Combinatorics, Modeling, Elementary Number Theory

From Basic to Advanced

Combinatorics, Modeling, Elementary Number Theory

From Basic to Advanced

Ivan Cherednik

The University of North Carolina at Chapel Hill, USA

World Scientific

NEW JERSEY • LONDON • SINGAPORE • BEIJING • SHANGHAI • HONG KONG • TAIPEI • CHENNAI • TOKYO

Published by

World Scientific Publishing Co. Pte. Ltd.
5 Toh Tuck Link, Singapore 596224
USA office: 27 Warren Street, Suite 401-402, Hackensack, NJ 07601
UK office: 57 Shelton Street, Covent Garden, London WC2H 9HE

Library of Congress Cataloging-in-Publication Data
Names: Cherednik, Ivan, author.
Title: Combinatorics, modeling, elementary number theory : from basic to advanced /
Ivan Cherednik, the University of North Carolina at Chapel Hill, USA.
Description: New Jersey : World Scientific, [2023] | Includes bibliographical references.
Identifiers: LCCN 2023015058 | ISBN 9789811265396 (hardcover) |
ISBN 9789811265402 (ebook)
Subjects: LCSH: Combinatorial analysis--Textbooks. | Combinatorial number theory--Textbooks. |
Graph theory--Textbooks.
Classification: LCC QA164 .C53 2023 | DDC 511--dc23/eng/20230605
LC record available at https://lccn.loc.gov/2023015058

British Library Cataloguing-in-Publication Data
A catalogue record for this book is available from the British Library.

Cover image: Mahabharata, The Game of Dice. Lithograph printed by Chore Bagan Art Studio, 1895.

For any available supplementary material, please visit
https://www.worldscientific.com/worldscibooks/10.1142/13116#t=suppl

Preface

This book is mainly based on about 25 years of teaching combinatorics at UNC Chapel Hill to two distinct sets of students: first-year students and seniors. The courses were intended for students from *all backgrounds*, not only for those majoring in mathematics and physics. The prerequisites were kept to a minimum: essentially only high school algebra was required, with a little linear algebra for seniors. The design was to go from zero knowledge to advanced themes and various applications during a 3 to 3.5 month semester; surprisingly it worked well.

These courses eventually developed into something different from what is covered by typical books in introductory combinatorics; a lot of modeling, statistics and related number theory were added, among other changes. Some topics were intended for the students as research projects, advanced assignments or additional reading. Accordingly, the plan was to write a book focused on the connections of combinatorics with other branches of mathematics and with wide range of applications, including dimers, random walks, games of chance, Catalan numbers and pricing options, magic and Latin squares, elements of cryptography, finite fields and Pisano periods, Pell's equations, Frobenius coin problem, Fibonacci numbers, population growth, fission reactors, epidemic spread, Eulerian circuits, Hamiltonian cycles, labeled trees and even something on neural networks.

These topics have deep historic roots. In its own unique way, mathematics reflects world history. I did my best to supply the readers with some historical context here and there, while keeping with the "condensed" spirit of the book. Algebra, number theory, combinatorics do require historic perspective.

However, this book contains only a few references. It is nowadays quite rare that Wikipedia is insufficient to find the basic ones; there are many other options too. When online searches seem not to be straightforward, some key names and links are provided. Aside from this, this book is a solid mathematical text. The exposition, which has been tested with dozens of cohorts of students, is accessible yet does not sacrifice mathematical rigor. This book can certainly be used for college courses in combinatorics, elementary number theory and modeling (mostly linear). It contains full definitions and comprehensive justifications.

The main challenges when writing this book were: creating a dynamic fully self-contained exposition with a focus on connections between the topics and cultivating an active learning style. There is quite a collection of problems and exercises (almost all with solutions, hints and answers). An additional challenge was to constantly provide links to statistics, statistical physics, computer science, economics and beyond. Also, this book was made convenient for remote teaching; it is close to the actual lectures, one-to-one in quite a few places. Success in these efforts, or not, will be judged by the readers.

Our Approach. The main objectives are as follows.

(1) *To make the book convenient in classrooms, including remote teaching.* Basically, one chapter (omitting starred sections and upon some other reductions) corresponds to approximately two weeks in real classes.

(2) *To focus on connections, links between the chapters and applications.* Quite a few topics are omitted or moved to the Exercises, but the core themes are exposed systematically and with many connections.

(3) *To provide additional material for motivated readers and projects.* Starred sections and some additional themes in the last chapter serve this purpose. About 25% of exercises are of advanced kind.

(4) *To try to incorporate almost all basic topics and standard problems,* though frequently in the Exercises. Hopefully, the book contains "almost everything" the users may expect to find in such a book.

All first-year students do projects in my classes, on topics such as dimers, tree growth and finite fields. For many of them, it was one of their first college mathematics classes. The culmination of their work comprised oral and/or written presentations with the primary purpose of assessing the students' ability to engage with the class material and conduct their own research. This mindset provided the motivation to structure the book's content and style for a blend of individual and pedagogical use and making it fully self-contained. I hope it is sufficient to learn the bunch of topics we consider from scratch and without other sources.

The main unifying themes of my classes and this book are *combinations and permutations, recurrences and modeling, randomness and games of chance* and *introductory number theory,* with constant attention to *statistical aspects* and applications. Perfect coverings by dominos are introduced on day one in my classes, and they reoccur many times. *Coin flipping and random walks* are the main examples of randomness throughout this book. *Lotteries, dice, poker, playing roulette* are systematically discussed. Population growth, dimers, resonances, epidemic spread, and a mini-introduction to pricing options are among the key modeling topics.

Elementary number theory is strongly present in the book. *Finite fields, Pisano periods, Pell's equations, Frobenius coin problem and Quadratic Reciprocity* are systematically exposed with full justifications and combinatorial aspects. These topics are closely related to the main themes. Let me mention that modular elliptic curves and their usage for digital signatures, *the ECDSA algorithm,* can be added to this list, but this is beyond the scope of this book.

Introductory texts in combinatorics and number theory are frequently too formal, sometimes unnecessarily so. This book is close to real notes: less formal and hopefully more readable. This is not always good; the exposition is bumpy in some places and some references are to the online notes; errors and misprints are possible, especially in the exercises. We hope that fixing them will not be a problem for the readers; their input will be greatly appreciated!

The justifications mostly follow the standards of research works in mathematics. Some are compressed. If they can be made one-line, they will be such. Actually, the proofs are mostly constructive (some calculations) in this book and should not be a problem for almost any reader.

There is a great variety of topics in combinatorics. I did not hesitate to reduce this Everest to minimum-minimorum. Top priority was given to the topics and problems that have, from my perspective, the most universal value and are not too specialized for most readers. The sections and problems marked by stars can be omitted during a first reading or used for projects and advanced assignments.

However, I did my best within the themes of this book to include major types of examples, problems and exercises that a reader can expect to find in introductory books on combinatorics, elementary number theory, and related modeling. They were frequently modified and developed to match the aims of this book.

Combinatorics expands now in many directions, which requires new approaches to the textbooks, especially when the applications are of high priority.

PROBLEMS AND EXERCISES. The number of key propositions and major formulas is limited in this book. A lot of material is provided in the form of relatively self-contained problems and many topics are covered in the Exercises. The active learning approach is absolutely the best way to teach and study combinatorics. Developing combinatorial skills requires practice.

This book contains about 135 problems and 275 exercises. Creating such a pool was not its main purpose, but this was appreciated by my colleagues: an update of the problem-oriented classical books in combinatorics and recreational mathematics from the 20th century (Gardner, Pólya, ...). The exercises significantly complement the topics in the main body of the book. I hope that the readers will appreciate this collection, as well as the twists, upgrades, multiple links and various angles I try to provide.

The exercises marked by $^\heartsuit$, $^\diamondsuit$ are generally suitable for home assignments and exams. The exercises with $^\diamondsuit$ are somewhat more involved. The ones with $^\clubsuit$ and $^\spadesuit$ are mostly quite doable too (sufficient hints are provided), but they are generally more difficult or beyond the core topics of this book. The distribution of the exercises is approximately 5:4:2:1 for $^{\heartsuit\diamondsuit\clubsuit\spadesuit}$ (*hearts: diamonds: clubs: spades*); about 75% are $^\heartsuit$and $^\diamondsuit$. Many can be solved without reading much of the related general material in the corresponding chapter.

The hints provided for exercises are frequently almost complete solutions. However they are condensed and following them will require effort. The students constantly request help with combinatorial problems; the answers and hints constitute an important part of the Exercises. Quite a few exercises complement the core topics. I favor the most the ones with applications and connections, but their choice of course reflects my preferences.

The problems in this book are generally instructional, closely connected with the main topics and suitable in the classroom; their solutions are provided. The exercises are for practising and to demonstrate that combinatorics has many dimensions and deeply penetrates mathematics, natural sciences, economics, sociology and life sciences. Every opportunity is used to show that combinatorics is an excellent training for statistics, probabilities and modeling.

The topics and problems marked by * are those I consider more advanced and/or not strictly necessary for the core themes of this book. This material can be omitted during a first reading and in the classroom. The stars and the difficulty "suits" for the exercises are designed to provide greater flexibility for the readers and teachers to adjust this book to their interests and needs.

This depends. For instance, number theory can be reduced and graph theory increased, or the focus can be on enumerative combinatorics, or modeling can be the main line. The book is made convenient for such adjustments.

DOMINOS, DIMERS AND MAGIC SQUARES. My courses on combinatorics always begin with Fig. 1 below: its major connections. Combinatorics is generally defined as the study of combinations and arrangements of any kind, but this hardly does justice to its actual role. The neighbors of combinatorics in this figure will be very much present in the corresponding chapters.

The first major topic of Chapter 1 is *perfect coverings by dominos*, equivalently, *tilings* or *dimer configurations*. Tilings are very important in combinatorics. The basics of their theory are a great introduction to recurrence relations, modeling, and basic statistical physics. I even introduce *partition functions*, the key in the physical study of dimers. They are quite understandable to motivated students; only the notion of *plaquets* is needed. The connections of dimers to graph theory and computer science, including *neural networks*, are discussed.

The 2nd half of Chapter 1 is devoted to magic and Latin squares, the 1st instance of the interplay between combinatorics and number theory in the book with deep historical roots. Various classical instances of magic squares are provided. A complete *modular* proof of the fact that the so-called *Siamese squares* are really magic is given, basically due to Leonard Euler.

Number theory, introductory and advanced, is inevitable in combinatorics for quite a few topics. The relation of magic squares to *mutually orthogonal Latin squares*, *MOLS*, in Chapter 1 and *modular split* are quite elementary: only the notion of division with a remainder is needed. Modular *MOLS* are of use to cryptography (the Advanced Encryption Standard), designs and error correcting codes. An interesting example is the Indian 4×4 magic square (around 12th century CE), which actually represents the field $\mathbb{F}_4$ with 4 elements.

Almost all students in my combinatorics classes enjoyed the number theory part and understood well *finite fields* and related topics, including applications to cryptography, designs and Pisano periods. Interestingly, those majoring in computer science were among those who liked this kind of number theory the most: they know that cryptography is a must (and that the universe is finite).

Anyway, the connection between combinatorics and number theory is fundamental; the students deserve to know this. Also, number theory always adds new dimensions to any courses and makes the exposition much more colorful.

PERMUTATIONS, COMBINATIONS. A systematic exposition of *enumerative combinatorics* begins in Chapters 2 and 3: permutations (P-numbers), combinations (C-numbers), combinations with repetition (R-numbers), the Pigeonhole Principle and the Inclusion-Exclusion Principle. Many problems and exercises are provided. Among the applications considered in these chapters, the analysis of the New Jersey Pick-6 Lotto is used to practice combinations and the *Poisson distribution*, discrete and continuous. The game "14 vs. 1" in Chapter 2 is an excellent model for practicing the *binomial distribution* and basics of statistics. The binomial distribution and its Stirling approximation are very important in this book, though the latter is mostly used to establish connections with statistics. For instance, the formula $2^{2n}/\binom{2n}{n} \approx \sqrt{\pi n}$ is frequently used, and we deduced the *normal distribution* from the binomial one.

This book revolves around the topics covered in my classes, but quite a few themes were added. For instance, the *symmetric group* was only marginally presented in the classroom, but the book contains its basic theory and applications. The algorithm for solving the *15-puzzle* is one of them. The composition of permutations does occur in some problems and exercises. However, it is *not* really significant in this book, even in the section on *higher derangements* in Chapter 3, which is directly related to the *symmetric group*.

Let me give examples of how the exercises and problems complement the main topics. The *Lah numbers, Bell numbers, necklace-counting, generalized continued fractions and q-binomial coefficients* occur in the book as (quite doable) exercises; the hints are provided, sufficiently complete when necessary.

The *Stirling numbers* are the subject of several exercises in Chapter 3 too, but there is a special Section 7.1.3 in Chapter 7 with further details and justifications. They are important but not absolutely necessary in this book. This section is mostly for additional reading or for extra assignments.

Needless to say that Chapters 1-3 contain a lot of history of combinatorics, continued in the next chapters. These include historical remarks on magic squares, the section "Historical perspective" (Fibonacci, Omar Khayyam and more), the discussion of games of chance, where the main line for us is dice-faro-poker-roulette, some basic history of Pascal's triangle, and many more.

RECURRENCES AND NUMBER THEORY. Chapter 4 on recurrences and modeling is a direct continuation of a lot of material on recurrences in prior chapters. This chapter is when the readers and students receive powerful and universal tools: recurrences and their algebraic solutions. Quadratic equations are actually sufficient in this chapter. Cubic equations occur in my classes, for instance when modeling population growth. The *resonance conditions*, modeling *tree growth* and the spread of *Covid-19* are of more advanced kind: important examples of linear-type and power-type growth.

Chapter 5 is closely related to Chapters 1-4. Cryptography, Latin and magic squares, nonattacking queens, *Pisano periods* and *Pell's equations* are clear transitional topics. I omit Pell's equations when teaching combinatorics; however, Pisano periods are always present. Finite fields are quite suitable and understandable for my audience, which includes the first-year students.

This book can be readily used for introductory courses in number theory. The beginning can be Section 1.2 on magic squares and their *modular split*, combined with the related exercises. Then one can switch to Chapter 5, including the Section "Pell's equations". Some additional material on *continued fractions*, usual and generalized, is in the Exercises (with justifications presented as hints). *Pick's Theorem* is Exercise 6.4.27; a sketch of the proof is provided. Section 7.3 of Chapter 7 contains more material on finite fields and field extensions. Also, it includes *Quadratic Reciprocity* with its full justification and various additional aspects. The finale can be Section 7.5.3, "Frobenius coin problem".

I tried to improve the exposition vs. "usual" textbooks in elementary number theory, especially with Pell's equations and the Frobenius coin problem. The proof of Quadratic Reciprocity is probably the best among "elementary ones" (quite combinatorial). Note that the proof of the cyclicity of the group $\mathbb{Z}_p^*$, used in Chapter 5, is "moved" to Chapter 7; we do this for any multiplicative finite subgroups in fields. The existence of *primitive roots* for $\mathbb{Z}_n^*$ (for $n = p^m, p \neq 2$) is readily deduced from that for $n = p$ in the Exercises.

GRAPHS AND CATALAN NUMBERS. Graph theory is considered in Chapter 6 and the corresponding portion of Chapter 7: *Eulerian circuits, Hamiltonian cycles, planar graphs, line graphs, counting labeled trees* and other themes. Some of them were "moved" to the Exercises. This is really an intro; we omit many "standard" topics, for instance Ramsey theory; though Ramsey numbers occur in the Exercises. *Catalan numbers* are exposed in detail in this chapter. They and their generalizations are among the key topics of this book. "Pricing options" is presented as a natural application of extended Catalan numbers.

More advanced topics on Catalan numbers are in Section 7.5, which includes *rational Catalan numbers*, Catalan numbers of type A_n and the *Poisson-Catalan distribution*. The Frobenius coin problem in Chapter 7 is directly related to rational Catalan numbers too.

Chapter 7 is exceptional; it is a comprehensive review of all topics. It contains problems *really* used for practicing in review classes and advanced themes suitable for projects and additional reading. The tools from various places of the book are freely used, assuming that the reader's skills reached some maturity.

The following is a natural sequence of sections *to use this book for introductory courses in modeling and data science.* A starting point can be Section 1.1 on dimers and the corresponding recurrence relations, possibly upon some reductions of the material. Then, Chapter 4 and the exercises there provide the core material, followed by Section 7.2. Modeling *dimers, population growth, tree growth, resonances, spread of epidemics* are systematically considered, including statistical aspects. See also modeling reinfections in Chapter 7. One can add some topics from Section 6.1, devoted to graphs. For instance, the usage of *neural networks* is discussed in Section 6.1.4 in the context of *knight's tours* and *Hamiltonian cycles* for dimer configurations.

FOCUS ON CONNECTIONS. Unless at a very introductory level, combinatorics can be difficult to learn and teach: more difficult, for instance, than college calculus. The 1st reason is that there are many different and seemingly disconnected combinatorial topics. The 2nd: too many puzzles can be a distraction for readers and students. The 3rd: books on introductory combinatorics are usually very sizable and not sufficiently adjusted to teaching real classes. This book is not small, but it contains a lot of themes beyond introductory combinatorics. The 4th: it is almost impossible to write a reasonably complete book on combinatorics without a lot of statistics and number theory. The 5th: the standard texts do not reflect well applied aspects of combinatorics. That said, we of course used some books; "Introductory combinatorics" by Richard Brualdi was one of them.

Applications of combinatorics attract a lot of attention, including relatively recent ones in artificial intelligence, data science and machine learning. Games of chance are important in this regard. The line *lottery-dice-poker-roulette* is well-exposed in this book, with a lot of implications and deviations.

The end of the book is a description of a new card game designed to model "momentum decision-making"; it was inspired by "momentum trading" in stock markets and presumably reflects well quite a few similar processes, including those in our brain. Poker is a natural starting point, but the risks there are too simplistic to model real trading; it is enhanced by the bridge-type auction and playing.

To conclude, combinatorics has many dimensions. It uses tools from various branches of mathematics and physics. In this sense, it is mostly an *eclectic science*: not really focused on creating its own tools. The methods can be from representation theory, linear algebra, special functions, number theory, geometry-topology, physics, graph theory, etc. This is the nature of combinatorics: *con-*

nected with everything and connecting everything. It is right in the middle of "everything" and under significant revision now.

Fig. 1 sketches some of these connections. It is also a preview of the applications of combinatorics we consider in this book.

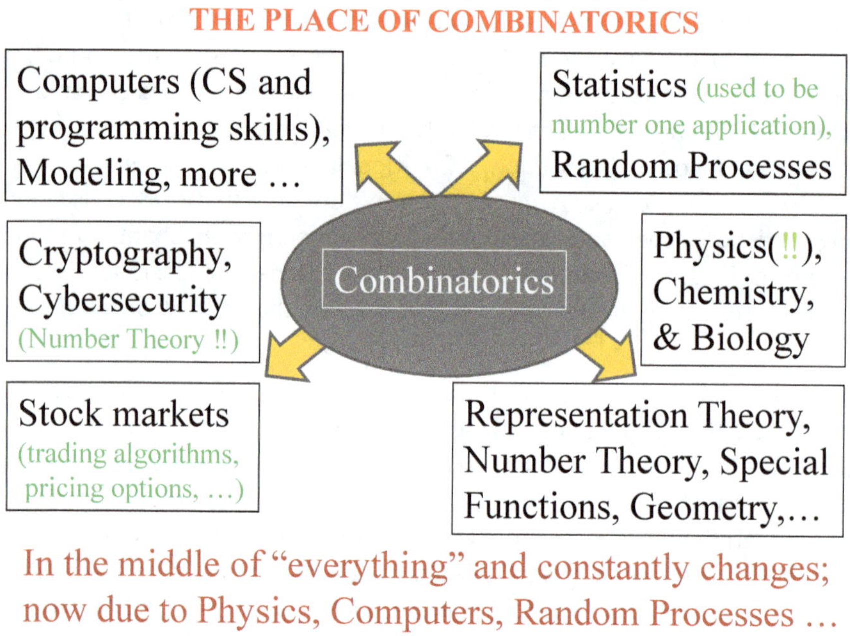

Figure 1: Combinatorics: practically and theoretically.

I use every opportunity to discuss in the book the "calculus limits" of combinatorics. Recurrence relations in combinatorics replace, in a sense, differential equations, the binomial distribution replaces the normal distribution, and so on.

Calculus is fundamental in mathematics, but college calculus is mostly from the 17-18th centuries. Combinatorics is very ancient, but it was significantly upgraded in the 19-20th centuries and is now undergoing rapid change, theoretically and via applications. Anyway, we all live in a very combinatorial environment.

Acknowledgements. Support from the National Science Foundation for all these years and from the University of North Carolina is gratefully acknowledged. I thank very much Rémi Lodh for his invaluable help in directing this project and improving the manuscript. Many thanks to Alexei Borodin, Pavel Etingof and Victor Kac. My thanks go to Bradley Hicks, Franklin Rea and Lai Fun Kwong for their help in editing this book.

Ivan Cherednik, Chapel Hill, North Carolina, 2023

Contents

Chapter 1

DOMINOS AND MAGIC SQUARES

This chapter introduces some basic combinatorial objects that are very interesting on their own and will constantly reappear throughout the book: *perfect coverings by dominos and magic and Latin squares*. The exposition is systematic and with many examples, problems and exercises. Unless the topics are marked by stars, we provide all of the details, though these themes will be revisited later in the text. We discuss the historical origins of magic squares, which are ancient, and the physical origins of perfect coverings by dominos, and *dimers* in the physics literature. The latter is sketchy, but we need to provide some perspective; only very basic physical concepts are needed.

1.1 Perfect coverings, dimers

A classical puzzle is to count the number $\mathcal{N}$, more exactly $\mathcal{N}^{\square\square}_{M\times N}$, of perfect coverings by dominos of the $M \times N$ chessboard, where M is the number of rows and N is the number of columns. The usual chessboard is, of course, 8×8. Generally, the product MN must be even, since dominos consist of 2 squares. At least one perfect covering therefore exists: with all horizontal dominos if N is even or all vertical if M is even. The exact definitions are:

chessboard: an $M \times N$ (rectangular) shape with MN squares;
domino: two adjacent squares, horizontally or vertically;
perfect covering: (i) no overlapping of the dominos,
(ii) every square must be covered.

Perfect coverings are also called ***tilings***. See [1] for a good introduction. We will call the unit squares boxes, cells or fields depending on the context. The problem is to determine *the number* $\mathcal{N}^{\square\square}_{M\times N}$ *of perfect coverings for the*

[1] F. Ardila, R. Stanley, *Tilings*, Math. Intelligencer 32:4 (2010), 32–43.

$M \times N$ *shape (chessboard)*. Here $\square\square$ means that the covering is by dominos, but more general tiles can be used.

It is very natural to consider dominos where one box is white and the other is black. Let $\widetilde{\mathcal{N}}^{\square\square}_{8\times 8}$ be the number of perfect coverings by such two-colored dominos. The calculation of such a tiling may seem more difficult than that without colors, but there is an almost immediate link.

For any uncolored perfect covering, i.e. that without any black and white boxes, one can perform the following procedure. There is no natural first domino, second domino, and so on in a perfect covering, but let us order them somehow. Then, we color the first domino, the second domino, and so on. The number of ways to color them is 2^{32} as there are 32 dominos used. Thus, $\widetilde{\mathcal{N}}^{\square\square}_{8\times 8} = 2^{32}\mathcal{N}^{\square\square}_{8\times 8}$.

Combinatorially, the process of coloring 32 dominos assuming that they are ordered is no different from tossing a coin 32 times as generally, there are 2^N possible sequences of heads and tails for N coin flips.

We now have for the first instance in this book one of various *product formulas*. One of the major principles of combinatorics and statistics is that *the number of outcomes for a sequence of consecutive choices is the product of the number of outcomes for each choice.* Here the choices can depend on the prior ones; the assumption is that the number of outcomes for a single choice depends only on that current step (regardless of the prior choices).

1.1.1 The cases $2 \times n$ and $3 \times n$

Let us consider the $2 \times n$ shapes, the simplest possible after "$1 \times n$". Any problem or computer program always begins with convenient notation. Let

$$d_n = \mathcal{N}^{\square\square}_{2\times n}, \text{ for instance, } d_1 = \mathcal{N}^{\square\square}_{2\times 1} = 1, d_2 = \mathcal{N}^{\square\square}_{2\times 2} = 2.$$

Calculating d_3, the number of perfect coverings by dominos of the shape 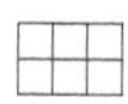, already requires a little consideration, which we provide next.

Let us consider the possibilities for the box marked by $*$ in 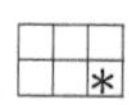. It can be covered vertically or horizontally by a domino. In the first case, we have: 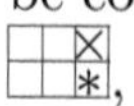, where the second box of the placed domino is marked by $\times$. Then we simply proceed with the remaining 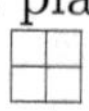, where the number of perfect coverings is $d_2 = 2$. It is even simpler in the second case: we arrive at 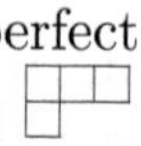 and *must* put a horizontal domino above the "removed" horizontal one. This gives only 1 covering. We conclude that $d_3 = 2 + 1 = 3$.

This method can be obviously extended to any value of n. For the sake of concreteness, let $n = 5$:

$\mathcal{N}($ 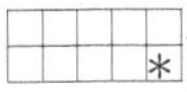 $) = \mathcal{N}($ 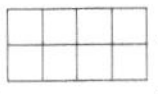 $) + \mathcal{N}($ 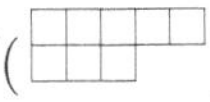 $)$, so $d_5 = d_4 + d_3$.

We will use below the notation $\mathcal{N}(\mathit{shape})$ for the number of perfect coverings by dominos for other shapes, not only rectangles.

Generally, we obtain $d_n = d_{n-1} + d_{n-2}$ for $n \geq 3$. This is the classical *Fibonacci recurrence*. However for d_n, the initial conditions are $d_1 = 1, d_2 = 2$. One thus has $d_n = f_{n+1}$ for the Fibonacci numbers f_n, (which will be discussed in the chapter on recurrences.)

This is a good example of the ***tree of possibilities***, a standard technique in statistics, computer science, and in writing practical computer programs.

Generally, we begin with some initial choice, then determine and analyze the second choices, then the third ones, and so on with the expectation that this procedure will eventually converge. For $\mathcal{N}$ we perform this by
(i) the analysis of placing dominos on some starred (marked) boxes, and then
(ii) reducing the "dimension" of the problem according to the analysis in (i).

After the recurrence is obtained, a computer can give you an answer for any n, but there is one more step: *solving recurrence relations algebraically*, if possible. For d_n, the algebraic formula (given in Chapter 4) is

$$d_n = \frac{\left(\frac{1+\sqrt{5}}{2}\right)^{n+1} - \left(\frac{1-\sqrt{5}}{2}\right)^{n+1}}{\sqrt{5}} = \text{Round}\left[\left(\frac{1+\sqrt{5}}{2}\right)^{n+1} / \sqrt{5}\right], \ n = 1, 2, \ldots .$$

Here, Round$[x]$ is the closest integer to x; for example, Round$[n+0.5] = n+1$ for integers n.

Thus the growth of d_n is *exponential*. Generally, the exponential function is $const\, a^t$. In this example, t is n and the base a is irrational: $a = (1+\sqrt{5})/2$. This will be clarified in full in Chapter 4. As we noted previously, $d_n = f_{n+1}$ for the celebrated Fibonacci numbers f_n counting rabbits (among other "things"); rabbits multiply exponentially.

The shapes $3 \times n$. Composing the recurrence relation is still simple in this case, with some interesting twists. As above, we begin with convenient notations. Let

$g_n = \mathcal{N}_{3\times n} = \mathcal{N}\Big($ 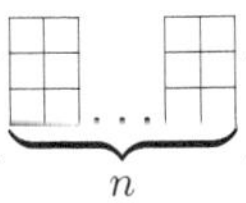 $\Big)$.

Here we omitted the domino symbol in $\mathcal{N}$ ; we will do this constantly below unless other *unit shapes* are considered.

Obviously, $g_{2m+1} = 0$, since $3(2m+1)$, the total number of boxes in the shape, is not divisible by 2, so the recurrence can be expected to be a formula for g_{2m} in terms of g_{2m-2}, g_{2m-4}, etc. Such a recurrence indeed exists, but the most natural approach requires one more "variable," which is defined as the number of perfect coverings of the following *non-rectangular* shape: the $3 \times n$ shape with 1 domino removed in the last column. Let $s_n = \mathcal{N}\Big(\underbrace{\square\ldots\square}_{n}\Big)$. Notice that n is the total number of columns, including the short one. Similarly to g, one has: $s_{2m+1} = 0$ for any $m = 0, 1, 2, \ldots$ due to the parity argument.

To calculate g_2, one can use that $g_2 = d_3 = 3$. With s_2 it is obvious: $s_2 = 1$, since there is only one way to cover the corresponding shape. Directly finding g_4, the next g, is a doable exercise, but it already presents some challenges. Let us take the starred box in the middle of the last *but one* column. One has for arbitrary $n \geq 4$:

$$
\begin{aligned}
g_n &= \mathcal{N}_{3\times n} = \mathcal{N}\Big(\underbrace{\square\ldots\square}_{n}\Big)\\
&= \mathcal{N}\Big(\underbrace{\square\ldots\square}_{n}\Big) + 2\mathcal{N}\Big(\underbrace{\square\ldots\square}_{n}\Big) + \mathcal{N}\Big(\underbrace{\square\ldots\square}_{n}\Big)\\
&= \mathcal{N}\Big(\underbrace{\square\ldots\square}_{n-2}\Big) + 2\mathcal{N}\Big(\underbrace{\square\ldots\square}_{n-2}\Big) + 2\mathcal{N}\Big(\underbrace{\square\ldots\square}_{n-2}\Big)\\
&= 3g_{n-2} + 2s_{n-2}.
\end{aligned}
$$

We mark the remaining box of the dominos covering $*$ by $\times$. We use that $\mathcal{N}\Big(\underbrace{\square\ldots\square}_{n}\Big) = \mathcal{N}\Big(\underbrace{\square\ldots\square}_{n}\Big)$. These two possibilities result in the same number of coverings because of the reflection symmetry, so only one is sufficient to consider. This gives the coefficient 2 of $\mathcal{N}\Big(\underbrace{\square\ldots\square}_{n}\Big)$.

Similarly,

$$
s_n = \mathcal{N}\Big(\underbrace{\square\ldots\square}_{n}\Big) = \mathcal{N}\Big(\underbrace{\square\ldots\square}_{n-1}\Big) = g_{n-2} + s_{n-2}.
$$

The reflection symmetry is used for the second "=".

If possible, it always makes sense to present solutions graphically in combinatorics; this recurrence is the case. In Fig. 1.1, we show all possibilities for the $\star$-box in the middle of the last but one (penultimate) column. The blue dominos are all choices for covering "star"; the yellow boxes are the corresponding arrangements after the placement of the blue domino. There are two possible variants for the yellow boxes in the first line of this figure.

Dominos: the recurrences for g_n and s_n

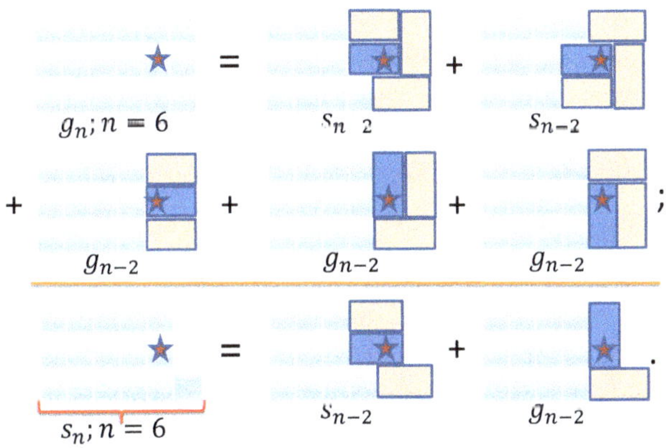

$g_n = 2s_{n-2} + 3g_{n-2}$, $s_n = s_{n-2} + g_{n-2}$, $g_2 = 3$, $s_2 = 1$;
$g_4 = 2 + 3{*}3 = 11, s_4 = 3 + 1 = 4, g_6 = 2{*}4 + 3{*}11 = 41$.

Figure 1.1: Recurrence for $3 \times n$.

The recurrences we obtained are sufficient to find any g_n and s_n, though only g_n are of actual interest of course; s_n are needed in the intermediate calculations. One has: $g_4 = 3g_2 + 2s_2 = 9 + 2 = 11$, which was the initial question. Continuing, $s_4 = g_2 + s_2 = 4$ and $g_6 = 33 + 8 = 41$, and so on. Computers can find "all of them" in no time (addition is fast).

General remarks. The procedure of composing the recurrence, briefly described in (i),(ii) above, always begins with finding as many *initial values* as you can: (1) to double-check that the problem is fully understandable, and (2) to look for possible "reduction steps". Anyway, the recurrences absolutely require knowing initial values by their very definition.

The next step is the composition itself. By analogy with the "$2\times n$" case, the most natural marked box to begin the recurrence can be expected to be one of the corners. Surprisingly, placing the "star" in the middle of the second to last column works much(!) better. The middle row makes some sense due to the obvious reflection symmetry of the $3 \times n$-problem. However, choosing the middle box in the second to last column is "art"; this requires experience.

The definition of *recurrences* is very relaxed. The number of sequences $g_n, s_n, \ldots$ can be as many as needed. Such sequences are discrete counterparts to functions, where n plays the role of t or x in calculus. The recurrence allows us to calculate the values "now" (at n) in terms of previous values for $g, s, \ldots$ and so on, i.e. at $n-1$ or smaller. This can be by a single algebraic formula, or by many formulas not only of algebraic nature. Whatever can be interpreted by computers is admissible, i.e. virtually everything.

Further analysis. Given n, the formula for g_n we obtained is ideal for computer calculations; the fastest possible. However, this is not the end. There are more aspects to consider:
(a) a matrix reformulation of the recurrence for g_n, s_n above;
(b) making the recurrence in terms of g_n only, i.e. without s_n;
(c) solving the recurrence: finding an algebraic formula for g_n;
(d) the analysis of the behavior of g_n as n approaches infinity.

This list is a preview of the methods we will develop in further chapters. Let us begin with a matrix reformulation. Using matrix multiplication, we can present these relations as a single one:

$$\begin{pmatrix} g_{2m} \\ s_{2m} \end{pmatrix} = \begin{pmatrix} 3 & 2 \\ 1 & 1 \end{pmatrix} \begin{pmatrix} g_{2m-2} \\ s_{2m-2} \end{pmatrix}, \quad \begin{pmatrix} g_{2m} \\ s_{2m} \end{pmatrix} = \begin{pmatrix} 3 & 2 \\ 1 & 1 \end{pmatrix}^{m-1} \begin{pmatrix} 3 \\ 1 \end{pmatrix}, \tag{1.1.1}$$

where the usage of $m-1$ is necessary to have g_2 and s_2 as $m = 1$ in the second formula. The latter is the result of the consecutive iteration of the first one, which gives $\begin{pmatrix} 3 & 2 \\ 1 & 1 \end{pmatrix}^{m-1}$. Powers of matrices are very basic in linear algebra; eigenvalues and eigenvectors can be used and there is a lot of software.

One can avoid using s_n and obtain a single recurrence relation in terms of g only. Namely, $g_{2m} = 4g_{2m-2} - g_{2m-4}$; see Problem 4.3.3(i). However, the usage of s_n is quite natural here, and it always makes sense to have *first-order recurrences*, even if they are in matrix form. This is parallel to the theory of ordinary differential equations.

A variant of this method works for the shape $4 \times n$. In the most natural approach, it requires 3 additional shapes. Finding the system of recurrence relations is not too difficult. See Exercise 1.4.12 and Problem 4.3.3(iii). Recall that almost everything in this chapter will be revisited later.

The matrix presentation from (1.1.1) gives the most universal way to justify that $g_{2m} = C_1\widetilde{\lambda}_1^m + C_2\widetilde{\lambda}_2^m$ for some constants $C_{1,2}$ and $\widetilde{\lambda}_{1,2} = 2 \pm \sqrt{3}$: these are the roots of the characteristic polynomial of this matrix. Such formulas are what we mean by "solving recurrences algebraically" in (c). Generally, matrices will not be needed much in this book, though we will use them to solve general recurrences and in some problems and exercises.

The relation with $\lambda_{1,2}$ at the very end of Section 1.1.3 is as follows: $\widetilde{\lambda}_{1,2} = (\lambda_{1,2})^2$, where $\lambda_{1,2} = \frac{1 \pm \sqrt{3}}{\sqrt{2}}$. The formula there reads: $g_{2m} = \frac{\lambda_1^{2m+1} - \lambda_2^{-2m-1}}{\lambda_1 - \lambda_2}$.

For instance, $g_2 = \lambda_1^2 + \lambda_1\lambda_2 + \lambda_2^2 = \widetilde{\lambda_1} + \widetilde{\lambda_2} - 1 = 3$, $g_4 = \lambda_1^4 + \lambda_1^3\lambda_2 + \lambda_1^2\lambda_2^2 + \lambda_1\lambda_2^3 + \lambda_2^4 = (\widetilde{\lambda}_1 + \widetilde{\lambda}_2)^2 - \widetilde{\lambda}_1\widetilde{\lambda}_2 + (\lambda_1\lambda_2)(\widetilde{\lambda}_1 + \widetilde{\lambda}_2) = 4^2 - 1 - 4 = 11$.

The growth of g_{2m} is exponential: $\lim_{m\to\infty} \frac{g_{2m}}{(2+\sqrt{3})^m} = C_1 = \frac{\lambda_1}{\lambda_1-\lambda_2} = \frac{1+\sqrt{3}}{2\sqrt{3}}$; λ_2 does not contribute since $|(2-\sqrt{3})| < 1$. This is (d).

The main purpose of this discussion is to preview the chapter on recurrences; we do not need (a-d) in this chapter.

1.1.2 An intro to dimers

In this section, we will use terms from physics without exact definitions, though only very basic concepts are needed. Readers unfamiliar with those terms are encouraged to believe the analogies to combinatorics without worrying about the exact connections.

In physics/chemistry, ***dimers*** refer to either diatomic molecules or pairs of molecules. Water dimers $(H_2O)_2$ are an important example of the latter. Combinatorially, *dimers are the segments connecting the centers of the domino boxes*, i.e. they just provide a different presentation for dominos. Perfect coverings by dominos are called ***states*** in the physics context. For instance, two parallel dimers $|\,|$ are translated to .

Two examples of perfect coverings of the 6×8 shape and the corresponding *states* (dimer configurations) are provided in Fig. 1.2; see also Fig. 6.7.

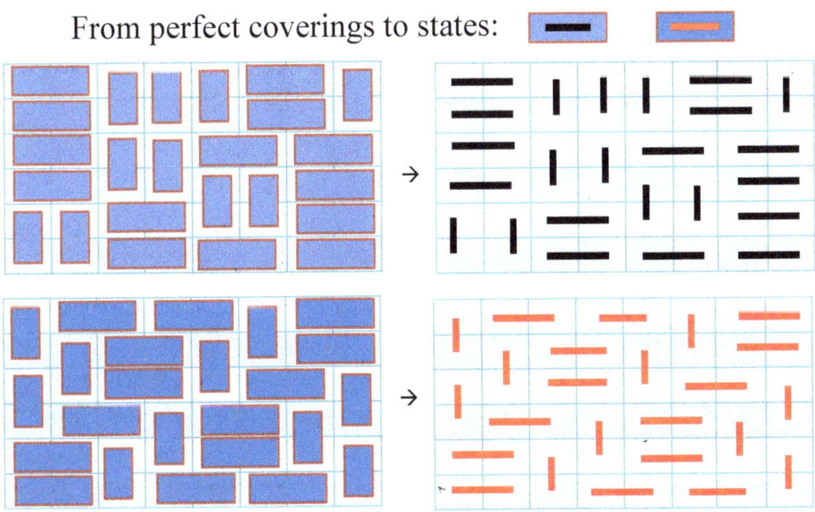

Figure 1.2: From dominos to dimers.

The general formula for $\mathcal{N}$ was found by Kasteleyn, Temperlay and Fisher in the early sixties; the so-called ***Kasteleyn's matrix*** is essential in the theory. See Section 1.1.3 below, Problem 7.2.13 and paper [2]. For the 8×8 shape, "8×8" for short: $\mathcal{N}^{\square\!\square}_{8\times 8} = 2^4(901)^2$, about 13 million.

Partition functions *. Is it necessary to know $\mathcal{N}^{\square\!\square}_{M,N}$ exactly? Not always. These numbers grow exponentially when M, N increase; see formula (1.1.3) below. For many applications, only large M, N make sense.

In the physics theory of dimers, ***partition functions*** are the key objects. They are defined as follows:

$$Z_{M\times N}(q) = \frac{1}{\mathcal{N}^{\square\!\square}_{M\times N}} \sum_{s=states} q^{E_s/(MN/2)},$$

of the $M \times N$ shape, where E_s is the *energy*, some number assigned to a given state s. The states are perfect coverings by dominos using the language of dominos. The division by $(MN/2)$ here is some natural normalization; q is a parameter of the theory. For $q = 1 + \epsilon$, the limit $\lim_{\epsilon\to 0} \epsilon\,(Z_{M\times N}(q) - 1)$ is the *average energy*.

If an *exact* algebraic-type formula for $Z_{M\times N}(q)$ can be found, the theory is called *integrable*. If we cannot calculate $\mathcal{N}$ exactly, the chances are slim that we can do this for $Z_{M\times N}(q)$. If there is no exact formula, physicists look for remarkable symmetries of $Z_{M\times N}(q)$: some relations between different values of M, N, and q.

What should the energy function E_s be here? Even relatively simple examples can lead to deep theories. For instance, let us assume that only vertical dominos "have energy". Then E_s is the number $\mathcal{N}_v$ of vertical dimers in a given state or, equivalently, of vertical dominos in a perfect covering. In this case, $Z_{M\times N}(q)$ can be calculated algebraically.

Plaquettes. In the theory of phase transitions, the following E_s is of importance: the number of ***plaquettes***, pairs of parallel neighboring dimers. In terms of dominos, it is the number of pairs 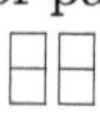 or 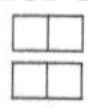. It is 13 and 4 for the states in Fig. 1.2. The case $q > 1$ is called the "aligning interaction of dimers"; the "columnar case" is when $q < 1$. See e.g. [3]. Strictly speaking, this theory is not integrable, but it is rich with symmetries.

In the case $q < 1$, the states with the lowest $Z_{M\times N}(q)$ are when all dimers (or dominos) are "columnar" indeed. To be exact, the maximal number of plaquettes is $\frac{(M-1)N}{2}$, where N is even and $M \geq N$ if both are even. See Problem 1.1.1. This is when all dominos are all horizontal: a "vacuum state".

[2] R. Kenyon, *An introduction to the dimer model*, arXiv:math/0310326, 2003.

[3] F. Alet, Y. Ikhlef, J.L. Jacobsen, G. Misguich, V. Pasquier, *Classical dimers with aligning interactions on the square lattice*, Phys. Rev. E 74 (2006), 041124.

Any perfect covering of the $M \times N$ shape with even MN and $M, N > 1$ has at least one plaquette, which results from the following general theorem of William Thurston (1990). By ***flips***, we mean transformations  $\leftrightarrow$ 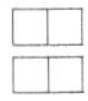 in any plaquettes. Thurston proved for configurations of dominos with no holes (empty regions surrounded by dominos) that the "connectivity" holds: *all perfect coverings can be obtained from any given one by sequences of flips.* Since "$M \times N$" for $M, N > 1$ has at least 2 perfect coverings, we conclude that any perfect covering must have at least one plaquette. This is checked directly in Exercise 1.4.6; actually, all shapes without holes that have no plaquettes are described there, i.e. those with $Z_{M\times N}(q) = 1$.

Problem 1.1.1 * *Prove that the maximal number of* plaquettes *in a perfect covering of "$M \times N$", is $\frac{(M-1)N}{2}$, where N is assumed even and $M \geq N$ if both are even. This maximum occurs only when all dominos are horizontal if $M > N$. If $M = N$, the other possibility is when all dominos are vertical.*

Solution. Consider a perfect covering with the greatest possible number of plaquettes. We represent it as a disjoint union of rectangle blocks of size $k \times 2$ or $2 \times m$ for $k, m \geq 2$ that are maximal possible sequences inside this covering of *consecutive* parallel dimers, horizontal or vertical correspondingly. For instance, 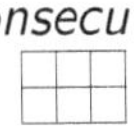 covered vertically is such for $m = 3$. Every such block contributes $m - 1$ or $k - 1$ to the total number of plaquettes.

If the first column of "$M \times N$" is covered only vertically or only horizontally we can remove this column or two of them in the horizontal case: the 1st and the 2nd. Thus, we diminish N and can proceed by *induction* with respect to the size of the rectangle, which means the assumption that the claim holds for "$(M-1) \times N$" or "$M \times (N-1)$" (or smaller). We can obviously assume this since our claim holds for $N = 2$. Mathematical induction is discussed in detail in Section 3.3.3. *Recall that "$M \times N$" denotes the $M \times N$ shape.*

For instance, if $M \geq N$ and M is odd, then the maximal number of plaquettes for "$M \times N$" will be $(M-1) + \frac{(M-1)(N-2)}{2} = \frac{(M-1)N}{2}$ by induction. Check by induction that if M is even and $M > N$, the maximum number of plaquettes can be reached only for the horizontal covering of the first two columns in "$M \times N$", continued then horizontally in "$M \times (N-2)$". We leave the case $M = N$ to the readers.

Now we begin a *proof by contradiction*. Let us assume that the first column is covered by $k_1 \times 2, \ldots k_r \times 2$ and $2 \times m_1, \ldots 2 \times m_s$ as above in some order. They contribute $(k_1 - 1) + \cdots + (k_r - 1) + (m_1 - 1) + \cdots + (m_s - 1)$ to the number of plaquettes. Here $M = (k_1 + \cdots + k_r) + 2s$. Now let us change this covering as follows. We cover the first 2 columns of our shape horizontally, and leave the remaining $m_i - 2$ vertical dominos in every $2 \times m_i$

block unchanged, the total number of the corresponding plaquettes becomes $M-1+(m_1-2)+\cdots+(m_s-2) = (k_1+\cdots+k_r)+2s+(m_1-2)+\cdots+(m_s-2) = M-1+k_1+\cdots+k_r+m_1+\cdots m_s$, i.e. greater than it was before, since $r+s$ was assumed greater than 1. We come to a contradiction with the maximality of the number of plaquettes; the end of the proof, denoted further by ❑.

To conclude, the formula for $Z_{M\times N}(q)$ is very reasonable mathematically. We divide one big number by another big number, which is $\mathcal{N}^{\square\square}_{M\times N}$. In a sense, we divide infinity by infinity, so the ratio can be expected to be finite.

1.1.3 Any *M* and any *N*

We use the formulas in this section very little, and only in some advanced context; see e.g. Section 4.3.4. However, we need to provide some general perspective. For any $M \times N$ shape, the following remarkable formula holds:

$$\mathcal{N}^{\square\square}_{M\times N} = \Big(\prod_{1\le m\le M}\prod_{1\le n\le N}\big(2\cos(\frac{m\pi}{M+1})+2\imath\cos(\frac{n\pi}{N+1})\big)\Big)^{1/2}.$$

Use your computer to obtain $2^4(901)^2$ for $N = 8 = M$, which value was mentioned above. Yes, here $\imath = \sqrt{-1}$, so complex numbers are used.

This formula is too involved to use practically. Even verifying the simple fact that there are no perfect coverings by dominos for odd MN, the number of boxes, requires some consideration. It is so because the term in this product with $m = (M+1)/2, n = (N+1)/2$ is $\cos(\pi/2)+\imath\cos(\pi/2)$, where $\cos(\pi/2) = 0$. In this form, this relation seems unrelated to the recurrence relations for $\mathcal{N}^{\square\square}_{M\times N}$, which we consider systematically for small M in this book. The method of ***transfer matrices*** is generally better for this; see Problem 7.2.12.

A somewhat simpler formula holds (at least without complex numbers) for even M and any N:

$$\mathcal{N}^{\square\square}_{M\times N} = 2^{\frac{MN}{2}}\prod_{1\le m\le \frac{M}{2}}\prod_{1\le n\le N}\big(\cos^2(\frac{m\pi}{M+1})+\cos^2(\frac{n\pi}{N+1})\big)^{1/2}. \qquad (1.1.2)$$

Both formulas are beyond our book, as is the following result:

$$\mathcal{N}^{\square\square}_{M\times N}/e^{GMN/\pi} \to 1, \quad \text{for } M\to\infty, N\to\infty, \qquad (1.1.3)$$

where $G = 1 - \frac{1}{3^2} + \frac{1}{5^2} - \frac{1}{7^2} + \ldots$ is ***Catalan's constant***. This formula is important and relevant to physics because there it is mostly assumed that M and N, both tend to infinity. The shapes like "$2\times n$" are of limited interest to physicists. A convenient reference for this subsection is [4].

[4] J.L. Jacobsen, *Algebra, integrability and exactly solvable models*, 2021, `http://www.phys.ens.fr/~jacobsen/AIMES/Dimers.pdf`.

Kasteleyn's formula.* We use it only to obtain some explicit recurrences. It is more suitable in this book than the product formulas above; our recurrence relations are certainly connected to it. See formula (2) in [5]. Let MN be even, $L=$Floor$[M/2]$, for the integer part of $M/2$, and:

$$\lambda_{1,2}^{(\ell)} = \cos\left(\frac{\ell\pi}{M+1}\right) \pm \sqrt{\delta_\ell}, \text{ where } \delta_\ell = 1+\cos^2\left(\frac{\ell\pi}{M+1}\right), \text{ and } \lambda_1 \text{ is for } +.$$

Here $\ell = 1, 2, \ldots, L$. Note that $\lambda_{1,2}^{(\ell)}$ are two solutions of the quadratic equation $\lambda^2 - 2\cos(\frac{\ell\pi}{M+1})\lambda - 1 = 0$.

We set $\Delta_L = \prod_{\ell=1}^{L} \delta_\ell$. It can be calculated in terms of the Pell numbers: Exercise 4.6.28. For any subset S from the set $[L] = \{1, 2, \ldots, L\}$, including the empty set and $[L]$, let $\lambda_S = \prod_{\ell \text{ in } S} \lambda_1^{(\ell)} \prod_{\text{otherwise}} \lambda_2^{(\ell)}$. We denote by $|S|$ the number of elements in S; the indices ℓ are from the set $[L]$. Then it turns out that:

$$\mathcal{N}_{M\times N}^{\square\square} = \prod_{\ell=1}^{L} \frac{(\lambda_1^{(\ell)})^{N+1} - (\lambda_2^{(\ell)})^{N+1}}{\lambda_1^{(\ell)} - \lambda_2^{(\ell)}} = \frac{1}{2^L \Delta_L^{1/2}} \sum_S (-1)^{L-|S|} \lambda_S^{N+1}. \quad (1.1.4)$$

The summation is over all subsets S of $[L]$; so 2^L is the number of S.

This formula is convenient for computations, but M, N are not on equal grounds here. The symmetry $\mathcal{N}_{M\times N}^{\square\square} = \mathcal{N}_{N\times M}^{\square\square}$ is far from obvious in (1.1.4).

For $M = 2$ and $N = n$, we obtain that $d_n = \mathcal{N}_{2\times N}^{\square\square} = \frac{\lambda_1^{n+1} - \lambda_2^{n+1}}{\lambda_1 - \lambda_2} = \frac{\left(\frac{1+\sqrt{5}}{2}\right)^{n+1} - \left(\frac{1-\sqrt{5}}{2}\right)^{n+1}}{\sqrt{5}}$. We omit the super-index (1) in $\lambda_{1,2}^{(1)}$.

For $M = 3$ and even $N = 2n$: $L = 1$ and $g_{2n} = \mathcal{N}_{3\times 2n}^{\square\square} = \frac{\lambda_1^{2n+1} - \lambda_2^{2n+1}}{\lambda_1 - \lambda_2}$, where $\lambda_{1,2}$ solve the equation $\lambda^2 - \sqrt{2}\lambda - 1 = 0$.

1.1.4 Two non-existence problems

The focus in the examples above was on finding the exact number of possible configurations. However, there are quite a few combinatorial problems where none exist, and you are supposed *to prove* this, i.e. to prove that $\mathcal{N} = 0$.

The non-existence problems are mostly of mathematical interest, but they are very useful to sharpen and develop theoretical and practical tools. Some are famous and involve very advanced methods.

---

[5] R. Stanley, *On dimer coverings of rectangles of fixed width*, Disc. Appl. Math. 12 (1985), 81–87.

A basic one is to *prove that* $\mathcal{N}_S^{\square\square}=0$ *for the following* S *: the* 8×8 *chessboard "minus" its* 2 *opposite corners.* Since the number of squares in S is even, there is no immediate contradiction. We need to apply something more advanced. The "existence problems" in combinatorics are mostly some explicit constructions, but "non-existence" always requires some proof.

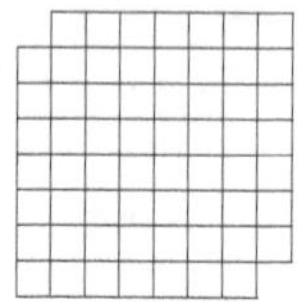

It is important to note that *proofs are not that formal in combinatorics*; pictures, verbal arguments, anything that seems relevant can be used. However, the sequences of arguments must, of course, be complete: all possibilities have to be considered. For instance, if you are asked to prove that something holds for arbitrary n, considering some examples is insufficient. In this problem, the following reasoning will do:

Proof. Let us use that the chessboard has two colors in the classic pattern: say, black and white. Then we removed 2 boxes of the same color in making S. However, any single domino covers both colors. Thus S cannot be covered perfectly, as well as any shape where there is no match between black and white boxes (by the same reasoning). ❑

Adding colors is part of the solution but, obviously, not in the claim: "8×8" is just a shape (no color). In physics, we would say that the colors are *hidden symmetries.* This is also parallel to *additional constructions*, sometimes very ingenious, in Euclid's geometry.

Coverings by 4-minos. A variant of the argument we used for the problem above is the key to the following non-existence problem.

Problem 1.1.2 *Prove that the shape* $T_6 =$ "6×6" *cannot be perfectly covered by 4-minos:* $\square\square\square\square$, *i.e.* $\mathcal{N}_{6\times 6}^{\square\square\square\square}=0$ *in our notation.*

Solution. Generally, you are supposed to begin with some experiments. Here 36 is divisible by 4, so potentially such coverings can exist. Though you will seriously doubt this after several attempts.

It may not be a simple problem to manage. However, the previous one about the shape S provides a strong hint: some coloring seems the best if not the only way for this problem. The usage of 4 colors is natural here. The *coloring scheme* is more subtle. We want the following lemma to hold:

Lemma. *There exists a coloring scheme for* T_6 *with* 4 *colors such that any 4-mino there covers all* 4 *colors.*

We use "Lemma" here to pay tribute to Euclid's geometry. The *axiomatic method* has its unique origin: Ancient Greece. Though, as we already noted, the proofs are much less formal in combinatorics than in "Euclid" or, say,

Hadamard's "*Lessons in Geometry*". There are no actual axioms in combinatorics; we have "principles" instead, which can be readily justified.

Theorems in this book will be mostly about obtaining explicit formulas. Mathematical induction is useful sometimes, but even this is not strictly necessary. However, combinatorial problems for very simple and explicit objects and situations can be very difficult.

Let us prove the lemma. The colors will be $1, 2, 3, 4$. Restricting ourselves to the 4×4 shape T_4, the following coloring does the job:

$$\begin{pmatrix} 1 & 2 & 3 & 4 \\ 4 & 1 & 2 & 3 \\ 3 & 4 & 1 & 2 \\ 2 & 3 & 4 & 1 \end{pmatrix}.$$

The rows and columns are cyclic permutations of the first ones. Our lemma obviously holds for T_4. Let us extend it to "6×6". We will do this *cyclically*, first horizontally and then vertically.

Lemma 1.1.3 *Any 4-mino covers all 4 colors for the following coloring:*

$$\left(\begin{array}{cccc|cc} 1 & 2 & 3 & 4 & 1 & 2 \\ 4 & 1 & 2 & 3 & 4 & 1 \\ 3 & 4 & 1 & 2 & 3 & 4 \\ 2 & 3 & 4 & 1 & 2 & 3 \\ \hline 1 & 2 & 3 & 4 & 1 & 2 \\ 4 & 1 & 2 & 3 & 4 & 1 \end{array}\right).$$

Proof. Since colors are cyclic in every row and column, any 4-mino, regardless of its position, in a row or in a column, covers all 4 colors. ❑

This one-line proof is perfectly acceptable; adding more words or sentences will only make it worse. Recall that the square marks the end of the proof.

To conclude the proof of the claim from Problem 1.1.2, we observe the following. There are two 1's (and no 3) in the lower-right 2×2 corner, so covering "6×6" by 4-minos is impossible; more precisely, because there are more 1's than any other color and because only shapes which have an equal number of boxes of each color (from 4) are perfectly coverable. The latter match is necessary but not always sufficient to ensure that some covering exists, but this is not part of this problem. ❑

1.2 Magic squares

Magic squares, MS for short, are $n \times n$ square matrices with n^2 entries that are *all* numbers from 1 to n^2 (each of them occurs once) such that all vertical sums Σ_v, all horizontal sums Σ_h and 2 diagonal sums Σ_d are the same number, ***the magic sum*** Σ. The figure below is for $n = 5$:

$$\begin{array}{cccccccccc} 65 & & 65 & 65 & 65 & 65 & 65 & & 65 \\ & \nwarrow & \uparrow & \uparrow & \uparrow & \uparrow & \uparrow & \nearrow & \\ 65\leftarrow & & 20 & 8 & \boxed{21} & 14 & 2 & & 65 \\ 65\leftarrow & & 11 & \boxed{4} & 17 & 10 & 23 & \nearrow & \boxed{65} \\ 65\leftarrow & & \boxed{7} & 25 & 13 & 1 & 19 & \nearrow & 65 \\ 65\leftarrow & & 3 & 16 & 9 & 22 & \boxed{15} & \nearrow & \\ 65\leftarrow & & 24 & 12 & 5 & \boxed{18} & 6 & & \end{array} \qquad (1.2.5)$$

The value of $\Sigma = \Sigma_v = \Sigma_h = \Sigma_d$, the magic sum, is simple to determine. One has: $n\Sigma_h = 1 + 2 + \cdots + n^2 = \frac{n^2(n^2+1)}{2}$, so $\Sigma = n(n^2 + 1)/2$.

We used that $1 + 2 + \cdots + N = \frac{N(N+1)}{2}$. For even N, this is immediate: $(1 + N) + (2 + N - 1) + \ldots = (N + 1)\frac{N}{2}$. This is a simple example of "combinatorial justifications," where we prove something by recombining the terms. For odd N, we have: $\frac{N-1}{2}(N + 1) + (N + 1)/2 = (N + 1)\frac{N}{2}$.

The magic sum for $n = 5$ is $\Sigma = 26 \cdot 5/2 = 65 = (1 + \cdots + 25)/5 = 65$. The arrows in (1.2.5) show the corresponding sums. This square is also ***pandiagonal magic***: the sums of the entries in any *pandiagonals* and *anti-pandiagonals* coincide with the magic sum Σ. Their definition is as follows.

A ***pandiagonal*** in the $n \times n$ square is a partial diagonal with k boxes (not only the main diagonal) combined with the complementary one with $n - k$ boxes and in the same direction. Pandiagonals are in the direction $\searrow$. The definition of ***anti-pandiagonals*** is the same for the direction $\nearrow$.

For instance, the boxed numbers in (1.2.5) form an anti-pandiagonal. The sums of the numbers in this and *any* pandiagonals or anti-pandiagonals are indeed 65 for this particular square. Pandiagonal MS were also called *diabolic*, which name can be of interest to investigate. They do not exist only for $n = 3$ and $n = 2 + 4k$ for integer $k \geq 0$, i.e. for $n = 2, 3, 6, 10, \ldots$.

Pandiagonals algebraically. A natural question is how to present pandiagonals and anti-pandiagonals algebraically; computers understand only this language. Let i be the row number, j the column number; mostly we will assume that $0 \leq i, j \leq n-1$. Then the subdiagonals or anti-subdiagonals are those with $i - j = C_1$ or $i + j = C_2$. The (full) diagonal and anti-diagonal are for $C_1 = 0$ and $C_2 = n + 1$.

So $i - j$ is taken for diagonals and $i + j$ for anti-diagonals, but we will frequently call *both* diagonals, abusing the strict definition.

The complementary ones are for $i' - j' = C'_1$ or $i' + j' = C'_2$ for $C_1 - C'_1$

and $C_2 - C_2'$ divisible by n. Combinatorially, C_1' and C_2' are chosen to make the total number of boxes of the union exactly n.

For the boxed pandiagonal in (1.2.5) above: the relations are $i + j = 2$ and $i' + j' = 7$, combined. Indeed, $7 - 2$ is divisible by 5.

1.2.1 Examples of MS

The origin of the following MS is India (12th century, the Parshvanath Temple). It is a (rare) 4×4 *pandiagonal* MS. The magic sum is $4(16+1)/2 = 34$ for $n = 4$. For instance, $2 + 12 + 15 + 5 = 34$:

$$\begin{pmatrix} 7 & \underline{12} & 1 & 14 \\ \underline{2} & 13 & 8 & 11 \\ 16 & 3 & 10 & \underline{5} \\ 9 & 6 & \underline{15} & 4 \end{pmatrix}.$$

The next one to discuss will be *Dürer's square* (1514), from his engraving "Melencolia I", the first MS printed in Europe: Fig. 1.3 from [6]. See also Section 1.3.3. It is actually equivalent to Yang Hui's square (about 250 years before Albrecht Dürer). This one has many symmetries, but it is not pandiagonal. For instance, the boxed pandiagonal below has the sum 30 ($\neq 34$):

$$\begin{pmatrix} 16 & \boxed{3} & 2 & 13 \\ 5 & 10 & \boxed{11} & 8 \\ 9 & 6 & 7 & \boxed{12} \\ \boxed{4} & 15 & 14 & 1 \end{pmatrix}.$$

By the way, 15,14 from the bottom line is the year 1514 of Dürer's engraving; also, 1 = A, 4 = D in this line are his initials. Magic squares were used in astrology and beyond. For instance the idea of this particular MS was that "it engages with the protective astral influences of Jupiter as a way to combat the dark anxiety and melancholic effects of the planet Saturn." See e.g. [7].

Lo Shu square. One of the exercises below is that there is only one 3×3 MS up to the symmetries. The symmetries are the rotations by $\pi/2, \pi$, and $3\pi/2$ around the center, and the reflections in the central lines of the middle rows and columns and the diagonals. Generally, one can send $x \mapsto n^2 - x + 1$ too for the entries x, but we do not need this for $n = 3$. We will check it now, omitting the justification of the fact that 5 must be in the center. So the proof of the uniqueness will be *conditional*: using that 5 is in the middle.

---

[6] Albrecht Dürer - Melencolia I - Google Art Project (AGDdr3EHmNGyA).jpg.

[7] C. Beyer, *Planetary magical squares*, January 2019, `https://www.learnreligions.com/planetary-magical-squares-4123077`.

Figure 1.3: Dürer's Melencolia I (public domain)

Proof. Since 5 is in the middle, we can use the rotations to make 1 in the last row as follows: $\begin{pmatrix} & & * \\ & 5 & \\ 1 & & \end{pmatrix}$ or $\begin{pmatrix} & * & \\ & 5 & \\ & 1 & \end{pmatrix}$. The magic sum is 15, so we then have two possibilities: $\begin{pmatrix} & & 9 \\ & 5 & \\ 1 & & \end{pmatrix}$ or $\begin{pmatrix} & 9 & \\ & 5 & \\ & 1 & \end{pmatrix}$.

Continuing, we can eliminate the first choice since 9 can only be included in 2 different "triples": $1+5+9 = 15$, $2+4+9 = 15$; the third one is $3+3+9 = 15$, using 3 twice, which is not allowed in magic squares. Using the reflection in the column $\{9\,5\,1\}$, we may assume that 4 is in the upper-left corner. Then we can readily complete the square: $\begin{pmatrix} 4 & 9 & 2 \\ & 5 & \\ & 1 & \end{pmatrix} \Rightarrow \begin{pmatrix} 4 & 9 & 2 \\ & 5 & \\ 8 & 1 & 6 \end{pmatrix} \Rightarrow \begin{pmatrix} 4 & 9 & 2 \\ 3 & 5 & 7 \\ 8 & 1 & 6 \end{pmatrix}$.

This is not the end of the proof. We only checked that if a 3×3 MS exists, then it must be this square up to the symmetries, so we still need to be sure that all 8 sums are really 15. We checked some of these, but possibly not all. Mathematically speaking, we proved the *uniqueness*, but not the *existence*. However, this is very simple to fix by inspection. ❑

We arrived at the oldest and the most famous ***Lo Shu square*** from China (circa 650 BCE). It is not pandiagonal (diabolic) magic. This is good, because it was used and is used for many spiritual things.

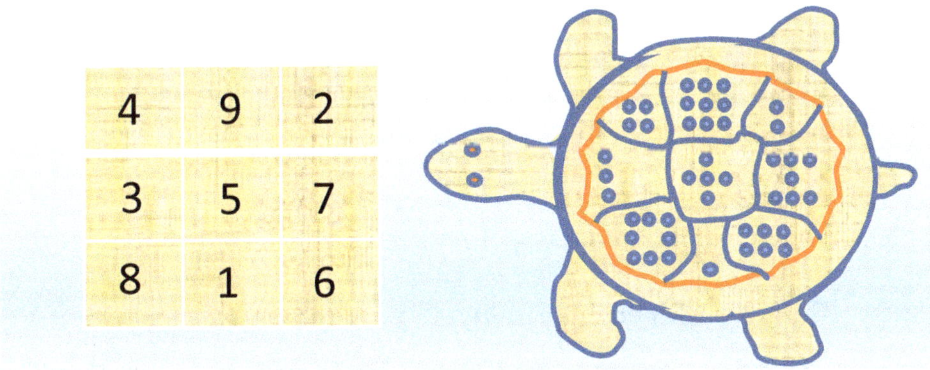

Figure 1.4: Lo Shu magic square.

The legend was that this square was on the carapace of a magical tortoise during a huge flood. The carapace of a turtle was also present in the name of a 4×4 MS in Varāhamihira's book (India, 550 CE). Interestingly, he used his magic square to prescribe combinations and quantities of ingredients for perfume. Possibly it was close in spirit to the usage of mutually orthogonal Latin squares (*MOLS*) for designs, which will be discussed later.

Semi-magic squares. Let us remove the requirement on the diagonal sums (both!) from the definition of MS. We will call such "relaxed" squares SMS, ***semi-magic squares***. Any permutations of rows or columns send SMS to SMS. The reflections in the vertical and horizontal middle lines are actually of this kind, as well as the rotation by π around the center (check). Generally, SMS form a much bigger class than MS. However, this is not so for $n = 3$.

Problem 1.2.1 *Check that any SMS of size $n = 3$ can be transformed into "Lo Shu" by permutations in the rows/columns and rotations by $\pi/2$ if necessary.*

Solution. Given an SMS, the permutations allow us to place 1 in the middle. Let us use that 1 is involved only in 2 sums $a+b+c = 15$ with pairwise distinct a, b, c from 1 to 9; they are $1 + 5 + 9 = 15 = 1 + 6 + 8$. Similarly, 9, 3, and 7 have this property; the remaining such presentation for 9 is $2 + 4 + 9 = 15$. Rotating and reflecting the SMS with entry 1 in the middle, we can assume that it is $\begin{pmatrix} 4 & 9 & 2 \\ & 1 & \\ & 5 & \end{pmatrix}$. Using the remaining presentation $1+4+8 = 15$ with 1, the square can be only $\begin{pmatrix} 4 & 9 & 2 \\ 6 & 1 & 8 \\ & 5 & \end{pmatrix}$ or $\begin{pmatrix} 4 & 9 & 2 \\ 8 & 1 & 6 \\ & 5 & \end{pmatrix}$. The first is impossible, since the last row of it would contain 5 twice. The second results in $\begin{pmatrix} 4 & 9 & 2 \\ 8 & 1 & 6 \\ 3 & 5 & 7 \end{pmatrix}$. Permuting the last 2 rows we obtain the Lo Shu square. ❑

For $n > 3$, the diagonal conditions in the definition of MS *significantly* reduce their number. For instance, the number of SMS is 549504 for 4×4 vs. $7040 = 880 \times 8$ for MS. This is totally, without the identifications due to the symmetries. See Section 7.1.4. If there exists an n-permutation such that the sum of corresponding entries is magic, then at least one of the diagonals can be made magic using permutations of rows or columns; see Exercise 1.4.29.

1.2.2 Siamese Method: odd n

This method became known in Europe in 1687 thanks to De la Loubère, a French diplomat, who shared it with his friends interested in mathematics upon his return from Siam (Thailand). He learned this construction from mathematicians there. It is called *Siamese/De la Loubére method.*

It works for any odd n, but provides essentially one MS from their great variety. There are 275305224 magic squares for $n = 5$ up to the rotations around the center, and reflections (horizontal, vertical and diagonal). However, this method is interesting, and not so many MS are needed ... for magic. There are several such constructions; see Exercise 1.4.36.

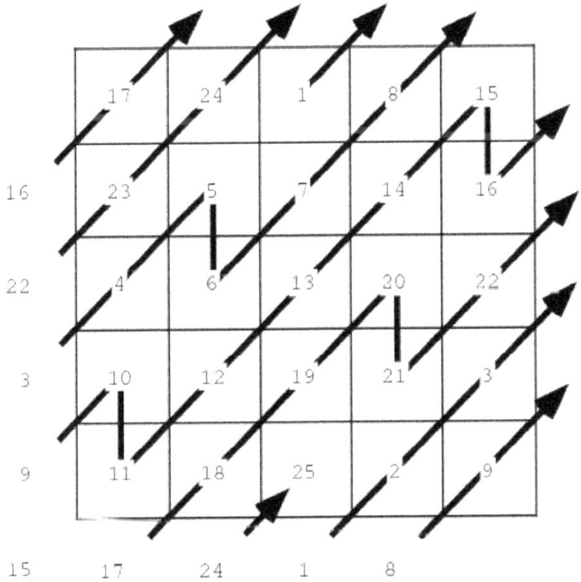

Figure 1.5: Siamese (de la Loubère) method.

The procedure is presented in Fig. 1.5; see [8]. As an algorithm:

(1) place 1 in the center of the first row; (2) inscribe consecutively the remaining numbers diagonally $\nearrow$; (3) if impossible, switch to the complementary portion of the corresponding pandiagonal; (4) if impossible, i.e. if the next box has been already taken, place the next number one step down: $\downarrow$.

It is not obvious that this procedure will reach the end, i.e. that these 4 rules are really sufficient; potentially, there can be other "corrections." This is some combinatorial challenge, which is obviously not about the magic sums. We will *prove* this below together with the justification of the fact that we really obtain an MS. The Siamese construction gives magic squares, but not magic pandiagonally (diabolically). Some other constructions are in [8].

1.2.3 Modular split of MS

Let us justify that this algorithm works well, i.e. all boxes will be occupied and the resulting square is really a magic one. We will need a little something from number theory, namely, *division with remainder*.

Let $n = 1, 2, \ldots$; the usual notation is $n \in \mathbb{N}$. (By the way, $\mathbb{Z}$ is used for integers, $\mathbb{Q}$ for rational numbers, $\mathbb{R}$ for real numbers, $\mathbb{C}$ for complex numbers,

[8] E. Weisstein, *Magic square*, `https://mathworld.wolfram.com/MagicSquare.html`.

and $\mathbb{Z}_p$ or $\mathbb{F}_q$ for finite fields.)

Any integer m can be uniquely represented as $m = qn + r$ where the remainder r satisfies the inequalities $0 \le r < n$. There can be some variants with choosing r; for instance $-n/2 \le r < n/2$ is sometimes convenient. However, in number theory, we typically take the remainder from 0 to $n-1$.

The proof of the existence and uniqueness is immediate: we simply remove from m the greatest possible multiple of n. This will be one of the key definitions in the number theory chapter.

Now, for any number $1 \le k \le n^2$, we represent $k-1 = qn+r$ for $0 \le r < n$ and $0 \le q < n$, and this presentation is unique. The inequalities for r are granted automatically, because it is the remainder. The inequalities for q follow from the restriction $k-1 < n^2$ due to the definition of MS.

For instance, let $n = 5$. Then $k = 18$ corresponds uniquely to $q = 3, r = 2$: indeed, $k - 1 = 17 = 3 \cdot 5 + 2$. The subtraction of 1 is a technicality. Historically, 0 was introduced in mathematics by Indian mathematicians after magic squares appeared. For any modular constructions, the numbers $\{0, \ldots, n^2-1\}$ are somewhat more convenient than $\{1, \ldots, n^2\}$, but this is insignificant: we simply subtract 1 from k to address this.

Starting with a magic square M, we will now create two squares, Q and R of the same size but with different entries: given k, we switch to $(k-1)$, calculate q, r as above, put q in Q and r in R in the boxes corresponding to that occupied by k. For the Siamese magic square for $n = 5$:

$$M = \begin{pmatrix} 17 & 24 & 1 & 8 & 15 \\ 23 & 5 & 7 & 14 & 16 \\ 4 & 6 & 13 & 20 & 22 \\ 10 & 12 & 19 & 21 & 3 \\ 11 & \boxed{18} & 25 & 2 & 9 \end{pmatrix} \Longrightarrow$$

$$Q = \begin{pmatrix} 3 & 4 & 0 & 1 & 2 \\ 4 & 0 & 1 & 2 & 3 \\ 0 & 1 & 2 & 3 & 4 \\ 1 & 2 & 3 & 4 & 0 \\ 2 & \boxed{3} & 4 & 0 & 1 \end{pmatrix}, R = \begin{pmatrix} 1 & 3 & 0 & 2 & 4 \\ 2 & 4 & 1 & 3 & 0 \\ 3 & 0 & 2 & 4 & 1 \\ 4 & 1 & 3 & 0 & 2 \\ 0 & \boxed{2} & 4 & 1 & 3 \end{pmatrix}. \qquad (1.2.6)$$

We see that the squares Q, R are ***Latin squares***, LS for short. The definition of Latin squares is as follows.

An $n \times n$ square with entries $0, 1, \ldots, n-1$ is called Latin *if all entries are different in every column and every row.*

Equivalently, all numbers $0, 1, \ldots, n-1$ must be present in any column and any row. Indeed, this means that there are no repetitions there (due to the Pigeonhole Principle from the next chapter).

We readily obtain that $\Sigma_v = \Sigma_h = 65$. For instance, for column 2 has: $24+5+6+12+18 = 5+5\cdot(4+0+1+2+3)+(3+4+0+1+2) = 5+5\cdot(0+1+2+3+4)+(0+1+2+3+4) = 5+50+10 = 65$. We simply use the definition of Q, R and that $1+2+3+4 = 10$ regardless of the order of terms. Of course the same holds for any column and any row.

It is equally applicable to the diagonal $\{17, 5, 13, 21, 9\}$ in M, since there are no repetitions in the corresponding diagonals of Q, R. The anti-diagonal $\{11, 12, 13, 14, 15\}$ has the sum 65 too, but we need to use here that $2+2+2+2+2 = 10$ (still 10) in the corresponding anti-diagonal of Q; its counterpart in R is a permutation: $\{0, 1, 2, 3, 4\}$. So the sum is 65 also.

This proves "without calculations" that M is really a magic square. Well, the *modular split* is some calculation; however it is systematic, and a clear improvement over checking every sum. This is essentially due to the Leonard Euler work (1776) published as [9].

1.2.4 Siamese method: justification

The analysis above shows clear patterns for the squares (matrices) Q and R, which can be readily extended to any odd n. For Q, we begin with $\{\frac{n-1}{2}, \ldots, \frac{n-1}{2}\}$ at the anti-diagonal (from the lower-left corner to the upper-right one). Then, we continue from the entry $\frac{n-1}{2}$ on this anti-diagonal by inserting $\frac{n+1}{2}$ and so on *cyclically!* in all rows or columns (go right or down). Note that the anti-pandiagonals in Q are all constant.

This is similar for R. Now the main anti-diagonal is $\{0, 1, \ldots, n-1\}$, which is not constant as for Q, and we cyclically inscribe the consecutive numbers in the columns starting with the ones at this anti-diagonal (go down). The step will be 2 if you go right in any row. Equivalently, the middle column in R is $0, 1, \ldots, n-1$ and we continue these numbers in the anti-pandiagonals in the direction $\nearrow$ by inserting the corresponding consecutive numbers cyclically.

Combining the latter pattern for R with the one for Q, we obtain an entirely modular reformulation of the Siamese procedure. This makes the combinatorial description of this procedure, including the 4-step algorithm above, unnecessary.

This is not only simpler theoretically; practically, a computer program for the Siamese method becomes much simpler. An almost immediate *proof* that this procedure really gives a magic square follows from the modular split. Here we say "almost" because we still need to justify a little something. The sums will be exactly the magic sum Σ because Q, R are Latin, and the diagonals

[9] L. Euler, *De quadratis magicis*, Commentationes arithmeticae 2 (1849), 593–602; see Jordan Bell's English translation (2004).

are fine; this is obvious. However, we still need to check that *all* numbers $1, 2, \ldots, n^2$ occur in the corresponding square.

Lemma *If we begin with Q, R defined via the modular patterns determined above and construct M by placing $m = qn + r + 1$ in the corresponding boxes, then all numbers $1, \ldots, n^2$ will occur in M.*

Proof.* The anti-pandiagonals of R contain all distinct entries due to the second modular description of R, i.e. they are some permutations of $\{0, 1, \ldots, n-1\}$ for every anti-pandiagonal. Let us use that the entries in every anti-pandiagonal of Q (the direction is $\nearrow$) are constant and they are different in different anti-pandiagonals. If it is $0 \leq a < n$, then the entries of M in this pandiagonal will be $1 + an$ plus those from the corresponding pandiagonal of R (all are different).

Thus, any partial quotient $0 \leq q < n$ will be coupled with any remainder $0 \leq r < n$, which gives that all $1, 2, \ldots, n^2$ will really occur in our M. ❑

The star * here and in several similar places in this book is not really the difficulty level. It mostly means that this is optional. It is sufficient (for now) if the readers are comfortable with the Siamese method *practically* and "see" the patterns of Q, R in examples. Recall that this is an introduction; the exposition will be more systematic later. The link [10] is a convenient one for some extra material on magic squares, related matters and references.

1.3 Latin squares, 36 officers

1.3.1 Mutually orthogonal LS

We have already used ***Latin squares***, LS for short, when doing MS. They are generally defined as *squares of size $n \times n$ with the entries $0, 1, \ldots, n-1$, and with no repetitions in their rows and columns.* They have many symmetries. Given an LS, one can:

(a) rotate them around the center and reflect (as for MS);

(b) permute the rows and columns by any n-permutations;

(c) rename the entries: permuting $\{0, 1, 2, \ldots, n-2, n-1\}$.

The rotations are by $\pm\pi/2$ and π around the center of the square. The reflections are along its 2 middle lines and the diagonals. The renamings are any permutation of the set $\{0, 1, \ldots, n-1\}$ *applied to the entries.* For instance, the first row can be always made $0, 1, \ldots n$ by "renaming," which determines the corresponding permutation uniquely.

---

[10]Mathematische Basteleien, *Magic square*, `http://www.mathematische-basteleien.de/magsquare.htm`.

Definition. *(i) We call two $n \times n$ Latin squares L_1, L_2 mutually orthogonal Latin squares, or* ***MOLS*** *for short, if the union $L_1 \vee L_2$ of L_1, L_2 contains all pairs $\langle u, v\rangle$ for $0 \le u, v \le n-1$. We use a special notation $\vee$ for the component-wise union: it is the* **bi-square** *with the entries $\langle u, v\rangle$ (or simply uv) placed at (i, j), where u is from L_1 at (i, j) and v from L_2 at (i, j).*

(ii) If this property holds, then $L_1 \vee L_2$ contains each and every such pair exactly once. We then write $L_1 \& L_2$, so that & indicates the combinatorial orthogonality. The notation $\langle u, v\rangle$ will be frequently used for the bi-entries uv if misunderstanding is possible; they are ordered pairs.

The row and column numbers i, j naturally range from 1 to n. However, we will mostly switch to the range $0 \le i, j \le n-1$ in Chapter 5, and in some exercises below. See e.g. Exercise 1.4.21. Note that it is not true that any Latin square L has some ***orthogonal mate***, which is defined as a Latin square L' such that $L\&L'$ may be formed. See Exercise 5.6.45.

The pair Q, R from (1.2.6) forms a pair of MOLS. Let us provide an example for $n=3$:

$$\begin{pmatrix} 0 & 1 & 2 \\ 2 & 0 & 1 \\ 1 & 2 & 0 \end{pmatrix} \vee \begin{pmatrix} 0 & 1 & 2 \\ 1 & 2 & 0 \\ 2 & 0 & 1 \end{pmatrix} = \begin{pmatrix} 00 & 11 & 22 \\ 21 & 02 & 10 \\ 12 & 20 & 01 \end{pmatrix}.$$

The second LS is obtained here from the first simply by permuting the last two rows. All pairs $\langle u, v\rangle$ are present in the resulting bi-square.

Given any MOLS, one can rename the entries *independently* in L_1 and L_2. Thus the first row can be assumed $\{0, 1, \ldots, n-1\}$ in both Latin squares; see Exercise 1.4.15(i). However, permuting the columns and the rows must be done with *the same permutation* in both, L_1 and L_2.

1.3.2 The problem of 36 officers

It is simple to see that there are no MOLS for $n = 2$. Leonard Euler tried hard to find at least one MOLS for $n = 6$; this was the famous puzzle of "36 officers" (around 1782). He stated it as follows:

Arrange 36 officers to be drawn from 6 different regiments so that they are arranged in a square so that in each line (both horizontal and vertical) there are 6 officers of different ranks and different regiments.

It was proved "by exhaustion" that this is impossible by Gaston Tarry in 1901 (120 years later!). Euler conjectured that the problem is not solvable for $2, 6, 10, \ldots$, i.e. for $n = 2+4k$ and $k = 0, 1, 2, \ldots$. He is regarded as one of the best mathematicians throughout the whole history of mathematics, but his conjecture turned out to be incorrect; only $n = 2, 6$ are "bad," as was finally proven in 1959.

There are various applications of MOLS and the sequences of pairwise orthogonal LS. The well-known ones are *designs*, say fabric designs; randomness is enjoyable for our eyes, which they provide.

Design of experiments must be mentioned here. For instance, suppose you need to check the efficiency of a combination of 4 factors, each in the range from 1 to n, when their order matters. Instead of going through all n^4 their possible sequences, one can take a MOLS of size n and consider its bi-entries extended by the corresponding column and row numbers as such sequences. This will require "only" n^2 experiments, and any possible (ordered) pairs of these 4 factors will occur; see Exercise 1.4.21(ii). The usage of sequences of pairwise orthogonal LS (not only pairs) for *error correcting codes* in broadband internet over powerlines and similar purposes is discussed in Section 5.2.3.

1.3.3 From MOLS to MS

Let us reverse the modular split construction. We begin now with MOLS $L_1\&L_2$ of size $n \times n$. We set $L_1 = (a_{ij}), L_2 = (b_{ij})$ using the standard matrix notation. This means that a_{ij} and b_{ij} are the entries in L_1 and L_2 at the intersection of row i and column j.

The inversion of the "modular split" is the following formula for M:

$$M = (c_{ij}), \quad \text{where} \quad c_{ij} = 1 + n \cdot a_{ij} + b_{ij}, 1 \le i, j \le n.$$

All horizontal and vertical sums *automatically* coincide with the magic sum for n. For instance, the sum of the entries in the first column of M is

$$c_{11}+c_{21} + \cdots + c_{n1} = n+n \cdot (a_{11}+a_{21} + \cdots + a_{n1}) + (b_{11}+b_{21} + \cdots + b_{n1})$$
$$= n+n(1 + \cdots + (n-1))+(1 + \cdots + (n-1)) = n + (n+1) \cdot \tfrac{n(n-1)}{2} = \tfrac{n(n^2+1)}{2}.$$

For the 3×3 MOLS considered above, we obtain the following:

$$\begin{pmatrix} 00 & 11 & 22 \\ 21 & 02 & 10 \\ 12 & 20 & 01 \end{pmatrix} \Longrightarrow$$

$$\begin{pmatrix} 1+0\cdot 3+0 & 1+1\cdot 3+1 & 1+2\cdot 3+2 \\ 1+2\cdot 3+1 & 1+0\cdot 3+2 & 1+1\cdot 3+0 \\ 1+1\cdot 3+2 & 1+2\cdot 3+0 & 1+0\cdot 3+1 \end{pmatrix} = \begin{pmatrix} 1 & 5 & 9 \\ 8 & 3 & 4 \\ 6 & 7 & 2 \end{pmatrix}.$$

The entries of the latter are indeed $1, 2, \ldots, 9$, and the vertical and horizontal sums are all 15 by construction, but it is not MS. *We did not impose any diagonal conditions in the definition of Latin squares*, so there is no reason to expect the diagonal sums to be magic. However, if you transpose the first and the second lines in the square above and rotate, it will be the "Lo Shu"!

Actually, "not so many" MS come from MOLS, even if we do something with the diagonals. For instance, this is not the case with the Dürer square. Let us provide its modular split:

$$M=\begin{pmatrix} 16 & 3 & 2 & 13\\ 5 & \boxed{10} & 11 & 8\\ 9 & 6 & 7 & 12\\ 4 & 15 & 14 & 1 \end{pmatrix} \implies Q=\begin{pmatrix} 3 & 0 & 0 & 3\\ 1 & \boxed{2} & 2 & 1\\ 2 & 1 & 1 & 2\\ 0 & 3 & 3 & 0 \end{pmatrix}, R=\begin{pmatrix} 3 & 2 & 1 & 0\\ 0 & \boxed{1} & 2 & 3\\ 0 & 1 & 2 & 3\\ 3 & 2 & 1 & 0 \end{pmatrix}.$$

The formula for Q, R is $m = 1+q\cdot 4+r$, where q, r are placed in the same location as m in Q, R. For instance, $10 = 1+2\cdot 4+1$ for the boxed entries in this figure. Obviously, Q, R are not Latin, but they have other interesting features. The proof of the MS-property using Q, R is immediate: all vertical, horizontal, and diagonal sums in Q, R are 6, so the corresponding sum in M is $4+4\cdot 6+6=34$ indeed.

This MS has many other remarkable 4-element subsets with the sum equal to 34. For instance, it is a natural union of four 2×2 squares (at the corners), and all of them are such; (use the modular split to see this!) The 2×2 square in the middle has this property too.

The modular split of the MS of size 4×4 which we began with in Section 1.2.1 does result in MOLS. However, the patterns for the corresponding Q, R are more involved than those in the Siamese construction. The field with 4 elements will be necessary to obtain them "naturally." See Section 5.3.2.

Using the tables in [11], one can check that Q, R for all 880 MS of size 4×4 have the vertical and horizontal sums 6, however this is not always true for the diagonals. It is of interest to justify this without the usage of the list of all MS of this size.

By the way, there are 48 pandiagonal magic squares of order 4 (ibid.). Pandiagonal magic squares have a greater number of symmetries, not only the generic ones, which are the rotations and reflections. One can move rows or columns from one side of the square to the other, though these transformations can coincide with the generic ones. See Exercise 1.4.35. Up to the symmetries, there are 3 different types of pandiagonal MS for $n = 4$.

Finally, let us provide MOLS for $n = 7$ using colors for "regiments" and letters for "ranks". In Fig. 1.6, $0 \le i, j \le 6$ are the row and column numbers of boxes (i, j); the one for $(0, 0)$ is the bottom-left. We color box (i, j) with respect to the remainder $4i+j$ modulo 7 and inscribe the letter corresponding to the remainder of $5i+j$. The first column is the key. This is a *pandiagonal*

[11] Harvey Heinz, *Order-4 magic squares*, `http://recmath.org/Magic%20Squares/order4list.htm`.

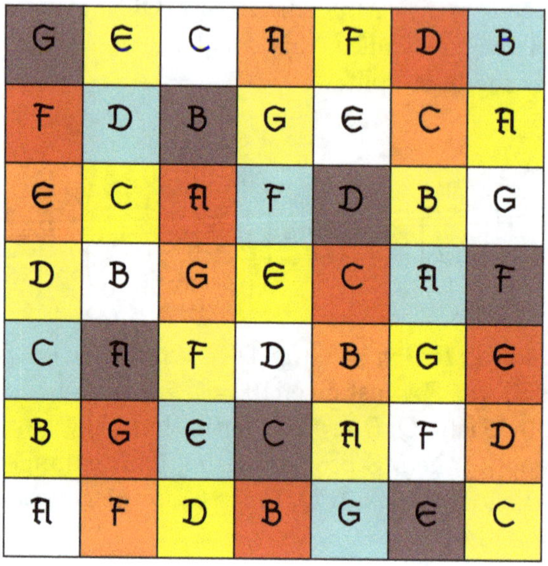

Figure 1.6: Pandiagonal (diabolic) MOLS for $n = 7$.

MOLS: there is no repetition of colors and of letters in any pandiagonal and any anti-pandiagonal. Generally, one can use here any $ai+bj, ci+dj$ provided that $a, b, c, d, a \pm b, c \pm d$ and $ad - bc$ are all relatively prime with n. Such *modular MOLS* are the subject of Proposition 5.2.3.

1.4 Exercises

Exercise 1.4.1$^{\heartsuit}$ *(i) Prove that any $m \times n$ chessboard with $n > 1$ and even m can be perfectly covered by dominos if any domino (any pair of squares that can be covered by a domino, to be exact) is removed.*

(ii) Deduce from (i) that if two dominos separated by a column or a row in the chessboard are removed then it is perfectly coverable; find an example when these dominos cannot be separated this way and no perfect covering exists.

(iii) Similarly, if m and n are both odd numbers, then prove that removing any square of the same color as the one in any corner makes the remaining shape perfectly coverable by dominos.

Hint for (i): Include this domino in a 2×2 square, and remove the greatest number of even rows $m \times 1$ to reduce is to $2 \times n$.
Hint for (iii): Similarly, reduce it to the case of 1×1 or 3×3 squares.

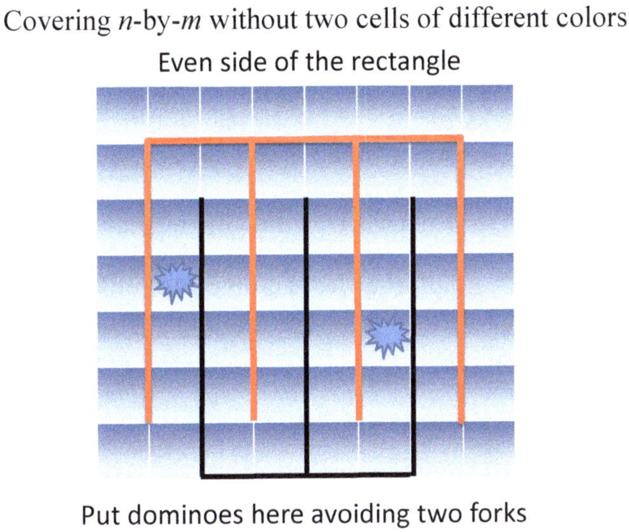

Figure 1.7: Proof of Gomory's Theorem.

Exercise 1.4.2$^\heartsuit$ *Use Fig. 1.7 to prove* Gomory's Theorem: *any $m \times n$ chessboard with even n and arbitrary $m > 1$ that has had any 2 squares with different colors removed is perfectly coverable by dominos.*

Exercise 1.4.3$^\heartsuit$ *(i) Construct a perfect covering of the chessboard of the size $(2m-1) \times (2n)$ for $m, n \geq 1$ with two opposite corners removed. This is a particular case of Gomory's Theorem.*

(ii) Deduce from (i) that the $m \times n$ chessboard for even m, n without 2 squares with different colors is perfectly coverable by dominos.

Hint: Consider the minimal rectangle containing these 2 squares and check that exactly one of its sizes is even, let it be horizontal. Then cover maximal vertical subrectangles in the chessboard $m \times n$ before and after the one containing these squares, and continue with the remaining rectangle, which is even-by-even.

Exercise 1.4.4$^\heartsuit$ *Find a connected subset of the board "2×6" with the coinciding numbers of black and white squares that has no perfect covering by dominos. By connected, we mean that any 2 squares are connected by a sequence of neighboring squares, ◫ or ◫◫, inside this subset.*

Exercise 1.4.5$^\diamond$ *A* ***Young diagram*** *is a collection of boxes (i, j) with $1 \leq i \leq n$ and $1 \leq j \leq k_i$ for $k_1 \geq k_2 \geq k_3 \geq \cdots \geq k_n > 0$. Assume that it has the same number of black and white boxes under the standard coloring (induced from the chessboard "$n \times k_1$"). Prove that it can be perfectly covered by dominos. You can use this to reprove Gomory's Theorem.*

Hint: Any Young diagram contains a pair of neighboring boxes such that their removal will result in another Young diagram, unless it is , 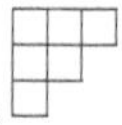 and so on. However the latter have different numbers of black and white boxes and therefore are not allowed. Thus, cover the pair above by a domino and continue by induction.

Exercise 1.4.6$^\clubsuit$ *(i) Check that the shapes* 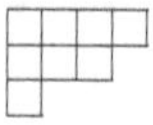 *and* 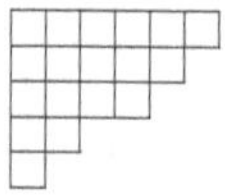 *can be* **uniquely** *covered perfectly by dominos. Extend this claim to the shapes where the first column contains $(2m-1)$ boxes for $m \geq 1$ and the numbers of boxes in the corresponding rows are $2m, 2m{-}1, \ldots, m{+}1, m{-}1, \ldots, 2, 1$. Check by induction that these coverings are as follows: the dominos covering boxes (i, j) for $j \geq i$ are horizontal and they are vertical otherwise.*

(ii) For perfect coverings, **plaquettes** *are defined as pairs of neighboring parallel dimers: a 2×2 subsquare covered vertically or horizontally. Check that the coverings of the shapes from (i) have no plaquettes. [Generally, the uniqueness provides this due to the Thurston Theorem; see Section 1.1.2.] Prove that the number of plaquettes is always nonzero for shapes "$M \times N$" with even MN and $M, N > 1$. For this, check that if there are no plaquettes and box $(1, 1)$ (the first one) is covered horizontally in such a perfect covering of "$M \times N$" then this covering can be embedded into the shape from (i) for sufficiently large m (where box $(1, 1)$ is the first).*

(iii) Generalize the claim from (ii) as follows. Any perfect covering of a union of boxes (i, j) with $1 \leq i \leq n$ and $1 \leq j \leq k_i$ for $k_1 = k_2 \geq k_3 \geq \cdots \geq k_n > 1$ boxes, i.e. that of a **Young diagram** *with at least 2 boxes in the last row and the last column, has at least one plaquette. See also Exercise 1.4.5.*

Hint for (ii)-(iii): Assume that a shape S without plaquettes contains $D =$ 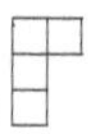 covered by 2 dominos. Check that it must contain its continuation 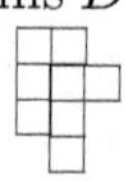 unless $\times$ in 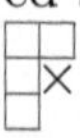 does not belong to S. This can be equally applied to the reflections or rotations of D, for instance, to $D' =$ 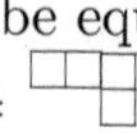. Assume that the top of S, which begins with 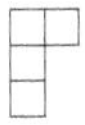 or with 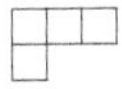 covered by 2 dominos, contains D' with its first row in the first row of S. Then construct the corresponding maximal chain of copies of D' in S as described in the direction $\swarrow$. The last D' in this sequence cannot belong to any rectangle Young diagram. So the first row of S is covered horizontally, we can delete it and proceed by induction.

Exercise 1.4.7 $^\diamond$ *(i) Use* $d_n = \mathcal{N}^{\square\square}_{2\times n}$ *and* $g_{2m} = \mathcal{N}^{\square\square}_{3\times 2m}$ *to check that the number of perfect coverings of the* 4×4 *chessboard is* 18 *assuming that the top-left corner is covered horizontally. Thus, the total number of coverings of the* $4{\times}4$ *shape is* 36.

(ii) Check that there are only 6 *coverings of "*4×4*" with a horizontal domino covering the top-left corner, if they are considered up to rotations. In other words, any of the 18 coverings can be obtained from 6 (no smaller) different coverings by some rotation of the chessboard.*

Hint for (i): If the top line is covered by 2 horizontal dominos, then the number is $g_4 = 11$; otherwise there are 7 coverings.
Hint for (ii): There is only 1 covering that is invariant under all 4 rotations, and 3 coverings invariant under the rotation by π; so the required number is $(12-1-3)/4+1+3=6$.

Exercise 1.4.8 $^\diamond$ *Prove that perfect coverings of the* $m{\times}n$ *chessboard by* k*-minos (the shapes of size* $1{\times}k$ *or* $k{\times}1$*) exist if and only if* k *divides both,* m *and* n.

Hint: Color the chessboard following Problem 1.1.2.

Exercise 1.4.9 $^\diamond$ *Let* $N(a_1,a_2)_n$ *be the number of coverings of "*$1\times n$*" for* $n\geq 1$ *by non-overlapping black* a_1*-minos (rectangles "*$1\times a_1$*") and white* a_2*-minos when the a-mino containing the last (right) square in "*$1\times n$*" is allowed to go beyond it. Justify that* $N(a_1,a_2)_n = N(a_1,a_2)_{n-a_1} + N(a_1,a_2)_{n-a_2}$*, where* $N(a_1,a_2)_n = 2$ *for* $1\leq n\leq \min(a_1,a_2)$ *and* $N(a_1,a_2)_n = 1$ *for* $0\geq n > -\max(a_1,a_2)$*. Use this recurrence to check that* $N(2,3)_{10}=21$.

Hint: For $a_1{=}2, a_2{=}3$, $N_{10} = N_8 + N_7 = \cdots = 3N_3 + 4N_2 + 2N_1 = 9{+}8{+}4$.

Exercise 1.4.10 $^\diamond$ *(i) Prove that for any perfect covering of the* $6{\times}6$ *chessboard, it is always possible to find a* ***fault line****, that is a horizontal or vertical line between the squares of this chessboard that does not cut through dominos.*

(ii) Find a perfect covering of the 6×8 *board without fault lines. Which step of your proof of Part (i) fails in this case?*

Hint: Assume that there are no fault lines and check that then all such lines, horizontal and vertical, would cross 20 dominos, ... when only 18 are possible.

Exercise 1.4.11 $^{\heartsuit}$ *(i) Consider the shapes* 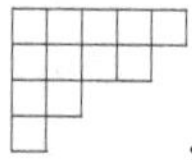 , 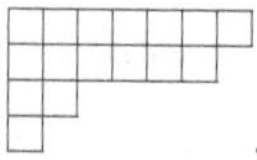 *, . . . , generally the union of the* **hook** *of size* (2m)×(2k+1) *and that of size* (2m–2)×(2k–1) *inside it for* $m, k \geq 2$. *Check that such shapes can be* **uniquely** *covered perfectly by dominos. Check that this covering can be extended (not uniquely) to a perfect covering of the rectangle "(2m)×(2k+1)" with* □□ *at the right-bottom corner.*

(ii) Continuing, take any covering of the rectangle "(2m)×(2k+1)" from (i) and add "2×(2k+1)" to its bottom covered horizontally for "2×(2k−2)" and with 3 *vertical dominos at the end of these* 2 *lines. Then redo the covering of* 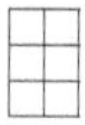 *at the right-bottom corner of "(2m+2)×(2k+1)" as follows:* 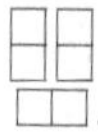. *For* $k = 2$, *the bottom part will be* 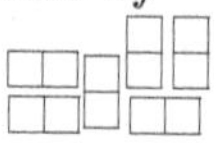 . *Check that the resulting covering has no* **fault lines** *in the sense of Exercise 1.4.10.*

Exercise 1.4.12 ♠ *Find a system of recurrence relations for* $h_n = \mathcal{N}^{\square\square}_{n\times 4}$, *the number of perfect coverings by dominos of the shape* $n \times 4$. *Use additional shapes with the following last rows:* □□■■, □■■□, ■□□■, *where the white boxes are those removed. Check that* $h_4 = 36$, *which coincides with that from Exercise 1.4.7.*

Exercise 1.4.13 $^{\heartsuit}$ *Check that there are* 12 *different Latin squares of size* 3×3 *with the entries* $0, 1, 2$ *(or any given set of* 3 *symbols).*

Exercise 1.4.14 $^{\diamondsuit}$ *Let* L' *be a* 4×4 *square with one empty box and the entries* $0, 1, 2, 3$ *in the others.*

(i) Check that if there are no repetitions in all columns and all rows in L', *then it can be uniquely extended to a Latin square* L.

(ii) Assuming additionally that the diagonal and anti-diagonal in L' *have no repetitions, prove that* L *then has no repetitions there too.*

Hints: (i) If 3 rows have no repetitions, then the 4th row must contain $0, 1, 2, 4$ and cannot have repetitions too; the same for columns.
(ii) Extend L' to L as in (i). Assume that the diagonal of L' contains the empty space and it contains in L some x twice. There are 6 places for x;

it suffices to consider $\begin{pmatrix} x*** \\ *x** \\ **** \\ **** \end{pmatrix}, \begin{pmatrix} x*** \\ **** \\ **x* \\ **** \end{pmatrix}, \begin{pmatrix} x*** \\ **** \\ **** \\ ***x \end{pmatrix}, \begin{pmatrix} **** \\ *x** \\ **x* \\ **** \end{pmatrix}$. In the 1st and the 2nd cases, there must be at least one x in the anti-diagonal, which is impossible because L is a Latin square. In the 3rd and the 4th cases, it can be only $\begin{pmatrix} x*** \\ **x* \\ *x** \\ ***x \end{pmatrix}, \begin{pmatrix} ***x \\ *x** \\ **x* \\ x*** \end{pmatrix}$. This is impossible because the empty space cannot belong to the main anti-diagonal, which gives that it contains x exactly once by the assumption.

Exercise 1.4.15 ♡ *(i) Construct a pair of orthogonal Latin squares of order* 4 *as follows. Using the renamings, the first rows can be assumed* $\{0, 1, 2, 3\}$ *in both squares, and the first column in the first LS can be assumed* $\{0, 1, 2, 3\}$ *too (why?). Then permute the remaining* 3 *rows provided that they are "sufficiently general" to obtain the second LS.*

(ii) More exactly, check that there are exactly 4 *LS with* $\{0, 1, 2, 3\}$ *as the 1st row and the 1st column and only one of them forms MOLS with at least one LS obtained from it by permutations of the remaining* 3 *rows; find it.*

Hint for (ii): There are 3 possibilities for 0 in the second row, and totally 4 their extensions to LS. Their list is in Exercise 1.4.18(iii). The 3-permutations for obtaining MOLS from them can be only cyclic, so there are 8 possibilities to examine. The zeros must be at the main diagonal for this and any cyclic permutation will do in this case.

Exercise 1.4.16 ♣ *A group G is a set of elements x with multiplication* $(x, y) \mapsto xy$ *and the unit* e, *satisfying the defining relations* $ex = xe$ *for any* x, *provided the associativity* $x(yz) = (xy)z$ *for any* x, y, z *and the existence of the inverses* x' *for any* x, *defined as follows:* $x'x = x'x = e$. *We do not assume that* $xy = yx$. *Then the multiplication table* $(x, y) \mapsto xy$ *and the division table* $(x, y) \mapsto y^{-1}x$ *are mutually orthogonal Latin squares for any finite G of odd order (the number of elements).*

Hint: Assume that $xy = \tilde{x}\tilde{y}$ and obtain from this that $y^{-1}x = \tilde{y}^{-1}\tilde{x}$ and $x^2 = \tilde{x}^2$. Then use the Lagrange Theorem: Problem 5.3.1 below.

Exercise 1.4.17 ♣ *Justify that the size of a set of* $n \times n$ *Latin squares that are pairwise MOLS is no greater than* $n - 1$.

Hint: The first rows in *all* squares can be assumed to be $\{0, 1, \ldots, n-1\}$, and the first entry in the second row cannot be 0 in any of these LS; see also Problem 5.2.4.

Exercise 1.4.18 ♣ *Latin squares are called* ***isotopic*** *if they can be obtained from each other by (a) permuting the rows, (b) permuting the columns, and (c) renaming the symbols. You can use below that any permutation is a composition of transpositions (ij), exchanging i and j.*

(i) Check that operations of different types commute with each other, i.e. their composition does not depend on the order. Use that it suffices to verify that applying the transposition of rows $i_1 < i_2$ after or before the transposition of columns $j_1 < j_2$ gives the same output; see also Section 7.1.4.

(ii) Check that any LS can be isotopically transformed to a ***reduced*** *one, which is defined to be with the sequence $\{0, 1, \ldots, n-1\}$ in the first row and $\{0, 1, \ldots, n-1\}$ in the first column.*

(iii) Prove that there are exactly 4 different reduced LS of size 4×4:

$$\begin{pmatrix} 0123 \\ 1032 \\ 2301 \\ 3210 \end{pmatrix}, \begin{pmatrix} 0123 \\ 1302 \\ 2031 \\ 3210 \end{pmatrix}, \begin{pmatrix} 0123 \\ 1032 \\ 2310 \\ 3201 \end{pmatrix}, \begin{pmatrix} 0123 \\ 1230 \\ 2301 \\ 3012 \end{pmatrix}.$$

Exercise 1.4.19 ♣ *(i) Use (i), (ii) from Exercise 1.4.18 to check that any Latin square of size 4×4 is isotopic either to the 1st or to the 2nd LS in (iii) there.*

(ii) Check that among these 4 LS above the first 2 are not isotopic to each other, and the last 2 are isotopic to the second.

Hint for (ii): To prove that the first two LS from (iii) are non-isotopic, count the number of 2×2 Latin subsquares there, which is an isotopy invariant (check!). To obtain, for instance, the 4th square from the 3rd one, apply (23), "the middle transposition" in the columns, then that in the rows and then that for the entries (i.e. all three types will be used).

Exercise 1.4.20 ♣ *Prove that there are only two orthogonal mates with the first row 0, 1, 2, 3 for the first square in Exercise 1.4.18(iii), which is* $L = \begin{pmatrix} 0123 \\ 1032 \\ 2301 \\ 3210 \end{pmatrix}$.

Namely, they are: $M_{1,2} = \begin{pmatrix} 0123 \\ 3210 \\ 1032 \\ 2301 \end{pmatrix}, \begin{pmatrix} 0123 \\ 2301 \\ 3210 \\ 1032 \end{pmatrix}.$

Hint: We use that the diagonal and anti-diagonal of L are constant. This gives that its orthogonal mate can be only of the following 2 types: $M_1 = \begin{pmatrix} 0\ 1\ x\ 3 \\ *\ x\ *\ * \\ *\ *\ *\ x \\ x\ *\ ** \end{pmatrix}$, or $M_2 = \begin{pmatrix} 0 & x & 2 & 3 \\ * & * & * & x \\ * & * & x & * \\ x & * & * & * \end{pmatrix}$, where $x = 2$ in M_1 and $x = 1$ in M_2. Then, the diagonal must be $0, 2, 3, 1$ for M_1 and $0, 3, 1, 2$ for M_2; the rest is straightforward.

Exercise 1.4.21 $^\diamondsuit$ *Given an $n \times n$ Latin square, we construct a set of ordered triples $\langle i, j, a\rangle$, where $0 \le i, j, a \le n-1$, i is any row number, j is any column number and a is the corresponding entry.*

(i) Given LS, check that any 2 such triples may have no greater than 1 coincidence entry-wise; for instance, $\langle 3, 1, 3\rangle$ and $\langle 3, 2, 3\rangle$ (2 coincidences) cannot appear together in this set.

(ii) Form the sets of pairs $\{\langle i, j\rangle\}$, $\{\langle i, a\rangle\}$ and $\{\langle j, a\rangle\}$ from $\{\langle i, j, a\rangle\}$. Deduce from (i) that given LS, all possible pairs occur in any of these 3 sets, and that an arbitrary set of n^2 triples satisfying the condition from (i) can be uniquely lifted to a Latin square.

(iii) Given a 3-permutation σ, apply it to $\langle i, j, a\rangle$ for all triples associated with a given Latin square. Then Latin squares remain Latin, and MOLS go to MOLS if the same 3-permutation is used for both LS in MOLS.

Hint for (iii): The pair LS, LS' is MOLS if and only if no two entries coincide in the corresponding ordered quadruplets $\langle i, j, a, a'\rangle$.

Exercise 1.4.22 $^\clubsuit$ *As in Exercise 1.4.21, we represent Latin squares as collections of triples $\langle i, j, a\rangle$, where i and j are the column and row numbers, and a is the corresponding entry at (i, j) for $0 \le i, j, a < n-1$. Verify that the 4th Latin square from Exercise 1.4.18(iii) remains isotopic to itself upon any of the 3-permutations of i, j, a. See Exercise 1.4.19 and Exercise 1.4.21(iii).*

Hint for (ii): The 4th square, L, is symmetric; it is the addition table modulo 4 (below). Therefore it suffices to examine only the cyclic permutation $\langle i, j, a\rangle \mapsto \langle a, i, j\rangle$. The corresponding LS will be the subtraction table modulo 4; its reduced form coincides with L.

Exercise 1.4.23 $^\heartsuit$ *Extend each of the partial squares below to a* ***mini-Sudoku****, which is a Latin square where $0, 1, 2, 3$ (all) are present in its 4 natural 2×2 blocks. The partial squares are:* $\left[\begin{array}{cc|cc} 0 & 3 & & \\ & & & \\ \hline & & & \\ & & 1 & 3 \end{array}\right]$, $\left[\begin{array}{cc|cc} & & 0 & \\ 2 & & & \\ \hline & 2 & & \\ & & & 1 \end{array}\right]$.

Exercise 1.4.24 ♢ *Partial Latin squares are when numbers* $0, 1, \ldots, n-1$ *(or for any* n *symbols) are placed in a subset of boxes in the* $n \times n$ *square such that there are no repetitions in columns and rows within this subset. Find at least two completions of the partial Latin square below to a* 6×6 *Latin square:* $\begin{bmatrix} 0 & 1 & 2 & 3 & 4 & 5 \\ 3 & 4 & 1 & 5 & 2 & 0 \\ 5 & 2 & 3 & 0 & 1 & 2 \end{bmatrix}$. *[We note that generally, the existence is guaranteed for any partial subrectangles of size* $m \times n$ *with* $m < n$ *by* Hall's Theorem*; the non-uniqueness holds for any* $m \leq n-2$*. Do not use this fact.]*

Exercise 1.4.25 ♣ *By analyzing* 4 *or* 5 *as possible entries at the bottom-right corner, prove that the following partial Latin squares cannot be completed to (full) Latin squares:*

$$\begin{bmatrix} 1 & 2 & 3 & 4 & 5 & 6 \\ 2 & 6 & 1 & 5 & 4 & 3 \\ 3 & 5 & 4 & & & \\ 4 & 3 & 5 & & & \\ 5 & 4 & 6 & & & \\ 6 & 1 & 2 & & & \end{bmatrix}, \quad \begin{bmatrix} 1 & 2 & 3 & 4 & 5 & 6 & 7 \\ 2 & 1 & 7 & 6 & 4 & 5 & 3 \\ 3 & 7 & 2 & & & & \\ 4 & 5 & 6 & & & & \\ 5 & 6 & 4 & & & & \\ 6 & 4 & 5 & & & & \\ 7 & 3 & 1 & & & & \end{bmatrix}.$$

[See [12] *for further information. Conjecturally, the completion of such* $\{n \times 3\} \cup \{2 \times n\}$ *partial Latin "fat hooks" is always possible for* $n \geq 8$*.]*

Exercise 1.4.26 ♠ *A* ***critical set*** *is a partial Latin square such that it has a unique completion to a full Latin square but deleting any of its entries makes such a continuation not unique.*

Prove that there is a unique continuation of the following partial square to one of size 6×6*, which will actually be the addition table modulo* 6*, and verify that this set is critical:* $\begin{bmatrix} 0 & 1 & 2 & & & \\ 1 & 2 & & & & \\ 2 & & & & & \\ & & & & & \\ & & & & & 3 \\ & & & & 3 & 4 \end{bmatrix}$.

Exercise 1.4.27 ♡ *Show that a magic square of order* 3 *must have* 5 *in the middle. Then prove that there exists only one MS of this size up to the rotations and reflections. Finally, justify that there are* 8 *magic squares of order* 3.

[12] C. Casselgren, H. Göransson, *Completing partial Latin squares with two filled rows and three filled columns*, Journal of Combinatorics, 14:1 (2023), 139–153.

Exercise 1.4.28 $^\heartsuit$ *Check that among all pandiagonals and anti-pandiagonals in the Lo Shu MS, only the (main) diagonal and anti-diagonal have 6 as the sum of entries. Deduce from it that any transposition of two neighboring rows or columns in this MS, does not change the vertical and horizontal sums, but changes the sums for the (main) diagonal and anti-diagonal, both.*

Hint: The diagonal is transformed to some anti-pandiagonals (not the main anti-diagonal) under such transpositions.

Exercise 1.4.29 $^\diamondsuit$ *A quasi-magic square,* ***quasi-MS***, *is when the anti-diagonal has any sum of entries, but the sum remains magic for the diagonal.*

(i) Check that any permutation of the rows followed by the same *permutation of the columns sends quasi-MS to quasi-MS, and they can be used, given a quasi-MS, to perform any permutation of the entries at the diagonal.*

(ii) For $n = 3$, check that the diagonal of a quasi-MS can be $\{2, 5, 8\}$ or $\{4, 5, 6\}$ upon a permutation from (i). For $\{4, 5, 6\}$, check that the resulting quasi-MS is "Lo Shu" or its reflection in this diagonal.

(iii) Apply all 6 simultaneous permutations as in (i) to "Lo Shu" and verify that there only 2 different quasi-MS, excluding "Lo Shu", can be obtained up to the rotation by π.

Hint: The numbers $1, 3, 9, 7$ are involved only in 2 different sums $a+b+c = 15$ for pairwise distinct $1 \le a, b, c \le 9$.

Exercise 1.4.30 $^\heartsuit$ *(i) Use de la Loubère's (Siamese) method to construct a magic square of order 9. (ii) Calculate its modular split, i.e. the matrices Q, R.*

Exercise 1.4.31 $^\diamondsuit$ *Associate to a given SMS, semi-magic square, of order n a physics system treating each entry m_{ij} as the mass of the particle positioned at the center of the corresponding unit square. Check that the center of mass is the middle point of SMS. The momentum of inertia I_z about this center (about the z-axis through it) is the sum of all $m_{ij}r_{ij}^2$, where r_{ij} is the distance from the corresponding particle to the center of mass. Justify that I_z depends only on n (Peter Loly), $I_z = 60$ for $n = 3$ and $I_z = 340$ for $n = 4$.*

Hint: $I_z = I_x + I_y = 2I_x$ for $I_x = \sum_{i,j} m_{ij}(i - \frac{1}{2} - c)^2 = \sum_{i=1}^{n}(i - \frac{1}{2} - c)^2 \sum_{j=1}^{n} m_{ij} = \Sigma_n \sum_{i=1}^{n}(i - \frac{1}{2} - c)^2$, for the center (c, c) and $\Sigma_n = \frac{n^2(n^2+1)}{2}$. To simplify a bit: $I_x = I_{edge} - n\Sigma_n\left(\frac{n-1}{2}\right)^2$ for $I_{edge} = \sum_{i,j} m_{ij}(i-1)^2$. Using Exercise 3.7.10, $I_z = n^2(n^4 - 1)/12$ (a surprisingly nice formula due to P. Loly).

Exercise 1.4.32 $^\diamondsuit$ *Prove that the following* partial square *cannot be completed to a full magic square of order* 4×4*:* $\begin{bmatrix} 2 & 3 & \cdots & \cdots \\ 4 & \cdots & \cdots & \cdots \\ \cdots & \cdots & \cdots & \cdots \\ \cdots & \cdots & \cdots & \cdots \end{bmatrix}$.

Hint: The first row $2, 3, x, y$ satisfies $2+3+x+y = 34$. The possibilities for $\{x, y\}$ are $\{13, 16\}$ and $\{14, 15\}$. On the other hand, the first column $2, 4, a, b$ is with $2+4+a+b = 34$. So $\{a, b\}$ can be $\{12, 16\}$ or $\{13, 15\}$. Thus $\{a, b\} = \{12, 16\}$, and $\{x, y\} = \{14, 15\}$ (up to transpositions). Otherwise, 13, 15, or 16 would appear twice. Next, b cannot be 16, since $b+y \geq 16+14 = 30$ in the anti-diagonal (through b) and $2, 3, 4$ were already used. Thus $a = 16, b = 12$. Then the anti-diagonal must contain 1: otherwise, $12 + 14 + 5 + 6 > 24$. So its 2 inner entries are $\{1, 6\}$ or $\{1, 7\}$. Thus we have either $\{3, 1\}$ in the 2nd column or $\{4, 1\}$ in the 2nd row. This is impossible, since their sum is 4 or 5, and $12, 14, 15, 16$ are already used: $5 + 13 + 11 = 29 < 34$.

Exercise 1.4.33 $^\spadesuit$ *(i) Prove that the following* partial square *cannot be completed to a full magic square of order 4:* $\begin{bmatrix} 3 & 4 & \cdots & \cdots \\ 5 & \cdots & \cdots & \cdots \\ \cdots & \cdots & \cdots & \cdots \\ \cdots & \cdots & \cdots & \cdots \end{bmatrix}$.

(ii) However, replacing 5 *by* 6 *here or* 3 *by* 2 *will make this doable; find the corresponding completions.*

Hint for (i): Only $\{16, 11\}, \{15, 12\}, \{14, 13\}$ can occur in the first row (with a possible transposition inside these pairs). Correspondingly, the 1st columns will contain $\{14, 12\}$, $\{16, 10\}$ and either $\{16, 10\}$ or $\{15, 11\}$ in the case of $\{14, 13\}$ above. Therefore, either the pairs (a) $14, 16$, or (b) $15, 16$, or (c) $14, 15, 13$ must be among the 4 entries in the 1st column. Next, the rows and columns containing $16, 15$ or 14 must also contain 1 or 2, because $3, 4, 5$ were already used. Thus, one of the 2 last entries in the diagonal through 3 must be either $v = 1$ or $v = 2$. However, then $3 + x + y + v = 34$ for the remaining unknown entries x, y at this diagonal, and $x + y$ is 30 or 29, which is possible only for $16 + 14$, $16 + 13$, or $15 + 14$. Neither possibility is compatible with (a), or (b), or (c) obtained above.

Answers for (ii): $\begin{bmatrix} 3 & 4 & 13 & 14 \\ 6 & 16 & 1 & 11 \\ 15 & 9 & 8 & 2 \\ 10 & 5 & 12 & 17 \end{bmatrix}, \begin{bmatrix} 2 & 4 & 15 & 13 \\ 5 & 14 & 3 & 12 \\ 16 & 7 & 10 & 1 \\ 11 & 9 & 6 & 8 \end{bmatrix}$.

Exercise 1.4.34 $^\heartsuit$ *Perform modular split for the first* 4×4 *magic square in Section 1.2.1 (the Indian one):* $\begin{pmatrix} 7 & 12 & 1 & 14 \\ 2 & 13 & 8 & 11 \\ 16 & 3 & 10 & 5 \\ 9 & 6 & 15 & 4 \end{pmatrix}$. *Check that both squares, Q and R, are Latin. Moreover, the diagonals and anti-diagonals of Q and R have all different entries, though some of their pandiagonals contain repetitions. Deduce from this "without calculations" that the initial square is magic.*

Exercise 1.4.35 $^\diamond$ *Given a pandiagonal MS of size n, present it as a union of the rectangular blocks "*$k \times n$*" and "*$(n-k) \times n$*" for* $1 \le k \le n-1$ *and transpose them (without changing the blocks). Check that this results in a pandiagonal MS, and the same holds for columns. Use this to check that the transpositions of rows* $1, 3$ *or* $2, 4$ *(or with the columns) transform pandiagonal MS of size* 4×4 *to other pandiagonal MS.*

Hint: The transposition of rows $1, 3$ is a composition of the reflection of a given MS in its middle horizontal line and then moving the resulting bottom row to the top.

Exercise 1.4.36 $^\diamond$ *The Pyramid method of constructing magic squares is described in Fig. 1.8. Follow this figure to define it for any odd n. Then construct such a square for* $n = 7$.

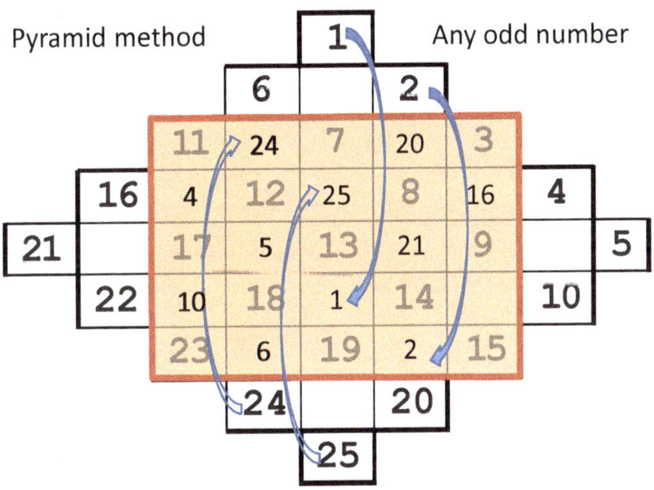

Figure 1.8: Pyramid method for odd MS.

Exercise 1.4.37 $^\clubsuit$ *(i) Demonstrate that the modular split provides a justification of the Pyramid method for* $n = 5, 7$*, which is explained in Fig. 1.8, i.e. check that the corresponding* Q, R *are Latin squares, and that their diagonal and anti-diagonal sums are* $0 + 1 + \cdots + (n-1)$.

(ii) Find the corresponding patterns for Q, R *and justify that they hold for any odd n. Then prove that the Pyramid method always results in MS.*

Exercise 1.4.38 ♢ *As we know, the rotations by* $0, \pi/2, \pi, 3\pi/2$ *around the center and the reflections, vertical, horizontal (in the middle lines) and 2 diagonal ones, transform* $n \times n$ *Latin or magic squares to other such ones for any* n*. Justify that the composition of any two of them is again one of these 8.*

For Latin squares, check that any of these operations can be obtained as compositions of the permutations in their rows and columns and the permutation $\langle i, j, a \rangle \mapsto \langle j, i, a \rangle$*; see Exercise 1.4.21.*

Hint: Check that the composition of a rotation and a reflection is a reflection, and the composition of 2 reflections is a rotation. The reflection in the diagonal is $\langle i, j, a \rangle \mapsto \langle j, i, a \rangle$. The rotation by $\pi/2$ is this reflection composed with $(0, 1, \ldots, n-1) \mapsto (n-1, n-2, \ldots, 1, 0)$ in the columns.

Exercise 1.4.39 ♡ ***Fleissner grille*** *is a cipher based on the* $2m \times 2m$ *square grid with* m^2 *perforations (holes) and the top of the grid marked "North". It is called* admissible *if the perforations, i.e. the* m^2 *boxes that are removed, satisfy the following property. No perforation (hole) can be rotated onto another around the center of the square by the angles* $\pi/2$, π *and* $3\pi/2$.

A message of length $4m^2$ *is written in these holes starting from top to bottom, then in the holes obtained by the rotations by* $\pi/2, \pi$ *and* $3\pi/2$*; see Fig. 1.9. If there are several holes in the same row, they are ordered naturally. The idea is old, but this cipher was used even during World War I.*

(i) Check that there are exactly 4^{m^2} *admissible* $2m \times 2m$ *squares with* m^2 *perforations and with one side marked by "North". Use that the orbit of any box with respect to such rotations must contain exactly 1 perforation.*

(ii) Instead of the usage of rotations in (i), consider the horizontal and vertical reflections about the center lines of a given square, and their composition, which is the rotation by π*. Then there are* 4^{m^2} *(the same number) admissible squares with perforations and with 1 corner marked by "NE" for the corresponding definition of admissibility.*

Hint: It is 4^{m^2} due to m independent choices of perforations in the orbits under the rotations in (i) or for the symmetries in (ii).

Exercise 1.4.40 ♢ *Fleissner grille for* $2m = 8$ *from Exercise 1.4.39 is called* balanced *if the corresponding* 8×8 *square* S *is as follows. Any row and any column of its 4 natural* 4×4 *subsquares* S_i $(i = 0, 1, 2, 3$ *for "NE", "SE", "SW", "SE") has exactly one perforation. Establish one-to-one correspondence between such* S *of size* 8×8 *and* 4×4 *Latin squares* L.

Hint: Given L, the corresponding perforations are the boxes in S_0 upon their rotations by $i\pi/2$ around the center of S to the corresponding S_i, where i are

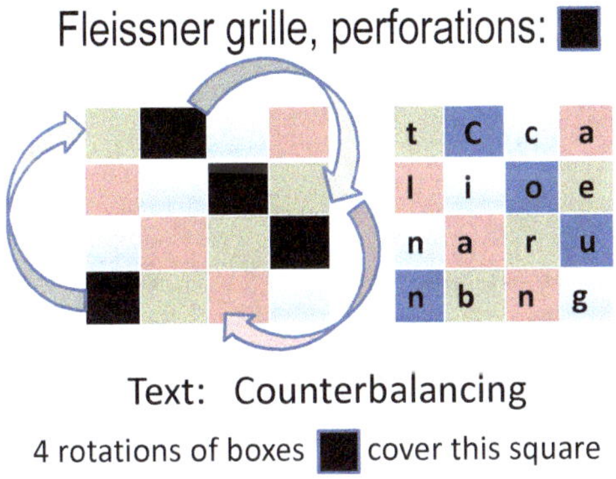

Figure 1.9: Using Fleissner grille.

the corresponding entries in L. For instance, the perforations in S_0 are the boxes of L with zero entries.

Exercise 1.4.41 ♣ *Calculate the number of $2m \times 2m$ squares with m^2 perforations admissible as in (i) from Exercise 1.4.39 (for the rotations) and, at the same time, admissible in the sense of (ii) there. In other words, count the squares with perforations that satisfy both definitions of admissibility. The example in Fig. 1.9 is such.*

Hint: The orbits of the boxes of S under all 8 symmetries of the square contain either (a) 4 boxes for the diagonal and anti-diagonal starting ones, or (b) 8 boxes otherwise. There are 4 choices for the perforation in every orbit for (a); there are m such orbits. For (b), the perforations suitable for (i) and (ii), for both, can be any pairs of boxes in the orbit obtained from each other either by the diagonal reflection of S or by the anti-diagonal one. This gives 8 choices for these perforations in every such orbit. There are $(4m^2 - 4m)/8 = \frac{m(m-1)}{2}$ orbits of type (b). Finally, the answer is $4^m \times 8^{\frac{m(m-1)}{2}}$.

Exercise 1.4.42 ♢ *The **Aztec diamond** of order n is the union of unit squares with integral corners (x, y) such that $|x|+|y| \le n+1$. See Fig. 1.10 for $n = 3$. Let A_n be the number of its perfect coverings (tilings) by dominos.*

(i) Check that the number of such coverings with the top 2 boxes covered vertically is A_{n-1}, and $2 + 2$ extreme left and right boxes will be covered vertically too in this case. Here $n > 1$, $A_1 = 2$. Obtain from this that $A_2 = 3 + d_5 = 8$ for $d_n = \mathcal{N}^{\square\square}_{2\times n}$.

(ii) Apply (i) to check that $A_3 = \mathcal{N}^{\square\square}_{4\times 4} + 2(2A_2 - A_1)$. Consider the cases when $2+2+2+2$ extreme top, bottom, left and right boxes are covered by 4

dominos (which gives $\mathcal{N}_{4\times 4}^{\square\square}$), 5 dominos and 6 dominos ($A_1 = 2$ coverings). Thus, $A_3 = 64$ due to $\mathcal{N}_{4\times 4}^{\square\square} = 36$ (do not prove this: see Fig. 7.1). [Generally, $A_n = 2^{n(n+1)/2}$ due to N. Elkies, G. Kuperberg, M. Larsen, and J. Propp, which is the **Aztec Diamond Theorem**. *See e.g.* [13]*.]*

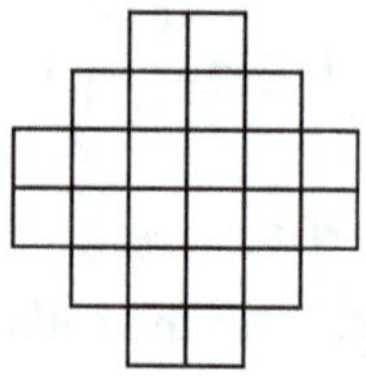

Figure 1.10: Perfect coverings of the Aztec diamond for $n = 3$.

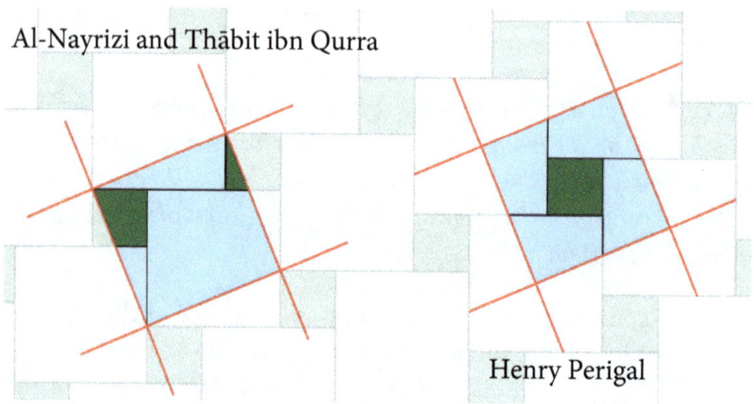

Figure 1.11: Pythagorean Theorem by dissections.

Exercise 1.4.43$^{\heartsuit}$ *Prove that there exists only one (full) tiling of the plane (up to translations) with 2 distinct squares such that no two squares have a common side, called* **unilateral**. *It is shown in Fig. 1.11. Use it to prove the* **Pythagorean Theorem** *in two ways: due to Al-Nayrizi and Thābit ibn Qurra (9th century) and due to Henry Perigal (19th century).*

Hint: Check that any small square must share a vertex with any neighboring large square. So it must be surrounded by 4 large squares. Then continue by "induction." [The classification of unilateral coverings of the plane with 3 distinct squares is significantly more difficult.]

[13] S-P. Eu, T-S. Fu, *A simple proof of the Aztec diamond theorem*, The Electron. J. Combin. 12: 1 (2005), R18.

Chapter 2

PERMUTATIONS AND COMBINATIONS

The key concepts in enumerative combinatorics are permutations and combinations. Groups of permutations are fundamental in the origins of representation theory, and combinations are the basis of the fundamental binomial distribution. Permutations have a lot of applications, but combinations have many more.

2.1 Pigeonhole Principle

A starting point of almost any course in combinatorics is the *Dirichlet Drawer Principle*, also called the ***Pigeonhole Principle***. It is as follows.

(I) *If n items are put into m containers, with $n > m$, then at least one container must contain more than one item.*

(II) *If $n = m$, then the following two statements are equivalent: "all containers are occupied" and "each one contains 1 item".*

See Fig. 2.1; the 2nd map there is a bijection (we use $\leftrightarrow$). This principle appeared in the literature about 200 years before Dirichlet. The name "Pigeonhole Principle" was introduced well after his work. Dirichlet himself called it the "drawer-shelf principle" (1834), having in mind a drawer used for storing and sorting mail, called (even now) a "pigeon-hole mailroom sorter."

This principle is of course obvious, but all of Euclid's axioms are "obvious" too, except for the one about parallel lines. The Pigeonhole Principle is used very frequently. For instance, we already applied (II) several times when doing Latin squares: *an $n \times n$ matrix with entries $\{0, 1, \ldots, n-1\}$ has no repetitions in columns nor rows if and only if each column/row contains them all.*

Figure 2.1: Pigeons $\rightsquigarrow$ pigeonholes, (I) and (II).

2.1.1 Some formalization

We will introduce the typical mathematical language employed for studying and proving enumerative facts. The set $\{1, 2, \ldots, n\}$ will be denoted by $[n]$. It is a standard set with n distinct elements. The names of the elements are irrelevant for many considerations, but we will mostly stick to $1, 2, 3, \ldots$ unless stated otherwise. Sometimes the usage of $\{0, 1, \ldots, n-1\}$ will be more convenient. Let $f : [n] \to [m]$ be a map from set $[n]$ (the domain of f) to set $[m]$ (its target). Maps in this book are between finite sets.

It will be convenient sometimes to write $f : x \mapsto y$ instead of $f(x) = y$. Notice a somewhat different kind of arrow here, convenient to distinguish between the maps of sets, and the images of individual elements. Such usage of different types of arrows is common in mathematics papers, but ignoring this will hardly lead to any misunderstanding.

In this framework, (I) means that at least two elements from $[n]$ will have coinciding f-image, i.e. $f(x) = f(y)$ for at least one pair of distinct elements x and y in $[n]$ (we write $x \neq y$) if $n > m$.

We will frequently use the notation $|A|$ for the number of elements of a set A. Then (I) can be compressed as follows. *If* $|A| > |B|$ *then for any map* $f : A \to B$ *there exist* $x \neq y$ *in* A *such that* $f(x) = f(y)$.

Claim (II) is that f is a *bijection* of sets if and only if $n = m$ and different elements from $[n]$ have different images in $[m]$.

We have already used bijections. A ***bijection*** or ***1-1 correspondence*** between two sets A and B is a pair of maps, $f : A \to B$ and $g : B \to A$ such that $gf = id_A$ and $fg = id_B$; then g is called the inverse of f denoted f^{-1}.

Here id_A is the ***identity map***: $a \mapsto a$ for any a in A. Correspondingly, id_B is the identity map on the set B. The product fg is the composition (consecutive application) of two maps, f and g (first g then f); it is also denoted by $f \circ g$, but we will mostly omit $\circ$. Namely, for any maps $g : A \to B$ and $f : B \to C$, $fg : a \mapsto c$ for $c = fg(a) = (f \circ g)(a) = f(g(a))$. See also Definition 2.4.2.

Using the standard notation $a \in A$ for elements a in A, the definition of bijection means that $f^{-1}(f(a)) = a$ and $f(f^{-1}(b)) = b$ for any $a \in A$ and any $b \in B$. The existence of a bijection implies that $|A| = |B|$.

We note that beyond combinatorics, the word "isomorphism" is common instead of "bijection"; it has the same meaning for maps between sets. The usage of the language of maps is very convenient for the Pigeonhole Principle. Let us demonstrate it in the following generalization of (I):

(III) *For a map $[kn + l] \to [n]$ with $l > 0$, at least $(k+1)$ pairwise distinct elements in the set $[kn+l]$ map to the same object in $[n]$. Equivalently, for any sets A,B such that $|A| > k|B|$ and any map $f : A \to B$, there exist pairwise distinct $a_1, \ldots, a_{k+1} \in A$ such that $f(a_i) = f(a_j)$ for any $1 \le i, j \le k+1$.*

Indeed, the set A would contain no greater than kn elements if no greater than k elements in A have image b for each $b \in B$. These three principles, (I), (II) and (III), are frequently used in combinatorics. They are somewhat similar to axioms in geometry, but their justification is immediate.

A basic example of Pigeonhole Principle (I) is the claim that in a company with more than 367 workers there are at least two people with the same birthday. Of course, there can be many more than two; this principle only gives a lower bound. As for (III), at least 31 have the same zodiac sign.

2.1.2 Sets and subsets

Problem 2.1.1 *For the set of numbers $A = \{1, 2, \ldots, 200\}$, let us randomly pick a subset B with 101 (different) elements from A. Prove that B will contain at least one divisible pair, which means the existence of two different numbers x and y in B such that x divides y.*

Solution. Before presenting the solution, let us discuss this problem a bit. First of all, if 1 is a member of B, then *any* other element will be divisible by 1. Obviously, if small numbers occur in B, then there will be many divisible pairs. The bound 101 for B is really necessary. Indeed, the set $B_{100} = \{101, ..., 200\}$

with 100 elements has no divisible pairs because $2 \cdot 101 > 201$; 2 is the smallest possible factor here.

Generally, the number of elements in the set $\{m, m+1, \ldots, n\}$ is $n-m+1$ (notice the $+1$); we will use this frequently. For instance, $|B_{100}|$, the number of elements in B_{100}, is indeed $200-101+1 = 100$, though this is of course obvious without any "$n - m + 1$".

So the claim of the problem is *sharp*: there are sets B without divisible pairs if $|B| < 101$. Good problems are supposed to be sharp.

We note that in principle, many combinatorial problems can be answered by computer programs. Computers are very helpful, but not in this case. The number of subsets with 101 elements from a set with 200 elements is $\binom{200}{101} \approx 10^{59}$ (discussed later), so a computer solution is infeasible.

As the last general remark related to this problem, its solution becomes easier to approach if the following hint is provided: examine x and y in B such that y/x is a power of 2. Using this hint, we define $op(m)$, *the odd part of* m: m divided by the greatest possible power of 2. This is not a standard map; its purpose is only to solve this problem. For example, $op(137) = 137$, $op(8) = 1$, $op(24) = 3$, and so on. Then the solution is as follows.

Applying op to every m of A gives us the set $op(A) = \{1, 3, 5, \ldots, 199\}$; let us denote it by C. To count the number of elements $|C|$ of C, we use a variant of the formula "$n - m + 1$" with the gap s between neighboring numbers; such gaps are frequently called steps in loop operators in computer languages. Let $n - m$ be divisible by s. Then the number of elements in the sequence $m, m + s, m + 2s, \ldots, n$ is $(n - m)/s + 1$. This gives that $|C| = (199 - 1)/2 + 1 = 200/2 = 100$.

Upon the restriction of the map op to B, we have a map $op : B \to C$, where $|B| = 101$ and $|C| = 100$. So, by the Pigeonhole Principle, at least two numbers $x < y$ from any such subset B will have coinciding images under op. In other words, $op(x) = op(y)$, which implies that y/x is a power of 2. ❑

Main set symbols. We will use the following notations from time to time. For subsets B of A: we write $B \subset A$; or $A \supset B$. Recall that $a \in A$ or $A \ni a$ are used for elements a of A. The difference of A and B for $B \subset A$, is denoted by $A \setminus B$, which means removing B from A, i.e. taking the complementary subset. Also, $\emptyset$ is the empty set, $A \cup B$ and $A \cap B$ are the union and the intersection of the sets A and B.

The following problem is good practice for working with sets, their subsets and maps.

Problem 2.1.2 *If A is any set of 10 different integers between* 1 *and* 106, *then*

A must have two different nonempty subsets X and Y for which the sum of the elements in X equals the sum of the elements in Y. Moreover, such a pair exists with $X \cap Y = \emptyset$.

Solution. Consider a function σ on subsets of A defined by $\sigma : X \mapsto \sum_{x \in X} x$ (f maps the set X to the sum of the elements of X). Its values are from 1 to $106+105+\cdots+97=(106+97)5=1015$. However, the total number of subsets of A is $2^{10}=1024$, which becomes 1023 if we disregard the empty set $\emptyset$. Indeed, we can decide independently whether an element in A is to be in X or not, which is equivalent to tossing a coin 10 times. Since $1023 > 1015$, there are different subsets X and Y with the same σ due to Pigeonhole Principle (I). By subtracting $X \cap Y$ from X and Y, we make them non-intersecting. ❑

The existence of 10-sets A satisfying the "subset-sum-distinctness" for all its subsets X under $\sigma(X)$ requires the bound much greater than 106, so this problem is not sharp. This notion is due to Paul Erdős; it is connected with *complexity* in computer science. See, for instance, [14].

To give an example, there are $31=2^5-1$ different sums of nonempty subsets in $\{6, 9, 11, 12, 13\}$. They are: $\{6, 9, 11, 12, 13, 15, 17, 18, 19, 20, 21, 22, 23, 24, 25, 26, 27, 28, 29, 30, 31, 32, 33, 34, 36, 38, 39, 40, 42, 45, 51\}$; employ your computer.

Problem 2.1.3 *Use the list above to check that 35 is the smallest number that can be added to $\{6, 9, 11, 12, 13\}$ so that all subsets in the resulting set have different (pairwise distinct) sums, i.e. the Erdős' property is satisfied.* ❑

There are quite a few problems based on the Pigeonhole Principle. The following problem due to Erdős and Szekeres is of a more involved kind; it has applications in geometry-topology and is related to the *Robinson-Schensted correspondence*.

Problem 2.1.4 * *Any sequence $a_1, a_2, \ldots, a_{n^2+1}$ of distinct real numbers for $n = 1, 2, \ldots$ contains either a decreasing* subsequence *of length $n+1$ or an increasing one of the same length. By definition, sequences are ordered sets and subsequences are sequences that are obtained from the initial one by omitting some of its terms.*

Solution. First, this claim is a sharp one. For example, consider the sequence $3, 1, 4, 2$ with n^2 numbers for $n = 2$, so it is not of length $n^2 + 1$. Any subsequence of it with 3 numbers is neither increasing nor decreasing; only increasing/decreasing subsequences of length 2 can be achieved. However, if

[14] T. Bohman, *A construction for sets of integers with distinct subset sums*, The Electron. J. Combin. 5 (1998), R3.

you add the number 5 to any position in the sequence, then there will always be a subsequence with 3 numbers that is increasing or decreasing (check it). For instance, $5, 3, 1, 4, 2$, has 3 decreasing subsequences, $5, 3, 1$ and $5, 3, 2$ and $5, 4, 2$ of length 3, but no such 4-element subsequences.

The proof goes as follows. For any m from $1, 2, \ldots, n^2 + 1$, let ℓ_m be the length of the maximal, i.e. the longest, possible increasing subsequence in the sequence $(a_i, 1 \le i \le n^2 + 1)$ that begins with a_m.

For the sequence $5, 3, 1, 4, 2$, the ℓ-numbers are $1, 2, 2, 1, 1$. For instance, $3, 4$ is such a maximal subsequence from $a_2 = 3$. Thus $\ell_2 = 2$. Similarly, $\ell_3 = 2$ because $1, 4$ and $1, 2$ will do.

We can assume that each ℓ_m is smaller than $n+1$; otherwise, we have what we need: the existence of some increasing subsequence of length $n+1$. Then ℓ can be considered as a map $[n^2 + 1] \to [n]$, where the inputs are the positions m for each a_m and the output (image) is ℓ_m, which is at most n by assumption. So there must be a subset of at least $n+1$ "starting points" $a_{m_1}, a_{m_2}, \ldots, a_{m_k}$ with coinciding ℓ-numbers, i.e. $\ell_{m_i} = \ell_{m_j}$, due to the Pigeonhole Principle; let this value be ℓ, abusing the notation. The value of ℓ is important; it guarantees the inequality $k \ge n + 1$, which is what we need. One can of course assume that $m_1 < m_2 < \cdots < m_k$. So it gives a subsequence $a_{m_1}, a_{m_2}, \ldots, a_{m_k}$, and we claim that it is decreasing.

Indeed, if this is not true and $a_{m_1} < a_{m_2}$ (the first 2 numbers in this subsequence), then a_{m_1} can be added at the start of any maximal increasing subsequence starting with a_{m_2}, which must have length ℓ. However, this gives an increasing subsequence starting from a_{m_1} of length $\ell+1$, which contradicts the assumption that there is no such subsequence from a_{m_1}. This argument is applicable to any neighboring a_{m_i}, so this subsequence decreases. Hence we found a decreasing subsequence of length at least $n+1$. ❑

For example, $5, 3, 1, 4, 2$, has no increasing subsequences of length 3, the value of ℓ from the proof is 1, and the corresponding subsequence of starting points is $5, 4, 2$. It is decreasing indeed! So this proof is constructive. If there are no increasing subsequences of length $n+1$, then you (or your computer) will find a decreasing one of this size (at least).

This problem has interesting aspects. See e.g. Exercise 2.6.9. The next application of the Pigeonhole Principle is one of the origins of number theory.

2.1.3 Chinese Remainder Theorem

The Chinese Remainder Theorem, from about the 3rd century CE, is a relatively simple example of the usage of the Pigeonhole Principle, but this application is of fundamental importance. Let us summarize what we need from number theory: only some basic facts (no justifications for now).

Division with remainder was used in the previous chapter. Two integers $m, n \geq 1$ are called ***relatively prime*** if they do not have common divisors, other than 1. Using the ***gcd***, the greatest common divisor, $\gcd(m, n) = 1$.

The ***Euclid's algorithm*** shows that $\gcd(m, n)$ is the smallest positive integer represented as $am + bn$ for some integers a and b. This readily results in the ***Prime Factorization Theorem***, frequently called *Fundamental Theorem of Arithmetic*, which states that every natural number $n > 1$ can be broken down into its prime number factors (whose product is the n). The set of these primes and the corresponding powers is uniquely determined by n. These 2 facts, coupled with the invertibility of any nonzero numbers modulo a prime number p (the claim that $\mathbb{Z}_p$ is a field) are the foundation of number theory. The justifications are in Section 5.1.3; also, see [15].

Let us state and justify the ***Chinese Remainder Theorem***. We set $r = m \bmod n$ if r is the remainder of m upon the division by n; $0 \leq r < n$.

Proposition 2.1.5 *Let m and n be relatively prime positive integers. Then for any integers $0 \leq a < m$ and $0 \leq b < n$ there exists a unique $0 \leq x < mn$ such that $a = x \bmod m$ and $b = x \bmod n$, which means that a and b are the remainders of x when divided by m and n, respectively.*

Proof. Let us justify that the sequence $b, b + m, b + 2m, \ldots, b + (n-1)m$ forms a complete set of remainders modulo n, which is $\{0, 1, \ldots, n-1\}$. By "modulo n", we mean upon the division by n. This gives the existence of x.

Using Pigeonhole Principle (II), it suffices to check that no $b + um$ and $b+vm$ can have coinciding remainders. In other words, no two pigeons, $b+um$ and $b + vm$, occupy the same hole, the remainder b. We can assume $v > u$. If they did coincide, then $b + vm - (b + um) = (v - u)m$ would be divisible by n. However, this is only possible if $(v - u)$ is divisible by n.

Generally, if ab is divisible by n and a is relatively prime to n, then a has no prime factor in common with n, and therefore n has to divide b. This is why we concluded that n divides $(v-u)$. However, $(v-u)$ cannot be divisible by n simply because $0 < (v - u) < n$.

Thus, the required x is found and we only need to check its uniqueness in the range $0 \leq x < mn$. This actually follows from what we have already proved (check it). More directly: if x, x' have coinciding remainders modulo m and n, then the difference $x - x'$ is divisible by m and n and therefore must be divisible by mn, since m and n are relatively prime. So $x - x'$ is divisible by mn and thus $x = x'$ because $0 \leq x, x' < mn$. ❑

[15] G. Andrews, *Number theory*, Dover Books on Advanced Mathematics, 1994.

2.2 16 Cards and fault lines

The Pigeonhole Principle is used in many forms. The following two exercises from the previous chapter are some examples.

2.2.1 4×4 MOLS and 16 cards

Old famous problem of 16 cards:

Arrange the 16 face cards of a deck of playing cards in a 4 x 4 array so that each denomination (Ace, King, Queen, Jack) and each suit (Clubs, Hearts, Diamonds, Spades) appears only once in each row and column.

♠A	♥K	♦Q	♣J
♦J	♣Q	♠K	♥A
♣K	♦A	♥J	♠Q
♥Q	♠J	♣A	♦K

Replacing the suits and denominations by 1,2,3,4:

MOLS:

1	2	3	4
3	4	1	2
4	3	2	1
2	1	4	3

&

1	2	3	4
4	3	2	1
2	1	4	3
3	4	1	2

MOLS → MS: $< a, b > \rightarrow$ 1+4(a-1)+(b-1)=4a+b-4 :

MS:

1	6	11	16
12	15	2	5
14	9	8	3
7	4	13	10

Figure 2.2: Old 16-card problem as MOLS.

Problem 2.2.1 *(i) Find an example of the arrangements of 16 cards with denominations A, K, Q, J and all 4 suits as a 4×4 table without repetitions of denominations and suits in the rows, columns, the (main) diagonal, and anti-diagonal.*

(ii) Justify that the absence of the repetitions of denominations and suits in columns and rows among any 15 cards taken from 16 always provides their absence for all 16 cards. This claim can be extended to include the diagonal and anti-diagonal.

Solution. **(i)** An example is presented at Fig. 2.2. It is convenient to switch to the corresponding MOLS, obtained there if we replace the suits and the denominations by 1, 2, 3, 4. There are no repetitions in the diagonals and anti-diagonals of these LS. Recall the passage detailing MOLS over MS.

Due to the absence of the repetitions in (anti-)diagonals, the map $\langle a, b\rangle \mapsto 4a+b-4$ in this figure for the corresponding pairs of entries results in a MS. It is not pandiagonal magic: examine the pandiagonals of the first LS.

(ii) Let us now remove 1 card from 16 from consideration and disregard the row and column that it was in. The assumption in (ii) gives that the remaining 3 rows have no repetitions of denominations and suits. Since all cards are used, the disregarded row must contain all 4 denominations and all 4 suits. The same argument is applicable to the columns. The Pigeonhole Principle (II) was used here. See Exercise 1.4.14(i). The extension to the (anti-) diagonals is Part (ii) of this exercise (follow the hint there). ❑

Permutations, our next topic, are heavily present in the theory of Latin squares. In the example from Fig. 2.2, we permuted rows 2, 3 and 4 to go from the first LS to the second one, as was suggested in Exercise 1.4.15. We will interpret this MOLS using the *field with* 4 *elements* $\mathbb{F}_4$ in Section 5.3.

2.2.2 Shields from dominos

Exercise 1.4.10 on the fault lines in the 6×6 chessboard is of some general interest. Its natural solution (the hint was provided) looks somewhat tricky. However it appears relevant when you try to extend it to more general shapes.

Problem 2.2.2 *(i) Any perfect covering by dominos of the* 6×6 *shape has a fault line, one drawn between the boxes vertically or horizontally that does not intersect any dominos. (ii) Construct examples of perfect coverings by dominos without fault lines for "*6×8*" and for "*6×7*".*

Solution. It is based on some version of Pigeonhole Principle (III). The proof goes by contradiction. If there is no fault line, then every horizontal and vertical line between boxes ($5+5=10$ of them) crosses at least one domino. A parity argument gives that each line must pass through an even number of dominos. Indeed, any such line splits the chessboard into 2 rectangles with an even number of boxes in each. If it intersects an odd number of dominos, then these rectangles without the corresponding halves are impossible to cover.

For any domino, there is a *unique* line through it. Thus, every covering results in a map $\{18\ \textit{dominos}\} \to \{10\ \textit{lines}\}$. Parallel dominos will have coinciding images under this map. For instance, $\boxminus\cdots\boxminus$ map to the same horizontal line, the one through them. All 10 lines will be in the image of this map if no fault lines exist for a given covering. But then the total number of dominos must be no smaller than 2 times the number of lines, which is $10\cdot2=20$; this contradicts that there are only 18 dominos.

There is no such contradiction for the 6×8 shape: $(5+7)\cdot 2 = 48/2$. So the previous argument does not work. Perfect coverings without fault lines do exist for this shape; see Fig. 2.3. They exist for "6×7" too. The corresponding count of dominos gives $(6+5)\cdot 2 = 22 > 42/2 = 21$ in this case, which seems a proof that such coverings cannot exist. A mistake here is that 3 lines from 11, those that split "6×7" into 2 rectangles with odd number of boxes, can intersect $1, 3, \ldots$ dominos (not ≥ 2 as we used above).

(ii) An example of the fault-free covering in this case is the one at Fig. 2.3 for "6×8" where we delete the 1st column and replace the 2 halves at the bottom by one vertical domino (check). See Exercise 1.4.11 for "6×5", the smallest rectangle with fault-free perfect coverings (check). ❑

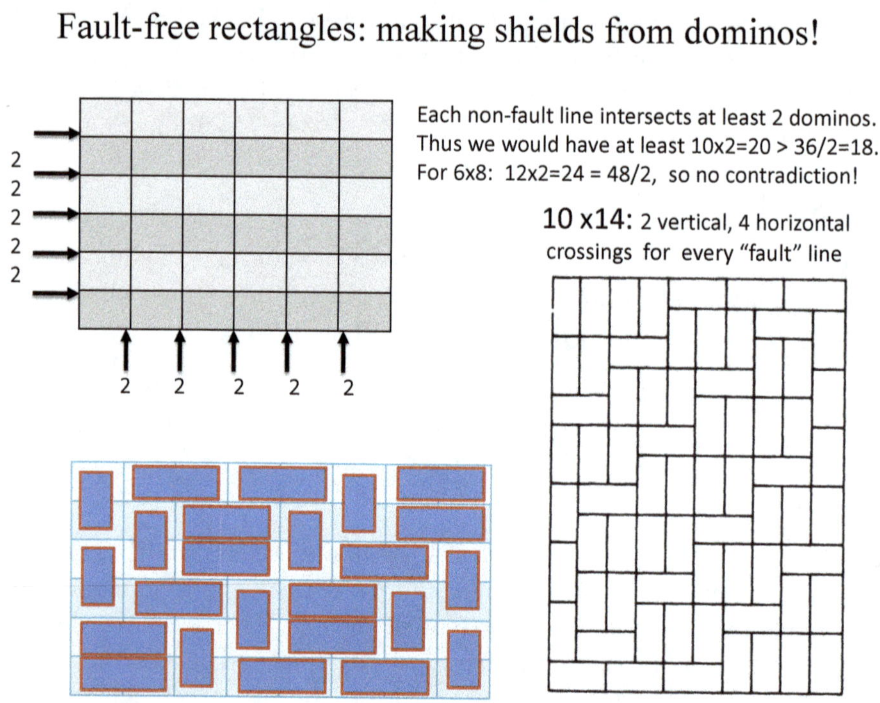

Figure 2.3: Fault-free coverings.

Shields. You want to create a shield against arrows, stones, even bullets (you name it) using "small" super-strong units, like dominos. Take a rectangle of plastic and glue the dominoes to it according to a perfect covering without fault lines. If there is at least one fault line you are not really protected: the shield can fold. If there are none, a shock may bend your shield, but at least it will not fold along horizontal or vertical lines.

If this is insufficient, make the number of intersections for every line greater than 2; see Fig. 2.3 and [16]. You can try to make them equally distributed, for instance, by maximizing the minimum of the distances between "parallel" dominos. Also, you can glue another set of dominos to the opposite side of your plastic. For a better protection, the second set must have no single domino in common with the first one; see Problem 6.1.5 and Fig. 6.7 concerning "non-intersecting" perfect coverings.

2.3 Permutations

This is the actual beginning of enumerative combinatorics. Permutations appear almost everywhere. The group of permutations, the *symmetric group*, is at the origin of the field of mathematics known as representation theory. Groups of symmetries are of major importance in mathematics and natural sciences. The symmetric group is discussed in Section 2.3.3 and is used in some topics and exercises, but we do not need it too much in this chapter.

Let us distribute n different objects in n different boxes (containers). The objects will mostly be the numbers $1, 2, \ldots, n$, but not always. The total number of such arrangements, ***permutations***, will be denoted $P(n)$.

Permutations can be of course defined without boxes, simply as any *sequences* (ordered!) of n different objects. However, the separation of objects and boxes greatly helps in applications and to define the composition of permutations, as we will see.

Using the set $[n] = \{1, 2, \ldots, n\}$, permutations are arbitrary sequences $(a_1, a_2, \ldots, a_n)$ where $a_i \neq a_j$ for $i < j$ and $1 \le a_i \le n$ for any indices i, j. Equivalently, *all* numbers from 1 to n must be used in such sequences.

Obviously: $P(1) = 1, P(2) = 2, P(3) = 6$. Generally, we have n choices for the first box, $(n-1)$ choices for the second box (from the remaining objects), and so on. Thus $P(n) = n(n-1)\cdots 3 \cdot 2 \cdot 1$.

Using the famous ***factorial function*** $n! = 1 \cdot 2 \cdots (n-1) \cdot n$, we obtain: $P(n) = n!$. Continuing: $P(4) = 4! = 24,\ P(5) = 5! = 120,\ P(6) = 720$.

The ***parity of a permutation***, 0 or 1, respectively *even* or *odd*, is the parity of the total number of ***inversions***: the pairs $i < j$ such that $a_i > a_j$. For $n = 3$, the permutations $(123), (231), (312)$ are even, the transpositions of any 2 elements, $(213), (132)$ and (321), are odd. Any transposition is odd (check). Here we omitted the commas between the numbers. Omitting the commas can be sometimes confusing when *cycles* are considered (later).

---

[16] M. Atkinson, W. Lunnon, *Regular fault-free rectangles*, The Mathematical Gazette, 64:428 (1980), 99–106.

More generally, *let $P(n,r)$ be the number of r-permutations from n objects*, the number of all ***sequences*** of r objects produced from a set of n different objects. Following the calculation of $P(n)$: $P(n,r) = n\cdot(n-1)\cdots(n-r+1)$, where there are n choices for the first box, $n-1$ choices for the second, ... , and, finally, $(n-r+1)$ choices for box r. The number of terms in this product is $n-(n-r+1)+1 = r$, as indicated.

One has: $P(n) = P(n,n), P(n,1) = n, P(n,2) = n(n-1)$. In terms of factorials, we obviously have: $P(n,r) = n!/(n-r)!$. This formula is compact and sometimes really convenient, but we will mostly use the direct product.

The latter formula requires some comment for $P(n,n)$: $0!$, which occurs in the denominator, is confusing. Permuting 0 objects might have been interpreted to be 0 because there is nothing to permute, or alternatively 1 because doing ***nothing*** is in a sense doing something. However, we ***must*** set $0! = 1$ here to "save" the formula for $P(n,n)$: *this is a mathematical necessity.* We will use $0! = 1$ constantly.

Similarly, $(-1)! = \infty, (-2)! = \infty$, and so on. Indeed, $P(n,n+m) = n!/(-m)!$ is obviously 0 for $m > 0$, because picking $n+m$ objects from n is impossible. Thus we ***must*** set $(-m)! = \infty$. We will use this too; it is especially convenient when we do not want to show explicit summation ranges.

Gamma function *. Extending the relation $n! = n(n-1)!$ to any number x instead of n, mathematicians (mostly Euler, 1729), found that $(1/2)! = \sqrt{\pi}/2$, $(-1/2)! = \sqrt{\pi}$ and so on. The recurrence above holds: $(1/2)! = (1/2)(1/2-1)!$. See any texts on the *Gamma function* $\Gamma(x)$. Its defining relation is $\Gamma(x+1) = x\Gamma(x)$ (plus some analyticity properties); the connection to the factorial function is $n! = \Gamma(n+1)$.

2.3.1 Game of Fifteen

This well-known game requires a 4×4 frame and 15 sliding units, 1×1-tiles, labeled $1,2,\ldots,15$. We begin with a random configuration of tiles in the frame, take one of the neighbors of the empty space and move (slide) this tile to the empty space. The objective is to arrange them in the order $1,2,\ldots,15,$ *space* in the frame from left-to-right and top to bottom.

Let us count the number of all possible configurations: it is 16! and not 15!. We can treat the empty space as a unit labeled by 16: then 16! is obvious.

The configurations can be obtained from each other if and only if they have the same ***15-invariant***, ξ_{15}, defined as the parity, 0 or 1, of the following sum: the parity of the corresponding permutation of $1,2,\ldots,15,16$ (left-to-right and down the frame) plus $dist_\square$, the distance from the empty space to the bottom-right corner. The distance is defined as $dist_\square = (4-i)+(4-j)$ if

Figure 2.4: Rearrange the tiles into order.

the empty space $\square$ is at row i and column j. Only the parity of $dist_\square$ matters: the top-left corner can be used instead of the bottom-right one in ξ_{15}.

This invariant does not change after an allowed move due to Exercise 2.6.38, so it turns out that $\xi_{15} = 0$ is necessary. The proof that $\xi_{15} = 0$ is sufficient for a configuration to be solvable is in Problem 2.3.7 at the end of Section 2.3.3. It essentially amounts to finding sufficiently many types of permutations that can be obtained by the moves.

To provide an example, the number of inversions is 45 in Fig. 2.4 (check), so this configuration cannot be transformed to $1, 2, \ldots, 15$, *space*. We will use ξ_{15} in the following variant of the 15-puzzle with cards instead of tiles.

Using cards as tiles ***.** Remove "♠A" from the 16 cards in Problem 2.2.1; its space will be treated as the empty field. Then begin with the configuration Fig. 2.5 (the "Richelieu" deck by A.G. Müller, 1935) or with any of its permutations with the empty field somewhere.

The moves will be as in the 15-puzzle: cards next to the empty space may move into it. The objective will be now to reach an arrangement of 15 cards without repetitions of denominations and suits in all rows and columns. We will also require the absence of repetitions in the (main) diagonal and the anti-diagonal. Exercise 1.4.14 provides that if no repetitions occur without "♠A", then there will be none after we put it (back) in the empty space. One may consider here avoiding repetitions in *all* pandiagonals or anti-pandiagonals, but this is impossible mathematically: no such Latin square exists for even n due to Exercise 2.6.39!

Now there are many "winning configurations" and any starting arrangement is "solvable," which is the following problem.

Figure 2.5: Use the empty (black) field to make it MOLS.

Problem 2.3.1 *Assign the numbers* $0, 1, \ldots, 15$ *in order left-to-right and down the frame to the* 16 *cards in the arrangement in Fig. 2.2. Perform the reflection of this configuration in the middle vertical line. It will become then* $3, 2, 1, 0, 7, 6, 5, 4, \ldots, 15, 14, 13, 12$. *Check that the parity of this* 16*-permutation is* 0. *Since the distance* $dist_0$ *from "♠A" (denoted by* 0*) to the top-left corner of "4×4" changes from* 0 *to* 3, *conclude that (exactly) one of these two configurations can be reached from any given configuration using the moves as in the 15-puzzle, where "♠A" is treated as the empty space.*

Solution. The moves are the same for 15 cards and 15 units, so we will use ξ_{15}. The permutation corresponding to this reflection results in the opposite ordering of the cards in every row. So the number of the corresponding inversions for this reflection is $6 \cdot 4$, an even number, and its parity is 0. See the definition of the parity of permutations above. Then, $dist_0$ changes after the reflection by an odd number, namely by 3. Thus, the invariant ξ_{15} defined above for the Game of Fifteen, will be different for these two configurations, 0 and 1, and (exactly) one of them can be reached from *any* starting arrangement. ❑

For instance, the configuration in Fig. 2.5 can be transformed to that in Fig. 2.2 upon removing "♠A". Assigning $0, 1, 2, \ldots, 15$ to the former, the latter corresponds to the permutation $(0, 13, 6, 11, 7, 10, 1, 12, 9, 4, 15, 2, 14, 3, 8, 5)$.

The number of inversions is 60, an even number; the empty spaces are at the same position in both.

Reducing the number of tiles. Let us consider now the 13-puzzle, where we take only 13 units for the same 4×4 shape. All configurations now are *solvable*: they can be transformed into any other configurations.

It is true even for the 14-puzzle, so it holds for any smaller numbers of tiles. For the 14-puzzle, use two empty spaces below units 11 and 12 in the initial configuration $(1, 2, \ldots, 11, 12, 13, 14)$ to transpose 11 and 12 without changing the positions of the other units. If we add unit 15 at its standard place, the corresponding permutation and $(1, 2, 3, \ldots, 15)$, will have different parities, so any 15-configuration can be transformed to one of them: use ξ_{15}. This obviously holds if 15 is removed.

For the 13-puzzle, let us count the total number of configurations. We will label the empty spaces $0_1, 0_2, 0_3$. The number of configurations will be $16!$ divided by $3! = 6$, the number of permutations within the set $\{0_1, 0_2, 0_3\}$. They are *indistinguishable*, so we need to divide $16!$ by the number of their permutations, which is $3!$.

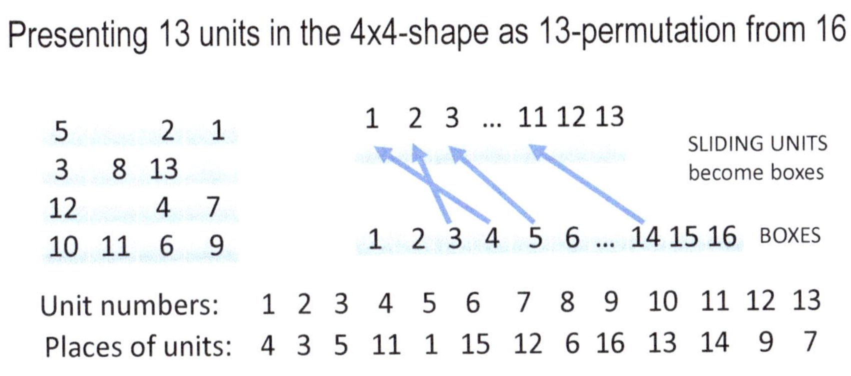

Figure 2.6: 13-puzzle in 4×4.

The mathematical meaning of the word "indistinguishable" is that only the number of such objects matters. For instance, "there are 10 children in the room" is a statement which does not distinguish between the separate qualities of the thing in question. The children have, of course, all different personalities, but we only want to count the number of elements of some type. Obviously, permuting indistinguishable objects does not change this *set*.

The calculation above yielded an explicit product: $\frac{16!}{3!} = 16 \cdot 15 \cdots 5 \cdot 4$. It coincides with $P(16, 13)$, which is hardly accidental. To explain this coincidence, let us restate the problem as follows. We now assign the corresponding place number (from 1 to 16) to every sliding unit, which is from 1 to 13. In a sense, the boxes become objects and the units become boxes! See Fig. 2.6 for an example of such a 13-sequence. So true!

This is a different approach to the same problem, which readily results in $P(16, 13)$. An informal definition of combinatorics is "a branch of mathematics where every problem has at least 2 different solutions."

2.3.2 Four basic examples

There are not too many types of *introductory* combinatorial problems and puzzles concerning permutations; the problems on combinations "dominate."

Problem 2.3.2 *Count the number $\mathcal{N}$ of* nice *4-digit numbers, where by nice, we mean that all digits must be different.*

Solution. First, 4-digit numbers have to begin with a nonzero digit; otherwise, they are 3-digit numbers. This is a matter of definition, but it is quite common. It is frequent when the statements of combinatorial problems rely on standard definitions (and common sense); this is the case.

Obviously, this problem is about $P(n, r)$, but with some deviations. Here and in *many* problems on permutations, it is better to *not* use our formula and "go back" to "boxes." Namely, we have □□□□, where there are 9 choices for the first box (0 is not allowed), 9 choices for the 2nd, then 8 and finally 7. So the answer is $\mathcal{N} = 9 \cdot 9 \cdot 8 \cdot 7$; the independent choices are multiplied. Calculate it numerically if you wish $(= 4536)$ or leave as such. ❑

Problem 2.3.3 *Count the number $\mathcal{N}$ of* nice *5-letter English words, where by nice, we mean that all letters are different and there are no consecutive consonants or consecutive vowels.*

Solution. There are 21 consonants and 5 vowels (a,e,i,o,u); we will use c, v for them. Thus the "pattern" □□□□□ can be of type $cvcvc$ (1) or $vcvcv$ (2). The consonants and vowels are totally independent, so we have $(21\cdot20\cdot19)(5\cdot4)$ for (1) and $(21\cdot20)(5\cdot4\cdot3)$ for (2). Finally, $\mathcal{N} = (21\cdot20\cdot19)(5\cdot4)+(21\cdot20)(5\cdot4\cdot3) = 21 \cdot 20 \cdot 5 \cdot 4(19 + 3) = 22 \cdot 21 \cdot 20 \cdot 5 \cdot 4$.

Such a nice product can be an indication of a different solution. Indeed, the words above are in 1-1 correspondence with pairs of a 3-word $C = c_1c_2c_2$ made from 22 consonants (add ð from Old English), and a 2-word $A = a_1a_2$ from 5 vowels. If there is no ð in C, then the corresponding nice 5-letter word is of type (1) as above. Otherwise, let $ð = c_i$, delete it from the word C, order the remaining 3 vowels ($\neq a_1, a_2$) alphabetically, and add the i_{th} vowel there to A as a_3. Finally, combine the remaining portion of C with the resulting

$a_1a_2a_3$ in a nice word of type (2). Justify that this is 1-1. Generalizing, use this argument to count the number of *nice* v-letter words as above in any alphabet with v vowels for odd v and at least $(v+1)/2$ consonants. ❑

Problem 2.3.4 *There are 3 ladies and 3 gentlemen to dine at a round table. What is the number $\mathcal{N}$ of arrangements, so that no 2 ladies sit together?*

Solution. By "a round table," we naturally mean that only your neighbors matter, not the position of your chair in the room (such exact definitions are frequently omitted in puzzles). If no 2 ladies sit together, then obviously no 2 gentlemen sit together. Using the rotational symmetry, we can assume that gentleman "A" occupies the first chair. Then we need to decide about the remaining 5 chairs, $\square\square\square\square\square$, where the first chair must be for a lady. Similar to the previous problem, the configuration after "A" is $lglgl$, which gives $3 \cdot 2 = 6$ (for ladies) and 2 (for gentlemen). Conclude that $\mathcal{N} = 12$.

There is another approach: we can "unfold" the table (break the symmetry), follow the previous problem, and then divide the resulting number by 6 (to go back to the round table). The line configuration then can be $glglgl$ or $lglglg$, with $(3!)^2$ possibilities for each. So we obtain $(2 \cdot 3! \cdot 3!)/6 = 12$. ❑

Problem 2.3.5 *How many different "words" can be obtained by permuting the letters in the word LITTLE?*

Solution. This is quite similar to the 13-puzzle. We can do any permutations of 6 letters, but permuting the 2 "L"s and the 2 "T"s will not change the resulting word. So the answer is $6!/(2 \cdot 2) = 6 \cdot 5 \cdot 3! = 180$. ❑

2.3.3 Symmetric group

Problem 2.3.5 is some introduction to the theory of the *symmetric group*, where the permutations are considered as *operations.* For this, we will begin with the so-called *two-line notation* for permutations. This allows us to define the *composition of permutations.* Actually, this definition is mostly used in some exercises and problems in this book (not too much).

The standard, one-line, presentation of a permutation u is $u = (a_1a_2 \dots a_n)$, where a_i are all numbers from 1 to n. The ***two-line mode*** is $\begin{pmatrix} 1 & 2 & \dots & n \\ a_1 & a_2 & \dots & a_n \end{pmatrix}$. The first line is always $1, 2, \dots, n$, so it does not add any information, but this presentation is convenient. We interpret u as an operation. For any sequence of objects in boxes $1, 2, \dots, n$, *the content* of box 1 will go to box a_1, the content of box 2 to box a_2, and so on.

The second line, the one-line presentation of u, is the result of the application of u to the sequence of symbols $1, 2, \dots, n$ as the initial objects in the boxes. However, the objects inside the boxes can be of any nature; they can be with repetitions. So we permute the boxes (together with their contents).

Given 2 permutations u, v, the ***composition*** of permutations is defined as their consecutive execution, *first* v *then* u. The notation is simply uv, the product. If the initial sequence of objects in the boxes is $1, 2, \ldots, n$, then uv is the resulting sequence upon the application of v and then u.

For instance, let $u = (213), v = (132)$ (in one-line notation). Then uv sends any sequence of objects abc to acb under v and then to cab under u; so $uv = (321)$ (we set $a = 1, b = 2, c = 3$). Similarly, $vu : abc \mapsto bac \mapsto bca$, so $vu = (231)$ (different from uv). Such calculations require some practice, but we can just follow the definition. The order in the products (compositions) matters: uv is generally different from vu, as in the example above.

The ***identity permutation*** is $id = (1, 2, \ldots, n)$: nothing is permuted. The ***inverse permutation*** u^{-1} is the one such that $uu^{-1} = u^{-1}u = id$. In the two-line notation, the columns of u^{-1} will be $\binom{a_i}{i}$, which must be ordered with respect to the first numbers to make the first line $1, 2, \ldots, n$. Say, $\begin{pmatrix} 1 & 2 & 3 & 4 & 5 \\ 4 & 1 & 5 & 2 & 3 \end{pmatrix}^{-1} = \begin{pmatrix} 1 & 2 & 3 & 4 & 5 \\ 2 & 4 & 5 & 1 & 3 \end{pmatrix}$.

We arrive at the definition of the ***symmetric group*** $\mathbf{S}_n$, permutations with the product operation above, which plays a fundamental role in representation theory and almost anywhere in mathematics and physics.

Generally, a group G is a set of elements g (all n-permutations for $\mathbf{S}_n$) with the multiplication operation $\{g, h\} \mapsto gh$, the identity element e, and the existence of g^{-1} for any g: such that $gg^{-1} = g^{-1}g = e$. The associativity $f(gh) = (fg)h$ for any elements f, g, h in G is required in this definition, which is obvious for the composition of permutations.

Transpositions. The ***transposition*** (ij) for $i < j$ transposes the contents of boxes i and j and preserves the objects in all other boxes. If $j = i+1$, it is called a *simple* transposition, denoted by s_i.

Any non-identity permutation can be obtained as a composition of simple transpositions s_i. Prove it by induction as an exercise.

For $n = 3$: $s_1(123) = (213), s_2(123) = (132), s_2s_1(123) = s_2(213) = (231)$, $s_1s_2(123) = s_1(132) = (312), s_1s_2s_1(123) = s_1(231) = (321)$. The products s_1s_2 and s_2s_1 were calculated above; we omit commas between the numbers in $(123), (213)$, etc.

There are some relations for the products of s_i. For $n = 3$, the basic ones are $s_1^2 = id = s_2^2$, and $s_2s_1s_2 = s_1s_2s_1$. All other relations follow from them. For instance, deduce the relation $(s_1s_2)^3 = id = (s_2s_1)^3$ from the basic ones. Also, $(uv)^{-1} = v^{-1}u^{-1}$ for any permutations u, v and in any groups, a simple exercise. Notice the change of the order. For instance, $(s_1s_2)^{-1} = s_2s_1$.

Let us rephrase the solution of Problem 2.3.5 using the notion of composition. There is nothing really new here vs. the solution provided above. So such formalization looks excessive. However, it is useful for more involved problems of this kind.

The 6-permutations v, w applied to the sequence of 6 objects L,I,T,T,L,E create coinciding words if and only if $w = vu$ where u preserves or transposes (the contents of) the boxes with "L" and the boxes with "T". Which means that u can be id, (15), (34), or the product (15)(34). So the number of such u is 4, and there are $6!/4$ words obtained from "LITTLE" by permutations. Formally, $vu_1 = vu_2$ if and only if $u_1 = u_2$, since $v^{-1}vu_1 = u_1 = v^{-1}vu_2 = u_2$.

Cycles. Let us define the so-called ***cycle decomposition*** of permutations. For $u = \begin{pmatrix} 1 & 2 & \cdots & n \\ a_1 & a_2 & \cdots & a_n \end{pmatrix}$, the first cycle will be $(1 \to x_1 \to x_2 \to \cdots x_\ell \to 1)$, where $x_1 = a_1$, x_2 is the bottom number in the column that begins with x_1, x_3 is the bottom number in the column that begins with x_2, and so on. We stop when 1 occurs again in this sequence. There can be no repetitions before this moment (check).

Then we take the smallest number that does not appear in the first one and perform the same procedure starting with this number, which gives the second cycle. This process is continued until all numbers $1, 2, \ldots, n$ occur in some cycles. Finally, u becomes a product (composition) of cycles.

Generally, a permutation sending $x_1 \to x_2 \to \cdots \to x_\ell \to x_1$ and preserving the other numbers is called a *cycle of length* ℓ if the last x_1 is the first repetition here. For instance, transpositions (ij) are 2-cycles: $i \to j \to i$. The notation is $(x_1 \to x_2 \to \cdots \to x_\ell \to x_1)$ or, simply, $(x_1 x_2 \ldots x_\ell)$, though the latter can be confused with the one-line notation for permutations. *To avoid misunderstanding it is better to switch to 2-line notation for permutations if the cycle decompositions are constantly considered.* We can write cycles starting with any x_i, but the minimal x_i is commonly taken as x_1.

The cycles in the *cycle decomposition* of u are uniquely determined by u. Their order in the product does not matter since they involve different numbers and commute with each other. Finally, permutations are products of *disjoint* cycles from $1, 2, \ldots, n$, and this presentation is a one-to-one correspondence; it is of fundamental importance in the theory of $\mathbf{S}_n$.

For instance, id becomes $(1)(2)\cdots(n)$ in the cycle presentation, though we almost always omit $(i) = (i \to i)$. Permutations that are n-cycles are frequently referred to as cycles, where it is assumed that all numbers $1, 2, \ldots, n$ are taken. We can begin with 1 for them, so their number is $(n-1)!$. For $n = 3$, $\begin{pmatrix} 1 & 2 & 3 \\ 2 & 3 & 1 \end{pmatrix} = (1 \to 2 \to 3 \to 1)$, and $\begin{pmatrix} 1 & 2 & 3 \\ 3 & 1 & 2 \end{pmatrix} = (1 \to 3 \to 2 \to 1)$ are the only cycles, denoted by (123) and (132) correspondingly.

Check that $u= \begin{pmatrix} 1 & 2 & 3 & 4 & 5 \\ 4 & 1 & 5 & 2 & 3 \end{pmatrix} = (142)(35)$. The latter product means the following composition of cycles: $(1 \to 4 \to 2 \to 1)(3 \to 5 \to 3)$.

The inverse permutations are obtained when we reverse the order of numbers in all cycles (and make the first entries in the cycles minimal possible). The cycle decomposition of u^{-1} is $(124)(35)$ for the u above.

Parity. The parity, 0 or 1, of a permutation $u = (a_1 a_2 \ldots a_n)$ has been already used; it is the parity of the number of ***inversions***, the pairs $i < j$ such that $a_i > a_j$. The *parity* of the product of (any number of) transpositions is then the parity of their number (0 for an even number of transpositions, 1 for odd). See Exercise 2.6.37. The parity does not depend on a particular presentation of u as such a product. The parity of an r-cycle is the parity of $(r-1)$ (as a number); so *parity*(u) is the parity of the number of *even cycles* (with even r) in its cycle decomposition.

Problem 2.3.6 *Count the number D_n of* **derangements** *for $n = 4$, the permutations that preserve no elements, i.e. such that $a_i \neq i$ in the two-line notation. See Section 3.6 for an approach based on the Inclusion-Exclusion Principle.*

Solution. They are 4-cycles $(1 \to a \to b \to c \to 1)$ or compositions of two disjoint transpositions: $(1 \to a \to 1)(b \to c \to b)$ for pairwise different $a, b, c > 1$; $(c \to b \to c) = (b \to c \to b)$. Thus, the answer is $3! + 3 = 9$. ❑

Solving 15-puzzle. * See Section 2.3.1. Practically, American Chess Grandmaster Bobby Fischer was an expert at solving the 15-puzzle (≤ 25 seconds). Theoretically, let us prove that if $\xi_{15} = 0$ then such a configuration is solvable, i.e. can be obtained from the starting one by admissible moves. Equivalently, it is the following problem.

Problem 2.3.7 *Any* even *15-permutation. i.e. that having parity 0, is solvable if the empty space $\emptyset$ is at its starting bottom-right position.*

Solution. We mainly follow W. Story (1879) and A. Archer. See a note by Robert Beeler [17].

Any composition of admissible moves can be presented as a sequence of letters u, d, r, l describing the moves of $\emptyset$. For instance, *drul* means that $\emptyset$ is moved left, up, right, down (first "left"), so it goes back to its starting position. The corresponding 15-permutation is the following 3-cycle: $\tau = (15, 12, 11) = (15 \to 12 \to 11 \to 15)$, i.e. *the content* of box 15, denoted by $\langle 15 \rangle$, goes

[17] R. Beeler, *The fifteen puzzle and related puzzles* (2022), `https://faculty.etsu.edu/beelerr/fifteen-supp1.pdf`.

to box 12, $\langle 12\rangle$ goes to box 11, and $\langle 11\rangle \mapsto 15$. So the transformation is $\begin{bmatrix} 11 & 12 \\ 15 & \end{bmatrix} \mapsto \begin{bmatrix} 12 & 15 \\ 11 & \end{bmatrix}$; the other boxes are unchanged.

Second, consider the following sequence of the 24 moves, which sends the empty field $\emptyset$ to its original position: $druldddrurdruuuldllurdlu$. It is the following 13-cycle: $\rho = (1,5,9,13,14,10,15,8,4,3,7,6,2)$. Note that 11, 12 occupy their initial positions under the action of ρ.

The configuration becomes: $\begin{bmatrix} 2 & 6 & 4 & 8 \\ 1 & 7 & 3 & 15 \\ 5 & 14 & \mathbf{11} & \mathbf{12} \\ 9 & 13 & 10 & \end{bmatrix}$. This construction is very geometric. It is similar to that in Fig. 1.7 (Gomory's Theorem). One has $\rho = drul\cdot\rho'\cdot rdlu = \tau\rho'\tau^{-1}$ for $\rho' = dddrurdruuuldllu$ presented at Fig. 2.7.

Figure 2.7: $\rho' = dddrurdruuuldllu\,.$

Third, for any k such that $11 \neq k \neq 12$, a proper power ρ^m maps the content of box k to box 15. Thus $\tau_k = \rho^{-m}\tau\rho^m$ sends the content $\langle k\rangle$ of box k to box 15 and then to box 12. Also: $\langle 12\rangle \mapsto \langle 11\rangle$ (only τ matters since ρ preserves boxes 11, 12), and τ_k sends $\langle 11\rangle \mapsto 15 \mapsto k$. We obtain that τ_k is the 3-cycle $(k, 12, 11)$; for instance, $m = 1$ for $k = 10$ and $m = -1$ for $k = 8$.

The following basic fact is Exercise 2.6.37(iv). The ***alternating group*** $\mathbf{A}_n$, which is a subgroup of $\mathbf{S}_n$ formed by all even permutations, is generated by all 3-cycles for $n \geq 3$. Moreover, the 3-cycles (k, a, b) for fixed $a \neq b$ and all k such that $a \neq k \neq b$ are sufficient here. Indeed (check this directly):

$$(x,y,z) = (x,a,b)^{-1}((z,a,b)(y,a,b)^{-1})(x,a,b),$$
$$(x,y,a) = (y,a,b)(x,a,b)^2, \text{ and } (x,y,b) = (y,a,b)^2(x,a,b),$$

assuming that x, y, z, a, b are pairwise distinct. We see that 3-cycles τ_k for all $k \neq 11, 12$ generate the whole $\mathbf{A}_{15}$, which gives the required solvability claim. ❑

2.4 Combinations

2.4.1 The Main Formula

We defined $P(n,r)$ as the number of r-permutations (r-sequences) of r different objects taken from n distinct ones (different numbers, people, etc.). The passage from P to C is simply disregarding the order in such selections, i.e. the passage from sequences to sets. Accordingly, $C(n,r)$ will be the number of ***r-combinations***, i.e. all r-element subsets, obtained from a set of n distinct objects, *distinguishable* to be a bit more exact. Informally, we buy r items and put them "randomly" in a basket. The objects are all distinguishable, so it is better to think of selecting a team with r players from a pool of n available players.

The most celebrated formula in the *whole* of combinatorics is as follows:

$$C(n,r)=\frac{P(n,r)}{r!}=\frac{\overbrace{n(n-1)\cdots(n-r+1)}^{r}}{r!}=\frac{n!}{(n-r)!r!}. \tag{2.4.1}$$

It readily results from the definition: we need to divide $P(n,r)$ by the number of permutations within the r-sequence, within the basket. As above, $0! = 1$, $(-m)! = \infty$ for $m > 0$. This allows us to take any $n, r \geq 0$ in the latter formula, but $C(n,r)$ is nonzero only if $0 \leq r \leq n$.

The tremendous role of combinations is actually surprising. Permutations are supposed to be the key because multiple events are ordered in time, in space, or both (quantum physics is somewhat exceptional). Disregarding this order is due to our ways of processing the information.

Lotteries are typical examples of "human-induced" combinations. There is no mechanical-electrical device that can *simultaneously* pick all lottery balls, but the winning combination is without any order of the corresponding numbers; the lottery slips are designed this way. Nature has many faces, but lotteries are what *we* (people) invented. Processing information is supposed to be secondary vs. events, but as a matter of fact probability theory, statistics, and statistical physics are heavily based on combinations.

2.4.2 Binomial coefficients

Instead of writing $C(n,r)$ the ***binomial coefficients*** $\binom{n}{r} = \frac{n!}{(n-r)!r!} = C(n,r)$ are almost always used now. We read these symbols as "n *choose* r"; this is their definition, the name "binomial" is due to *Newton's binomial* (below).

In computer languages, they are denoted by $Binomial[n,r]$, or in some similar form. The computers do not really use the factorials here; the triangle recurrence is one of the main tools (below) and there are other methods. It takes a couple of seconds to calculate $Binomial[10^8, 5\cdot 10^7]$ ... unless you want this number to be printed. Computers do not calculate $\binom{n}{1}$ as $\frac{n!}{(n-1)!}$, which is a ratio of huge numbers for big n; they "know" that it is n. The following "small" binomial coefficient is very common: $\binom{n}{2} = \frac{n(n-1)}{2}$, the number of games/tournaments between n teams/players.

Problem 2.4.1 *Consider the usual deck with 52 cards, which has 13 denominations $(2, \ldots, 10, J, Q, K, A)$ and 4 suits:$\heartsuit\diamondsuit\clubsuit\spadesuit$. Count the number of distinct* full-house hands *in poker, which are hands with 5 cards (you can permute them in a hand) with 3 cards of the same denomination and the remaining 2 of the same denomination too.*

Solution. We choose one of the 13 denominations for the 3 matching cards ("three of a kind") and one of the 12 remaining denominations for the remaining two ("pair"). They cannot coincide since this would require 5 suits. Then there are $\binom{4}{3}$ options for the suits of the triple and $\binom{4}{2}$ for the pair. Finally, the number of such hands is $13 \cdot 12 \cdot \binom{4}{3} \cdot \binom{4}{2} = 13 \cdot 12 \cdot 4 \cdot 6 = 3744$. ❑

The symmetry of binomial coefficients. One has $\binom{n}{r} = \binom{n}{n-r}$. The first proof: use the formula via factorials, which is obviously $r \leftrightarrow (n-r)$-symmetric. The second proof: an r-element subset from a set with n elements is fully determined if you know the *complementary subset*, which is of size $n-r$.

More precisely, we have a *1-1 correspondence* between r-element subsets and $(n-r)$-element subsets. We already used 1-1 (one-to-one) correspondences; see also Definition 2.4.2 below for a somewhat more formal definition. It is the map $[n] \supset X \mapsto [n] \setminus X$, where $[n] = \{1, 2, \ldots, n\}$. Recall that $A \setminus B$ is *the difference of two sets*: removing B from A. It is assumed that $B \subset A$, i.e. that B belongs to A (one can also write $A \supset B$). The usage of 1-1 correspondences here and in similar situations is not just a formality. It is always important to interpret combinatorial identities as correspondences.

Definition 2.4.2 *If the elements of two sets can be paired so that each element is paired with exactly one element from the other set, then this gives a* one-to-one correspondence, *1-1 correspondence for short, between the two sets. Such a map between two sets is also called a* bijection *or an* isomorphism.

The definition remains the same for infinite sets. For instance $f : x \mapsto x^3$ is a bijection from $\mathbb{R}$ to $\mathbb{R}$, the set of all real numbers. Generally, $f : X \to Y$ is an isomorphism between two sets X, Y if there exists a map $f^{-1} : Y \to X$ such that $f^{-1}(f(x)) = x$, $f(f^{-1}(y)) = y$ for all x, y in X, Y.

2.4.3 Pick-6 Lotto

Let us discuss one of the oldest USA lotteries, the New Jersey (NJ) Pick-6: pick 6 numbers from 49. The winner must guess all 6 numbers. We will call such a person a ***super-winner***, since all lotteries reward those who match some winning numbers but not all of them. "Pick-6" has a ***progressive jackpot***, added to until a super-winner hits it, as almost all lotteries have. We discuss the probability of no super-winners in Section 2.4.4.

Pick-6 Lotto. A bet slip is divided into five separate panels. Each panel contains a field of numbers from 1 to 49. Select 6 numbers from each panel. The jackpots start at $2M and keep growing until someone wins. Prizes upon $1 wager:

As of Monday, July 27, 2020

Match	Prize Amount	Odds
Match 6	$4.5M* - Jackpot	1 in 13,983,816
Match 5	$2,713*	1 in 54,201
Match 4	$56*	1 in 1,032
Match 3	$3	1 in 57

(* means the average prizes, $3 for "match 3" is fixed)

PAY BACK (FROM $14M "RECEIVED"): 4,500,000
+ Binomial[6,3]*Binomial[43,3] * 3
+ Binomial[6,4]*Binomial[43,2] * 56
+ Binomial[6,5]*Binomial[43,1] * 2713 = $ 6,698,934.

For "3-winners", indeed, the odds are 1 in 56.6599 = Binomial[49,6] / (Binomial[6,3] * Binomial[43,3]).

Figure 2.8: NJ Pick-6 Lotto.

This lottery is essentially from 1970; its current form is from 1980 (it initially had 39 numbers). It is not the most "profitable" for the players among all lotteries but it has quite a good reputation. The more popular the lottery, the better the chances for thrilling jackpots.

Generally, lotteries pay back to the players about 55c (cents) per $1. The operating costs are about 12c; the rest, $100-55-12 = 33$c, is the profit of the organizers, which mainly goes to the state's budget and programs. The "return" for the players can be as high as 70c per $1, so generally higher than the average return with "Pick-6." This data is of course conditional: we are just outlining how this industry works.

Obviously, lotteries are far from "zero-sum games." However if you think that roulette or stock markets are more fair, think twice. You play "only" twice a week with "Pick-6"; 100 rounds (spins) of roulette take "just" about 3-6 hours! One "round" can be in seconds with stock markets.

Universe of possibilities. Any calculations in combinatorics and statistics begin with the definition of the "universe of possibilities," the sample space, and calculating its size. This is basic for understanding "profitability." Let us discuss the profitability of "Pick-6" for us, the players: what we receive per \$1. The universe here is the total number of *panels*. For "6 from 49", its size is $\mathcal{N} = \binom{49}{6} = 13983816 \approx 14\text{M}$ (million), so your chances to correctly guess all 6 numbers of the winning combination, are about $1 : 14\text{M}$.

Accordingly, if your \$1-panel wins, you expect a return of \$14M, assuming that there are no charges, i.e. that the lottery is a kind of coin flipping with the odds $1 : 14\text{M}$. Obviously, it is not! Importantly, your mathematical chances to win with 1 panel do not depend on the total number of players. This is between you and the lottery (as a business model).

Let us calculate the odds to become a *3-winner*, i.e. to match exactly 3 numbers from 6. We can assume that the winning panel is $\{1, 2, 3, 4, 5, 6\}$ for this calculation. To become a 3-winner, you need to pick 3 numbers from these 6, and, independently, 3 from the remaining $49 - 6 = 43$. Thus the chances are $\binom{6}{3}\binom{43}{3}$ to $\mathcal{N}$, which is almost $1 : 57$ as in Fig. 2.8 (for 2020).

To determine the profitability, we need only to know how much the lottery receives from the "universe" and how much it pays back. It receives $\mathcal{N}$ dollars ($\approx$ \$14M), and pays back what is provided in Fig. 2.8. The prize \$3 for the "3-winners" is fixed; the other prizes are statistical (determined by the Pick-6 lottery organizers). It is perfectly possible to use average figures in this and similar calculations, assuming that someone (the lottery?) calculated them. Generally, they are in the annual reports of the lotteries.

The main component is of course the *average* top prize. See [18]. Not all lotteries publish it. Using data from the link above, we take the average to be \$4.5M as in Fig. 2.8. This is conditional of course; the fluctuations are significant here. We need to know this to proceed.

Taking \$4.5M, the lottery pays back to "the universe" \$6,698,934 in total, as you see. Thus, they pay back about $48\% = 0.48 \approx 6698934/13983816$. Importantly, in this and similar calculations, we need only to know the actual or average prizes *per one panel*; for instance, the size of the jackpot is irrelevant.

Our 0.48 is very close to 0.5: officially, the *total prize pool* is set to be 50% of the gross dollars wagered for a lottery drawing. Also, from the rules: the amount of money in the prize pool is divided among the winners so that 6

[18] New Jersey Lottery Results and News, 9/6/2021.

out of 6 winner(s) will equally share 68% of the total prize pool, 5 out of 6 winners will equally share 10% of the total prize pool, 4 out of 6 winners will equally share 11% of the total prize pool and 3 out of 6 winners will be paid a fixed prize of \$3.00 (approximately 11% of the total prize pool). Indeed, $1032/57 \approx \$56/\$3$ in the table (4,3-winners).

There are many other lotteries. For instance, you can play multi-state *Powerball*, where you pick 5 from 69 and 1 (powerball) from 26. The corresponding size of the universe of panels is $\binom{69}{5} \cdot 26 = 292,201,338$. It is not by chance that the creators of this lottery made the size of their universe roughly comparable with the population of the USA; for "Pick-6", it is about the population of New Jersey plus "some of its neighbors." Then the frequency of super-winners makes the game "optimal" for the players, as we have seen.

2.4.4 Poisson distribution

A natural question is: "what are the chances that there are no super-winners," so that the jackpot is not triggered and continues to grow. On one occasion, it reached \$36M for "NJ Pick-6." This obviously depends on the number of people playing the lottery. Let us assume that $\mathcal{N}x$ *panels* play "Pick-6" for one *drawing*, where x is a given or unknown fraction and, again, $\mathcal{N} = \binom{49}{6} \approx 14\text{M}$.

For every panel, the probability to *not* become a super-winner is $(1 - 1/\mathcal{N})$, almost 1. Generally, any probability is the ratio *good/all*. "Good" here depends on the problem. In this particular calculation, by *good*, we mean non-super-winning panels (i.e. "bad"). The number of *all* panels is $\mathcal{N}$ and the probability equals $\frac{\mathcal{N}-1}{\mathcal{N}}$.

People play independently, so the probability of the event "no-super-winners" is $P_0 = (1 - \frac{1}{\mathcal{N}})^{\mathcal{N}x} = \big((1 - \frac{1}{\mathcal{N}})^{\mathcal{N}}\big)^x \approx e^{-x}$ for sufficiently large $\mathcal{N}$. We use one of the most celebrated formulas in calculus:

$$e^x = \lim_{N\to\infty} \left(1 + \frac{x}{N}\right)^N, \quad e = \lim_{N\to\infty} \left(1 + \frac{1}{N}\right)^N \approx 2.71828\ldots.$$

If the number of playing panels equals $\mathcal{N}$, i.e. $x = 1$, then P_0 is approximately $1/e \approx 0.37$: about 1 in 3 games will have no super-winners. So the jackpot will be frequently used, but not too frequently in this case. This explains why it makes sense to keep $\mathcal{N}$ comparable with the number of playing panels. The jackpot is unreasonable for small x; for instance, about 9 in 10 games will have no super-winners for $x = 0.1$, i.e. if $\mathcal{N}/10$ panels are sold.

Problem 2.4.3 *Find the probability P_m of the event "exactly m super-winners." What is its "calculus limit" for fixed m and $\mathcal{N} \to \infty$?*

Solution. First, "we" need to pick the super-winners from the players; actually, *Fortune*, a daughter of *Jupiter*, is in charge. The number of choices is $A = \binom{\mathcal{N}x}{m} = \frac{\mathcal{N}x(\mathcal{N}x-1)\cdots(\mathcal{N}x-m+1)}{m!} \approx \frac{(\mathcal{N}x)^m}{m!}$. We assume here that m is much smaller than $\mathcal{N}x$: using the notation $m \ll \mathcal{N}x$.

The probability of a given group of m people to become the winners (all of them) is obviously $B = \frac{1}{\mathcal{N}^m}$. Then, for every such choice, we need to make all other players non-winners. The chances are $C = (1 - \frac{1}{\mathcal{N}})^{\mathcal{N}x-m} \approx e^{-x}$, where $\ll \mathcal{N}x$ as above.

Finally, $P_m = ABC = \frac{\mathcal{N}x(\mathcal{N}x-1)\cdots(\mathcal{N}x-m+1)}{\mathcal{N}^m m!}(1 - \frac{1}{\mathcal{N}})^{\mathcal{N}x-m}$, which can be simplified a bit:

$$P_m = \frac{\mathcal{N}x(\mathcal{N}x-1)\cdots(\mathcal{N}x-m+1)}{(\mathcal{N}-1)^m m!}\left(1-\frac{1}{\mathcal{N}}\right)^{\mathcal{N}x}.$$

Respectively, the "calculus limit" of P_m is $\frac{(\mathcal{N}x)^m}{m!\mathcal{N}^m}e^{-x} = \frac{x^m}{m!}e^{-x}$, the famous ***Poisson distribution***. See also Exercises 2.6.1, 2.6.2, 2.6.3 on the chances of hurricanes and reinfections, and Problem 7.2.14. ❑

By construction, $P_0 + P_1 + \cdots + P_m + \ldots = 1$; we counted all possible events and the total probability must be 1. Setting $M = \mathcal{N}x$, this results in the following *identity*:

$$\sum_{m=0}^{\infty} \frac{M(M-1)\cdots(M-m+1)}{(\mathcal{N}-1)^m m!} = \left(\frac{\mathcal{N}}{\mathcal{N}-1}\right)^M.$$

We do not need to assume that M is an integer here. If it is, then the left-hand side is also, indeed: $(1 + \frac{1}{\mathcal{N}-1})^M = \left(\frac{\mathcal{N}}{\mathcal{N}-1}\right)^M$ due to *Newton's binomial* from Section 3.2.1.

Under the "calculus approximation," we obtain the second (and much more convenient) formula for e^x:

$$\sum_{m=0} \frac{x^m}{m!}e^{-x} = 1, \text{ and } e^x = \sum_{m=0} \frac{x^m}{m!}.$$

This is actually a rigorous "statistical" proof of this formula.

Problem 2.4.4 *If* 2 *drawings out* 7 *statistically end without super-winners, estimate the number of* panels *sold by "Pick-6" per one drawing.*

Solution. We denote the number of panels that play by $\mathcal{N}x$ for $\mathcal{N} = \binom{49}{6}$, i.e. those sold by the lottery per one drawing. Then $e^{-x} \approx 1/3.5$, $x = \log 3.5 \approx 1.25$, and about $1.25 \times 14\text{M} = 17.5\text{M}$ panels "play" this lottery for one drawing. This is the answer. ❑

Using this calculation, the lottery receives about \$1.68B in this year ($= 8 \cdot 12 \cdot 17.5$M) under the assumption of this problem; there are 2 drawings per week. The total sales of the New Jersey Lottery (they offer several kinds) were about \$3.2B in 2020. So this estimate for "Pick-6" sounds realistic. The total lottery sales in the United States reached about \$80.5B in 2016.

2.5 Binomial distribution

2.5.1 Random walks

What we will call ***random walks*** are the key to understanding why ***combinations*** play such a fundamental role. One can expect ***permutations*** to be of greater value to model "natural processes." Indeed, unless in the microscopic world, any processes in nature are "ordered" in time and in space. Permutations play their role of course, but the combinations appear to be of greater importance.

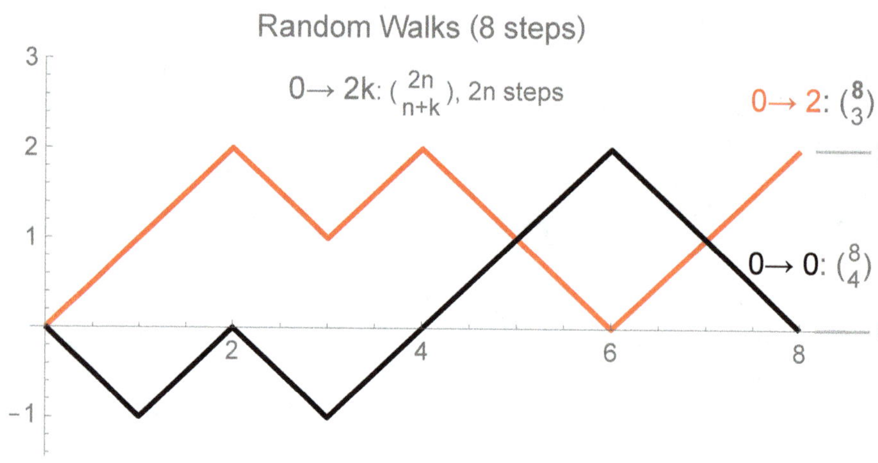

Figure 2.9: Random walks.

The most basic stochastic process is the random walk with two equally probable outputs: *one step up* and *one step down*, i.e. ± 1. This is exactly coin flipping, where we count a total number of heads minus the total number of tails. It is convenient to present this as a graph; see Fig. 2.9.

The paths (walks) from 0 to 0 require an even number of steps: $2n$ steps for some n. We will mostly stick to even numbers of steps below. To recover such a path, we need only know when it goes up (by $+1$); then the other steps will be down (by -1). Thus the total number of paths $0 \to 0$ with $2n$ steps is ***the middle binomial coefficient*** $\binom{2n}{n}$: from $2n$ steps we choose n of them to be up. For an odd number of moves, there are two coinciding middle binomial coefficients: $\binom{2n+1}{n}$, and $\binom{2n+1}{n-1}$. Accordingly, the outcome, the last value in the graph, must be *odd*.

Similarly, the paths from 0 to $2k$ for $2n$ steps require $n+k$ "ups" and $n-k$ "downs". Their number will then be $\binom{2n}{n+k}$. This formula works for any k, including negative ones, but we will mostly use it for $0 \le k \le n$. Indeed, k and $-k$ result in the same number of paths due to the symmetry of binomial coefficients; geometrically, it is due to the reflection in the x-axis in Fig. 2.9.

Distributions. A distribution is generally any sequence of non-negative numbers with sum 1; these numbers can be interpreted as the *probabilities* of some process. The binomial distribution is absolutely the key.

The ***binomial distribution*** is $\{p_k = \binom{2n}{n+k}/2^{2n}, -n \le k \le n\}$. It gives the probabilities of the outcome $2k$ for $2n$ random steps of size ± 1. We use that 2^{2n} is the total number of paths: we can move up or down for each step. Thus, $p_{-n} + p_{1-n} + \cdots + p_{n-1} + p_n = 1$. More generally, the (non-centered) *binomial distribution* can be defined as $\{\binom{n}{r}/2^n\}$, where n is fixed and any r is taken in the range $0 \le r \le n$. This one is for n random steps of size $0, 1$. We note that the *Poisson distribution* for random walks of this kind with the probability α/N to go up for $N \gg 0$ steps.

The identity $\sum_{r=0}^{n} \binom{n}{r} = 2^n$, which is the key here, can be restated combinatorially as follows. The number of subsets with r elements in the set $[n] = \{1, 2, \ldots, n\}$ is $\binom{n}{r}$ for $0 \le r \le n$. Therefore the total number of subsets, which is 2^n, equals the sum of all these numbers. Importantly, we include the empty set $\emptyset$ and the whole set. This identity will be revisited in Section 3.2.1 when discussing Newton's binomial.

Let $2n = 8$. Then, for instance, p_1 = probability of "$0 \to 2$" $= \binom{8}{3}/2^8 = \frac{1}{2^8}\frac{8\cdot 7\cdot 6}{3!} = \frac{56}{256}$. The simplest are $p_{-4} = p_4 = 1/256$: there is only one path from 0 to ± 8. For $k = 0$, $\binom{8}{4} = \frac{8\cdot 7\cdot 6\cdot 5}{24} = 7\cdot 2\cdot 5 = 70$, so $p_0 = \frac{70}{256}$. The middle binomial coefficient $\binom{2n}{n}$ and the corresponding p_0 are the greatest. This follows from the interpretation via random walks.

2.5.2 Stirling approximation

The key link to calculus is the following ***Stirling approximation***:

$$\binom{2n}{n+k}/2^{2n} \approx \frac{e^{-k^2/n}}{\sqrt{\pi n}}, \quad \binom{2n}{n}/2^{2n} \approx \frac{1}{\sqrt{\pi n}} \quad \text{for } n \gg 0,\ k \sim \sqrt{n}. \tag{2.5.2}$$

Generally, the famous ***normal distribution*** $\dfrac{e^{-(x-\mu)^2/(2\sigma^2)}}{\sigma\sqrt{2\pi}}$ is the limit of $p^r(1-p)^{n-r}\binom{n}{n+r}$, describing the random walk with the probabilities $p, 1-p$ for steps of size $1, 0$. Here $x = r$, the *mean* is $\mu = np$ and the *standard deviation* is $\sigma = \sqrt{np(1-p)}$. This is justified in Exercise 3.7.47, a special case of the *Central Limit Theorem*. The limit of $p^{n+k}(1-p)^{n-k}\binom{2n}{n+k}$, generalizing (2.5.2), is obtained for $\mu=(2p-1)n$, $\sigma^2=2np(1-p)$, $x=k$. The 2nd relation in (2.5.2) follows from the *Wallis formula*; see Exercise 2.6.5.

To give some examples, $\binom{200}{100}/2^{200} \approx 0.056349$ matches well $\frac{1}{\sqrt{100\pi}} \approx 0.056419$. Also, $p^{n+k}q^{n-k}\binom{2n}{n+k} \approx 0.0596057$, for $p=\frac{2}{3}$, $q=1-p=\frac{1}{3}$, $n=100$, $k=33$, and $\frac{e^{-(k-\mu)^2/(2\sigma^2)}}{\sigma\sqrt{2\pi}} \approx 0.059766$, where $\sigma=\sqrt{2npq}=\frac{20}{3}$, $\mu=(2p-1)n=\frac{100}{3}$. Changing here $k=33$ to $k=30$, we obtain 0.0521722 combinatorially and 0.0528098 via the normal distribution.

The role of the Stirling formula (and the Wallis one) for us is mostly to establish connections with calculus and statistics. Binomial coefficients can be calculated in no-time in the age of computers, however, the Stirling formula is very important when trying to understand why many of our formulas make sense as $n \to \infty$, which limit is standard in statistics and statistical physics.

Practically, the Stirling formula works better near the middle binomial coefficient, i.e. when $k \approx 0$ for $p=\frac{1}{2}$ and $k \approx \mu$ for any $0<p<1$. Then its accuracy is very reasonable. Surprisingly, it works well even for relatively small n. For $n=4$ ($2n=8$) and $p=\frac{1}{2}$, the probabilities from the 1st line of the table below are close to their approximations via (2.5.2) in the 3rd line:

$\binom{8}{4+k}/2^8$	$k=0$	$k=1$	$k=2$	$k=3$	$k=4$
exactly :	70/256	56/256	28/256	8/256	1/256
decimal :	0.273	0.219	0.109	0.031	0.004
Stirling :	0.282	0.22	0.104	0.03	0.005

Let us provide the Stirling approximation for $n!$. For $n \ge 1$:

$$n! \sim \sqrt{2\pi n}\left(\frac{n}{e}\right)^n, \quad \sqrt{2\pi n}\left(\frac{n}{e}\right)^n e^{\frac{1}{12n+1}} < n! < \sqrt{2\pi n}\left(\frac{n}{e}\right)^n e^{\frac{1}{12n}}. \qquad (2.5.3)$$

For instance, let $\nu(m)$ be the number of digits of integer $m \ge 1$. Then $\nu(n!)$ is approximately $n(\nu(n)-1)$; see Exercise 3.7.46.

Problem 2.5.1 *We will model shooting hoop in basketball (from some distance) by the 8-step random walk as above. Due to the backboard and the rim the goal is when $0, 2$ or -2 is reached (not only 0). What is the probability of the goal and what happens if the number of steps is 2 times greater? Calculate the latter probability exactly and then using the Stirling approximation.*

Solution. Using the table, the probability is $\frac{56+70+56}{256} \approx 0.71$. For $2n=16$, the probability becomes $(\binom{16}{8}+2\binom{16}{7})/2^{16} = \frac{35750}{65536} \approx 0.5455$, which is ≈ 0.5515 via (2.5.2). The steps here are actually about the shot preparation and the shot itself, but their number reflects the distance to the basket. ❑

2.5.3 Game "14 vs. 1"

You are on a plane and have already watched all of the movies in their selection. Your neighbor suggest to play with you a very simple game: *"14 vs. 1"*. It requires a chessboard (he has one), a coin, and a token to be used as a chess piece (another coin will do). See Fig. 2.10.

Place the token at the bottom-left field, then toss the coin. For tails the token moves up by 1; for heads it moves right by 1. If the token reaches the opposite corner of the chessboard (red), then he wins. Otherwise it reaches one of the green fields (not within the board!) and you win. It looks like the chances are 14 : 1. Moreover, reaching "his field" always requires 14 moves, but only 8 to 14 moves for "your fields", which is clearly in your favor. *Will you play if you have to pay him 14-fold when he wins?*

Warning 1: Do not play any games with strangers based on tossing a coin! There are many ways to influence the outcome (for con artists).

Warning 2: If you are persuaded to play, then divide their suggested bet by 2! They know the game; you do not.

Concerning tossing the coin, just "coincidentally" he has 3 dice with him ("I just played yesterday with my kids ..."). Using dice under "odd-even" is of course mathematically equivalent to tossing the coin, but this is better protected against possible manipulations. So will you now play using 3 dice and under the reduced bet 7 : 1?

Warning 3: "Do not trust anyone who proposes a deal too good to pass up!" This is a quote from some FBI report on con games.

Warning 4: Do not play with a guy who has a chessboard in one pocket and dice in the other! Actually, the best advice: do not play with strangers! And the very best: do not play!

At this point, the teacher may interrupt the airplane story and ask the students to "vote" for different bets; the readers are welcome to participate too. All of course understand that this game (any game) is a "business model." So it is instructional to estimate the odds following our instincts.

Statistically, some agree to play with 7 : 1 odds, and almost all (who voted, not everyone) accept 5 : 1, i.e. "agree" to pay the neighbor 5-fold if he reaches his (red) corner: you receive $1 from him after you reach one of your 14 (green) fields and you pay him "just" $5 for his red corner.

Let us now calculate the actual ***fair bet*** rigorously. Obviously, this game is about random walks. Generally, the bet in this or any similar situation is the number of "your games" divided by the number of "his games". Here games are paths. The problem is that the paths have different lengths, from 8 to 14. If you assume that all paths (until the corresponding green/red field) are equally likely in the game, the ratio $\beta = \frac{\text{your paths}}{\text{his paths}}$ will be very different from the *fair bet* $b = \frac{\text{your games}}{\text{his games}}$. This results in a very different universe of events; any path is counted as 1, regardless of the number of steps.

The ultimate check is of course to write a program for counting your/his games as the author of this book did. It must randomly produce $\uparrow$ and $\rightarrow$ and move the (virtual) token accordingly. There is no need to continue of course after a green field is reached.

After 100-1000-10000 games (here 100 is a bit too few), it will calculate the fair bet b by the latter formula with good accuracy without any "mathematics." This is the famous ***Monte Carlo method.***

The discrepancy $b \neq \beta$ is because the endpoints closer to the starting one are faster to reach and the paths (to green/red fields) are *not* equally likely for this game. Making them equally likely events would require a device that produces the whole path *at once*. If all such paths, short and long, are assumed equally likely, the corresponding bet β is calculated in Exercise 3.7.9. One has: $\beta = 2\binom{14}{6}/\binom{14}{7} = \frac{7}{4} = 1.75$. The true bet b is predictably higher, significantly higher as we will see.

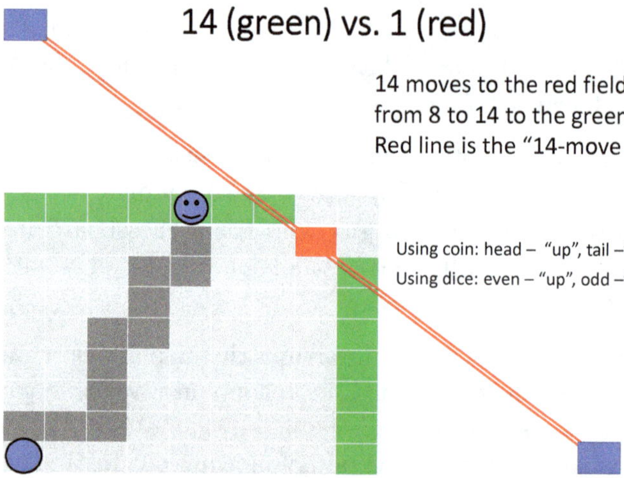

Figure 2.10: Playing "14 vs. 1".

Generally, the rule of thumb is *do not compare apples with oranges!* In this context: *you cannot generally compare paths of different lengths.* The correct approach is to always play 14 times, even if this would be beyond the chessboard. So we play until the (red) anti-diagonal through "his box" (the top-right corner) in Fig. 2.10 is reached, which always requires 14 moves. The corresponding "green boxes" there (when you win) will be all but the red one (still 14 vs. 1). See also Exercises 2.6.23 and 2.6.24.

The ratio *(your games)/(his games)* remains the same of course; your computer will just work a bit longer. With this understanding, we really have the coincidence: $\frac{\text{your 14-paths}}{\text{his paths}} = \frac{\text{your games}}{\text{his games}}$. It must be obvious for those with some "statistical experience," but people make mistakes here all the time. Anyway, your computer will confirm that b (not β) is correct.

The calculation of b is as follows. The number of *your games* will be a sum over the green fields, which is not too simple to calculate. However, *his games* + *your games* $- 2^{14}$, and *his games* $- \binom{14}{7}$; we did this before. The switch from some set to its complement frequently works well in combinatorics. Thus:

$$\text{bet} = b = \frac{\text{your 14-paths}}{\text{his paths}} = \left(2^{14} - \binom{14}{7}\right)/\binom{14}{7}$$
$$= 2^{14}/\binom{14}{7} - 1 \approx 3.774;$$
$$\text{using calculus: bet} \approx \sqrt{7\pi} - 1 \approx 3.689.$$

So, if you do not have your computer with you, $\sqrt{7\pi} - 1$ is good enough.

Despite the very significant reduction of the initial "14:1" to "5:1", the latter appeared not in your favor. His average gain per 1 round (for the \$1 bid) is approximately $(5\,\frac{1}{4.774} - 1\frac{3.774}{4.774})\$ \approx 25.684c$. The game is fast, so your hope is that the flight is short. The Monte Carlo method indeed gives about 26c for $\sim$ 100000 runs. See Section 7.1.1.

Adding tailwind. If you are getting increasingly disappointed (you will), your neighbor may offer to change the roles: green fields will now be his, and you will "own" the corner (red), which appears much more "profitable" for "red" than one could expect. He assumes that you already understand (experimentally or theoretically) that the fair bet b for this game is smaller than 5; about 20 games are basically sufficient to see that it is from 3.5 to 4.

He suggests changing roles with a twist: *3 equally probable outputs instead of 2.* Roll the dice and consider the remainder of the resulting sum of the faces modulo 3, i.e. upon division by 3. The token will now move $\rightarrow, \uparrow, \nearrow$, *king's moves*, with equal chances (1/3). Adding $\nearrow$ obviously creates a strong "wind," a double move, toward the corner (red), and a significant reduction of the fair bet 3.774 above can be expected, say to 3 or smaller. Anyway, his suggestion is to pay you \$3 for reaching the corner, assuming that you pay him \$1 for the other (green) fields. What is the exact new fair bet?

Problem 2.5.2 * *Find the formula for the* new fair bet, *when the diagonal move is added and the probabilities of $\rightarrow, \uparrow, \nearrow$ are 1/3. It will be via some sum of* *binomial coefficients. Check that the* new fair bet *is about* 3.48*; if you need a hint, formula (3.2.9) below addresses this.* ❑

Thus, the new fair bet is significantly greater than 3. Actually, there is little change versus 3.774 calculated without the "wind." Your neighbor will win big again, assuming that you accepted playing under 1 : 3.

Conclusion. Our intuition can fail us when *binomial distribution* is involved. The red field appeared much more likely to be reached than one could expect. Also, the "tailwind" contributes only a little. The middle field is the key factor ... just because it is in the middle. We (all creatures) are good with linear approximations, but the binomial distribution and its calculus limit, the celebrated *normal distribution*, are very nonlinear.

2.6 Exercises

Exercise 2.6.1♡ *There were* 12 *hurricanes from 1851 to 2020 that brought winds of Category 3 status to North Carolina, about* $\alpha = 12/17.5$ *such hurricanes in a decade. What are the probabilities* P_0, P_1, *and* P_2 *that there will be* $0, 1$ *and* 2 *such hurricanes during any* 10 *years? Follow Section 2.4.4 and calculate these probabilities* p_i *under the calculus limit, i.e. those for the corresponding* Poisson distribution. *The rationale is that the probability of a hurricane on a certain day is* α/N *for the total number* N *of days during one decade (or only for the days during the hurricane seasons); it must be large for the calculus limit.*

Answers: $p_0 = e^{-\alpha} \approx 0.50$, $p_i = \alpha^i e^{-\alpha}/(i!)$, $p_{1,2} \approx 0.35,\ 0.12$.

Exercise 2.6.2♡ Reinfections. *(i) During an epidemic,* 0.1% *of the population of some country is infected every day, and there is no immunity. Find the probability* p_r *that an individual is infected* r *times during* 1000 *days and the calculus limit* p'_r *of* p_r*, i.e. the corresponding* Poisson distribution. *What are* p'_r *for* $r = 0, 1, 2, 3$*?*

(ii) Assume that the probability of being infected r *times during an epidemic was* $p'_r = \frac{\alpha^r}{e^{\alpha} r!}$ *for some* α*, and* 50% *of the country was non-infected, i.e. the probability of being non-infected for an individual was* 0.5*. Find* α*, and calculate* p'_r *for* $r = 1, 2, 3$.

Answers: (i) When calculated exactly, $p_r = (0.001)^r (0.999)^{1000-r} \binom{1000}{r}$. Approximately, $p'_r = \frac{1}{r!\,e}$; $p'_0 = p'_1 = \frac{1}{e} \approx 0.37$, $p'_2 = \frac{1}{2e} \approx 0.18$, $p'_3 = \frac{1}{6e} \approx 0.06$. (ii) $\alpha \approx 0.69$, $p'_1 \approx 0.35$, $p'_2 = 0.12$, $p'_3 \approx 0.03$.

Exercise 2.6.3 ♣ *Reinfections and immunity.* *During an epidemic lasting N days, assume that people are exposed to the infection with probability β per day. However this exposure results in the infection only after day $x + L$, where x is the day of the previous infection and L is the* immunity period. *Let $P(r, N, \beta) = \binom{N-Lr}{r}\beta^r(1-\beta)^{N-Lr-r}$.*

(i) Prove that $\pi_r = P(r, N, \beta) + \beta\sum_{i=1}^{L} P(r-1, N-i, \beta)$ is the probability of r infections for one individual. This results in the identity $\sum_{r=0}^{\infty}\pi_r = 1$, where $r \le Floor[\frac{N+L}{L+1}]$ are sufficient. For $N = 4, L = 2$ and $x = 1-\beta$, it becomes: $x^4 + \beta x(2 + x + x^2) + \beta^2 = 1$ (check).

(ii) Following (i), take $\beta = 0.001$ and $N = 1000$, as in Exercise 2.6.2, and use your computer to check that $\pi_0 \approx 0.37$ (it equals p_0), $\pi_1 \approx 0.42$, $\pi_2 \approx 0.18$, $\pi_3 \approx 0.03$ when $L = 150$ (the sum of these approximations is 1).

(iii) Assuming that $\beta = \frac{\alpha}{N}$ and $L = \nu N$ for some α, ν, calculate that the calculus limit of π_1 provided $\nu < 1$ is $\pi_1' = (1 + (1-\nu)\alpha)e^{(\nu-1)\alpha} - e^{-\alpha}$. For $\alpha = 1$ and $\nu = 0.15$ as in (ii), check that $\pi_1' \approx 0.42$, which matches (ii).

(iv) The parameter $\beta = \alpha/N$ is not supposed to be given; it can be found using $\pi_0' = e^{-\alpha}$ as in Exercise 2.6.2 (ii). Assuming that 50% of the population appeared infected (as there), find α and then (numerically) π_r for $r = 0, 1, 2, 3$ when $N = 1000, L = 150$ as above.

Hint for (i): If $1 \le x_1 < x_2 < x_3 < \ldots < x_r \le N$ are the infection days, then $x_1 < x_2 - L < x_3 - 2L \ldots x_r - (r-1)L$ and one needs to consider 2 cases: (1) $x_r + L \le N$, and (2) $x_r + L > N$. Any number of *potential* infections can occur for days $x_i + 1, \ldots, x_i + L$ for $1 \le i < r$ and during the end period $x_r + 1, \ldots, \min\{x_r + L, N\}$. These periods do not contribute to π_r (justify!). So the probability of the event $x_1 < x_2 < \ldots < x_r$ is $\beta^r(1-\beta)^{N-rL-r}$ for (1) and $\beta^r(1-\beta)^{(x_r-(r-1)L-r)}$ for (2).
For (iv): $\alpha \approx \log(2) \approx 0.69$, $\pi_0 \approx 0.50$, $\pi_1 \approx 0.38$, $\pi_2 \approx 0.10$, $\pi_3 \approx 0.01$.

Exercise 2.6.4 ♠ *From Exercise 2.6.3(i), let π_r be the probability of r infections during N days of an epidemic with the immunity period $L < N$ and the probability β of being exposed to the infection during 1 day. Then $\pi_r = P(r, N, \beta) + \beta\sum_{i=1}^{L} P(r-1, N-i, \beta)$ for $P(r, N, \beta) = \binom{N-Lr}{r}\beta^r(1-\beta)^{N-Lr-r}$. Let $F_r(X) = d^{r-1}\left(X^{N-Lr}\,\frac{1-X^L}{1-X}\right)/dX^{r-1}$ (the $(r-1)$th derivative, $r, L, N-Lr > 0$).*

(i) Prove that $\pi_r = \binom{N-Lr}{r}\,\beta^r\,(1-\beta)^{N-Lr-r} + \frac{\beta^r}{(r-1)!}\,F_r(X \mapsto 1-\beta)$. In particular, $\pi_1 = (N-L)\,\beta\,(1-\beta)^{N-L-1} + (1-(1-\beta)^L)\,(1-\beta)^{N-L}$.

(ii) Simplify the formula for π_r in (i) when $r = 2$ and $2L \le N$:

$$\pi_2 = \beta^2\binom{N-2L}{2}(1-\beta)^{N-2L-2} + \beta(N-2L)\left(1-(1-\beta)^L\right)(1-\beta)^{N-2L-1}$$
$$+\left(1-(1-\beta)^L - \beta L(1-\beta)^{L-1}\right)(1-\beta)^{N-2L}.$$

(iii) Let $\beta = \alpha/N$, $L = \nu N$ and $N \to \infty$, as in Exercise 2.6.3(iii). Check that the limit of π_2 provided the inequality $2\nu < 1$ becomes

$$\pi_2' = e^{(2\nu-1)\alpha}\left(1 + \alpha(1-2\nu) + \frac{\alpha^2(1-2\nu)^2}{2}\right) - e^{(\nu-1)\alpha}\left(1 + \alpha(1-\nu)\right).$$

Check that for $\alpha = 1$, $L = 150$, $N = 1000$: $\pi_2 = 0.175193$, $\pi_2' = 0.175141$.

Exercise 2.6.5 $^\diamondsuit$ *Rewrite the formula $\lim_{n\to\infty}\sqrt{\pi n}\binom{2n}{n}/2^{2n} = 1$, a particular case of the Stirling approximation from (2.5.2), as follows:*

$$\frac{\pi}{2} = \prod_{n=1}^{\infty}\frac{(2n)^2}{(2n-1)(2n+1)} = \frac{2\cdot 2}{1\cdot 3}\frac{4\cdot 4}{3\cdot 5}\frac{6\cdot 6}{5\cdot 7}\cdots.$$

Then deduce the latter from the Wallis formula*:*

$$\frac{\sin(x)}{x} = \prod_{n=1}^{\infty}\left(1 - \frac{x^2}{\pi^2 n^2}\right);$$

perform the substitution $x \mapsto \pi/2$.

Exercise 2.6.6 $^\heartsuit$ *In the card game* faro *played with the usual deck (52 cards, 13 ranks and 4 suits), the dealer earnings depend on the probability p_2 that 2 random cards have coinciding ranks. See Section 3.5.1. Calculate p_2. This probability remains the same if a partial deck is used with at least 2 cards, if it is created* randomly *(why?). Calculate the average number $\mathcal{N}_2$ of such coincidence events during 2 rounds, which means that 51 random pairs are selected (this is adjusted to real faro). The chances of such coincidences can be manipulated. For instance, find p_2' assuming that the deck became, in the process of play (manipulated by the dealer), to have 7 ranks and 4 suits; which subset maximizes p_2 among all 28-subsets of the 52-deck (explain this).*

Answers: $p_2 = 1/17 < p_2' = 1/9$; $\mathcal{N}_2 = 51/17 = 3$.

Exercise 2.6.7 $^\diamondsuit$ *There are 7 boys and 7 girls sitting around a round table. Check that at least 1 person is surrounded by girls, and at least 1 person is surrounded by boys.*

Hint: If no one is surrounded by girls, then the maximal groups of neighboring girls can be g or gg and the minimal group of neighboring boys is bb. However, these groups are alternating, which excludes "g" and contradicts "7".

Exercise 2.6.8 ◇ *(i) Check that $N_p = (k^p - k)/p + k$ is the total number of* ***necklaces*** *of length p for prime p ($p = 2, 3, 5, 7, \ldots$) when there are k types of beads. Here the rotations are allowed, i.e. necklaces are considered up to the rotational symmetry. The divisibility of $k^p - k$ by p, which follows from this calculation, is Fermat's Little Theorem.*

(ii) For necklaces of length p^s for prime p, check that their number is $N_{p^s} = k + \frac{k^p - k}{p} + \frac{k^{p^2} - k^p}{p^2} + \frac{k^{p^3} - k^{p^2}}{p^3} + \cdots + \frac{k^{p^s} - k^{p^{s-1}}}{p^s}$. This is a particular case of the Redfield-Pólya Theorem: $N_n = \frac{1}{n}\sum_{d|n} \varphi(d) k^{\frac{n}{d}}$ for necklaces of length $n \geq 1$, where the summation is over all divisors d of n, including 1 and n, and $\varphi(d)$ is the number of m in the range $0 < m < d$ relatively prime to d, the Euler function; see Exercise 5.6.13.

Hints: (i) By breaking the symmetry, a necklace can be presented by a string $a_1 a_2 \cdots a_p$ for some types of beads a_i, possibly coinciding. Its rotations are $a_p a_2 a_3 \cdots a_{p-1}$ and so on. This string remains unchanged upon some non-trivial rotation, only if it is $aa \cdots a$, i.e. all a_i coincide; here we use that p is a prime number. Indeed, if the string is invariant for the rotation by m, where $1 \leq m \leq p-1$, then the rotations by $m, 2m, \ldots, (p-1)m$ preserve it too, but they constitute all non-trivial rotations. Therefore all a_i must coincide. Thus, the total number of necklaces is k for $aa \cdots a$ plus $(k^n - k)/p$ for all other ones up to the rotations.
(ii) For the necklaces of length p^s, the rotation invariance can be modulo $1, p, \ldots, p^s$; this number must divide p^s. The number of strings $a_1 a_2 \cdots a_{p^s}$ that have the period exactly p^e is $k^e - k^{e-1}$ for $1 \leq e \leq s$, which must be divided by p^e; we subtract k^{e-1}, since the period is smaller than p^e for them. These strings are $aa \cdots a$ with any a for $e = 0$, i.e. for the period 1.

Exercise 2.6.9 ♣ *Use the Erdős and Szekeres Theorem from Problem 2.1.4 to check that any set of different $n^2 + 1$ points in a plane contains a subset with $(n+1)$ points without a single equilateral triangle.*

Hint: Rotating the plane makes the x-coordinates of these points all distinct, and the same for their y-coordinates. Then pick an $(n+1)$-subsequence of the naturally-ordered sequence of the x-coordinates with all increasing or all decreasing y-coordinates.

Exercise 2.6.10 ◇ *A part-time employee must work at least one day during every week for 52 weeks. If the total number of days he (actually) worked was 90, prove that there were some sequences of consecutive weeks (from 52) when he worked (totally) exactly 1 day, exactly 2 days, …, exactly 13 days.*

Hint: Combine the increasing 52-sequence of total days he worked during the first week, the first 2 weeks, ..., all 52 weeks with this sequence shifted by any x (days) such that $1 \le x \le 13$. The resulting sequence is bounded by the number $90 + x$; employ the Pigeonhole Principle.

Exercise 2.6.11$^{\heartsuit}$ *Prove that the decimal expansion of a rational number* $\frac{m}{n}$, *where* $\gcd(m,n) = 1$, *is finite if and only if* n *is a product of powers of* 2 *and* 5. *Otherwise this expansion becomes repeating with the period no greater than* $n - 1$ *(use the Pigeonhole Principle). Find an example when it is* $n - 1$.

Exercise 2.6.12$^{\diamondsuit}$ *In any set of* 101 *integers, there are* a *and* b *such that either* $a - b$ *or* $a + b$ *is divisible by* 197.

Hint: Consider the map from x in this set to $i = x \bmod 197$ (the remainder modulo 197), and then to the set of unordered pairs $\{i, 199 - i\}$ for $i = 0, 1, 2, \ldots, 99$.

Exercise 2.6.13$^{\heartsuit}$ *Assume that all pairwise intersections are nonempty for* m *different subsets* $S_1, \cdots S_m$ *of a set with* n *elements. Prove that* $m \le 2^{n-1}$.

Hint: Consider unordered pairs of subsets $\{S, [n] \setminus S\}$.

Exercise 2.6.14$^{\diamondsuit}$ *In how many ways can* 4 *girls and* 4 *boys sit around a round table together with (i) one parent and (ii) two parents, provided that there are no neighboring girls and no neighboring boys? The parents separate children and can sit anywhere (possibly, together).*

Answers: For one parent, it is $2 \cdot (4!)^2$; for two, $24 \cdot (4!)^2$.

Exercise 2.6.15$^{\spadesuit}$ *Let* $N_m(n)$ *be the number of circular arrangements of* n *girls,* n *boys and* $2m$ *parents for* $n \ge 1, m \ge 0$, *provided that there are no neighboring girls and no neighboring boys.*

(i) Setting $N_m(0) = 0$, *obtain the following recurrence relation:* $N_{m+1}(n) = (2n + 2m + 1)(2n + 2m)N_m(n) + 2\big(n(n+1)\big)^2 N_m(n-1)$.

(ii) Check that $N_m(1) = (2m+1)!$, $N_0(n) = n!(n-1)!$, $N_1(n) = 6n(n!)^2$, *and* $N_2(n) = 12(3n^2 + 3n - 1)(n+1)!n!$.

(iii) For the probability of a circular arrangement as above, which is $P_m(n) = \frac{N_m(n)}{(2n+2m-1)!}$, *deduce from (i) that* $\lim_{n\to\infty} \frac{P_m(n)}{P_0(n)} = \left(\frac{3}{2}\right)^m$.

Exercise 2.6.16$^{\heartsuit}$ *In how many ways can* 5 *children sit around a round table with* 7 *chairs?*

Answer: $4!(5+\binom{5}{2})$.

Exercise 2.6.17$^\diamondsuit$ *(i) In how many ways can* 5 *children sit around a round table with (a)* 11 *chairs, and with (b)* 12 *chairs, assuming that they are non-neighboring (separated by empty chairs) for (a) and (b).*

(ii) Generalizing, in how many ways can m non-neighboring children sit around a round table with n chairs? Consider the chairs next to the children.

Answers. For 11 chairs: $5!$; for 12: $5!+\binom{5}{2}4!$; for m,n: $(m-1)!\binom{n-m-1}{m-1}$.

Exercise 2.6.18$^\heartsuit$ *(i) In how many ways can* 14 *apples and* 2 *oranges be distributed among* 3 *children in such a way that each child gets at least* 2 *pieces?*

(ii) What is the answer if at least 1 *piece for each child is required?*

Answer to (i): $66\cdot 6$.

Exercise 2.6.19$^\heartsuit$ *One panel in* ***Powerball*** *is when you pick* 5 *from* 69 *and* 1 *(powerball) from* 26. *Check that the number of panels is* $\binom{69}{5}\cdot 26$.

(i) Check that the odds to match 5 *but not the powerball are* $1:11688053.52$.

(ii) Check that the odds to match 4 *and the powerball are* $1:913129.18125$.

(iii) Find the odds to match 3 *from* 5 *numbers but not the powerball.*

Answers: (ii) $1:\binom{69}{5}\cdot 26/(\binom{5}{1}\cdot\binom{64}{1})$; (iii) $1:\binom{69}{5}\cdot 26/(\binom{5}{2}\cdot\binom{64}{2}\cdot 25)$.

Exercise 2.6.20$^\heartsuit$ *Let $(\binom{69}{5}\cdot 26)x$ panels play the powerball from Exercise 2.6.19. Assume that only* 1 *from* 3 *drawings results in super-winners. What is x?*

Answer: In the *calculus limit*, $x=\log(3/2)\approx 0.4$.

Exercise 2.6.21$^\heartsuit$ *The dice game in* Mahabharata *was (presumably) played with* 2 *square cuboids with the labels* $\{1,3,4,6\}$ *and* $\{1,2,5,6\}$ *on their long faces (each possibility is equally likely). An additional rule (contributed to ruining* Yudhisthira: *see Section 3.5.1) was that a player declares the expected value σ of the sum Σ and makes one more throw if it coincides with the actual one. Assume that no more than two consecutive throws are allowed.*

(i) Check the probability of the sum Σ to become $5,6,7,8,9$ *is* $1/8$, *and it is* $1/16$ *for* $2,3,4,10,11,12$. *Check that the average value of Σ is* 7.

(ii) Player A always declares $\sigma=7$, B always declares $\sigma=9$, C declares 11. *What are the corresponding average total sums including the* 2*nd throw (if any)? The greater this sum, the better for the player.*

(iii) What will be the average total sum for player A from (i) if any number of consecutive throws is allowed under the additional rule above?

Answer for (ii): $(7-\frac{\sigma}{8})+\frac{1}{8}(\sigma+7)=7\frac{7}{8}$ for A, B ($\sigma=7,9$), and $7\frac{7}{16}$ for C.
Answer for (iii): $7(1+\frac{1}{8}+\frac{1}{64}+\ldots)=8$.

Exercise 2.6.22$^\heartsuit$ **Lady tasting tea.** *(Based on a historical account.) A lady claimed the ability to distinguish whether the tea or the milk was added first to a cup.* Ronald Fisher *gave her randomly 8 cups, four of each kind, and asked her to determine which 4 were of the 1st kind. Assuming the* null hypothesis, *that the subject has no ability to distinguish the teas, calculate the probability that (exactly) 2 cups of the 1st kind and 2 of the 2nd kind were identified correctly. Is it 1/2?*

Answer: $\binom{4}{2}^2/\binom{8}{4}\approx 0.514\neq 0.5$.

Exercise 2.6.23$^\diamondsuit$ *(i) In the game "14 vs. 1" from Fig. 2.10, you make from 8 to 14 moves $\uparrow,\rightarrow$ with equal probabilities until you reach a green/red field. What is the probability of reaching the field at $(9,5)$ (shown there with ☺) from $(1,1)$? The fields are counted from the bottom-left one.*

(ii) Generally, let the square be "$(n+1)\times(n+1)$". You begin from $(1,1)$ and play until you reach the fields $(n+2,m)$ or $(m,n+2)$ for $1\le m\le n$, or $(n+1,n+1)$. Check that the probability of reaching $(n+2,m)$ or $(m,n+2)$ is $P_m(n)=(\frac{1}{2})^{n+m}\binom{n+m-1}{n}$. Obtain from this that $2\sum_{m=1}^{n}P_m(n)+\binom{2n}{n}/2^{2n}=1$.

(iii) Interpret the identity $2^n\binom{n}{n}+2^{n-1}\binom{n+1}{n}+\cdots+2^{n-m+1}\binom{n+m-1}{n}+\cdots+\binom{2n}{n}=2^{2n}$ from (ii) as follows. For $1\le m\le n+1$, the mth term there counts the number of subsets X in $[2n]=\{1,2,\ldots,2n+1\}$ with at least $(n+1)$ elements, which are assumed ordered naturally (as integers), such that the $(n+1)$th consecutive element in X is $n+m$.

Answer for (i): you need 12 moves, where the last must be $\uparrow$; thus, the probability is $\binom{11}{4}/2^{12}$.

Exercise 2.6.24$^\clubsuit$ *(i) Generalizing part (ii) of Exercise 2.6.23, assume that the probabilities of $\uparrow,\rightarrow$ are p,q $(p+q=1)$ for the square "$(n+1)\times(n+1)$". Obtain the following identity for such $p,q\ge 0$ and any $n\ge 0$:*

$$\sum_{m=1}^{n}(p^{n+1}q^{m-1}+q^{n+1}p^{m-1})\binom{n+m-1}{n}+(pq)^n\binom{2n}{n}=1,\ \ where\ 0^0=1.$$

(ii) Furthermore, consider a path $\{(i_m,j_m)\mid 0\le m\le n,\ i_n=0=j_0\}$ obtained by $\downarrow$ or $\rightarrow$ $(i\mapsto i-1$ or $j\mapsto j+1)$. The probability to go beyond *this path from $(0,0)$ is 1 using $\uparrow,\rightarrow$ with probabilities p,q. Interpret this as the identity $\sum_{m=0}^{n}\pi_m\,p^{i_m}\,q^{j_m}\binom{i_m+j_m}{i_m}=1$, where $\pi_m=p,q,1,0$ if, respectively, (i_m+1,j_m) or (i_m,j_m+1) or both or neither become beyond this path.*

(iii) Check that the identity from (ii) becomes $p\frac{1-q^{n+1}}{1-q}+q^{n+1}=1$ *for the segment* $\{(0,0),(0,1),\ldots,(0,n)\}$. *Calculate it for* 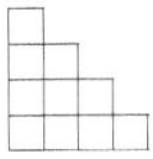, 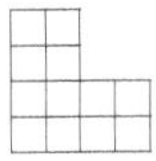 *and* 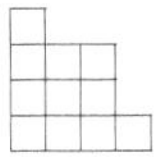.

Hint for (iii): Newton's binomial will be obtained for the "triangle" (the 1st diagram).

Exercise 2.6.25$^\heartsuit$ *For the deck of* 13 *cards that has no suits and only denominations,* $1,2,\ldots,13$, *count the number of* 4*-hands without any triples with consecutive denominations. Cards can be permuted in a hand.*

Exercise 2.6.26$^\heartsuit$ *Among* 3*-digit numbers, how many have no neighboring digits with the difference* ± 1? *The numbers with* <3 *digits are not counted.*

Answer: 600.

Exercise 2.6.27$^\diamondsuit$ *In a card variant of the game* Bingo, *you win absolutely if the denominations of the cards in your* 5*-hand belong to the set of those in the dealer's* 10*-hand. The hands are from one standard deck with* 52 *cards. What is the probability,* P^k_m, *that your hand contains* exactly $k=0,1,\ldots,5$ *cards of denominations different from those in the dealer's hand, assuming that the latter is known and contains* m *different denominations* $(3\le m\le 10)$?

Hint: There are $\binom{4(13-m)}{k}$ choices for the cards that do not match the denominations of the dealer's hand and (independently) $\binom{4m-10}{5-k}$ choices for the remaining $(5-k)$ cards. So $P^k_m=\binom{4(13-m)}{k}\times\binom{4m-10}{5-k}/\binom{52-10}{5}$.

Exercise 2.6.28$^\diamondsuit$ *For* Bingo *in Exercise 2.6.27 and probability* P^k_m *there, assume that the dealer's* 10*-hand is known, that it contains* m *different denominations, and there are* 2 *players with* 5*-hands. All hands are from the same deck. Calculate the probability that neither of the* 2 *players wins absolutely.*

Answer: $\sum_{k=1}^{5}P^k_m\left(1-\binom{4m-10-(5-k)}{5}/\binom{52-15}{5}\right)$.

Exercise 2.6.29$^\heartsuit$ *Count the number of* 4*-digit numbers (from* 1000 *to* 9999*) such that increasing consecutive triples* 012, 123 , . . . , 789 *do not appear (somewhere) in the number. For instance,* 1263 *has no such triples.*

Answer: 8864.

Exercise 2.6.30 ♡ *License plates in North Carolina consist of four digits (and can begin with 0) as well as some letters. Car owners can pay some constant fee for personalized numbers. However, some numbers are rarer than others; we suggest the fee to be based on "rarity."*

(i) If A paid \$100 for a special number with 2 pairs of consecutive increasing digits (say, 6712), but without 4 consecutive increasing digits (way too expensive), then what is B supposed to pay for a consecutive increasing triples (say, 6456 or 2459) without 4 consecutive increasing digits?

(ii) If A paid \$100 for a special number with 2 pairs of coinciding digits somewhere in the plate (say, 6116), but without all 4 coinciding digits (too expensive), then what is B supposed to pay for 3 coinciding digits somewhere (say, 6166) but without all 4 coinciding digits?

Answer for (i): For A: $9 \cdot 9 - 7 = 74$ possibilities; for B: $2 \cdot 80 - 7 - 7 = 146$; B must pay $74/146 \approx 50.685$ \$. Answer for (ii): $3!/(2!2!) = 150$ \$.

Exercise 2.6.31 ♣ Two doors and 2 guards. *There are two doors, A and B: one of them is the exit, the other is a dead end. There are two guards, X and Y, and only Y knows which of these doors is the exit. As many times as you wish, you can request guard X to ask guard Y the question* {is A the exit?} *and then answer your own question* {was Y's response "yes"?}. *Let us assume that the probability that X tells the truth is p,* which is not known to you. *However, you know (can calculate) the probability q of X's answer "yes" to your question (about Y's response). Check that this q is sufficient for finding the right door in the following 2 cases.*

(i) When the probability that Y tells the truth is $1-p$. In particular, if $p=0$ or $p=1$ in this case, then the exit is always A if X responds "no". Assume that $p \neq \frac{1}{2}$; otherwise, asking X any questions is pointless (why?).

(ii) When the probability that Y tells the truth is $\alpha - p$ for some α such that $0 \le \alpha < 1/2$ or $2 \ge \alpha > 3/2$, where p is arbitrary satisfying $\alpha - 1 \le p \le \alpha$. As an example: if both guards are truth-tellers, i.e. $p = 1$ and $\alpha = 2$, or always lie ($p = 0, \alpha = 0$), then the exit is A if X responds "yes".

Hint for (i): Assume that A is the exit. The probability that the answer is "yes" to your question is $q_1 = p(1-p) + (1-p)p = 2p(1-p)$. Assume that B is the exit, then this probability is $q_2 = p^2 + (1-p)^2 = 1 - 2p(1-p) = 1 - q_1$. One has: $q_1 < 1/2$. Thus, if the probability q of "yes" is $q < 1/2$, then the exit is A. It is B, if $q > 1/2$.

Hint for (ii): Now the probability q_1 take the values from $|1-\alpha|$ to $1-\alpha+\frac{\alpha^2}{2} \ge |1-\alpha|$ for different p. As above, $q_2 = 1 - q_1$. Therefore, if the probability q of "yes" is $q > 1/2$, then the exit is A, provided that $|1-\alpha| > 1/2$.

Exercise 2.6.32 $^\heartsuit$ *Begin with an initial sequence of r groupings of identical objects:* $\overbrace{A, A, \ldots, A}^{m_1}, \overbrace{B, B, \ldots, B}^{m_2}, \ldots$*; their total number is* $m_1 + m_2 + \cdots + m_r$. *Justify that the number of all different permutations of this sequence is* $\frac{(m_1+m_2+\cdots+m_r)!}{(m_1)!(m_2)!\cdots(m_r)!}$. *How many different "words" can we obtain by permuting the letters in the word SLEEPLESSNESS?*

Exercise 2.6.33 $^\heartsuit$ *How many different "words" can we obtain by permuting the letters in the word BOOKKEEPER, assuming that the first letter is a consonant? What is the probability P that such a word begins with a consonant?*

Answer: $3{\cdot}9!/(2!3!2!){+}9!/(2!3!) = 5{\cdot}9!/(2!3!2!)$, $P = \frac{5\cdot 9!/(2!3!2!)}{10!/(2!3!2!)} = \frac{1}{2}$. Actually, $P = 0.5$ is obvious because of the number of consonants in this word $(5 = \frac{10}{2})$.

Exercise 2.6.34 $^\heartsuit$ *(i) How many different "words" can we obtain by removing 3 different (pairwise distinct) letters from the word BOOKKEEPING?*

(ii) How many different "words" can we obtain by removing any 2 letters (possibly coinciding) from the word BOOKKEEPING?

Answers: (i) $\binom{8}{3}$; (ii) $\binom{8}{2} + 3$.

Exercise 2.6.35 $^\diamondsuit$ *(i) How many "words" of any length can be obtained from the word COUNT using its letters only once?*

(ii) Given any word with $n \geq 1$ different letters, justify that the number of words with different letters of any length that can be made from the letters of this word equals Floor$[n!e] - 1$, where Floor$[x]$ is the largest integer before x, e is the Euler number.

Answer for (i): 325. For (ii), use the remainder formula for the series for e.

Exercise 2.6.36 $^\heartsuit$ *Assuming that an English word with n letters you have typed is not in the dictionary, the level-one spelling program automatically deletes one of its letters and then inserts $0, 1$ or 2 letters in any 2 places (possibly neighboring) before/after/between the remaining letters. Then it checks such corrections against the dictionary and provides suggestions to you. How many corrections will it check? The program ignores that some of the potential corrections obtained this way coincide or may coincide with the initial word; it checks them all.*

Answer: $n\left(1 + 26n + 26^2\binom{n+1}{2}\right)$.

Exercise 2.6.37 ♣ *(i) Check that any transposition (lm) for $1 \le l < m \le n$ can be presented as a product (composition) of $2(m-l)-1$ simple transpositions $s_i = (i\,i{+}1)$, defined for $1 \le i < n$.*

(ii) Deduce from (i) that the parity *$\sigma(u)$ of an arbitrary n-permutation $u = (a_1, a_2, \ldots, a_n)$, which is the parity of the number of pairs $i < j$ such that $a_i > a_j$, coincides with the parity of the number of transpositions in* any *presentation of u as a product of transpositions.*

(iii) Deduce from (ii) that $\sigma(uv) = \sigma(u) + \sigma(v)$ modulo 2 for any n-permutations u, v and their product uv; in particular, $\sigma(uvu^{-1}) = \sigma(v)$. Check that the map $u \mapsto (lm)u$ is a 1-1 correspondence between the sets of odd and even permutations for any fixed transposition (lm).

(iv) Deduce from (iii) that even *permutations (of parity 0) form a subgroup of $\mathbf{S}_n$, denoted by $\mathbf{A}_n$ (called the* alternating group*); it is generated by 3-cycles in $\mathbf{S}_n$ for $n \ge 3$. Also, $u\mathbf{A}_n u^{-1} = \mathbf{A}_n$ for any u in $\mathbf{S}_n$, and $|\mathbf{A}_n| = n!/2$. Recall that a subgroup H of a group G is a subset such that xy^{-1} belongs to H if x, y belong to H (then it contains id and xy for any x, y in H).*

Hint for (i): $(lm) = s_l s_{l+1} \cdots s_{m-2} s_{m-1} s_{m-2} \cdots s_{l+1} s_l$.
Hint for (iv): For $i < n$, $s_i s_{i+1}$ is a 3-cycle. If $n \ge j > i+1$, then $s_i s_j = (s_i s_{i+1})(s_{i+1} s_{i+2}) \cdots (s_{j-1} s_j)$.

Exercise 2.6.38 ♡ *In the 15-puzzle from Section 2.3.1, neighboring tiles labeled by $1, 2, \ldots, 15$ can be moved to the free space in the 4×4 frame. Assigning number 16 to the empty field, ξ_{15} for any arrangement of 15 tiles is the parity of the sum of the number of inversions of the corresponding 16-permutation and the distance, $dist_\square$, from the empty field to the bottom-right corner. The distance between tiles is the length of any path in terms of $\downarrow, \rightarrow$ between them. Check that ξ_{15} does not change upon the admissible moves of the tiles.*

Hint: The distance $dist_\square$ changes by ± 1. For horizontal moves, the number of inversions changes by ± 1 too, which make the parity of their sum unchanged. For vertical moves, the change is from $xabc$ to $abcx$ for some segment $[xabcd]$ in the permutation or in the opposite direction: from $abcx$ to $xabc$. This changes the number of inversions by ± 3: it will be for the pairs ax, bx, cx. So ξ_{15} remains unchanged as well.

Exercise 2.6.39 ♢ *Use Problem 7.3.15, which states that the addition table modulo n has no orthogonal mate for even n, to prove that there are no Latin squares L of even order n without repetitions of symbols in all its anti-pandiagonals (pandiagonals are not necessary to consider). The addition table* *is the following LS:* $M = \begin{pmatrix} 0 & 1 & 2 & \ldots & n-1 \\ 1 & 2 & 3 & \ldots & 0 \\ * & * & * & \ldots & * \\ n-1 & 0 & 1 & \ldots & n-2 \end{pmatrix}$.

Hint: Any such L must be orthogonal to M. Indeed, its anti-pandiagonals, permutations of $\{0,1,2,\ldots,n-1\}$, will be coupled with the corresponding constant in M. Thus all bi-entries $\{u,v\}$ for $0 \le u,v \le n-1$ will occur in $L \vee M$, which contradicts Problem 7.3.15.
Alternatively, follow Exercise 5.6.50. Take any k such that $0 \le k \le n-1$. Let (i_m, j_m) for $1 \le m \le n$ be the indices of the boxes in L with the entry k. Then $\sum_{m=1}^{n} i_m = \sum_{m=1}^{n} j_m = \sum_{m=1}^{n}(i_m + j_m) = X = n(n-1)/2 \bmod n$. We use that $i_m + j_m$ are all distinct because no anti-pandiagonal can contain k twice. Thus, $X + X = X$ and $X = 0 \bmod n$, which is impossible for even n. Note that we need only one such k for this argument.

Exercise 2.6.40 ♣ *(i) Follow Problem 2.3.6 to calculate the number D_5 of 5-* ***derangements****, the 5-permutations that preserve no elements.*

(ii) Calculate D_6 in the same way and confirm that $D_6 = 5(D_5 + D_4)$.

Hints: (i) There can be only 5 cycles (4! of them) or 3-cycles composed with the transpositions of the 2 remaining elements ($2\binom{5}{2}$ of them). Thus $D_5 = 24 + 20 = 44$.
(ii) For $n = 6$, we obtain 5! (6-cycles) $+\binom{6,2}{3}!$ (4-cycles combined with transpositions) $\binom{6}{2}2^2$ (pairs of transpositions). Thus $D_6 = 120 + 15\cdot 6 + 20\cdot 4 = 280$.

Exercise 2.6.41 ♡ *For the usual deck with 52 cards (13 denominations and 4 suits), count the number of 5-hands for the following consecutive poker events.*

(i) Count the number of "straights": when 4 cards in your hand are consecutive (of sequential rank) of any suit, but not all 5 are consecutive.

(ii) Count the number of hands that are "three-of-a-kind": 3 of a kind from any suit, but no better than this (no "3 + 2", "4 + 1").

Answers: (i) $10 \cdot 9 \cdot (4^5 - 4)$, where 10 is for the number of 4 consecutive denominations, 4^5 is the number of "applying" suits to them. (ii) $(13 \cdot \binom{12}{2}) \cdot (4 \cdot 4^2)$.

Exercise 2.6.42 ♢ ***Bell numbers*** *B_n count different ways to partition a set with n distinguishable elements: to split it into groups (subsets) without ordering these groups. For instance, $(13)(24) = (24)(13)$ is one partition of $[4]$. Check that $B_1 = 1, B_2 = 2, B_3 = 5$, and $B_{n+1} = \sum_{k=0}^{n} \binom{n}{k} B_k$, where $B_0 = 1$. Use the latter formula to calculate B_4.*

Hint: The element $(n+1)$ added to $[n] = \{1, 2, \cdots, n\}$ will join a subset with $n-k$ elements for $k = 0, 1, \ldots, n$. So we have $\binom{n}{n-k} = \binom{n}{k}$ choices here, multiplied by the number of partitions of the remaining k elements, which is B_k. Using the formula: $B_4 = 15$.

Exercise 2.6.43 ♣ *Let us consider coverings of "$1\times n$" by non-overlapping black "$1 \times a_1$" and white "$1 \times a_2$" when the one containing the nth square is allowed to go beyond it. Let $N^n_{a_1,a_2}(r_1, r_2)$ be the number of such coverings where r_i is the number of segments of type i. For instance, $N^3_{2,1}(2, 0) = 1 = N^3_{2,1}(0, 3)$. Compare with Exercise 1.4.9.*

(i) Prove that $N^n_{a_1,a_2}(r_1, r_2)$ equals the sum of $\binom{r_1+r_2-1}{r_i-1}$ for all possible $r_1, r_2 \geq 0$ and $i = 1, 2$ such that $a_1r_1 + a_2r_2 \geq n$ and $a_1r_1 + a_2r_2 - a_i < n$. At least one such i must exist for $\{r_1, r_2\}$ to occur in this sum.

(ii) Let $a_1=2, a_2=3, n=10$. Check that there are 6 possible pairs $\{r_1, r_2\}$. Calculate the probabilities of these pairs if the probability of any particular combination of type $\{r_1, r_2\}$ is $\left(\frac{1}{2}\right)^{r_1+r_2}$. Verify that their sum is really 1.

Answer for (ii): For $\{0, 4\}, \{1, 3\}, \{2, 2\}, \{3, 2\}, \{4, 1\}, \{5, 0\}$, the corresponding N are $1, 4, 6, 4, 5, 1$ (the total is 21, matching Exercise 1.4.9), and the probabilities are $1/16, 1/4, 3/8, 1/8, 5/32, 1/32$. Their sum is 1.

Exercise 2.6.44 ♣ *Given a Latin square of size 4×4, justify that the number of different Latin squares that can be obtained from it by permutations of the columns followed by permutations of the rows can be either $(4!)^2$, or $(4!)^2/2$ or $(4!)^2/4$. Check that it is $(4!)^2/4$ for* $\begin{matrix} 0 & 1 & 2 & 3 \\ 1 & 2 & 3 & 0 \\ 2 & 3 & 0 & 1 \\ 3 & 0 & 1 & 2 \end{matrix}$ *and* $\begin{matrix} 0 & 1 & 2 & 3 \\ 1 & 0 & 3 & 2 \\ 3 & 2 & 1 & 0 \\ 2 & 3 & 0 & 1 \end{matrix}$.

Hint: The total number of such symmetries is $(4!)^2$, which must be divided by the number of those preserving a given Latin square L. A composition of a permutation of the columns, c, and then a permutation of the rows, r, fixes L if and only if the rows of L are obtained from each other by c. If c is such, then there is a unique r that sends the entries back to L. Unless $c = id = r$, c cannot leave unchanged any column number: the entries of L do not coincide in columns. So c can be: $\{1 \to x \to 1, y \to z \to y\}$ or $\{1 \to x \to y \to z \to 1\}$ for different x, y, z $(\neq 1)$. Accordingly, the total number of c permuting the rows of the corresponding L can be 2 or 4 in the 1st case, and it is 4 in the 2nd case (including $c = id$). The examples at the end are straightforward.

Chapter 3

Binomial Coefficients and Applications

3.1 *R*-numbers

3.1.1 The *R*-formula

We defined the P, C-numbers and then calculated them as follows:

$P(n,r) = r$-permutations from n objects $= \frac{n!}{(n-r)!}$

$C(n,r) = r$-combinations from n objects $= \frac{n!}{(n-r)!r!}$.

The next and last in this classical sequence will be:

$R(n,r) = r$-combinations from n *types* of objects.

This is the number of combinations with repetition allowed.

For instance, if we purchase 10 items from 3 types, ("apples", "oranges" and "peaches") then $R(3,10)$ is the total number of possible "baskets." The formula for R deserves to be designated a "theorem"; its proof is interesting.

Theorem 3.1.1

$$R(n,r) = \binom{n+r-1}{n-1} = \binom{n+r-1}{types-1}$$
$$= \binom{n+r-1}{r} = \binom{n+r-1}{objects}.$$

Proof. The contents of the basket with n objects of types $A, B, C, \ldots$ can be presented as the line

$$\overbrace{\square\cdots\square}^{a}\overbrace{\square\cdots\square}^{b}\overbrace{\square\cdots\square}^{c}\ldots,$$

where $a, b, c, \ldots$ are the numbers of As, Bs, Cs and so on; $a+b+c+\ldots = r$. Let us insert separators ‡ between the groups:

$$\overbrace{\square\cdots\square}^{a}\ ‡\ \overbrace{\square\cdots\square}^{b}\ ‡\ \overbrace{\square\cdots\square}^{c}\ ‡\ \ldots.$$

We need $(n-1)$ separators. If their positions (between the boxes) are known, we obviously know the contents of the basket: the group before the first separator is A, the one between the first and the second ‡ is B, and so on.

Let us interpret the separators in this line as additional boxes with ‡ inside; so ‡ is considered a new type of object. We obtain $(n+r-1)$ boxes. If we know those with ‡, we know everything: then As are between the 1st and the 2nd ‡, Bs are between the 2nd and the 3d, etc. Thus the number of different baskets equals the number of choices of $(n-1)$ boxes for ‡ from all $(n+r-1)$ boxes: initial n and added $(r-1)$. This is $\binom{n+r-1}{n-1}$. ❑

3.1.2 Reformulations

There is an entirely mathematical reformulation:

$R(n,r)$ is the number of presentations $r = \overbrace{a+b+c+\ldots}^{n}$, where we assume that $a,b,c,\ldots \geq 0$ and the order of $a,b,c,\ldots$ matters.

Here a gives the number of As, b gives the number of Bs, and so on: the order matters! For instance, let $r = 12$ and $n = 3$ (12 objects from 3 types), then $R(3,12) = \binom{14}{2} = 7 \cdot 13 = 91$.

We are allowed to select zero items from any type; for instance, $a = 0$ means that there are no As in the basket. Using the separators, this would be shown by ‡ ‡ □ ⋯ in the beginning of the presentation with boxes.

Interestingly, the interpretation of $R(n,r)$ as $r = \sum_i^n a_i$ does not help too much to justify the formula for $R(n,r)$; the proof we presented is the simplest possible. Though an entirely algebraic proof is not difficult to find.

Note that the "numerator" $n+r-1$ in the binomial formula for $R(n,r)$ is unchanged when n and r are swapped, but the "denominator" is not preserved. Practically, we need to be careful here: n is the number of types, and r is the number of objects. This is a frequent confusion, especially because r can be greater than n in $R(n,r)$.

For instance, $R(3,12) \neq R(12,3) = \binom{14}{11} = \binom{14}{3}$, where we used the symmetry of binomial coefficients in the last equality.

Let us demonstrate that the latter can be readily calculated without the R-formula. One has:

$$R(12,3) = |\{x_1 + \cdots + x_{12} = 3\}| \quad \text{for} \quad x_i \geq 0,$$

and there are 4 possibilities here: $1+1+1 = 3$, $1+2 = 3$, $2+1 = 3$, and $3 = 3$ (plus zeros). Accordingly, we need to pick 3 nonzero x_i in the first case, 2 in the 2nd and 3rd, and 1 in the 4th case. We arrive at the identity:

$$R(12,3) = \binom{14}{3} = \binom{12}{3} + 2\binom{12}{2} + \binom{12}{1}.$$

The next topic will be about the identities between binomial coefficients, including this one. Before going to this, let us discuss one related exercise.

3.1.3 Two prior problems

The R-numbers are not really necessary for Exercise 2.6.18, but let us employ them for the following variant of this exercise. *We use that $R(n,r)$ is the number of distributions of r indistinguishable objects among n people.*

Problem 3.1.2 *In how many ways can 17 "identical" (indistinguishable, to be exact) apples and 2 "identical" oranges be distributed among 3 children, provided that each child gets at least 3 pieces of fruit?*

Solution. The requirement for at least 3 pieces each is easy to satisfy as follows. Give 2 oranges and 1 apple to one of them (3 choices) and $3+3$ apples to the remaining 2, or give 2 children one orange and 2 apples each (3 choices) and then 3 apples to the remaining one, so there are 2 scenarios and 6 choices totally. Then we will continue with the remaining $17-7=10$ apples, which gives $R(10,3)=66$ choices. Without the usage of the R-formula, let A be the first child. Then: $R(3,10)=1(A \text{ has } 10)+2(A \text{ has } 9)+\cdots+11(A \text{ has } 0)=66$. The final answer is $6\cdot 66$. ❑

Two-parent problem. Let us provide a solution for a somewhat modified Exercise 2.6.14. This one is not really connected with combinations.

Problem 3.1.3 *In how may ways can 7 girls, 7 boys, and 2 parents be seated at a round table if no boy sits next to a boy and no girl sits next to a girl?*

Solution. Without parents the answer is $6!7!$; we solved a similar problem about 3 ladies and 3 gentlemen. The answer is $2(7!)^2$ for 1 parent, since the parent breaks the rotational symmetry and the arrangements can be $p\,\overbrace{g\cdots b}^{14}$ or $p\,\overbrace{b\cdots g}^{14}$, where we use p,b,g for parents, boys, and girls.

With 2 parents, the answer is $2\cdot(7!)^2\cdot 15+2\cdot(7!)^2\cdot 6=42(7!)^2$. The first term, $2\cdot(7!)^2\cdot 15$, means that we may add the second parent, P_2, anywhere between the 14 children and P_1 in the arrangement with 1 parent. The second term is 2 times the number of all circular arrangements of $7g+7b$ with exactly one pair (bb) and (accordingly) exactly one pair (gg). Then there are 2 ways to "use" the parents A,B for "splitting" (bb) and (gg): bAb, gBg or bBb, gAg. Let us fix the order of the parents, and obtain $(7!)^2\cdot 6$.

Using rotations, the arrangements with 2 pairs can be assumed to be $\{(b_1b_2)G_1b_3G_2b_4G_3b_5G_4\}$, where (b_1b_2) sit together, and G_i is a either a girl or a pair of girls sitting together. Thus, we have 7! for boys and $6 \cdot 7!$ for girls, since $\{G_1G_2G_3G_4\}$ can be represented as a permutation $\{g_1g_2g_3g_4g_5\}$ where we add (gg) to two neighboring girls (6 possibilities). The total is $(7!)^2 \cdot 6$.

A somewhat shorter way here is as follows. We consider (gg) as one "girl", and (bb) as one "boy". Thus, there will be 6 "boys", 6 "girls", and $6! \cdot 5!$ configurations for them around a table. Then we multiply it by $(7 \cdot 6)^2$, the number of choices for such (ordered) pairs of boys and girls.

This approach is more convenient to manage an arbitrary even number of children $(n+n)$, and an even number of parents. The answer for 2 parents is $6n(n!)^2 = (2n+1)(2n)n!(n-1)! + (n(n-1))^2(n-1)!(n-2)!$, where the 1st term is for adding parents and the 2nd is for using them as separators.

There is one more solution. Let us consider P_1 as a "boy" and P_2 as a "girl" and then switch their roles. Then there are $2(n+1)!n!$ admissible arrangements. However, the situations when the parents sit *separately* between (gb) or (bg) are not counted here, which gives $(2n-1)(2n)n!(n-1)!$ additionally. We obtain $2(n+1)!n! + (2n-1)(2n)n!(n-1)! = 6n(n!)^2$. ❑

3.2 Combinations algebraically

3.2.1 Basic identities

The algebraic theory begins with the ***Newton's binomial formula***:

$$(x+y)^n = \sum_{r=0}^{n} \binom{n}{r} x^r y^{n-r} = y^n + nxy^{n-1} + \binom{n}{2} x^2 y^{n-2} + \cdots + x^n. \tag{3.2.1}$$

The proof is immediate. To calculate the coefficient of $x^r y^{n-r}$ in the product $\overbrace{(x+y)(x+y)\cdots(x+y)}^{n}$, we need to take x from r binomials $(x+y)$ in this product; the remaining ones will then be y. This gives $\binom{n}{r}$ choices.

This was known well before Newton; when $n = 2$ it was "always" known. This is actually an alternative *algebraic* definition of binomial coefficients.

We have used many times the symmetry $\binom{n}{r} = \binom{n}{n-r}$. In the algebraic approach, it results from the $x \leftrightarrow y$ symmetry in $(x+y)^n$. Recall that its combinatorial proof is the switch from r-element subsets in $[n]$ to the complementary ones, which is a 1-1 correspondence. See Definition 2.4.2.

Mainly we will need the following special case of (3.2.1):

$$(1+x)^n = \sum_{r=0}^{n} \binom{n}{r} x^r = 1 + nx + \binom{n}{2} x^2 + \cdots + x^n, \text{e.g.} \tag{3.2.2}$$

$$(1+x)^5 = 1 + 5x + 10x^2 + 10x^3 + 5x^4 + x^5 \text{ (check this).}$$

Here $y = 1$. Reducing it further by making $x = 1$, we obtain:

$$\sum_{r=0}^{n} \binom{n}{r} = 1 + n + \binom{n}{2} + \cdots + \binom{n}{n-1} + 1 = 2^n.$$

The sum here is the total number of *all subsets* from $[n]$. We counted it combinatorially in a very simple way. If the size (r) of subsets is not fixed, then there are 2 choices for every object in $[n]$, "in" or "out"; totally, 2^n.

3.2.2 Even and odd subsets

The next identity is the key for the *Inclusion-Exclusion Principle*:

$$\sum_{r=0}^{n} (-1)^r \binom{n}{r} = 0, \text{ or } \sum_{even\, r} \binom{n}{r} = \sum_{odd\, r} \binom{n}{r}. \tag{3.2.3}$$

It is not a mistake to omit here and in similar formulas the range of the summation; the irrelevant binomial coefficients automatically vanish. This first formula is immediate from (3.2.2): make $x = -1$. Let us find a proper 1-1 correspondence between even and odd subsets of $[n]$.

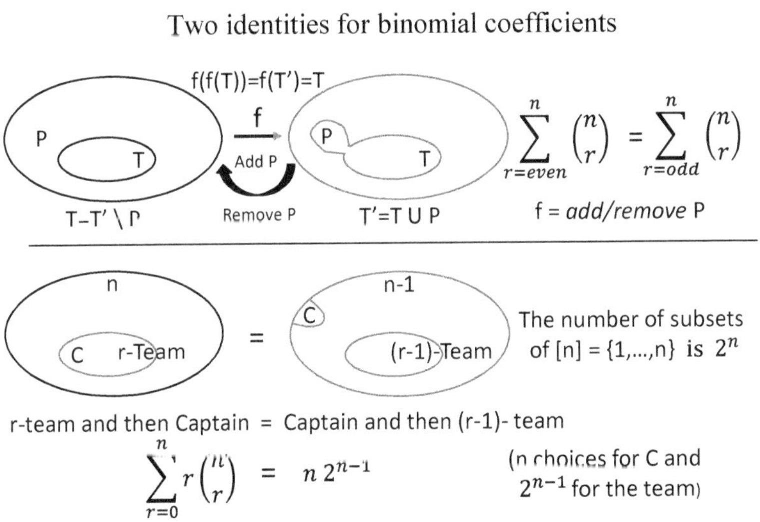

Figure 3.1: Two combinatorial identities.

Let T be an r-element subset formed from $[n]$, say, an r-team from a pool of n players. Let us select a player p. Then we have a map $T \mapsto T \cup p$, where $\cup$ is the *union*. Strictly speaking, we need to write $T \cup \{p\}$, since here we consider p as a set $\{p\}$ with 1 element, but we will mostly disregard this in this book. This can be of importance. For instance, a common mistake when writing computer programs is when "Union" or "Join" are used for elements instead of arrays. Join[$\{a, b, \ldots\}, p$] is not allowed; it must be $\{p\}$ here.

The map $T \mapsto T \cup p$ does the job (it changes the parity of T) with one important reservation. This map is formally well-defined if $p \in T$, but changes the parity only if $p \notin T$ (not in T), which is of course possible for some T. This can be fixed as follows: we add p if it is not in T or remove p otherwise. The correct definition is

$$f : T \mapsto \{T \cup p \text{ if } p \notin T\} \text{ and (otherwise) } \{T \setminus p \text{ if } p \in T\}. \quad (3.2.4)$$

Recall that $A \setminus B$ is subtracting a subset $B \subset A$ from A.

The number of players on the resulting team upon applying f will now be $r \pm 1$, which gives the correct parity. So even teams will "go" to odd ones, and odd teams will "go" to even ones. This is not the end: We need to find the inverse of f. In this case, it is simply f. Obviously, f^2, applying f twice, is the identity map *id*: $T \mapsto T$ for any T. So f establishes a bijection between the set of all even subsets of $[n]$ and the set of all odd subsets.

This construction is shown in the first half of Fig. 3.1. The bijection we established proves the second formula in (3.2.3), but everyone would agree that the algebraic proof is simpler. It is unusual that f depends on some choice of p, so it is not some "canonical" map.

3.2.3 Further identities

Let us begin with the *Multinomial Theorem*. For $m, n \geq 1$,

$$(x_1+x_2+\cdots+x_m)^n = \sum_{k_1+k_2+\cdots+k_m=n} \frac{n!}{k_1!k_2!\cdots k_m!} x_1^{k_1} x_2^{k_2} \cdots x_m^{k_m}. \quad (3.2.5)$$

Here $k_i \geq 0$, x_i are variables. This is simple to justify combinatorially. Consider an arbitrary permutation of $1, \ldots, n$; there are $n!$ of them. Given such a permutation, the first k_1 numbers are interpreted as the numbers of the quantities $(x_1 + \cdots + x_m)$ where we pick x_1. The following k_2 numbers then give those where we pick x_2, and so on. The permutations within the first group of k_1 numbers, the second one with k_2 numbers and so on will not affect the count of monomials, since we need only to know the *set* of quantities where we take x_i for $1 \leq i \leq m$, not the *sequence*. Thus, we need to divide $n!$ by $(k_1)!(k_2)!\cdots(k_m)!$, which gives the required identity.

One can of course simply apply Newton's binomial $m-1$ times to justify this formula. As a demonstration, let us do this for $m = 3$:

$$(x_1+x_2+x_3)^n = \sum_{k_1=0}^{n} \frac{n!}{k_1!(n-k_1)!} x^{k_1}(x_2+x_3)^{n-k_1}$$

$$= \sum_{k_1=0}^{n} \frac{n!}{k_1!(n-k_1)!} x^{k_1} \sum_{k_2=0}^{n-k_1} \frac{(n-k_1)!}{k_2!(n-k_1-k_2)!} x_2^{k_2} x_3^{n-k_1-k_2}.$$

Then we perform the multiplication, delete $(n-k_1)!$ in the numerator and denominator, and rewrite the summation using $k_3 = n - k_1 - k_2$. See also Problem 2.5.2 and Section 7.1.3.

Letting $p_{k_1,\ldots,k_m} = \frac{n!}{k_1!k_2!\cdots k_m!}/(m^n)$, we arrive at the ***multinomial distribution***. The sum of all $p_{k_1,\ldots,k_m}$ is 1 due to (3.2.5) where we set $x_1 = x_2 = \cdots = x_n = 1$.

The next level of formulas is based on the derivatives of $(1+x)^n$:

$$\frac{d}{dx}(1+x)^n = n(1+x)^{n-1} = \sum_{r=1}^{n} r\binom{n}{r} x^{r-1}, \tag{3.2.6}$$

$$\text{which gives} \quad n2^{n-1} = \sum_{r=1}^{n} r\binom{n}{r} \quad \text{for} \quad x = 1. \tag{3.2.7}$$

Here we used that $d(x^r)/dx = rx^{r-1}$ on the right-hand side of (3.2.2), and the chain rule on the left-hand side. The derivatives of polynomials belong to algebra; this is "pre-calculus." Though the chain rule is generally not "pre-calculus." It is certainly better to have an entirely combinatorial proof of (3.2.7), without any derivatives.

Consider r-teams T' from a pool of n players with captains. Given a team T, the captain is a distinguished player in T. The number of such "enhanced" teams T' is $\sum_{r=1}^{n} r\binom{n}{r}$ (check); note that we can begin the summation here with $r = 0$ (it will be the same sum). This is the right-hand side. The left-hand side is based on the following alternative way of "creating" T'. First, we select the captain, any player from the pool of n, and, second, the captain picks from the remaining $(n-1)$ players to make the team (of any size). This gives $n2^{n-1}$, and we arrive at (3.2.7). See the second half of Fig. 3.1.

The combinatorial proof is very suitable here: it is from scratch and can be readily generalized. For instance, consider teams T'' with captain C and vice-captain vC. Check that we arrive at the identity

$$\sum_{r=2}^{n} r(r-1)\binom{n}{r} = n(n-1)2^{n-2}.$$

Two differentiations of (3.2.2) give of course the same result, but it is not really necessary to use some calculus here.

A similar identity is that from Exercise 3.7.2:

$$\sum_{r=1}^{n}(-1)^r r\binom{n}{r} = 0, \text{ or } \sum_{\text{even } r} r\binom{n}{r} = \sum_{\text{odd } r} r\binom{n}{r}.$$

One can differentiate here or modify the map f : {even subsets} → {odd subsets} from (3.2.4) as follows. Pick 2 different players, p and q, and add one line to the definition of f: {use q if $p = C$ in T'}. Check this. The following identity seems complicated, but its proof is simple.

Problem 3.2.1 *Prove that* $\sum_{r=0}^{n}\binom{n}{r}^2 = \binom{2n}{n}$.

Solution. We split $2n$ players into two halves: $\overbrace{\square\cdots\square}^{n}\overbrace{\square\cdots\square}^{n}$. Then composing a team with n players requires taking r from the first half and $n-r$ from the second. Thus:

$$\binom{2n}{n} = \sum_{r=0}^{n}\binom{n}{r}\binom{n}{n-r} = \sum_{r=0}^{n}\binom{n}{r}\binom{n}{r},$$

where we use that $\binom{n}{n-r} = \binom{n}{r}$. ❑

3.2.4 Generating functions I

If $\mathcal{N}_r$ is the number of outputs of value r for a certain process, then $G(x) = \sum_r \mathcal{N}_r x^r$, where the sum is over all r, is called the corresponding ***generating function***. See also Section 4.2.4. If $r = r_1 + r_2$ where r_1 and r_2 are for two *independent* processes with the generating functions $G_1(x)$ and $G_2(x)$, then the corresponding generating function will be $G(x) = G_1(x)G_2(x)$. A variant is when we switch to the probabilities p_r of "r" and calculate $g(x) = \sum_r p_r x^r$. Then the probability of $r = r_1 + r_2$ will be the coefficient of x^r in $g_1(x)g_2(x)$.

To give an example, the generating function of the Poisson distribution $\{p_r^\alpha = e^{-\alpha}\frac{\alpha^r}{r!}\}$ from Section 2.4.4 is $g_\alpha(x) = e^{-\alpha}\sum_{r=0}\frac{\alpha^r x^r}{r!} = e^{\alpha(x-1)}$. For instance, let us assume that there are 2 sources of hurricanes in some area with the Poisson distributions $\{p_r^\alpha\}$ and $\{p_r^\beta\}$ for r hurricanes per decade; see Exercise 2.6.1. Then the generating function for the probability of r hurricanes of any type in one decade is $e^{x(\alpha+\beta)(x-1)}$. It is the same as for 1 source of hurricanes with the corresponding parameter $\alpha + \beta$.

This "multiplicativity" is actually the defining property of the Poisson distribution; it is directly connected with that for the *binomial distribution*. For coin flipping, the generating functions are $G(x) = 1+x$ and $g(x) = (1+x)/2$. Accordingly, they become $G_n = (1+x)^n$ and $g_n = (\frac{1+x}{2})^n$ for n flips. This is immediate: we flip a coin n times, which gives that $G_n = (G_1)^n = (1+x)^n$.

R-numbers. Let us provide the generating function for $R(n,r) = \binom{n+r-1}{r}$, where we fix $n \geq 1$ and consider the sum over all $r \geq 0$. It is:

$$\frac{1}{(1-x)^n} = \sum_{r=0}^{\infty} \binom{n+r-1}{r} x^r, \; n = 1, 2, \ldots . \tag{3.2.8}$$

The justification is very similar to that for Newton's binomial. We need the identity $(1-x)^{-1} = (1+x+x^2+\ldots)$. This has nothing to do with the convergence; it simply means that $(1-x)(1+x+x^2+\ldots) = 1$, which is an entirely algebraic identity. Then (3.2.8) is a particular case of the product formula for generating functions stated above. Explicitly,

$$\overbrace{(1+x+x^2+\ldots)(1+x+x^2+\ldots)\cdots}^{n} = \sum_{r=0}^{\infty} \binom{n+r-1}{r} x^r.$$

We take any x^a from the first factor, then any x^b from the 2nd factor and so on. The coefficient of x^r coming from this product will then be the number of presentations $r = \overbrace{a+b+\ldots}^{n}$, which is $\binom{n+r-1}{r}$. This is parallel to Newton's binomial (the infinite series do not change anything). Zeros are allowed for $a, b, \ldots$, which corresponds to picking 1 from $(1+x+x^2+\ldots)$.

Counting king's paths. Let us solve Problem 2.5.2, which was about the chances to start in the bottom-left corner of the 8×8 chessboard and reach the top-right corner using the moves ↑, →, ↗, *king's moves*, each with equal probability 1/3. The challenge was to find the corresponding *fair bet*.

For $2n$ moves with an unlimited chessboard, the number of paths K_n from box $(1,1)$ to box $(n+1, n+1)$, and the corresponding bet is:

$$K_n = \sum_{m=0}^{n} 3^m \binom{2n-m}{m} \binom{2n-2m}{n-m} = \sum_{m=0}^{n} 3^m \frac{(2n-m)!}{m!((n-m)!)^2}, \tag{3.2.9}$$
$$\text{Bet}_n = (3^{2n} - K_n)/K_n, \; \text{Bet}_7 = 3.47875.$$

We can restrict ourselves to the $(n+1) \times (n+1)$ board since any of these paths will be inside it. For the classical chessboard, $n = 7$. We will always play $2n$ times (which can be beyond this chessboard).

Here m is the number of diagonal moves, which can be from 0 to n. There will be $2n - 2m$ remaining moves then, and $A = \binom{2n-2m}{n-m}$ possible paths made from them, assuming that the diagonal moves are fixed. Since the total number of moves is $2n-m$, there are $B = \binom{2n-m}{m}$ possibilities to pick the diagonal moves. However, with $2n-m$ moves, we must multiply AB by 3^m, the number of *all* their continuations from $2n-m$ moves to $2n$ moves.

We obtain that the probability $p_{n,n}$ to reach box $(n+1, n+1)$ is $K_n/3^{2n}$. Thus, $p_{7,7} = 0.223277$ for $n = 7$, and $\mathrm{Bet}_7 = (1/p_{7,7}) - 1 = 3.47875$.

It is convenient to use the generating function $g(x,y) = \sum_{a,b=0}^{\infty} p_{a,b} x^a y^b$, where $p_{a,b}$ is the probability to reach box $(a+1, b+1)$ from box $(1,1)$. One has: $g(x,y) = 1/(1 - \frac{x+y+xy}{3})$; check this following (3.2.8). Accordingly, $K_n/3^{2n}$ is the coefficient of $x^n y^n$ in this series. More explicitly, it is the coefficient of $x^n y^n$ in $\sum_{m=n}^{2n} \left(\frac{x+y+xy}{3}\right)^m$. Then use your computer!

3.3 Pascal's triangle

3.3.1 Triangle recurrence

Actually, all relations above follow from only one, *the ultimate recurrence relation*, which is

$$\binom{n}{r} = \binom{n-1}{r} + \binom{n-1}{r-1} \quad \text{for } n = 0, 1, 2, \ldots, \ r \le n. \tag{3.3.10}$$

This is justified algebraically by expressing $(1+x)^n = (1+x)^{n-1}(1+x)$ and examining the coefficients. Combinatorially, the justification is nicer. Let us fix some element p in the set $[n] = \{1, 2, \ldots, n\}$. The binomial coefficient $\binom{n}{r}$ counts r-element subsets $X \subset [n]$. Any p can be outside or inside X. This splits the set of all subsets X into two groups with the sizes $\binom{n-1}{r}$ and $\binom{n-1}{r-1}$ respectively. This is shown in Fig. 3.2; if $p \in X$ then the remaining subset $X \backslash p$ is with $r-1$ elements to be picked from $[n] \backslash p$. This is one more example of "the tree of possibilities."

The triangle relation is absolutely the best for computer calculations, especially if all binomial coefficients for a given n are needed. It is additive!

3.3.2 Heat equation

The discrete ***heat equation*** on a line with an initial *point heat source* is an example of fundamental importance, not only in physics. It is a straightforward interpretation of (3.3.10). The classical heat equation is a partial differential equation. However the discrete setting is quite natural here ..., and computers understand only recurrence relations.

We consider a set of points $\ldots, -\frac{3}{2}, -1, -\frac{1}{2}, 0, \frac{1}{2}, 1, \frac{3}{2}, \ldots$. The integer and half-integer points are convenient for the discrete heat equation in dimension one. We will use them alternately when calculating the averages. The

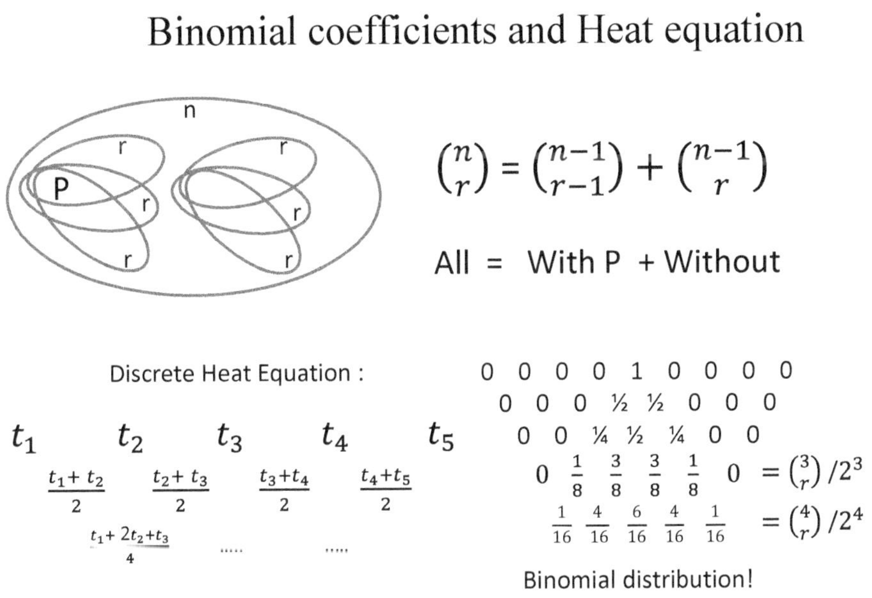

Figure 3.2: One-dimensional heat equation.

corresponding "physics law" is: if the temperatures at time n are $t_n(m/2)$ at points $x = m/2$, where integers m are assumed all even or all odd, then they become at the next moment the corresponding averages of the temperatures of the neighbors. The average is "placed" at $(m+1)/2$: $t_{n+1}(\frac{m+1}{2}) = \big(t_n(\frac{m}{2}) + t_n(\frac{m}{2}+1)\big)/2$. So we switch here from integer points x to half-integers and then back to integers: $\{\ldots,-1,0,1,\ldots\} \to \{\ldots,-\frac{1}{2},\frac{1}{2},\ldots\} \to \{\ldots,-1,0,1,\ldots\}$. This is easier this way. See the bottom part of Fig. 3.2.

One can take here any initial distribution of temperatures. To obtain the binomial coefficients, let us assume that we have a single point heat source at $x=0$ at moment $n=0$: $t_0(0)=1$ and $t_0(m)=0$ for any integers $m \neq 0$. Then the temperatures at moment $n=1$ will be $t_1(\pm 1/2) = (1+0)/2$ (zero otherwise), then, at moment $n=2$, $t_2(0)=(\frac{1}{2}+\frac{1}{2})/2=\frac{1}{2}$, $t_2(\pm 1)=(\frac{1}{2}+0)/2=\frac{1}{4}$ (zero otherwise), and so on.

We obtain that the distribution of temperatures at moment n becomes:

$$\ldots,0,\binom{n}{0}/2^n,\binom{n}{1}/2^n,\ldots,\binom{n}{r}/2^n,\ldots,\binom{n}{n}/2^n,0,\ldots\,.$$

The x-points with nonzero temperatures at time n will range from $-\frac{n}{2}$ to $\frac{n}{2}$ (with step $=1$): $\{-\frac{1}{2},\frac{1}{2}\}$ for $n=1$, then $\{-1,0,1\}$ for $n=2$, $\{-\frac{3}{2},-\frac{1}{2},\frac{1}{2},\frac{3}{2}\}$ for $n=3$, and so on. This is the *binomial distribution* from Section 2.5.1!

3.3.3 Mathematical induction

Since $\binom{n}{r}$ is fully determined by the initial values $\binom{n}{0} = 1 = \binom{n}{n}$ and the triangle relation, any relations for binomial coefficients can be deduced from the latter. ***Mathematical induction*** can be used here. It works as follows.

When checking that an identity holds for a certain n, we are allowed to use the identity for any $m < n$, provided that it was checked for sufficiently many initial n. The basic case is when we check the claim for $n = 1$ and then verify that "$(n-1)$ implies $n > 1$". However, the latter can be "$m < n$ implies n", i.e. the claim $(n-1)$ can be insufficient for the passage to n, and the number of necessary initial confirmations can not be only for $n = 1$.

The simplest example is the proof of the formula $1 + \cdots + n = n(n+1)/2$. One has $1 + \cdots + n = (1 + \cdots + (n-1)) + n = n(n-1)/2 + n = n(n+1)/2$, where we use this claim for $(n-1)$; the case $n = 1$ is obvious. Similarly, prove by induction that $\sum_{m=0}^{n} m^2 = \frac{n(n+1)(2n+1)}{6}$; see Exercise 3.7.10.

Let us reprove that $\sum_{r=0}^{n} \binom{n}{r} = 2^n$ *by induction*. This is immediate for $n = 1$. Then $\sum_{r=0}^{n} \binom{n}{r} = \sum_{r=0}^{n} \left(\binom{n-1}{r} + \binom{n-1}{r-1}\right) = \sum_{r=0}^{n-1} \binom{n-1}{r} + \sum_{r=1}^{n} \binom{n-1}{r-1} = 2 \cdot 2^{n-1} = 2^n$, where this identity for $(n-1)$ was used.

Similarly, $\binom{n}{r} = \binom{n-1}{r} + \binom{n-1}{r-1} = \binom{n-1}{n-1-r} + \binom{n-1}{n-r} = \binom{n}{n-r}$ by induction. Here we assume this symmetry for $n-1$ and prove it for n.

Mathematical induction is a particular instance of the general "reduction" from n to n–1. The following is an example. The proof is by a direct algebraic manipulation: certainly "induction" but not strict "mathematical induction".

Problem 3.3.1 *Let* $s_{n+1} = \binom{n}{0} + \binom{n-1}{1} + \binom{n-2}{2} + \ldots$ *for* $n = 0, 1, 2, \ldots$. *Check that* $s_1 = \binom{0}{0} = 1, s_2 = \binom{1}{0} + \binom{0}{1} = 1$, *and* $s_{n+1} = s_n + s_{n-1}$. *Graphically, these sums are the sums of* ***shallow diagonals*** *of Pascal's triangle; see Fig. 3.3 below. Finally,* $s_n = f_n$ *for the* Fibonacci numbers f_n, *defined via the recurrence* $f_{n+1} = f_n + f_{n-1}$ *subject to* $f_1 = 1 = f_2$.

Solution. To begin with, $s_3 = \binom{2}{0} + \binom{1}{1} = 2 = s_1 + s_0$. Generally, using (3.3.10): $s_{n+1} = \binom{n}{0} + \binom{n-1}{1} + \binom{n-2}{2} + \ldots$. Thus

$$s_{n+1} = \binom{n-1}{0} + \left(\binom{n-2}{1} + \binom{n-2}{0}\right) + \left(\binom{n-3}{2} + \binom{n-3}{1}\right) + \cdots$$
$$= \left(\binom{n-1}{0} + \binom{n-2}{1} + \binom{n-3}{2} + \ldots\right) + \left(\binom{n-2}{0} + \binom{n-3}{1} + \ldots\right),$$

which equals $s_n + s_{n-1}$. We simply employ the triangle formula and recombine the terms. Finally, $f_n = s_n$ for $n = 1, 2, \ldots$, since the Fibonacci numbers satisfy the same recurrence for the same initial conditions as for s_n. ❑

The equality $s_n = f_n$ is old; it was known to Indian mathematicians at least around 975. For them it was the recurrence for the sums of *shallow diagonals* of Pascal's triangle. Also, $s_{n+1} = d_n$ for the number d_n of perfect coverings of "$2\times n$" by dominos from Section 1.1.1. This is direct from the definition of d_n; see (4.1.4) below.

3.3.4 Graphic presentation

Due to the recurrence relation, it is quite natural to present $\{\binom{n}{r}\}$ as a triangle. We, actually, did this already for the heat equation. In Fig. 3.3, the "next" binomial coefficient $\binom{n}{r}$ is placed under its two "predecessors": $\binom{n-1}{r-1}$ and $\binom{n-1}{r}$. The number n of a horizontal line can be seen right after 1. The sum of all the terms in each such line is 2^n, as we know. We provide one example of the *shallow diagonals*: the sum $\binom{8}{1}+\binom{7}{2}+\ldots$ coincides with the Fibonacci number f_9.

The greatest entry in Fig. 3.3 is the middle binomial coefficient $\binom{16}{8} = 12870$; interestingly, the corresponding calculus approximation is not that bad: $2^{16}/\sqrt{8\pi} \approx 13073$. There are many interesting properties of Pascal's triangle. For instance, you can see 3003 twice in rows 14 and 15. Generalizing this phenomenon, David Singmaster proved in 1970 that the equation $\binom{n+1}{r} = \binom{n}{r+1}$ has infinitely many solutions n, r, where $n > r$.

Some history. The oldest instance of this triangle was probably in Pingala's work (5th-2nd century BCE!); the references can be found in later works of Indian mathematicians (10th century CE). They knew that the sums of the numbers in shallow diagonals satisfy the Fibonacci recurrence.

Newton's binomial and Pascal's triangle were certainly known to the famous Baghdad school, mathematicians of the "House of Wisdom." It is called *Khayyam's triangle* in Iran. Omar Khayyam was a great poet and a very good mathematician; he lived a long and very productive life (1048-1131), which could be equivalent to 100+ years now. We will say a little something about the "House of Wisdom" when discussing Fibonacci numbers.

The third instance of this triangle was *Yang Hui's triangle*. Its graphic presentation practically coincided with the modern one (but with rod numerals); this was the 13th century.

In Italy, it was called *Tartaglia's triangle* (16th century). It was well known to quite a few mathematicians in Europe around this time, a very fruitful period in the history of algebra. For instance, Petrus Apianus, an influential scholar, published it on the frontispiece of his work (1527); he mostly worked at the University of Ingolstadt (in Bavaria).

Finally, Blaise Pascal wrote his paper on this in 1655, the end of this

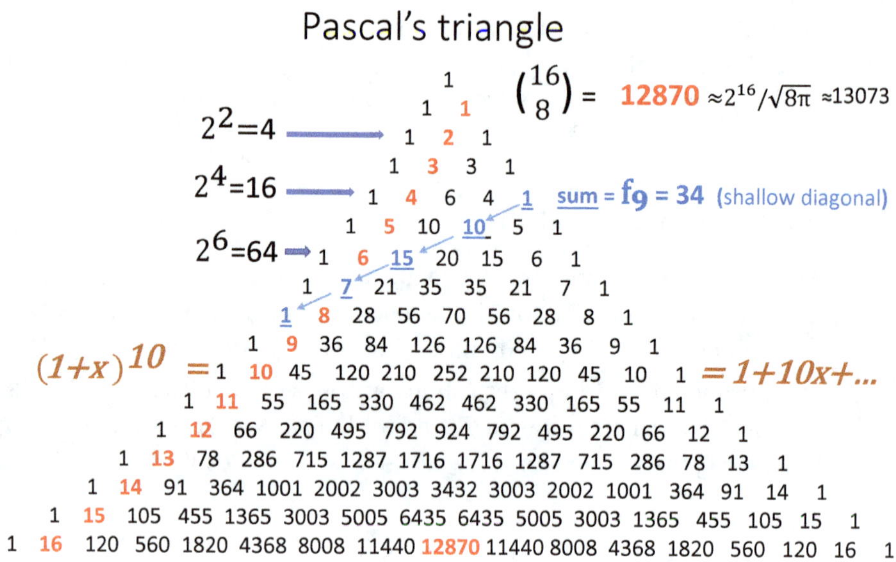

Figure 3.3: Pascal's triangle until $n = 16$.

ramified story through many centuries. After his work, it was impossible to rediscover this triangle again; it became a standard tool in mathematics.

The names of the latest contributors are frequently used as the names of mathematical facts and objects; this is not really fair but convenient.

Pascal's paper was of course not the end of this long story. For instance, q-binomial coefficients, satisfying the recurrence $\begin{bmatrix} n \\ r \end{bmatrix}_q = \begin{bmatrix} n-1 \\ r-1 \end{bmatrix}_q + q^r \begin{bmatrix} n-1 \\ r \end{bmatrix}_q$ for the initial conditions $\begin{bmatrix} n \\ 0 \end{bmatrix}_q = 1, \begin{bmatrix} n \\ r>n \end{bmatrix}_q = 0$, were added later. See Exercise 3.7.43. There have been other generalizations and developments.

3.4 Inclusion-Exclusion Principle

3.4.1 Overlapping sets

The problem is to find the *order* (the size), $| \cup_i A_i |$, the number of elements, of the union $\cup_i A_i$ of sets A_i, which generally can overlap. We continue to use here vertical bars for the orders of sets, $\cup$ for unions, and $\cap$ for intersections.

Obviously, we need to know at least the orders $a_i = |A_i|$ of these sets, and the orders of all intersections, which we will denote as follows:

$$A_{ij} = A_i \cap A_j, \ A_{ijk} = A_i \cap A_j \cap A_k, \ a_{ij} = |A_{ij}|, \ a_{ijk} = |A_{ijk}|, \ \text{etc.}$$

Here and from now on ij and ijk are all pairwise distinct. It is not obvious that the knowledge of the orders of all intersections is sufficient.

The following formula confirms this and in a very explicit way:

$$|A_1 \cup \ldots \cup A_n| = a_1 + \cdots + a_n - (a_{12} + a_{13} + \cdots + a_{1n} + \cdots + a_{n-1\,n}) + (a_{123} + \cdots) - (a_{1234} + \cdots) + \cdots + (-1)^{n-1} a_{123\ldots n}. \tag{3.4.11}$$

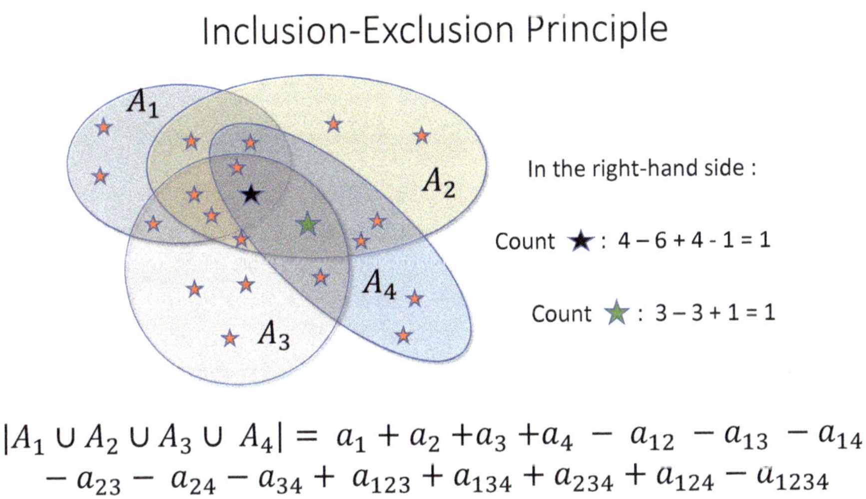

Figure 3.4: "Inclusion-Exclusion" for $n = 4$.

3.4.2 Proof via binomial coefficients

For instance, for $n = 2$, we begin with $a_1 + a_2$ and subtract a_{12}, because the objects in the intersection $A_1 \cap A_2$ are counted twice in $a_1 + a_2$. For $n = 3$, there will be *correction after correction*, namely, $+a_{123}$. Some consecutive corrections can be expected, but it is necessary to justify that the coefficients of such corrections will be ± 1. This is not obvious, though it seems quite likely after considering several small n.

To prove (3.4.11), let us count how many times a given object occurs on the right-hand side. It is counted once on the left-hand side. In Fig. 3.4, let us consider 2 objects marked by big stars.

We can assume that ★ belongs to all A_i; the A-sets that do not contain it obviously do not contribute to the alternating sum and therefore can be omitted. For instance, only A_2, A_3, A_4 are needed to consider for the second (green) big star in this figure.

Then, the numbers of intersections $\overbrace{A_\bullet \cap \ldots \cap A_\bullet}^{r}$ for r sets A_i containing ★ equals the number of subsets with r elements from the set of all n indices i ($n = 4$ in the figure). Thus, the multiplicity of ★ in this sum is $n - \binom{n}{2} + \binom{n}{3} - \cdots \pm \binom{n}{n}$. Formula (3.2.3) from Section 3.2.2 gives that this alternating sum equals 1, which proves the formula. See Fig. 3.4. ❑

3.4.3 Adding supply lines

We count all possible "baskets," when we purchase 10 objects from 3 types, $\triangle, \square, \odot$, assuming that the store has only $3\,\triangle + 4\,\square + 5\,\odot$. The latter is called the ***supply line***. It is more reasonable to assume that the store has a lot, but only limited amounts are allowed to be purchased (say, with discount prices); we will nevertheless use the name "supply."

We assume this when discussing the solution below: your basket can contain any number of objects of any type, but you will be stopped if it contains either $\geq 4\,\triangle$, or $\geq 5\,\square$, or $\geq 6\,\odot$. The following is a uniform and convenient procedure, which is helpful to avoid mistakes. We recommend sticking to it even in simple cases. Problems of this kind are very standard: practically "all" enumerative combinatorics can be practiced.

(1) First of all, check that the "free" problem, without the supply or other limitations, can be solved. For the problems on purchasing r objects from n types, $R(n, r)$ solves the free problem. Here $Free = R(3, 10) = \binom{12}{2} = 66$. The Inclusion-Exclusion Principle will "do the rest," i.e. formula (3.4.11) will automatically provide the solution subject to the supply line: steps (2) – (6).

(2) Assume that there are 3 devices checking that you are within the allowed amounts. They are "programmed" as follows. If your basket has $x_1\,\triangle + x_2\,\square + x_3\,\odot$, then you will be stopped as "Bad-1" if $B_1 = \{x_1 \geq 4\}$, or as "Bad-2" if $B_2 = \{x_2 \geq 5\}$, or as "Bad-3" if $B_3 = \{x_3 \geq 6\}$. Notice that $4, 5, 6$ are the first "bad" here; $3, 4, 5$ are allowed.

(3) The key in the usage of "Inclusion-Exclusion" is that if you are already "Bad" because your basket violated one of these conditions or any of their combinations, then you are "free" from this moment on to add to your basket any objects without any restrictions. So we are back to (1), but now for smaller numbers of objects: the remaining ones after you became "Bad".

(4) Then you calculate using the R-formulas for $b_i = |B_i|$ (the number of elements):

$b_1 = |B_1| =$|"10-4=6 objects of 3 types"|$= R(3, 6) = \binom{8}{2} = 28$,

$b_2 = |B_2| =$|"10-5=5 objects of 3 types"|$= R(3, 5) = \binom{7}{2} = 21$,

$b_3 = |B_3| = |$"10-6=4 objects of 3 types"$| = R(3,4) = \binom{6}{2} = 15.$

(5) Next, calculate the intersections of the B-sets. Only two are nonempty in the considered example:

$B_{12} =$ "you have 4 $\triangle$ and 5 $\square$ and 1 remaining object," $b_{12} = |B_{12}| = 3$,

$B_{13} =$ "you have 4 $\triangle$ and 6 $\odot$ and 0 remaining objects" $b_{13} = |B_{13}| = 1$.

Note that if you have 0 remaining objects in your basket, then $b = 1$ (not zero!). Obviously, $b_{12} = 3$ without any formulas. However, we advise using R-numbers for the sake of uniformity: $b_{12} = R(3,1) = \binom{3+1-1}{3-1} = 3$. Obviously, $b_{23} = 0$ and $b_{123} = 0$: for this you need more than 10 objects in your basket. If B_{ij} is empty, then B_{ijk} will be empty for any k.

(6) Finally, we will use the Inclusion-Exclusion Principle. Actually, it is rather "Exclusion-Inclusion" in this book: we calculate $Free$ and then begin with the exclusion of bad baskets. The calculation goes as follows:

$$\begin{aligned} \mathcal{N}_{10} &= Free - (b_1 + b_2 + b_3 + \ldots) \\ &+ (b_{12} + b_{13} + b_{23} + \ldots) - (b_{123} + \cdots) + \cdots \\ &= 66 - (28 + 21 + 15) + (1 + 3) - 0 = 6. \end{aligned} \tag{3.4.12}$$

A great thing is that "Inclusion-Exclusion" always works if the "free problem" is doable; a not-so-great thing is that it may result in large oscillations. Indeed, we began with 66 and finally obtained only 6. As a matter of fact, such a small number can be expected, since we "purchased" almost the entire supply. The contents of our basket can be uniquely recovered if it is known how many objects we left in the store. We left only 2!

Generally, let m be the total number of objects in the supply line, $3+4+5 = 12$ in this example. Then $\mathcal{N}_r = \mathcal{N}_{m-r}$, because allowed baskets are one-to-one with the remaining sets of objects after our purchase. Here, $\mathcal{N}_{10} = \mathcal{N}_2$, and the latter can be calculated of course without any "Inclusion-Exclusion," since 2 is below the minimal supply restriction, which is 3 for triangles. Thus, $\mathcal{N}_2 = \binom{3+2-1}{2} = 6$, indeed.

3.5 Rolling dice

Games serve many purposes, but the key is to create and use some artificially controlled environment to prepare us for taking real risks. This is not only for us, people. A kitten trying to catch a butterfly uses a natural zero-risk random environment to develop and sharpen its hunting skills.

"Randomness" as a concept is no smaller than "space" and "time". Combinatorics, statistics and statistical physics provide mathematical and physical

Figure 3.5: 5 dice: 7% to make it 14.

foundations of *randomness*. Quantum physics is almost entirely about randomness. From this perspective, we began with coin flipping and lotteries, and will now go to the "next level": *rolling dice*.

3.5.1 Some history of dice

Dice are the oldest known gaming implements. Egyptian tombs from 2000 BCE have been found containing dice, and some were discovered in older archaeological excavations. Fig. 3.6 shows a very old dice from those excavated from Harappa and Mohenjo-daro, Indus Valley Civilization (southern Pakistan). See "6. Stone Dice" from [19]. They can be as old as 4500 years. Dice from this civilization were quite prefect. One dice from Harappa was reported as marked in the same way as modern dice, with the opposite sides adding up to seven. There were also *stick dice*, compared with roughly similar finds from ancient Egypt (the First Dynasty).

Figure 3.6: Indus Valley Civilization, about 4500 years ago.

Casting dice was initially used for fortune-telling. Ancient people used stones, sticks, sea shells, and animal bones. The Arab word for knucklebones is the word they use for dice; we still say "rolling the bones." Romans and Greeks were keen dice players. The Roman goddess Fortuna was believed to determine the outcome of a throw. Europe during the Middle Ages was

[19] E. Mackay, *The Indus civilization*, L. Dickson and Thompson, London, 1935.

a continent of compulsive players. King Henry VIII lost the bells of old St Paul's church on a throw.

Figure 3.7: Mahabharata, The Game of Dice, Chore Bagan Art Studio, 1895.

Though the most emotionally-charged story among very many of this kind is probably in the *Mahabharata*, a famous Sanskrit epic: Fig. 3.7 and the cover of this book from [20]. One by one, Yudhisthira put at stake and lost prized jewel ornaments from his treasury, his brocades, vestments, his precious metal savings, his livestock and horses, his armies, his kingdom; then his brothers, his own self and his wife, all of whom became slaves. His wife, Draupadi, protested: "If he had already lost himself and he was already a slave, he had no right to put me at stake." However, it was decided not in her favor. The situation turned from civilized to brutal; Krishna's protection appeared necessary.

It was a "family game" when Yudhisthira could not really refuse or protest and anyway he was not skillful; his opponent was a grand master of this game and some magic was likely to be involved. Statistically, it is impossible to be so unlucky as it was described in the *Mahabharata*. See Exercise 2.6.21 for calculations with the dice used in similar games.

Dominos descended from dice in China and spread to the West during the 18th century, while standard dice became the vital component of a vast array of board games. Let us mention that in *Yahtzee*, 5 dice are a substitute for

[20]From Wikimedia Commons, the free media repository: `https://commons.wikimedia.org/wiki/File:Draupadi_s_presented_to_a_pachisi_game.jpg`.

the hand in poker; see Exercise 3.7.37.

Before poker, *faro* (Pharaoh, *le Pharaon*) was very popular. It was basically, betting on the 13 card denominations. Faro originated in France around 1713 as a revised form of the popular British pub game ***basset***. Now it practically vanished. There were special boards and tables designed for it, where all denominations were painted.

The cards are open in pairs by the "banker" (1 card from 52 was disregarded and there is some special rule for the remaining 3). The bank takes the bets for the 1st denomination and pays to the players with a 1 to 1 payout (even money) for the second. If the denominations coincide, the bank wins half of the stakes upon the denomination. Also, one can bet on "high card." The farther you get into the deck, the more interesting the game becomes. With fewer and fewer numbers to bet on, the bets tend to get larger.

It is relatively simple to cheat at *faro*. For instance, a "stacked deck" with many paired cards is obviously profitable for the banker. Exercise 2.6.6 describes what can happen if the deck is not really random.

Faro was the most widespread form of gambling in the United States from 1825–1915. So popular was the game that faro gaming tables could be found in just about every saloon in every frontier town. Draw poker was a rarity on the frontier until the late 1870s. However, books and western films of the 1940s disregarded faro in favor of poker.

The rage of play obsessed all classes of England in the 18th century, including high-stakes gambling (for the elite). Some gamesters used faro as a way of acquiring an income. Charles Fox earned about £30,000 in 3 months by operating a faro bank (at Brook's) in 1781. However, it was not unusual for him to lose a lot, for instance, about £20,000 over just 3 nights in February 1778. He was able to play for long periods without sleep, which did not have much effect on his duties. Some modeled their own conduct after that of him.

Charles Fox played a key role in the *Treaty of Paris* (1783) that formally ended the American Revolutionary War. He was mostly in "opposition" during his political career, a strong voice against slavery and for the freedom of press. It was Fox who championed the abolition of the slave trade in the House of Commons. He proposed the bill that passed into law in 1807 as the Slave Trade Act. He was a brilliant and powerful debater (his friends added "who ever lived"). Fox died on 13 September 1806, forgiven for his failings by many because of his defence of civil liberty and overwhelming charisma.

In 1774 his father, Lord Holland, paid the young man's gambling debts to the amount of £140,000, about £20M in 2020. He abandoned gambling and racing ... 20 years later. This was not unique; the Prince of Wales' debts reached £630,000 by 1795 (not only due to gambling).

Figure 3.8: James Gilray, Modern Hospitality (1792).

At the table in Fig. 3.8 are: the Prince of Wales (later King George IV) and Charles Fox, the 2nd and the 5th (the last). Here "Knave" can mean the card Jack (in Lady Archer's hand) or a deceitful fellow. See [21].

3.5.2 Formula for 3 dice

We will mostly discuss rolling three dice below, three standard cubes with dots from 1 to 6 on the faces. The outcome is the sum of the numbers at their top faces. From 2 to 5 dice are used in gambling. *Hazard* is an early English game (perhaps of Arabic origin) played with 2 dice, which is still popular. The outputs for 2 dice are in the following table.

$1+1=2$	$1+2=3$	$1+3=4$	$1+4=5$	$1+5=6$	$1+6=7$
$2+1=3$	$2+2=4$	$2+3=5$	$2+4=6$	$2+5=7$	$2+6=8$
$3+1=4$	$3+2=5$	$3+3=6$	$3+4=7$	$3+5=8$	$3+6=9$
$4+1=5$	$4+2=6$	$4+3=7$	$4+4=8$	$4+5=9$	$4+6=10$
$5+1=6$	$5+2=7$	$5+3=8$	$5+4=9$	$5+5=10$	$5+6=11$
$6+1=7$	$6+2=8$	$6+3=9$	$6+4=10$	$6+5=11$	$6+6=12$

Note that $2+1=3$ and $1+2=3$ are different outputs. The cubes

[21] From Wikimedia Commons: `https://commons.wikimedia.org/wiki/File:Modern_hospitality,-or-a_friendly_party_in_high_life._(BM_1868,0808.6173).jpg` .

are different "individuals": with different colors in Fig. 3.5 [22]. Rolling 2 dice is absolutely the same as rolling 1 cube twice. Using these outputs, the probability of 10 with 2 dice is $3/6^2 = 3/36 = 1/12$ (10 occurs 3 times).

Problem 3.5.1 *Calculate the numbers $\mathcal{N}_9, \mathcal{N}_{10}$ of all possible throws of 3 dice with the sums* $9, 10$.

Solution. This is directly about R-numbers and the usage of the Inclusion-Exclusion Principle, but with uniform inequalities. To follow the procedure from Section 3.4.3, we need to address that the dots on the dice cubes historically begin with 1, not with 0 as for the "supply problems."

The answer is the number of presentations $9 = y_1 + y_2 + y_3$ for $1 \le y_i \le 6$. To adjust the range, let us switch to $x_i = y_i - 1$; then $x_1 + x_2 + x_3 = 9 - 3 = 6$ and the inequalities are $0 \le x_i \le 5$. This is necessary since we always allow 0 in the sums for R-numbers. Note that we can readily incorporate any inequalities $y_i \ge a_i$ in the same way: by setting $x_i = y_i - a_i$. The "upper" inequalities $y_i \le b_i$ are significantly more involved than the "lower" ones.

The "free problem" is now to count all presentations $6 = x_1 + x_2 + x_3$ with $x_i \ge 0$, and $Free = \binom{6+3-1}{3-1} = \binom{8}{2} = 28$. Now, $\text{Bad}_i = B_i = \{\text{“}x_i \ge 6\text{”}\}$ for all i. This is uniform in contrast to purchasing $\triangle, \square, \odot$ above.

Accordingly, $b_i = |\{x_1 + x_2 + x_3 = 6 - 6 = 0\}| = 1$, which is obvious without any formula. The formula of course gives the same: $\binom{0+3-1}{3-1} = 1$. Obviously, $b_{ij} = 0$ for any i, j; recall that $i \ne j$. The sum of y_i must be at least 15 for $b_{ij} > 0$: $15 = 7 + 7 + 1$ is Bad_{12}, or would be $12 = 6 + 6 + 0$ in terms of x_i. This is significantly greater than 9 or 10 (in terms of y_i). Finally, we obtain that $\mathcal{N}_9 = 28 - 3 \cdot 1 = 25$.

Let us adjust this calculation to 10: $\mathcal{N}_{10} = Free - 3b_1 = \binom{9}{2} - 3\binom{1+3-1}{3-1} = 36 - 9 = 27$ (adding 1 to "numerators" of the binomial coefficients). ❑

The formula for any $\mathcal{N}_n$ is straightforward. Switching to $x_i = y_i - 1$ the free problem is $x_1 + x_2 + x_3 = m$ for $m = n - 3$. We obtain the following:

$$\mathcal{N}_n = \mathcal{N}_{m+3} = \binom{m+2}{2} - \binom{3}{1}\binom{m-4}{2} + \binom{3}{2}\binom{m-10}{2} - \binom{3}{3}\binom{m-16}{2}.$$

Actually, $n \le 6 \cdot 3 = 18$ and $m \le 15$, so the last term can be omitted here.

A challenge. Let us use that $\binom{n}{2} = \frac{n(n-1)}{2}$. Then $\mathcal{N}_n = \mathcal{N}_{m+3} =$

$$\frac{(m+2)(m+1)}{2} - 3\frac{(m-4)(m-5)}{2} + 3\frac{(m-10)(m-11)}{2} - \frac{(m-16)(m-15)}{2}.$$

Since $\mathcal{N}_n$ vanishes for sufficiently large n, the right-hand side must be identically zero for any m. Indeed, any polynomial of the fixed degree of one

[22] `https://www.peakpx.com/441488/5-dices.`

variable (here m) is identically 0 if it has infinitely many zeros. However, $\mathcal{N}_n$, the left-hand side, does not vanish for $2 \le n \le 18$, i.e. for $0 \le m \le 15$. Where was a mistake made?

3.5.3 Using computer algebra

Let us show how one can quickly calculate the probabilities with 3 dice or 5 dice using standard algebraic software. The simplest expansion operations for products of polynomials are needed here.

One die is represented by the polynomial $(x + x^2 + x^3 + x^4 + x^5 + x^6)$: the exponents are all possible outputs: $1, \ldots, 6$. Then the coefficient of x^n in the following product of these polynomials gives exactly $\mathcal{N}_n^3$, the number of throws with the sum m for 3 dice:

Expand$[(x + x^2 + x^3 + x^4 + x^5 + x^6)^3] = x^3 + 3x^4 + 6x^5 + 10x^6 + 15x^7 + 21x^8 + \mathbf{25x^9} + 27x^{10} + 27x^{11} + 25x^{12} + 21x^{13} + 15x^{14} + 10x^{15} + 6x^{16} + 3x^{17} + x^{18}$.

This polynomial is the *generating function* for 3 dice; see Section 3.2.4. We truncate the series $1 + x + x^2 + \ldots$ in (3.2.8) and multiply it by x to match the faces of a dice. The coefficient of x^n in this product is the number of presentations $n = \sum_i a_i$, where we pick x^{a_i} from the corresponding factors. The marked term $25x^9$ corresponds to $\mathcal{N}_9^3 = 25$ we calculated before, as well as the next one: $\mathcal{N}_{10}^3 = 27$. Notice that no alternating sums from "Inclusion Exclusion" occur in this approach.

This product formula can be readily extended to 5 dice. For instance, the number of presentations $a_1 + \cdots + a_5 = 14$ with $1 \le a_i \le 6$ is 540. The corresponding "Expand-formula" reads:

Expand$[(x+x^2+x^3+x^4+x^5+x^6)^5] = x^5+5x^6+15x^7+35x^8+70x^9+126x^{10}+205x^{11}+305x^{12}+420x^{13}+\mathbf{540x^{14}}+651x^{15}+735x^{16}+780x^{17}+780x^{18}+735x^{19}+651x^{20}+540x^{21}+420x^{22}+305x^{23}+205x^{24}+126x^{25}+70x^{26}+35x^{27}+15x^{28}+5x^{29}+x^{30}$. This was used in Fig. 3.5: chances of 14 are $540/6^5 \approx 7\%$.

The same approach works perfectly for any supply problems. For instance, the number of 10-baskets under the supply $3\triangle + 4\square + 5\odot$, which is 6, equals the coefficient of x^{10} in:

Expand$[(1 + x + x^2 + x^3)(1 + x + x^2 + x^3 + x^4)(1 + x + x^2 + x^3 + x^4 + x^5)] = 1 + 3x + 6x^2 + 10x^3 + 14x^4 + 17x^5 + 18x^6 + 17x^7 + 14x^8 + 10x^9 + \mathbf{6x^{10}} + 3x^{11} + x^{12}$.

The usage of computer algebra is a must for real cubes; the corresponding probabilities are never 1/6 (even due to the number of dots on the faces). Assuming that they are p_i^j for cube i and face j, one needs to calculate $\prod_{i=1}^m (p_1^i x + p_2^i x^2 + p_3^i x^3 + p_4^i x^4 + p_5^i x^5 + p_6^i x^6)$, where i is from 1 to the number

m of cubes ($m = 3$ for 3 dice). The coefficient of x^n will give the probability P_n of the event "$a_1 + \cdots + a_m = n$".

It is of course straightforward to find p_j^1 for 1 die: roll it 1000 times and count the probabilities of "$a_1 = n$". However, it becomes quite a challenge for $m > 1$ if the distribution of probabilities is known only for the sum of faces (not for individual cubes). This is "reverse engineering": assume that all P_n are known and find p_j^i, which requires solving algebraic equations with many variables. This is a basic problem of this kind. It is not unusual when we know some distribution of probabilities for a complex process and need to find the probabilities for its simple components.

Toward normal distribution *. The probabilities for 5 dice can be reasonably approximated by some binomial distribution. This happens because both distributions, the binomial distribution for n coin flips and that for n dice, in the limit of large n, become the celebrated ***normal distribution***, which is one of the most universal in nature. It is absolutely the key in statistics and statistical physics. The normal distribution is already "quite present" for 5 dice. Our conclusion will be that rolling 5 dice is approximately equivalent to tossing a coin 65 times. This is an example of *CLT*, the *Central Limit Theorem*; see Section 2.5.2 and Exercise 3.7.47.

Let $\mathcal{N}(m)$ be the number of ways to represent $m = x_1 + \cdots + x_5$ for $1 \le x_i \le 6$, i.e. using 5 dice. We will show numerically that it can be approximated by $780 \cdot \binom{65}{m+15} / \binom{65}{32}$ for m sufficiently close to the middle of the range, which is $17.5 = (5 \cdot 6 + 5)/2$, the average sum for 5 dice. The scaling coefficient 780 here is the number of ways to represent $m = 17$ or $m = 18$, the greatest possible for $5 \le m \le 30$. The number of tosses, 65, is taken to provide that $780/6^5 \approx \binom{65}{32}/2^{65}$. A more standard way is via *CLT* (below).

To adjust the usage of 5 dice to real problems, let us assume that the faces are $-2.5, -1.5, -0.5, 0.5, 1.5, 2.5$ with equal probabilities, instead of $1, \ldots, 6$ for 5 dice. For instance, the price changes of equities in stock markets are generally in both directions. Let $N(a)$ be the number of presentations of half-integers a (positive and negative) as $a = y_1 + \cdots + y_5$ with $y_i = \pm 0.5, \pm 1.5, \pm 2.5$. Then $N(a) = \mathcal{N}(a + 17.5)$; e.g. $N(\pm 0.5) = 780$.

Similarly, let $M(a)$ be the number of presentations $a = v_1 + \cdots + v_{65}$ for such a for $v_i = \pm 0.5$. It equals the number of presentations $a + 32.5 = u_1 + \cdots + u_{65}$ for $u_i = 0, 1$; so $M(a) = \binom{65}{a+32.5}$. To compare $M(a)$ with $N(a)$ we need to rescale it: $\widetilde{M}(a) = 780 \cdot M(a)/\binom{65}{32}$. Let $n(a) = 100\, N(a)/6^5$ and $m(a) = 100\, M(a)/2^{65}$, the corresponding probabilities in percent.

The following table shows that the values $\widetilde{M}(a)$ reasonably match $N(a)$ for $4.5 \ge a \ge 0.5$, but they begin to deviate for $a > 4.5$. Concerning matching the probabilities $n(a)$ and $m(a)$, 61-63 for $m(a)$ suit a bit better.

$a =$	9.5	8.5	7.5	6.5	5.5	4.5	3.5	2.5	1.5	0.5
$N(a)$	35	70	126	205	305	420	540	651	735	780
$\widetilde{M}(a)$	49	86	141	217	313	425	542	650	734	780
$n(a)$	0.45	0.90	1.62	2.63	3.92	5.40	6.94	8.37	9.45	10.03
$m(a)$	0.62	1.08	1.77	2.72	3.93	5.33	6.80	8.16	9.21	9.78

Using the *CLT*, the *variances* of these two processes are supposed to match. Generally, $var_x = \sum_{i=1}^{n} p_i(x_i - \overline{x})^2$ for the *mean* $\overline{x} = \frac{1}{n}\sum_{i=1}^{n} x_i$, where a random variable x takes its values x_i with probabilities p_i. For 5 dice, $var_x = 5(2.5^2 + 1.5^2 + 0.5^2)/3 \approx 14.6$; for 65 coin-tossing with $v = \pm 0.5$, $var_v = 65(0.5)^2 = 16.25$. Actually, the match of the variances becomes better for 59 instead of 65, but we focus on the shape of $N(a)$ near $a = 0$.

3.5.4 Poker via *R*-numbers

The last "station" before roulette and the stock market, will be ***draw poker***. The deck has 52 cards: 4 suits and 13 denominations. We will use $1, 2, \ldots, 13$ for the latter. A poker hand consists of 5 cards.

For the sake of simplicity, suits will be disregarded and the deck will have only 13 cards: $1, 2, \ldots, 13$. So the number of hands is $\mathcal{N} = \binom{13}{5} = 1287$; the hands are of course "combinations."

The "top event" in this deck is the *full straight*: 5 consecutive denominations. They are $\{1, 2, 3, 4, 5\}, \ldots, \{9, 10, 11, 12, 13\}$; so there are 9 of them. The chances of the "top event" are 9 to $\mathcal{N}$, about 0.7%. Let us calculate the chances of the event "no single *consecutive pair*"; we will simply say *pairs* instead of consecutive pairs. There are no reasons to continue playing in this case. To make the game playable this "non-event" must be relatively rare.

The hands are naturally the sequences $(1 \le a < b < c < d < e \le 13)$. Equivalently, we can represent them as $(a, b = a + x, c = a + x + y, d = a+x+y+z, e = a+x+y+z+t)$, where $a \ge 1, x, y, z, t \ge 1, a+x+y+z+t \le 13$. This makes their count very close to the usage of the R-numbers. The restrictions $a, x, y, z, t \ge 1$ are no problem for us, but the R-numbers require $a + x + y + z + t = const$, not some inequality here.

There is a very useful connection between equalities and inequalities, which is established by adding an extra variable. In our case, the following conditions are equivalent:

$$\{a + x + y + z + t \le 13\} \Leftrightarrow \{a + x + y + z + t + u = 13, u \ge 0\}.$$

Thus, all hands can be represented as sequences (a, x, y, z, t, u) such that $a, x, y, z, t \ge 1, u \ge 0$ and $a + x + y + z + t + u = 13$. Upon switching to

$a' = a-1, x' = x-1, y' = y-1, z' = z-1, t' = t-1, u' = u$ (as we did with dice), we have $a' + x' + y' + z' + t' + u' = 13 - 5 = 8$ for any non-negative a', x', y', z', t', u'. Here $n = 6, r = 8$ and the total number of hands is $R(6,8) = \binom{8+6-1}{6-1} = \binom{13}{5}$, ... which we know without any R-numbers.

The advantage of this approach is that it is simple to incorporate any events (or their absence) in terms of x, y, z, t. For instance, the event "no single pair" means that $x, y, z, t \geq 2$. Accordingly, we set $a' = a-1, x' = x-2, y' = y-2, z' = z-2, t' = t-2, u' = u$ and obtain $a'+x'+y'+z'+t'+u' = 13-5-4 = 4$. Thus the number $\mathcal{N}_\emptyset$ of such non-events is $R(6,4) = \binom{9}{5} = 126$, and its chances are $126/1287 \approx 9.8\%$.

Problem 3.5.2 *For the whole deck with 52 cards and the standard* 5*-hands, calculate the chances of the non-event "no single pair with coinciding or consecutive denominations."*

Solution. Since the coinciding denominations are excluded, we need to add 4 suits in all possible ways to the (ordered) "hands" with 13 denominations; their number is $\mathcal{N}_\emptyset$ counted above. This is $4^5\mathcal{N}_\emptyset$: 4 choices for the first card, 4 for the second and so on. Therefore the chances are $4^5\binom{9}{5}/\binom{52}{5}$, which is almost exactly (numerically) 5%. ❑

Problem 3.5.3 *How many* 5*-hands from 13 cards (the denominations only) have exactly* 1 *(increasing) pair?*

Solution. Represent the cards $a < b < c < d < e$ in terms of a, x, y, z, t as above: $b = a + x, \ldots, e = a + x + y + z + t \leq 13$, which gives $a + x + y + z + t + u = 13$ with $u \geq 0$. Set $a'a - 1$, $x = 1$ (i.e. ab form a pair), $y' = y-2$, $z' = z-2$, and $t' = t-2$ (we use that only 1 pair is allowed). Then $a' + x + y' + z' + t' + u = 13 - 6 - 1 = 6$ and $a' + y' + z' + t' + u = 5$. Thus, the number of hands with exactly 1 pair $b = a+1$ is $\binom{5+5-1}{5-1} = \binom{9}{4} = 126$, and the answer is $4 \cdot 126$, *four times greater* than the number of hands without pairs. Mostly people expect that it can be smaller or somewhat greater than for "no-pairs". The coefficient 4 is because the pair can be *ab, bc, cd, de*. Accordingly, the probability is $4 \cdot 126/1287 \approx 0.4$. See Exercises 3.7.35, 3.7.36 for further problems of this kind. ❑

3.6 Derangements etc.

3.6.1 Hat-check problem

This problem is from 1708. Several gentlemen went to a club and checked their hats. Accidentally their hats were returned in no specific order and with no specific intent. Assuming that they failed to notice this, what are the chances that no one received their own hat? Gentlemen's clubs are obviously considered here as some source of randomness; probably they were such in the early 18th century. In some variants, the hat-checker was to blame.

Generally, let D_n be the number of ***n-derangements***, which are defined as n-permutations $(p_1, \ldots, p_n)$ such that no element stays at its position, i.e. $p_i \neq i$ for any $1 \leq i \leq n$. It is convenient to use the *two-line notation* for permutations: $\begin{pmatrix} 1 & 2 & 3 \ldots (n-1) & n \\ p_1 & p_2 & p_3 \cdots p_{n-1} & p_n \end{pmatrix}$. The derangements are those with no repetition in columns, i.e. some counterparts of Latin squares for $2 \times n$-rectangles. Being random this way can be of some applied interest. Random permutations occur in many processes, including those in genetics. The meaning of derangements in biology is very different from that in combinatorics, but the concept of this kind of randomness is about the same: some repetitions must not be allowed.

The conditions $p_i = i$ are of course too simplistic for modeling real conditions preventing DNA assembly and similar processes. Anyway, if DNA are permutations in some universe and $p_i \neq i$ provide "successful assembly," then "life" based on such DNA will exist only if $D_n/n! > \epsilon > 0$ for large n and some $\epsilon > 0$. This is what we are going to examine using "Inclusion-Exclusion." Since $Free = n!$, some formula is granted.

One has: $D_1 = 0$, $D_2/2! = 1/2$, $D_3/3! = 2/6 = 1/3$. For $n = 3$, there are only two derangements: the cyclic permutations $(231), (312)$ and $D_3 = 2$ indeed. We use 1-line notation for permutations here and below. So far the trend is down; let us examine D_4.

For $n = 4$, the derangements are: (2,1,4,3), (2,3,4,1), (2,4,1,3), (3,1,4,2), (3,4,1,2), (3,4,2,1), (4,1,2,3), (4,3,1,2), (4,3,2,1). They were obtained in Problem 2.3.6. We have here $3! = 6$ cyclic permutations $(1 \mapsto i \mapsto \cdots \mapsto j \mapsto 1)$; generally, there are $(n-1)!$ of them. Additionally, the derangements for $n = 4$ can be any pairs of transpositions; their number is 3. Thus $D_4/4! = 9/24 = 3/8$, which is a bit greater than $D_3/3! = 1/3$.

A direct count like this can be performed for any n, but it quickly becomes involved when n grows. The formula for D_n will be a sum of contributions of n-cycles and other types of permutations. We need a better method.

Let B_i (Bad$_i$) be the set of all permutations such that $p_i = i$; B_{ij} be for $p_i = i, p_j = j (i \neq j)$, and so on. Informally, we need n "devices" programmed to prevent the events $p_i = i$ for $i = 1, 2, \ldots, n$. Then $b_i = |B_i| = (n-1)!$, $b_{ij} = |B_{ij}| = (n-2)!$, $b_{ijk} = (n-3)!$, and so on. We simply omit the columns with the coinciding entries. As always, $|B|$ is the size of a set B, and we assume that the indices ij, ijk and those in further intersections are pairwise distinct.

We arrive at the following formula:

$$\begin{aligned} D_n &= Free - nb_1 + \binom{n}{2} b_{12} - \binom{n}{3} b_{123} + \dots \\ &= n! \left(1 - 1 + \frac{1}{2} - \frac{1}{3!} + \dots + (-1)^n \frac{1}{n!}\right). \end{aligned} \tag{3.6.13}$$

We transformed here:

$$\binom{n}{k} b_{12\cdots k} = \frac{(n(n-1)\cdots(n-k+1))(n-k)!}{k!} = \frac{n!}{k!}.$$

Thus $D_n/n! \approx 1/e$ for the Euler number e and large n. We conclude that "life" may exist in a "universe" where DNA are permutations.

The convergence here is fast and the deviation from $1/e$ is smaller than $\frac{1}{(n+1)!}$, which is a simple general fact about alternating series. Thus $D_n =$ Round$[n!/e]$ for any n; Round$[x]$ is the closest integer to x, Round$[1.5] = 2$.

We already encountered $1/e$ when doing *Pick-6 Lotto*; it was the probability of the event "no-winner" if the number of panels playing this lottery coincides with the total number of panels, $\mathcal{N} = \binom{49}{6}$ for "Pick-6." There is a direct connection. Assume that p_i are independent variables with equally probable values $1, 2, \dots, n$. Then the chances that we have no derangement are $(1 - \frac{1}{n})^n \approx \frac{1}{e}$ for $n \gg 0$, which is what we obtained for Pick-6 Lotto with $\mathcal{N}$ instead of n. This deduction requires large n, since p_i are not really independent. For permutations, prior p_i diminish the possibilities for further p_i, but this can be disregarded for $n \gg 0$.

3.6.2 Recurrence for D_n

Practical usage of (3.6.13) is unreasonable due to the huge denominators; factorials grow very quickly. Almost always in combinatorics, recurrence relations (if they exist) provide the best computational tools. However direct formulas in terms of binomial coefficients and their various extensions and generalizations are equally important; for instance, they provide connections with the binomial distribution.

Problem 3.6.1 *Use (3.6.13) to check that* $D_n = nD_{n-1} + (-1)^n$ *for* $n \geq 0$, *where* $D_0 = 1$. *Deduce from this relation that* $D_{n+1} = n(D_n + D_{n-1})$ *for* $n = 1, 2, 3, \dots$.

Solution. Indeed, $D_n/n! - D_{n-1}/(n-1)! = (-1)^n/n!$ and then we can multiply this relation by $n!$. Then, consider the sum of $D_{n+1} = (n+1)D_n + (-1)^{n+1}$

and $D_n = nD_{n-1} + (-1)^n$. We have different ± 1, which cancel each other. Thus, $D_{n+1} + D_n = n(D_n + D_{n-1}) + D_n$, and $D_{n+1} = n(D_n + D_{n-1})$. ❑

The following combinatorial proof of the latter formula is elegant. In the two-line notation, any derangement has the columns $\binom{1}{x}$ and $\binom{y}{1}$ for some $x, y \neq 1$. If $x = y$, then we have n choices for x and can reduce the problem from $(n+1)$ to $(n-1)$ by removing these two columns. We obtain the term nD_{n-1}. If $x \neq y$, then let us replace y by 1 and 1 by y in the 1st row of our derangement. Then it will contain the columns $\binom{y}{x}$, $\binom{1}{1}$. By deleting $\binom{1}{1}$, we obtain an n-derangement, where one of the columns is moved to the 1st position (there are n choices for this column). This gives nD_n.

The first recurrence formula for D_n is of course sufficient for practical calculation. Let us use the second. Switching there to $\delta_n \stackrel{\text{def}}{=\!=} D_n/(n-1)!$, we obtain the following recurrence relation:

$$\delta_n = \delta_{n-1} + \frac{1}{n-2}\delta_{n-2},\ \delta_1 = 0, \delta_2 = 1. \tag{3.6.14}$$

Without the term $1/(n-2)$ this would be the Fibonacci recurrence (for rabbits, dimers in the shape $2 \times n$, and so on). This recurrence presumably models the growth of trees; with $\frac{c}{n-2}$ replacing $\frac{1}{n-2}$ it is related to the spread of epidemics. We will touch upon this in the next chapter.

3.6.3 Higher derangements *

The combinatorial proof of the 2nd order recurrence for D_n from Problem 3.6.1 can be simplified and significantly generalized using *cycle decompositions* of permutations from Section 2.3.3. Let us extend the definition of derangements:

Derangements of depth $\geq d$ (d or greater) are n-permutations w such that no group of $1 \leq k \leq d$ elements in $[n]$ remains fixed under w (as a group).

This means that w has no cycles $i_1 \to i_2 \to \cdots \to i_k \to i_1$ of length $1 \leq k \leq d$, i.e. its cycles have lengths $> d$. The number of such derangements will be denoted $D_n^{(d)}$; thus $D_n^{(1)} = D_n$. The notation $D_n^{(\geq d)}$ can be somewhat more consistent here, but the derangements of depth *exactly* d will not be used. The number $D_n^{(d)}$ of derangements of depth $\geq d$ can be expected to diminish quickly when d increases, but this does not happen.

Proposition 3.6.2 *Let $\mathcal{D}_n = D_n^{(d)}$ for $d \geq 1$. Then $\mathcal{D}_0 = 1$, $\mathcal{D}_1 = 0 = \mathcal{D}_2 = \cdots = \mathcal{D}_d$, and $\mathcal{D}_{d+k} = (d+k-1)!$ for $1 \leq k \leq d+1$. For any $n \geq d$:*

$$\mathcal{D}_{n+1} = n\mathcal{D}_n + n(n-1)\cdots(n-d+1)\mathcal{D}_{n-d}.$$

Proof. Derangements of depth $\geq d$ with $n+1$ elements are as follows. First, we can take any n-derangement v of such depth, and go from $[n]$ to $[n+1]$ by adding the element $n+1$ to one of the cycles of v; the number of possible places for it is n. The result is obviously a derangement of depth $\geq d$. This gives the term $n\mathcal{D}_n$.

Second, consider the derangements w in $[n+1]$ where $n+1$ is included in a cycle of length exactly $d+1$. They are not covered by the previous construction and can be obtained as follows. Let $x_1, \ldots, x_d$ be any *sequence* of d elements from $[n]$, and v a derangement of the remaining $n-d$ elements (of depth $\geq d$). Let w be the product of v and the $(d+1)$-cycle $((n+1) \to x_1 \to x_2 \to \cdots \to x_d \to (n+1))$. This gives all remaining derangements in $[n+1]$ and results in the term $n(n-1)\cdots(n-d+1)\mathcal{D}_{n-d}$. ❑

Similar to $D_n^{(1)}$, let $\delta_n = D_n^{(d)}/(n-1)!$. We obtain that $\delta_{n+1} = \delta_n + \frac{1}{n-d}\delta_{n-d}$, a recurrence with interesting applications. We have: $\delta_0 = 1, \delta_1 = 0 = \cdots = \delta_d$, $\delta_{d+1} = 1 = \cdots = \delta_{2d+1}$. It has an obvious solution n. The remaining $d-1$ *fundamental solutions* are interesting to find. Applications to modeling tree growth and beyond are expected. See Section 4.4.4 and also Exercise 4.6.43.

Problem 3.6.3 *Deduce the following formula from the definition of the cycle decomposition:* $D_n^{(d)} = \sum_{i_1,\ldots,i_{n-d}} \prod_{m=1}^{n-d} \frac{1}{(d+m)^{i_m}(i_m)!}$, *where the summation is over all* $\{i_1, i_2, \ldots, i_{n-d}\}$, i_m *is the number of cycles of length* $d+m$. *The conditions for them are:* $\sum_{m=1}^{n-d}(d+m)i_m = n$ *(the whole* $[n]$ *is covered by the cycles), and* $0 \leq i_m \leq$ *Floor*$[\frac{n}{d+m}]$. *This results from Exercise 3.7.18.* ❑

Depth two. Let us provide the calculation of $\mathcal{D}_n = D_n^{(2)}$ using "Inclusion-Exclusion," following the formula for the usual derangements (of depth ≥ 1). This can potentially be done for any depths, but we will consider only $d = 2$, not too technical. This is a great exercise for this chapter.

Problem 3.6.4 *Let* $\mathcal{D}_n = D_n^{(2)}$ *for* $n \geq 0$, *and* $[n/2] =$ *Floor*$[n/2]$. *Prove that*

$$\mathcal{D}_n = n! \sum_{m=0}^{n} \sum_{\ell=0}^{[m/2]} (-1)^{m-l} \frac{1}{(m-2\ell)!\,\ell!\,2^\ell} \tag{3.6.15}$$

$$= n! \sum_{\ell=0}^{[n/2]} \sum_{k=0}^{n-2\ell} (-1)^{k+\ell} \frac{1}{k!\,\ell!\,2^\ell}. \tag{3.6.16}$$

Obtain from this that $\lim_{n\to\infty} \mathcal{D}_n/n! = e^{-3/2}$ *for the Euler number* e.

Solution. For permutations $w = (a_1, a_2, \ldots, a_n)$, we need to monitor 2 types of basic "bad" (not allowed) conditions. The corresponding sets of "bad permutations" are as follows:

$$\mathcal{B}_i = \{w \text{ such that } a_i = i\},\ \mathcal{C}_{ij} = \{w \text{ such that } a_i = j, a_j = i\},$$

where $1 \le i, j \le n$ for $\mathcal{C}_{ij}$, and i, j are assumed unordered pairs: $\mathcal{C}_{ij} = \mathcal{C}_{ji}$. We obtain: $b_i = |\mathcal{B}_i| = (n-1)!$, $c_{ij} = |\mathcal{C}_{ij}| = (n-2)!$. Let us intersect ℓ sets $\mathcal{C}_{i_\alpha j_\alpha}$, where $\alpha = 1, 2, \ldots, \ell$, and $m - 2\ell$ sets $\mathcal{B}_{i_\beta}$ for $n \ge m \ge 2\ell$. Accordingly, $\beta = 1, 2, \ldots, m - 2\ell$. If $\ell = 0$ or $m - 2\ell = 0$, then the $\mathcal{C}$-sets or $\mathcal{B}$-sets will not be present. The total number of bad sets is $m - 2\ell + \ell = m - \ell$.

Here we naturally assume that $i_\beta \ne i_{\beta'}$ for $\beta \ne \beta'$, and the (unordered) pairs $(i_\alpha j_\alpha)$ are different for different α. Only the set (i_β) and the set of pairs $(i_\alpha j_\alpha)$ matter: the intersection of sets does not depend on their order.

One has: $i_\beta \ne i_\alpha$ and $i_\beta \ne j_\alpha$ for any α, β if the intersection under consideration is nonempty. Indeed, $\mathcal{B}_i \cap \mathcal{C}_{ij} = \emptyset$, because i is fixed under w due to the definition of $\mathcal{B}_i$ and must go to j due to the definition of $\mathcal{C}_{ij}$. Similarly, $\mathcal{C}_{ij} \cap \mathcal{C}_{kl} = \emptyset$ if the pairs (ij) and (kl) intersect. Indeed, if $i = k$, then w must send i to j due to $\mathcal{C}_{ij}$ and to l due to $\mathcal{C}_{kl}$, which is impossible.

The number of permutations in the intersection of $\mathcal{C}_{i_\alpha, j_\alpha}$ and $\mathcal{B}_\beta$ considered above depends only on n and m (not on ℓ): it is $k!$, where k is the number of elements not in these sets, which is $n - (m - 2\ell) - 2\ell = n - m$. So the size of this intersection is $(n - m)!$. This is why the usage of m is convenient.

Thus, we need only to calculate the number of all possible combinations of $\mathcal{B}_i, \mathcal{C}_{jk}$ with pairwise distinct indices for any given $n \ge m \ge 2\ell$. It is

$$N(m, \ell) = \binom{n}{m - 2\ell}\binom{n - m + 2\ell}{2\ell}(2\ell - 1)(2\ell - 3)\cdots 1.$$

Here $\binom{n}{m-2\ell}$ is the number of choices for the indices i_β in the $\mathcal{B}$-sets (from $[n]$). Then we pick a subset with 2ℓ elements from the remaining $n - m + 2\ell$ elements, which gives $\binom{n-m+2\ell}{2\ell}$. Then we split the last subset into pairs to form $\mathcal{C}_{ij}$ in all possible ways, which results in the product $(2\ell-1)(2\ell-3)\cdots 1$. Indeed, you pick the pair for 1 ($2\ell - 1$ choices), then take the smallest one from the remaining $2\ell - 2$ numbers and pick the pair for it ($2\ell - 3$ choices), and so on. This product can be simplified as follows:

$$N(m, \ell) = \frac{n!}{(n - m)!}\,\frac{1}{(m - 2\ell)!\ \ell!\ 2^\ell}.$$

We use that $(2\ell - 1)(2\ell - 3)\ldots 1 = (2\ell)!/(\ell!\,2^\ell)$. There is a simple combinatorial interpretation of the latter identity; see Exercise 3.7.17.

Finally, $\mathcal{D}_n = \sum_{m=0}^{n}(-1)^{m-\ell}N(m,\ell)(n-m)!$ due to "Inclusion-Exclusion," which is (3.6.15). Recall that the number of "bad" sets, $\mathcal{B}_i$ and $\mathcal{C}_{ij}$, is $m-\ell$ (not m), which gives $(-1)^{m-\ell}$. Changing the summation index m to $k = m-2\ell$, we arrive at (3.6.16). When $n \to \infty$, the upper summation limits in the latter formula become ∞: we obtain $e^{-3/2}$ (check). ❑

It makes sense to provide some small values of $\mathcal{D}_n = D_n^{(2)}$ and their corresponding $r_n = \mathcal{D}_n/n!$; the latter quickly approach $e^{-3/2} \approx 0.2231$:

$n =$	5	6	7	8	9	10
$\mathcal{D}_n =$	24	160	1140	8988	80864	809856 .
$r_n =$	0.2	0.2222	0.2262	0.2229	0.2228	0.2231

Let us provide some v_d such that $\lim_{n\to\infty} D_n^{(d)}/n! = e^{-v_d}$:

$d =$	1	2	3	4	5	6	7
$v_d =$	1	3/2	11/6	25/12	137/60	49/20	363/140 .

They are ***harmonic numbers***: $v_d = H_d = 1+\frac{1}{2}+\cdots+\frac{1}{d}$ (A. Weingartner, 2016). See e.g. [23]. This follows from Exercise 4.6.44, but a sequence of analytic steps is required.

Thus, the probability of obtaining a permutation with all cycles having length $\geq d$ for large d and $n \gg d$ is approximately $e^{-\log d - \gamma} = \frac{e^{-\gamma}}{d}$ for the Euler-Mascheroni constant $\gamma \approx 0.577215665$.

The argument we used to obtain the recurrence for $\mathcal{D}_n$ works well in the following problem, which we leave as an exercise.

Problem 3.6.5 *Prove that the number Q_n^m of n-permutations with exactly m cycles can be found from the following recurrence: $Q_{n+1}^m = nQ_n^m + Q_n^{m-1}$, where $n, m \geq 1$, $Q_n^0 = 0$, $Q_n^m = 0$ for $n < m$, $Q_n^n = 1$. For instance, $Q_n^1 = (n-1)!$, $Q_n^{n-1} = \binom{n}{2}$, $Q_n^{n-2} = 3\binom{n}{4} + 2\binom{n}{3}$.* ❑

3.7 Exercises

Exercise 3.7.1$^\heartsuit$ *For m, n and $0 \leq r \leq \min\{m, n\}$, prove combinatorially the Vandermonde identity: $\binom{m+n}{r} = \sum_{k=0}^{r}\binom{m}{k}\binom{n}{r-k}$ (actually, due to Zhu Shijie, Yuan Dynasty, 1303).*

[23] R. Petuchovas, *Asymptotic estimates for the number of permutations without short cycles*, Australas. J. Comb. 72(1) (2018), 1–18.

Exercise 3.7.2$^\heartsuit$ *Prove that for every integer* $n > 1$, $\binom{n}{1} - 2\binom{n}{2} + 3\binom{n}{3} + \cdots + (-1)^{n-1}n\binom{n}{n} = 0$. *Use the interpretation in terms of "teams with a captain" and follow our combinatorial proof that the alternating sum of binomial coefficients is zero.*

Hint: Alternatively, you can differentiate $(1+x)^n$ and substitute $x \mapsto -1$.

Exercise 3.7.3$^\heartsuit$ *Let the set* A *be a union* $\cup_{i=1}^{n} A_i$ *of its subsets* A_i *with* a_i *elements. We set* $a_{ij} = |A_i \cap A_j|$ *for any* $i < j$, $a_{ijk} = |A_i \cap A_j \cap A_k|$ *for any* $i < j < k$, *and so on. Here* $|X|$ *is the number of elements of* X, *the summations below are over all possible indices* $i < j < k < \ldots$.

(i) Using "Inclusion-Exclusion", check that the number of elements of A *that belong* **exactly** *to one of* A_i *is* $\sum_i a_i - 2\sum_{i<j} a_{ij} + 3\sum_{i<j<k} a_{ijk} - \ldots$, *the number of those exactly in two of* A_i *is* $\sum_{i<j} a_{ij} - \binom{3}{2}\sum_{i<j<k} a_{ijk} + \binom{4}{2}\sum_{i<j<k<l} a_{ijkl} - \ldots$, *and so on.*

(ii) Assuming that there exists 1 element in the total intersection of all A_i *count its contributions to* $a_{ijk\ldots}$ *and prove that (i) is* **equivalent** *to the following generalization of the identity from Exercise 3.7.2, which is for* $m = 0$. *Given* $0 \le m < n$, *one has:* $\sum_{r=m}^{n}(-1)^r\binom{r}{m}\binom{n}{r} = 0$.

Exercise 3.7.4$^\diamondsuit$ *Prove that for any integers* $n > r \ge 1$, $\binom{n}{1} - 2^r\binom{n}{2} + 3^r\binom{n}{3} - 4^r\binom{n}{4} + \cdots + (-1)^{n-1}n^r\binom{n}{n} = 0$. *For this, check that* m^r *is a linear combination of* $m(m-1)\cdots(m-r+1)$, $m(m-1)\cdots(m-r+2)$, $m(m-1)\cdots(m-r+3)$, $\ldots$, m *with the coefficients independent of* m, *and then use Exercise 3.7.2 with "captains, vice-captains, vice-vice-captains..." or follow the hint there. You can also use Exercise 3.7.3.*

Exercise 3.7.5$^\heartsuit$ *(i) Let* $M_n = \binom{n}{0} - \binom{n}{1}2 + \binom{n}{2}2^2 - \binom{n}{3}2^3 + \cdots + (-1)^n\binom{n}{n}2^n$. *Check that this sequence never becomes greater than 1.*

(ii) Let $M_n = \binom{n}{1} - 2\binom{n}{2}2 + 3\binom{n}{3}2^2 - 4\binom{n}{4}2^3 + \cdots + (-1)^{n-1}n\binom{n}{n}2^{n-1}$ *for* $n \ge 1$. *Check that* $M_n \le n$.

(iii) Check that the sequence $M_n = \sum_{r=1}^{n}\binom{n}{r}r^k(-1/2)^r$ *converges to 0 as* $n \to \infty$.

Hint for (iii): $r^k = C_1 r + C_2 r(r-1) + \cdots + C_k r(r-1)\cdots(r-k+1)$ for some constants C_i. Thus, $M_n = nC_1(-1/2)(1-1/2)^{n-1} + \cdots + n(n-1)\cdots(n-k+1)(-1/2)^k C_k(1-1/2)^{n-k}$, which converges to 0 because $n^k(1/2)^n \to 0$.

Exercise 3.7.6$^\heartsuit$ *In how many ways can 1 orange, 1 grapefruit and 12 "identical" apples be distributed among 3 children, assuming that each of them receives at least one piece of fruit?*

Answer: $3\binom{10+3-1}{3-1}+6\binom{11+3-1}{3-1}=666$.

Exercise 3.7.7$^{\heartsuit}$ *Distribute 1 orange, 1 grapefruit and 12 "identical" apples among 3 children, assuming that each of them receives something and nobody receives greater than k pieces. Let N_k be the number of such distributions. Find N_6 via the Inclusion-Exclusion Principle. What are N_4, N_5?*

Answers: $N_6 = 3\cdot 15 + 6\cdot 15 = 135$, $N_4 = 0$, $N_5 = 3\cdot 3 + 6\cdot 3 = 27$.

Exercise 3.7.8$^{\diamondsuit}$ *Prove the* Hockey Stick Identity: $\binom{r}{r}+\binom{r+1}{r}+\binom{r+2}{r}+\cdots+\binom{n}{r}=\binom{n+1}{r+1}$ *for* $n\ge r\ge 0$.

Hint: For $[n+1]=\{1,2,\ldots,n+1\}$, the maximal element $m+1$ of any subset with $r+1$ elements can be from $r+1$ to $n+1$. The number of such subsets with fixed max$=m+1$ is $\binom{m}{r}$, which gives that $\binom{n+1}{r+1}=\sum_{m=r}^{n}\binom{m}{r}$. A direct proof is as follows: $\binom{n+1}{r+1}=\binom{n}{r}+\binom{n}{r+1}=\binom{n}{r}+\binom{n-1}{r}+\binom{n-1}{r+1}=\ldots$.

Exercise 3.7.9$^{\heartsuit}$ *For any $n\ge 1$, use Exercise 3.7.8 or directly the triangle recurrence to check that $S_n=\binom{n}{0}+\binom{n+1}{1}+\cdots+\binom{2n-1}{n-1}$ equals $\binom{2n}{n-1}$. In Fig. 2.10, assume that all paths until the green/red fields are counted as 1, i.e. are considered equally probable. Then check that the fair bet β for this modification of the game "14 vs. 1" from Section 2.5.3 is $2S_n/\binom{2n}{n}$ for "$(n+1)\times(n+1)$" and the red square as for "14 vs. 1", which is $\beta=\frac{7}{4}$ for $n=7$.*

Hint: As in Section 2.5.3, use that the last move for the paths to green fields is always up. One has: $2\binom{14}{6}/\binom{14}{7}=\frac{7}{4}$ for $n=7$.

Exercise 3.7.10$^{\heartsuit}$ *Use Exercise 3.7.8 to check the steps in the following calculation: $\sum_{m=0}^{n} m^2=\sum_{m=0}^{n}(m+m(m-1))=\frac{n(n+1)}{2}+2\binom{n+1}{3}=\frac{n(n+1)(2n+1)}{6}$. Alternatively, prove the resulting identify using mathematical induction. Another way: calculate $\left(x(\frac{1-x^{n+1}}{1-x})'\right)'$ and make $x=1$. Use any approach.*

Exercise 3.7.11$^{\diamondsuit}$ *(i) Use the generating function for R-numbers from (3.2.8) to check that $\sum_{i=1}^{N-1} i(N-i)=\binom{N+1}{3}$ for $N\ge 1$.*

(ii) Generalizing, check that for $r,s,k\ge 0$, $n,m\ge 1$:

$$\sum_{r=0}^{k}\sum_{s=0}^{k-r}\binom{n+r-1}{n-1}\binom{m+s-1}{m-1}=\binom{n+m+k-1}{n+m-1}.$$

Hint: The corresponding $G(x)$ is $\frac{1}{(1-x)^{n+m}}$ for (ii) and $\frac{1}{(1-x)^4}$ for (i).

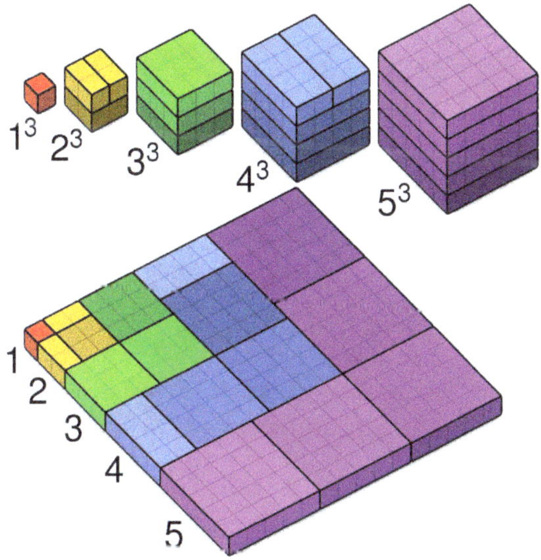

Figure 3.9: Sum of cubes geometrically.

Exercise 3.7.12 ♣ *Follow Exercise 3.7.10 to check that* $6\sum_{m=0}^{n}\binom{m}{3}=6\binom{n+1}{4}=\sum_{m=0}^{n}(m^3-3(m^2-m)-m)=S_3(n)-6\binom{n+1}{3}-\binom{n+1}{2}$, *where* $S_3(n)=\sum_{m=0}^{n}m^3$. *Finally,* $S_3(n)=6\binom{n+2}{4}+\binom{n+1}{2}=\left(\frac{n(n+1)}{2}\right)^2$. *Use mathematical induction to justify the latter identity directly. Also, check that Fig. 3.9 gives another proof* [24]. *Alternatively, justify the intermediate equalities in this calculation:* $S_3(n)=\sum_{m=1}^{n}\left((m^2-m+1)+(m^2-m+3)+\cdots(m^2+m-1)\right)=\sum_{k=1}^{(n^2+n)/2}(2k-1)=\left(\frac{n(n+1)}{2}\right)^2$.

Exercise 3.7.13 ♡ *By considering pairs of subsets* $A\subset B\subset[n]$, *prove combinatorially that* $\sum_{r=0}^{n}\binom{n}{r}2^r=3^n$, *where* $[n]=\{1,2,\ldots,n\}$.

Hint: Every element in $[n]$ can belong to $[n]\setminus B$, or $B\setminus A$, or A, where by $B\setminus A$ we mean the complement of A in B. This gives 3^n in the right-hand side; the left-hand side is the count of all r-element subsets B of $[n]$, which is $\binom{n}{r}$, times the number of their subsets, which is 2^r.

Exercise 3.7.14 ♢ *Follow Exercise 3.7.13 in the following.*

(i) By considering the triples of subsets $A\subset B\subset C\subset[n]$, *prove combinatorially that* $\sum_{r=0}^{n}\sum_{m=0}^{r}\binom{n}{r}\binom{r}{m}2^m=4^n$.

[24] `https://commons.wikimedia.org/wiki/File:Nicomachus_theorem_3D.svg`, Cmglee, CC BY-SA 3.0, via Wikimedia Commons.

(ii) By considering the triples of subsets $A \subset B \supset C$ inside $[n]$, prove combinatorially that $\sum_{r=0}^{n} \binom{n}{r} 2^{2r} = 5^n$.

Hint for (ii): Similar to Exercise 3.7.13, consider the stratification of $[n]$: $[n] \setminus B$, $B \setminus (A \cup C)$, $A \setminus (A \cap C)$, $C \setminus (A \cap C)$, $A \cap C$.

Exercise 3.7.15 ♢ ***Stirling numbers*** *of the 2nd kind are defined from the expansion $z^n = \sum_{m=1}^{n} z(z-1)\cdots(z-m+1)S(n,m)$. Check by a direct calculation that $S(n,1) = 1 = S(n,n)$, $S(n,2) = 2^{n-1} - 1$ for $n \geq 1$, and that*

$$S(n+1,m) = mS(n,m) + S(n,m-1) \text{ for } 2 \leq m \leq n.$$

Hint: Use the evaluation at $z = 1, 2$ for $S(n,1) = 1, S(n,2) = 2^{n-1} - 1$, and consider the leading z-term for checking $S(n,n) = 1$. For the recurrence formula, express $z = (z-m) + m$.

Exercise 3.7.16 ♡ *Calculate directly the number of ways to place 7 distinct objects into 5 nonempty identical bins. The order of objects in a bin and the order of bins do not matter.*

Hint: It can be either (a) one bin with 3 objects and the remaining 4 bins with single objects, or (b) 2 bins with $2+2$ objects with single objects in the remaining 3 bins. We need to add the number of 3-element subsets from 7 and the number of 4-element subsets from 7 multiplied by the number of pairs in $[4]$ containing 1. The answer is $\binom{7}{3} + 3\binom{7}{4} = 140$. This is Stirling number $S(7,5)$.

Exercise 3.7.17 ♡ *Check that the number of ways to split the set $[2n]$ into n pairs, i.e. to represent it as a union of 2-element subsets, is $(2n)!/(2^n\, n!)$. Use this observation to prove the following formula for the* **odd factorials**: *$(2n-1)(2n-3)\cdots 3 \cdot 1 = (2n)!/(2^n\, n!)$.*

Hint: Permute $[2n]$ in any way and consider the first 2 numbers as the first pair and so on. However, you can permute n resulting pairs, and the order of those in pairs does not matter. Alternatively, pick the pair for 1, then the pair for the smallest number among the remaining $2n-2$ ones, and so on.

Exercise 3.7.18 ♢ *For some set of integers $1 \leq k_1 < k_2 < \ldots < k_m$ and arbitrary positive $a_1, a_2, \ldots, a_m$, let $n = \sum_{i=1}^{m} a_i k_i$.*

(i) Check that there are $\frac{n!}{\prod_{i=1}^{m} (a_i)!(k_i!)^{a_i}}$ ways to place n distinct objects in a_1 bins with k_1 objects in each, a_2 bins with k_2 objects in each, and so on until m. The order of the objects in the bins and the order of the bins are disregarded.

(ii) Use (i) to calculate the number of ways to put 7 distinct objects into 4 nonempty identical (indistinguishable) bins.

Hint: Order n objects and put them in the (ordered) sequence of bins naturally. The number of such arrangements must be divided by the product of the number of permutations inside all bins, which is $\prod_{i-1}^{m}((k_i)!)^{a_i}$ and the number of permutations of bins with the same numbers of objects, which is $\prod_{i-1}^{m} a_i!$. To see this use that any composition of such permutations changes any arrangements with the ordering as above.

(ii) There are 3 types of distributions in this case: 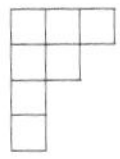 , 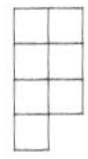 , 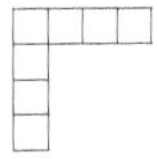 , where the rows give the numbers of objects in the corresponding bins. The answer is $\frac{7!}{3!(2!)^2} + \frac{7!}{3!(2!)^3} + \frac{7!}{3!4!} = 10\binom{7}{3} = 350$. This is Stirling number $S(7,4)$.

Exercise 3.7.19 $^{\diamondsuit}$ *In how many ways can 4 girls, 4 boys and 4 parents sit at a round table if no parents are allowed to sit together?*

Hint: Let the first seat be for parent P_1; this breaks the rotational symmetry of the problem. Then there are 8! permutations of boys and girls (only the total number 8 is relevant), 3! permutations of the remaining parents, and 9 potential places for the remaining 3 parents; the seat after P_1 and the last one (before P_1) are not allowed. The number of choices for the parents' seats is then $\binom{10-3}{3} = \binom{7}{3} = 35$, which is the number of coverings of the shape 1×10 by 3 dominos and 4 monominos, where the parents' seats "correspond" to the second boxes of the dominos. Follow the formula (4.1.4) for d_n in the next chapter. Also, it equals the number of presentations $y_1 + y_2 + y_3 \le 9$ with $y_1 \ge 1, y_{2,3} \ge 2$, which gives $\binom{9-5+4-1}{4-1} = \binom{7}{3}$. Here $y_1, y_1+y_2, y_1+y_2+y_3$ are the seat numbers (from 9) for the remaining (3) parents. Finally, the answer is $35 \cdot 3! \cdot 8!$.

Exercise 3.7.20 $^{\diamondsuit}$ ***Fleissner grille*** *from Exercise 1.4.39 is for a $2m \times 2m$ square S with m^2 perforations, where no two perforations can be obtained from each other by rotations by the angles $\pi/2, \pi, 3\pi/2$ around the center of the square. Given $k_0, k_1, k_2, k_3 \ge 0$ such that $\sum_{i=0}^{3} k_i = m^2$, count the number $N(m; k_0, k_1, k_2, k_3)$ of such squares S satisfying the following conditions. Represent S as a union of its 4 natural quarters, S_0, S_1, S_2, S_3, which are $m \times m$ squares: $S = \begin{pmatrix} S_0 S_1 \\ S_2 S_4 \end{pmatrix}$. We require that S_i has k_i perforations for $0 \le i \le 3$.*

Answer: $N(m; k_0, k_1, k_2, k_3) = \frac{(n^2)!}{\prod_{i=0}^{3}(k_i)!}$, where we partition S_0 into subsets K_i with k_i elements and then rotate K_i from S_0 to the corresponding S_i.

Exercise 3.7.21 $^{\heartsuit}$ *Let us stack bricks in the following manner. We begin with a row of $n \ge 1$ bricks, then place the next row on top of it leaving at least the leftmost and rightmost places empty, and so on.*

(i) Count the number N_m of such stacks if the number of rows (including the initial one) is $m+1$ for $0 \leq m < (n-1)/2$.

(ii) Use the formula for the sum of even binomial coefficients to check that the total number of stacks N is 2^{n-2} (m is unrestricted).

Answer for (i): $N_m = \binom{n-1}{2m}$.

Exercise 3.7.22$^\heartsuit$ *A city lawyer goes to her office located 7 blocks east and 10 blocks north of her home. She either (a) takes a shortest walk of 17 blocks, or (b) first goes to her favorite coffee shop that is 2 blocks west and 4 blocks north. How many different walks are possible for (a) and for (b)?*

Answer for (b): $\binom{6}{2}\binom{15}{6} = 75075$.

Exercise 3.7.23$^\heartsuit$ *You can select a team with r players and its captain from a pool of n players in 2 different ways. Select the captain and then the remaining $r-1$ players. Alternatively, pick an r-team and then select its captain. Check that this results in the identity $n\binom{n-1}{r-1} = r\binom{n}{r}$.*

Exercise 3.7.24$^\heartsuit$ *Use Exercise 3.7.8 and the identity from Exercise 3.7.23 in the form $r\binom{n+r}{r} = (n+1)\binom{n+r}{n+1}$ for $r \geq 0$ to check the following identity: $\sum_{r=0}^{n} \binom{n+r}{r} r = (n+1)\binom{2n+1}{n-1}$.*

Exercise 3.7.25$^\heartsuit$ *Use Exercise 3.7.23 to prove that the average of $\max[S_r]$ over all r-element subsets S_r in the set $[n] = \{1, 2, \cdots, n\}$ is $\frac{r(n+1)}{r+1}$, where the maximum is over the numbers in S_r.*

Hint: Count the number of subsets S_r with fixed $\max[S_r]$.

Exercise 3.7.26$^\heartsuit$ *Use binomial coefficients to check that the average number of elements in a subset of $[n]$ is $n/2$. Obtain the same answer in a more direct way using the duality $\binom{n}{r} = \binom{n}{n-r}$.*

Exercise 3.7.27$^\heartsuit$ *How many ways are there to arrange 10 books on a bookcase with 3 shelves? The books are distinguishable and the order of the books on each shelf matters.*

Answer: $10!\binom{12}{2}$.

Exercise 3.7.28$^\diamondsuit$ *The so-called unsigned* **Lah number** $L(n,k)$ *counts the ways of stacking* n *distinct books into* k *identical (indistinguishable) boxes so that there is a nonempty stack in every box. Their order in each box matters, but not the order of the boxes. Check that* $L(n,k)=\binom{n-1}{k-1}\frac{n!}{k!}$.

Hint: Permute the books ($n!$) and then use the R-numbers to split them into boxes. The boxes are assumed unordered, which gives $1/k!$.

Exercise 3.7.29$^\diamondsuit$ *(i) Prove that* $F_k(x)=\sum_{n=0}^{\infty}L(n,k)\frac{x^n}{n!}$ *equals* $\frac{1}{k!}\left(\frac{x}{1-x}\right)^k$ *for the* Lah numbers $L(n,k)$ $(k\ge 1)$ *from Exercise 3.7.28. This is the corresponding* modified generating function*: modified by* $x^n\mapsto\frac{x^n}{n!}$.

(ii) Prove that the modified generating function $B(x)=\sum_{n=0}^{\infty}B_n\frac{x^n}{n!}$ *for the* Bell numbers B_n *from Exercise 2.6.42 equals* $e^{(e^x-1)}$.

Hints. (i) $\frac{k!L(n,k)}{n!}=R(k,n-k)$. Then use $\frac{1}{(1-x)^n}=\sum_{r=0}^{\infty}R(n,r)x^r$.
(ii) The recurrence from Exercise 2.6.42 gives that $dB(x)/dx=e^xB(x)$.

Exercise 3.7.30$^\clubsuit$ *(i) For the usual deck with* 52 *cards, count the number* H_m *of* 10*-hands containing exactly* m *different denominations* $(3\le m\le 10)$*; apply the Inclusion-Exclusion Principle.*

(ii) Combine (i) with Bingo Exercise 2.6.27 to count the probability p_k *that your* 5*-hand contains exactly* k *cards of the denominations different from those in the dealer's* unknown 10*-hand, from the same deck. Calculate numerically your chances to win, which is the probability of the event "*$k=0$*".*

Answers. $H_m=\binom{13}{m}\left(\binom{4m}{10}-\binom{m}{m-1}\binom{4(m-1)}{10}+\binom{m}{m-2}\binom{4(m-2)}{10}-\ldots\right)$. Using P_m^k from Exercise 2.6.27: $p_k=\sum_{m=3}^{10}H_mP_m^k/\binom{52}{5}$, $p_0\approx 0.03$.

Exercise 3.7.31$^\heartsuit$ *Find the number* N *of integers among* $1,\ldots,1000$ *that: (i) are divisible by* 6*, or by* 10*, or by* 15*, (ii) have the remainder* 4 *upon the division by* 6*, or by* 10*, or by* 15*, (iii) have the remainder* 1 *upon the division by* 6*, or* 2 *upon the division by* 10*, or* 3 *upon the division by* 15*.*

Answers. (i): $N=266$, (ii): $N=266$, (iii): $N=344$.

Exercise 3.7.32$^\heartsuit$ *Find the number* N *of integers among* $1,\ldots,200$ *that are divisible by* 6*, or by* 7*, or by* 8*, or by* 10*.*

Answer: $N=80$.

Exercise 3.7.33 ♡ *For 5 dice, the probability* P_S *of the corresponding sum* S *is the number of presentations* N_S *of* $x_1 + x_2 + x_3 + x_4 + x_5 = S$ *with positive* (> 0) *integers* x_1, x_2, x_3, x_4, x_5 *not exceeding* 6*, which is divided by the total number of ways to roll 5 dice; i.e.* $P_S = N_S/6^5$.

(i) What are the number N_{14} *of presentations of* $S = 14$ *and the corresponding probability* P_{14}*?*

(ii) What are the number $N_{\le 14}$ *of presentations and the probability* $P_{\le 14}$ *for the sum to be* $S \le 14$*?*

Answers. For (i): $N_{14} = 540, P_{14} \approx 7\%$; for (ii): $N_{\le 14} = 1722, P_{14} \approx 22\%$.

Exercise 3.7.34 ♡ *Calculate the number* N_{16} *of integral solutions of the equation* $x_1{+}x_2{+}x_3{+}x_4{=}16$ *satisfying*

$$1 \le x_1 \le 6,\ 0 \le x_2 \le 4,\ 2 \le x_3 \le 8,\ 2 \le x_4 \le 9.$$

Calculate N_{25} *via the* duality*, the connection between* N_S *for large and small* S*. Check your answers using a computer: "Expand*$[\,(x + \ldots)(1 + \ldots)\cdots]$*".*

Answers: $N_{16} = 170, N_{25} = 10$.

Exercise 3.7.35 ♡ *For* 5*-hands from the deck with 13 cards* $(1, 2, \ldots, 13)$ *(denominations only), calculate the number of hands:* (a) $\mathcal{N}_3$ *with only one triple* $\ldots a, a+1, a+2, \ldots,$ *and* (b) $\mathcal{N}_{2,2}$ *with strictly* 2 *pairs. It is supposed to be no better than this: any pair can be only inside the triple for (a) or coincide with one of the 2 pairs for (b). Follow Problem 3.5.3.*

Hint: Represent the cards $a < b < c < d < e$ in terms of a, x, y, z, t as follows: $b = a + x, \ldots, e = a + x + y + z + t \le 13$, which gives $a + x + y + z + t + u = 13$ with $a \ge 1, u \ge 0$. Then follow Problem 3.5.3: $\mathcal{N}_3 = \mathcal{N}_{2,2} = 3\cdot\binom{6+4-1}{4-1} = 3\cdot 84$.

Exercise 3.7.36 ♣ *Pick randomly a* subset *of* k *numbers from* $[n] = \{1, 2, \ldots, n\}$*. Let* $\mathcal{N}_n^k(m)$ *be the number of such subsets that contain exactly* m *pairs of consecutive numbers, where a number can be used in different pairs (no greater than 2). For instance, a subset* $\{1, 2, 3, 5\}$ *contains 2 such pairs. Accordingly, the corresponding probabilities will be* $\mathcal{N}_n^k(m)/\binom{n}{k}$*. Fix* $k > 0$ *and check that* $\sum_{m=0,k=0}^{\infty} \mathcal{N}_n^k(m) x^n y^m = x^k \big(y + \frac{x}{1-x}\big)^{k-1} \big(\frac{1}{1-x}\big)^2$*. Then use this formula to calculate* $\mathcal{N}_{13}^5(m)$ *for* $0 \le m \le 4$*; check that* $\mathcal{N}_{13}^5(2) = \mathcal{N}_3 + \mathcal{N}_{2,2}$ *in the notation from Exercise 3.7.35, which is really* $6 \cdot 84$.

Answer: $\sum_{m=0}^{4} \mathcal{N}_{13}^5(m) y^m = 9\,(y^4 + 16y^3 + 56y^2 + 56y + 14)$.

Exercise 3.7.37 ◊ *When playing* Yahtzee, *one of the events is "three of a kind," which means that at least 3 dice give coinciding numbers from the total of 5 dice. Count the number of all rolls of this kind.*

Answer: $6\binom{5}{5} + 5 \cdot 6\binom{5}{4} + (5 \cdot 5) \cdot 6\binom{5}{3}$.

Exercise 3.7.38 ◊ *A company of n gentlemen walk into a party and give their hats to the coat and hat check guy. When the party finishes, their hats are returned in no specific order and with no specific intent. Consider the events: (a) "no one gets his own hat," (b) "exactly one receives his own hat." Denote the corresponding numbers of possibilities by D_n and D'_n.*

(i) Check that $D'_n = nD_{n-1}$ and calculate $D_n - D'_n$. When is $D_n > D'_n$?

(ii) Count the number $N_n^{(2)}$ of possibilities for the event "at least two people receive their own hats."

(iii) What is the probability $P^{(2)}$ of the event from (ii) as $n \to \infty$?

(iv) Use (iii) and the formula $D_n = Round[n!/e]$ to verify the inequality $N_n^{(2)} < D_n$ for any $n > 2$. In particular, $P^{(2)} < \lim_{n\to\infty} D_n/n! = 1/e$.

Answers. For (i): $D_n - D'_n = (-1)^n$; for (ii): $N_n^{(2)} = n! - D_n - n \cdot D_{n-1} = n! - 2D_n - (-1)^n$. For (iii,iv): $P^{(2)} = 1 - 2/e < 1/e$.

Exercise 3.7.39 ◊ *(i) In the $n\times n$ chessboard and any subset S of boxes there, let N_S be the number of configurations of n nonattacking* rooks *that are all not in S. Rooks can attack vertically or horizontally. Use the Inclusion-Exclusion Principle to justify that $N_S = n! - s_1(n-1)! + s_2(n-2)! - s_3(n-3)! + \ldots$, where s_k for $k = 1, 2, \ldots$ are the numbers of configurations of k nonattacking rooks strictly within S. For instance, this formula gives that $N_S = 0$ if S is a union of rows (why?).*

(ii) Apply this formula for $n = 8$ to calculate N_S for S that is a 4×4 subsquare inside this chessboard. If your calculation is correct, then it must be 24^2 regardless of the position of this subsquare (why?).

Hint for (ii): $s_1 = 4^2 = 16$, $s_2 = \binom{4}{2}4 \cdot 3 = 72$, $s_3 = \binom{4}{3}4 \cdot 3 \cdot 2 = 96$, $s_4 = 4! = 24$; $N_S = 8! - 16 \cdot 7! + 72 \cdot 6! - 96 \cdot 5! + 24 \cdot 4! = 576 = 24^2$.

Exercise 3.7.40 ◊ *For $n \geq 0$, check that $\sum_{r=0}^{n}(-1)^r\binom{2n}{2r} = \frac{(1+\imath)^{2n}+(1-\imath)^{2n}}{2} = (-1)^{[n/2]}(1+(-1)^n)2^{n-1}$, where $\imath^2 = -1$, $[x]$=Floor$[x]$ is the integer part of x.*

Hint: $(1+\imath)^2 = 2\imath$; treat $\imath$ algebraically as some symbol satisfying $\imath^2 = -1$.

Exercise 3.7.41 ♣ *Check that* $4\sum_{r=0}^{n}\binom{4n}{4r}x^{4r} = (1+\imath x)^{4n} + (1-\imath x)^{4n} + (1+x)^{4n} + (1-x)^{4n}$ *for* $\imath^2 = -1$. *Use that* $(1\pm\imath)^4 = 1 \pm 4\imath - 6 \mp 4\imath + 1 = -4$ *to establish that*

$$\sum_{r=0}^{n}\binom{4n}{4r} = 2^{4n-2} + (-1)^n 2^{2n-1}.$$

Hint: You do not need to know complex numbers for this calculation; treat $\imath$ algebraically as some "symbol" satisfying $\imath^2 = -1$.

Exercise 3.7.42 ♠ *Let* $q(n)$ *be the number of partitions on* $n \geq 1$ with distinct parts, *i.e. the number of presentations* $n = a_1 + \cdots a_r.$, *where* $1 \leq a_1 < a_2 < \ldots < a_r$ *and* r *is arbitrary. For instance,* $q(1) = 1 = q(2), q(3) = 2 = q(4), q(5) = 3$*:* $5 = 5 = 1+4 = 2+3$. *Check that* $1 + \sum_{n=1}^{\infty} q(n)x^n = \prod_{k=1}^{\infty}(1+x^k) = \prod_{k=0}^{\infty}\frac{1}{(1-x^{2k+1})}$ *and interpret it as follows:* $q(n)$ *equals the number of partitions of* n *with* odd parts, *possibly not distinct* (Euler). *For instance,* $5{=}5{=}3{+}1{+}1{=}1{+}1{+}1{+}1{+}1$, *which gives* 3 *odd partitions* $(= q(5))$.

Exercise 3.7.43 ♣ *Let us consider the lattice paths* p *of length* n *as in Fig. 2.10: from* $(0,0)$ *to* $(x = r, y = n - r)$ *using the unit steps* $\uparrow$ *or* $\rightarrow$, *where the standard plane coordinates* x, y *are used. So these paths are inside the rectangle* $(n-r)\times r$ *(the height is the first, the width is the second) and connect its opposite corners. The area* below *this path in this rectangle will be denoted by* $A(p)$. *For the following length 8 paths to* $(5,3)$, *which are with* $r{=}5, n{-}r{=}3$ *(inside "*$3{\times}5$*"):*

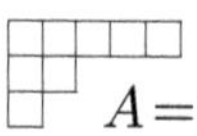 $A{=}7$, 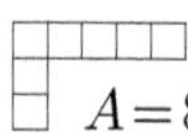 $A{=}8$, 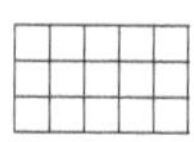 $A{=}0$.

The ***q-binomial coefficient*** $\begin{bmatrix} n \\ r \end{bmatrix}_q$ *counts* $q^{A(p)}$ *over all such paths.*

(i) Reflecting the paths in the anti-diagonals of the corresponding rectangles, check that $\begin{bmatrix} n \\ r \end{bmatrix}_q = \begin{bmatrix} n \\ n-r \end{bmatrix}_q$, *where* $\begin{bmatrix} n \\ 0 \end{bmatrix}_q = 1, \begin{bmatrix} n \\ r>n \end{bmatrix}_q = 0.$

(ii) Prove that $\begin{bmatrix} n \\ r \end{bmatrix}_q = \begin{bmatrix} n-1 \\ r-1 \end{bmatrix}_q + q^r\begin{bmatrix} n-1 \\ r \end{bmatrix}_q$ *for* $n \geq r \geq 1$.

(iii) Use (ii) to check that for $r \leq n$,

$$\begin{bmatrix} n \\ r \end{bmatrix}_q = \frac{(1-q^n)(1-q^{n-1})\cdots(1-q^{n-r+1})}{(1-q)(1-q^2)\cdots(1-q^r)}.$$

Exercise 3.7.44$^\diamondsuit$ *For m, n and $0 \le r \le \min\{m,n\}$, extend the* Vandermonde identity *from Exercise 3.7.1 to the q-case:* $\begin{bmatrix} m+n \\ r \end{bmatrix}_q = \sum_{k=0}^{r} \begin{bmatrix} m \\ k \end{bmatrix}_q \begin{bmatrix} n \\ r-k \end{bmatrix}_q.$ *Use directly the definition of q-binomial coefficients from Exercise 3.7.43.*

Exercise 3.7.45$^\diamondsuit$ *The **Möbius function** $\mu(n)$ is $(-1)^\nu$ if n is square-free positive integer with ν prime factors and 0 for other positive integers.*

(i) Prove that $\mu(mn) = \mu(m)\mu(n)$ if $\gcd(m,n) = 1$ and $\sum_{d|n} \mu(d) = 0$ for any $n > 1$, where the summation is over all divisors of n (indicated by $d|n$).

*(ii) The **Möbius transform** is defined for any function $f(n)$ for $n \ge 1$ as follows: $g(n) = \sum_{d|n} f(d)$. Deduce from (i) that the* inverse Möbius transform *is as follows: $f(n) = \sum_{d|n} \mu(d) g(\frac{n}{d})$ for $n \ge 1$.*

Hint for (i): If $n = p_1^{m_1} \cdots p_\ell^{m_\ell}$ is the prime factorization of $n > 1$, then $\sum_{d|n} \mu(d) = \sum_{r=0}^{\ell} (-1)^r \binom{\ell}{r} = 0$. Hint for (ii):

$$\sum_{d|n} \mu(d) g(\frac{n}{d}) = \sum_{d|n} \mu(d) \sum_{\delta|(n/d)} f(\delta) = \sum_{(\delta d)|n} \mu(d) f(\delta) = \sum_{\delta|n} f(\delta) \sum_{d|(n/\delta)} \mu(d) = f(n).$$

Exercise 3.7.46$^\diamondsuit$ *(i) Check that $n! = \psi(n) \left(\frac{n}{e}\right)^n$ with some $\psi(n)$ such that $\lim_{n\to\infty} \frac{\psi(n)}{n} = 0$. Use here the approximation $\log(n!) \approx \int_1^n \log(x) dx$. Formula (2.5.3) is more exact, but this one is sufficient below.*

(ii) The length of n!. *Check that $\nu(m) = Floor[\log_{10}(m)] + 1$ for the number $\nu(m)$ of digits of an integer $m \ge 1$, where $\log_{10}(x) = \frac{\log(x)}{\log(10)}$. Obtain from (i) that $\nu(n!) \approx n(\nu(n) - 1)$ when $n \approx e10^k$ for $k \gg 0$. Calculate that $\nu(n!)$ equals exactly $n(\nu(n)-1)$ for $n = 24, 268, 2712, 27176, 271821, 2718273$.*

(iii) The 1st digit of n!. *Let p_i be the probability that the 1st digit α_n of $n!$ is i for $1 \le i \le 9$ for sufficiently large n. For $n \sim 35000$, check that p_i are approximately: $\{0.3, 0.177, 0.127, 0.098, 0.078, 0.07, 0.055, 0.051, 0.044\}$. Due to* P.B.Sarkar *and* [25], *p_i match well **Benford's law**: $\pi_i \approx \log_{10}(\frac{1+i}{i})$ for $n!$ and many "natural sequences". Similarly, π_i for $10^{m-1} \le i \le 10^m - 1$ approximates the probability that $n!$ begins with i. Use this law to check that* average*$(\alpha_n) \approx 9 - \log_{10}(9!) \approx 3.44$, that $n!$ begins "statistically" with $Floor[m10^m - (m-1)10^{m-1} - \log_{10}(\frac{10^m!}{10^{m-1}!})] \approx Floor[\frac{9000\cdots}{\log(10)}]$ for $m \gg 0$, and $\nu(n!) \approx n\big(\nu(n) - 1 + \log_{10}(\frac{3.44}{e})\big)$, a bit better than in (ii). [In the United States, evidence based on* Benford's law *has been admitted in criminal fraud cases.]*

[25] P.Diaconis, *The distribution of leading digits and uniform distribution mod 1*, Annals of Probability (1977), 5:1, 72–81.

Hints for (i,ii). Use that $\log((n-1)!) < \int_1^n \log(x)dx < \log(n!)$. Note that the estimates for $\nu(n!)$ are $n(\nu(n)-1-\log_{10}(e))$ for $n = 10^k$ and $n(\nu(n)-\log_{10}(e))$ for $e = 10^k - 1$. So $\nu(n)(\nu(n)-1)$ is in between, and the coincidence $\nu(n!) = \nu(n)(\nu(n)-1)$ near $n = e10^k$ among n with fixed $\nu(n) = k+1$ k is generally possible, but not granted for any k.
Hint for (iii). If m begins with α, then $\nu(m) = \text{Floor}[\log_{10}(m) - \log_{10}(\alpha)] + 1$. Thus, $\log_{10}(m) \approx \nu(m) - 1 + \log_{10}(\alpha)$ and $\nu(n!) \approx n(\nu(n) - 0.9)$ is somewhat better statistically than the estimate $n(\nu(n)-1)$ from (ii).

Exercise 3.7.47 ◇ *Use Stirling's approximation* $n! \approx \sqrt{2\pi n}\left(\frac{n}{e}\right)^n$ *to justify that the binomial distribution approaches the normal one. Namely,* $p^r(1-p)^{n-r}\binom{n}{r}$ *for* $0<p<1$ *tends to* $\dfrac{e^{-(r-np)^2/(2np(1-p))}}{\sqrt{2\pi np(1-p)}}$ *for* $n \gg 0$ *and* $r-np = O(\sqrt{n})$. *This is basically due to* Abraham de Moivre *(1733). Here* $f(x) = O(g(x))$ *means that* $|f(x)| < Cg(x)$ *for some constant* $C>0$ *in the considered range.*

First, check by a direct algebraic manipulation that $p^r(1-p)^r\binom{n}{r} \approx U/V$ *for* $U = \left(\frac{np}{r}\right)^r\left(\frac{n(1-p)}{n-r}\right)^{n-r}$ *and* $V = \sqrt{\frac{2\pi r(n-r)}{n}}$.

Second, use the formula $\log(1+x) = x - x^2/2 + O(x^3)$ *for* $|x|<1$:

$$\log(U) = r\log\Big(\frac{np}{r}\Big) + (n-r)\log\Big(\frac{(n-r)(1-p)}{n-r}\Big)$$
$$= -r\log\Big(\frac{r}{np}\Big) - (n-r)\log\Big(\frac{n-r}{(n-r)(1-p)}\Big) \text{ and}$$
$$-\log(U) = r\log\Big(1+\frac{r-np}{np}\Big) + (n-r)\log\Big(1+\frac{np-r}{n(1-p)}\Big)$$
$$= r\left(\frac{r-np}{np} - \frac{1}{2}\Big(\frac{r-np}{np}\Big)^2\right) + (n-r)\left(\frac{np-r}{n(1-p)} - \frac{1}{2}\Big(\frac{np-r}{n(1-p)}\Big)^2\right) + O\Big(\frac{1}{\sqrt{n}}\Big)$$
$$= \frac{(r-np)^2}{2np(1-p)} + O\Big(\frac{1}{\sqrt{n}}\Big).$$

Here $r-np = O(\sqrt{n})$, $r = np + O(\sqrt{n})$ *and* $n-r = n(1-p) + O(\sqrt{n})$, *which result in* $\frac{r(n-r)}{n} = np(1-p) + O(\sqrt{n}))$. *Finally,* $V = \sqrt{\frac{2\pi r(n-r)}{n}} = \sqrt{2\pi np(1-p)}\left(1+O(\frac{1}{\sqrt{n}})\right)$.

Hint: Using $\frac{r}{p} - \frac{n-r}{1-p} = \frac{r-np}{p(1-p)}$ and $\frac{r}{p^2} + \frac{n-r}{(1-p)^2} = \frac{n}{p(1-p)} + \frac{(r-np)(1-2p)}{p^2(1-p)^2}$,

$$r\left(\frac{r-np}{np} - \frac{1}{2}\Big(\frac{r-np}{np}\Big)^2\right) + (n-r)\left(\frac{np-r}{n(1-p)} - \frac{1}{2}\Big(\frac{np-r}{n(1-p)}\Big)^2\right)$$
$$= \frac{(r-np)^2}{np}\left(1 - \frac{1}{2} + O\Big(\frac{1}{\sqrt{n}}\Big)\right) \approx \frac{(r-np)^2}{2np(1-p)}.$$

Chapter 4

LINEAR RECURRENCES

4.1 Fibonacci recurrence

4.1.1 Fibonacci rabbits

The Fibonacci sequence is a celebrated example of a recurrence relation. Fibonacci's assumptions were that 1 pair of rabbits gives birth to 1 new pair each month, and that young rabbits reach maturity after 1 month. We will always count *pairs of rabbits*. Let f_n be the number of pairs at month n, assuming that we purchase 1 pair of young (non-mature) rabbits. Thus $f_1 = 1, f_2 = 1, f_3 = 2$ and so on; $f_2 = 1$ because they have reached maturity but have not yet reproduced.

It makes perfect sense to split f_n according to the maturity: to set $f_n = m_n + y_n$, where m_n is the number of mature pairs, and y_n stands for the young pairs. So, $m_1 = 0, y_1 = 1, m_2 = 1, y_2 = 0$. Generally, $m_n = m_{n-1} + y_{n-1}$ and $y_n = m_{n-1}$, which recurrence relations are sufficient to calculate any f_n.

Algebraically, replacing y_{n-1} by m_{n-2}, we obtain that $m_n = m_{n-1} + m_{n-2}$. By substituting $m_n = y_{n+1}$, we arrive at $y_{n+1} = y_n + y_{n-1}$. One can of course present it as $y_n = y_{n-1} + y_{n-2}$, i.e. using n instead of $n+1$. Without the indices: the number of pairs of young rabbits now are those from last month plus the offsprings of the newborns from 2 months ago. Thus, $f_n = m_n + y_n = m_{n-1} + m_{n-2} + y_{n-1} + y_{n-2} = f_{n-1} + f_{n-2}$. Finally,

$$f_n = f_{n-1} + f_{n-2} \text{ for } n \geq 3, \; f_1 = f_2 = 1, f_3 = 2, f_4 = 3, \ldots. \tag{4.1.1}$$

This recurrence can be obtained without the usage of m_n, y_n. Indeed, the pairs "now", i.e. the pairs counted by f_n, are those that existed just before (they never die), which are counted by f_{n-1}, plus the newborns. The newborns are obviously 1-1 with all pairs, mature or newborn, at $n-2$, i.e. their number is f_{n-2}. This gives (4.1.1). The name "Fibonacci numbers" is due to Edouard Lucas; Fibonacci calculated them until $f_{14} = 377$.

4.1.2 Towards real rabbits

(1) We need to change the time-unit from 1 month to the real (time to) maturity, which is about 6 months, and use the actual average birth rate for 1 pair of mature rabbits for this time interval. Denoting it by b, the recurrence becomes $f_n = f_{n-1} + bf_{n-2}$.
(2) Real rabbits die, and so we need to add *life expectancy.* Let it be k time-units, where $k \geq 2$. So the age of rabbits can be 0 (newborn), $1, \ldots, k-1$, and no greater than this. Qualitatively, the recurrence $f_n = f_{n-1} + bf_{n-2} - f_{n-k}$ is what one would expect, but it must be corrected as follows: $f_n = f_{n-1} + bf_{n-2} - b\,f_{n-k-1}$, as we will show below.
(3) We also need to incorporate the duration of the reproduction period, which can be smaller than the life expectancy.

The justification for the changes caused by including (2) and (3) will be provided next. The reader may, for now, skip the next two problems without any loss of understanding until Section 4.4.3 on modeling human population.

Generally, *we count rabbits "right before" an "event"*: before they give birth, become mature, reach the end of the reproduction period or die. Recall that the "allowed" ages are $0, 1, \ldots, k-1$ if the life expectancy is k. The same will be assumed for the reproduction period: if it ends at age l, and they mature at age m, then the rabbits reproduce from age m to age $l-1$ (in months or for other time-units).

From now on, we will say that a recurrence $f_n = a_1 f_{n-1} + \cdots + a_m f_{n-m}$ is of ***depth*** m if $a_m \neq 0$. The depth of (4.1.1) is 2, the depth of (4.1.2) below is k, and the depth of (4.1.3) is $k+1$. *The recurrence depth equals the number of its free parameters*; this is simply because we can take arbitrary $f_1, f_2, \ldots, f_m$, and then the recurrence will be used to determine f_{m+1}, f_{m+2}, and so on. In general, the coefficients $a_1, \ldots, a_m$ are allowed to depend on n.

Problem 4.1.1 * *Assuming that the reproduction period coincides with the life expectancy $k \geq 2$ we get*

$$f_n = b(f_{n-2} + \cdots + f_{n-k}) \text{ for } n > k \quad \text{and "free" } f_1, f_2, \ldots, f_k. \qquad (4.1.2)$$

Solution. Let m_n^s be the number of pairs of rabbits at time n of age $0 \leq s < k$. Accordingly, $f_n = m_n^0 + \cdots + m_n^{k-1}$ for the number of pairs of rabbits at n. In terms of y_n and m_n used above: $y_n = m_n^0$ and $m_n = m_n^1 + \cdots + m_n^{k-1}$. The recurrence relations for m_n^s are as follows:

$$m_n^s = m_{n-1}^{s-1} \text{ for } 0 < s < k, \; m_n^0 = b(m_{n-1}^1 + \cdots + m_{n-1}^{k-1}), \text{ where } n > 1.$$

For instance, the relations are $m_n^1 = m_{n-1}^0$, $m_n^0 = b\,m_{n-1}$ for $k = 2$. Starting with one young pair of rabbits (newborn), the evolution of (m_n^0, m_n^1) will be:

$(1,0) \mapsto (0,1) \mapsto (b,0) \mapsto (0,b) \mapsto (b^2,0) \mapsto \ldots$. So $f_n = b^{[n/2]}$ in this case for $n = 1, 2, \ldots$, where $[n/2]$ is the integer part of $n/2$. We mostly use the notation Floor$[x]$, the largest integer $\le x$, instead of $[x]$.

Let us consider the general case. Upon the substitution $m_n^s = m_{n-s}^0$, one obtains that $m_n^0 = b(m_{n-2}^0 + \cdots + m_{n-k}^0)$ for $n > k$. The same recurrence holds for any m_n^s because the passage from $n \mapsto n - s$ (generally, any shift of the index n) does not change the recurrence relation. The same recurrence holds for f_n, which is a sum of m_n^s. Generally, any linear combination of solutions of *linear recurrences* are solutions too: see below. We obtain the required. ❑

We will demonstrate now that the recurrence $f_n = f_{n-1} + bf_{n-2} - bf_{n-k-1}$ follows from (4.1.2). Let us assume that only m_n, counting *all* mature pairs, and y_n are "available," i.e. the age stratification of m_n used above is not provided. Then the rabbits that die at time n are the newborns from time $n-k$, which are counted by y_{n-k}, and we obtain the system:

$$m_n = (m_{n-1} + y_{n-1}) - y_{n-k}, \text{ and } y_n = b\,m_{n-1} \text{ for } n > k.$$

Substituting as above, $m_n = (m_{n-1} + b\,m_{n-2}) - b\,m_{n-k-1}$ and $y_n = (y_{n-1} + by_{n-2}) - by_{n-k-1}$ for $n > k+1$. Finally, $f_n = m_n + y_n$ satisfies the same recurrence:

$$f_n = f_{n-1} + bf_{n-2} - bf_{n-k-1} \text{ for } n > k+1 \text{ and "free" } f_1, \ldots, f_{k+1}. \quad (4.1.3)$$

There is a logical mistake in this argument: the rabbits we began with will then never die according to the relation we obtain; they were not included in $-y_{n-k}$. Mathematically, we obtained an extension of (4.1.2) by adding constant solutions: exactly due to the initial rabbits that live forever. Indeed, $f_n = const$ solves (4.1.3); it solves (4.1.2) only for $b(k-1) = 1$.

Such an extension is insignificant for modeling population growth when $n \gg 1$. However, the straightforward recurrence with life expectancy is (4.1.2). Note that (4.1.3) becomes the Fibonacci recurrence with b, if we set $f_{n-k-1} = 0$; this is not immediate for (4.1.2). We will use (4.1.3) to "explain" the linear growth of human population in Section 4.4.3.

The duration of the *reproduction period* is actually the key here; it influences the rate of growth of f_n much more than life expectancy does. Let us address this in detail in terms of (4.1.3); the analysis of (4.1.2) is similar.

Problem 4.1.2 * *Let k be the end of the reproduction period for the rabbits with the birth rate b and let $k+\ell$ be their life expectancy for $\ell \ge 0$. Let f_n be now the number of pairs of rabbits at n counted from their birth until the end of their reproduction period. Then the recurrence (4.1.3) holds for such f_n and for y_n,* *which is the number of pairs of newborn rabbits as above. The total number of rabbits at n will now be denoted by $\widetilde{f}_n$. Then $\widetilde{f}_n = f_n + \sum_{i=k}^{k+\ell-1} y_{n-i}$, and $\widetilde{f}_n$ satisfies the same recurrence relation as f_n. Prove the same fact for (4.1.2) used instead of (4.1.3).*

Solution. Using m_n for mature pairs at time n that are of reproductive age, and y_n for young (newborn) pairs of rabbits, the system above for m_n and y_n remains unchanged:

$$m_n = (m_{n-1} + y_{n-1}) - y_{n-k}, \;\; y_n = b\, m_{n-1} \text{ for } n > k.$$

Thus $f_n = m_n + y_n$ and y_n satisfy (4.1.3). The same holds for any shifts of n in f_n and y_n. Therefore this holds for $\widetilde{f}_n$, which is f_n plus the number of pairs at time n after the end of their reproduction period. This addition equals $\sum_{i=k}^{k+\ell-1} y_{n-i}$ (which is 0 if $\ell = 0$). ❑

The natural question is: can maturity and mortality suppress the exponential growth of rabbits? The answer is negative for $b \geq 1$, except for $b = 1$ under the smallest possible $k = 2$, which will be addressed later.

4.1.3 Coloring "$1 \times n$"

A very different situation that resulted in the same recurrence was considered in Section 1.1.1 for "$2 \times n$". Namely, $d_n = \mathcal{N}^{\square\square}_{2\times n}$ for the number of perfect coverings of the $2 \times n$ shape by dominos. In contrast to rabbits, we were not free to choose the initial values in this and other such problems. Here $d_1 = 1$ and $d_2 = 2$. So $d_n = f_{n+1}$ and $\mathcal{N}^{\square\square}_{2\times n}$ coincides with the number of pairs of Fibonacci rabbits at the nth month with 1 *mature* pair to begin with.

There are no *a priori* reasons to expect connections between rabbits and perfect coverings. The fact that the Fibonacci recurrence describes such different processes is due to its fundamental nature. Compare this to the fundamental role of the functions $\cos(x), \sin(x)$. Originally they were from trigonometry, but they turn out to describe almost any periodic process. There is an important difference though: the Fibonacci numbers grow exponentially. The following problem provides another instance of the same recurrence.

Problem 4.1.3 *Find the recurrence for the number k_n of* colorings *of the $1 \times n$ shape $\overbrace{\square\square\cdots\square}^{n}$ with 2 colors, black and white, such that there are no adjacent black boxes. We will call these n-*colorings.

Solution. As in all "composing problems," we begin with small n. Obviously, $k_1 = 2$, $k_2 = 4 - 1 = 3$. For finding k_3, we use "Inclusion-Exclusion" with

$B_1 = \{\blacksquare\blacksquare\boxast\}$, $B_2 = \{\boxast\blacksquare\blacksquare\}$, where $*$ means "whatever." Thus, $b_1 = 2, b_2 = 2, b_{12} = 1$(all black), and $k_3 = Free - b_1 - b_2 + b_{12} = 8-2-2+1 = 5$. Obtaining the recurrence relation will be simpler than using "Inclusion-Exclusion." The first box can be either white or black, so

$$k_n = |\{\square\overbrace{\boxast\cdots\boxast}^{n-1}\}| + |\{\blacksquare\overbrace{\square\boxast\cdots\boxast}^{n-1}\}| = k_{n-1} + k_{n-2}.$$

❑

To double-check: $k_3 = 5$ was calculated above directly from the definition; it was discovered to be $k_2 + k_1$. One has $k_n = f_{n+2}$ in terms of the Fibonacci numbers. It looks like a new type of situation resulting in the same recurrence, but it can be connected with d_n as follows.

Let us consider the bottom rows of perfect coverings of the $2 \times n$ shape. For instance, it could be □□□□□□□□□□□, where □□ are horizontal dominos, □ are truncated vertical ones. Let us color the ***first*** boxes of □□ here with black; the map is $f : \square\square \mapsto \blacksquare\square$. Otherwise, they will be white. This coloring has no adjacent black boxes by construction.

Different lines of □□ and □ correspond to different colorings: these lines can be uniquely recovered from the corresponding colorings. So f embeds perfect coverings of "$2\times n$" into n-colorings with no adjacent black boxes.

It seems that we proved that $k_n = d_n$, but this cannot be true: we know that $k_n = d_{n+1}$. The mistake is that f has no inverse f^{-1}. It is necessary to check that f^{-1} exists to establish a 1-1 correspondence. The problem with f^{-1} is that if the last box is black then we cannot "lift" any such an n-coloring to a "$1\times n$" covered by □□ and □, and then to a perfect covering of "$2\times n$".

This mistake can be readily corrected: f^{-1} really identifies perfect coverings of "$2\times n$" and the admissible n-colorings ***with white last boxes***. The latter are one-to-one with all admissible $(n-1)$-colorings (without adjacent black boxes). Upon this adjustment, f and f^{-1} result in $k_{n-1} = d_n$. ❑

Since there is a "natural" 1-1 correspondence between the colorings with no adjacent black boxes and perfect coverings, we cannot really consider these two problems as independent sources of the Fibonacci numbers. Exercise 4.6.36 provides a really different instance of the Fibonacci recurrence, though it has some connections with d_n and k_n. See also Exercises 4.6.12, 4.6.13.

Shallow diagonals. Let us use the passage to the last row of the $2 \times n$ shape to provide a different solution of Problem 3.3.1, which states that the sums $\binom{n}{0} + \binom{n-1}{1} + \binom{n-2}{2} + \ldots$ satisfy the Fibonacci recurrence. They actually coincide with the corresponding Fibonacci numbers. These sums are those for the *shallow diagonals* in Pascal's triangle.

We use that from our analysis above, d_n equals the number of coverings of $\overbrace{\square\cdots\square}^{n}$ by dominos and monominos. If the number of dominos is k, then the total number of pieces is $n-k$: k dominos and $n-2k$ monominos. In terms of tossing a coin $n-k$ times, consider creating such configurations as the following process: in a sequence, for each coin toss place a domino if heads or place a monomino if tails. The total number of possibilities is $\binom{n-k}{k}$. Thus:

$$f_{n+1} = d_n = \binom{n}{0} + \binom{n-1}{1} + \binom{n-2}{2} + \dots \quad \text{for} \quad n = 1, 2, \dots . \tag{4.1.4}$$

4.1.4 Historical perspective

The numbers f_n are from Fibonacci's famous *Book of Calculations* (*Liber Abaci*, 1202). Its main purpose was to promote Hindu-Arabic numerals. Europe switched to digital numerics about 50 years later ... (rather what was "Europe" in the 13th century.) *Liber Abaci* was a very big handwritten book, and, as all books of this time, not too much interesting; the pages on rabbits made it famous (thanks to Lucas). Let us mention that the first known instances of the ***abacus***, a counting frame for addition and subtraction, are dated around 2500 BCE.

"Fibonacci" means from the family of Bonacci; in the book, his name was "Leonardo, son (from family) of Bonacci the Pisan." He was well-known in Italy, had some kind of distinguished professorship in Pisa. Little is known about his life; the dates 1175–1250 are approximate. The only autobiographical fragment can be found at the very beginning of *Liber Abaci.*

His father was a merchant, and Leonardo traveled with him a lot. His instructor of mathematics was presumably Moorish (it was in Algeria); he called his study "marvelous instructions." Let us try to put him and his book into perspective. The following sketch is very vague.

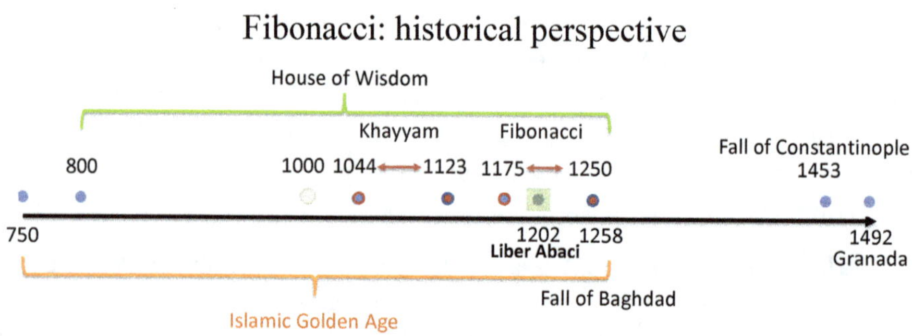

Figure 4.1: Fibonacci: before and after.

Fibonacci was born around 1175, about 50 years after Omar Khayyam died (1044-1123). He died (about 1250) several years before the Fall of Baghdad.

Both, Khayyam and Fibonacci, enjoyed very long lives, possibly equivalent to 100+ now. Omar Khayyam was a celebrated poet and also an excellent mathematician, who, while probably not the best in the *House of Wisdom*, was a real master of algebra. Let us mention *Khayyam's triangle*, and that he was able to solve some cubic equations, to the level that European mathematics reached and then surpassed only near the year 1500.

The Renaissance period was very productive for algebra. Let us mention the colorful history of the "Solution of the Cubic": the main actors were the Italian mathematicians Scipione del Ferro, Nicolo Tartaglia, and Hieronimo Cardano. Cardano was one of the best doctors in Europe (he treated the Pope), an astrologer, a keen gambler and much more (to say nothing of mathematics). His *Book of Games of Chance* is one of the first on mathematics of probabilities. He added a lot to Tartaglia's formula for solving cubics, which he received from him in 1539 ... with his oath to Tartaglia to never publish it. The usage of complex numbers was Cardano's great contribution to mathematics.

House of Wisdom. The interaction between people living in the Mediterranean region reached high levels in the 12-13th centuries. Baghdad in the time of Omar Khayyam was the cultural capital of this part of the world: it boasted a multitude of libraries and especially the House of Wisdom, which can be seen as a precursor of our Academies and Institutes of Advanced Studies. The universality of its "permanent members" was fantastic, and it existed for about 450 years, until the Fall of Baghdad in 1258. By the middle of the 13th century, the role of Baghdad significantly diminished. The expanding Mongol Empire was unstoppable. See e.g. [26].

It was probably not the plan of Hulagu Khan to ruin and depopulate the city in 1258, but it happened nonetheless. Among other cruelties, many scientists and philosophers of the House of Wisdom were killed. Those who survived recalled through their writings and the writings of their descendants that "Tigris water turned black from the ink". This was to emphasize the scale of the cultural disaster; Baghdad contained 36 libraries. The superb irrigation infrastructure in this region was destroyed and was never repaired.

The Fall of Baghdad came as a shock to the whole Muslim world. However, its cultural heritage did not vanish. The Ottoman Empire must be mentioned. The Moorish Kingdom of Granada was a great cultural center for almost 2 centuries until 1492, the Fall of Granada and the end of Moorish Spain. It is less known that Mali was an important cultural enclave (until about 1500 too): trading books was one of the most profitable businesses there.

The Fall of Constantinople in 1492 was another important turning point.

[26] *Islamic Golden Age*, `https://en.wikipedia.org/wiki/Islamic_Golden_Age`.

Many priceless books were accumulated in its libraries. Arab scholars were welcome there, despite permanent tensions and conflicts in this region (to say the least). We mention that the first printed version of Euclid's *Elements* was in Arabic: in 1594 by the Medicean Press in Rome. In the 16th century, the cultural interaction between people and countries in this region went down significantly, scientific contacts included.

4.2 Solving recurrences

4.2.1 Negative indices

Let us begin with the following observation. The Fibonacci numbers can be readily extended to negative n using the same recurrence: $f_n = f_{n+2} - f_{n+1}$. Here we changed n by $n+2$, but n is just an index; the recurrence remains the same, but the range of n to consider must be adjusted after such substitutions.

One obtains the following sequence:

$$\ldots, 13, -8, 5, -3, 2, -1, 1, 0, 1, 1, 2, 3, 5, 8, 13 \ldots. \qquad (4.2.5)$$

For instance, $f_0 = 0, f_6 = 8, f_7 = 13, f_{-6} = -8, f_{-7} = 13$. It is easy to prove that $f_{-n} = (-1)^{n-1} f_n$ for $n > 0$. To do this let us define $\widehat{f_n} \stackrel{\textbf{def}}{=} (-1)^{n-1} f_{-n}$ for any n, positive or negative, and check that (1) this sequence satisfies the Fibonacci recurrence, and (2) $\widehat{f_1} = 1 = \widehat{f_2}$. Indeed,

$$\begin{aligned}\widehat{f_n} &= (-1)^{n-1} f_{-n} = (-1)^{n-1}(f_{-n+2} - f_{-n+1}) \\ &= (-1)^{n+1} f_{-n+2} + (-1)^n f_{-n+1} = \widehat{f}_{n-2} + \widehat{f}_{n-1}.\end{aligned}$$

We used that $(-1)^n$ depends only on the parity of n.

Since f_n and $\widehat{f_n}$ satisfy the same recurrence and coincide at $n = 1, 2$, they coincide *identically* (for all n). Richard Stanley extended this symmetry from $d_n = f_{n+1}$ to perfect coverings of arbitrary rectangular shapes [27]. For a general recurrence of depth 2, this symmetry is as follows.

Problem 4.2.1 *Let f_n be a solution of the recurrence relation $f_n = af_{n-1} + bf_{n-2}$ for $b \neq 0$ and any integer n. Then $\widetilde{f_n} = f_n + \beta(-b)^n f_{-n}$ satisfies the same recurrence as for f_n (for any n and for any β). Accordingly, $\widetilde{f_n} = 0$ for all n and $f_n = -\beta(-b)^n f_{-n}$ if and only if $(1+\beta)f_0 = 0$ and $f_1 = \beta b f_{-1} = \beta(f_1 - af_0)$. These conditions hold when $f_0 = 0$ and $\beta = 1$, as in the case of the Fibonacci numbers, or when $\beta = -1$ and $2f_1 = af_0$.* ❑

[27] R. Stanley, *On dimer coverings of rectangles of fixed width*, Disc. Appl. Math. 12 (1985), 81–87.

The oscillation of f_n for negative n is of interest. Negative values of f are certainly not about rabbits, but they can be interpreted as follows.

Let $\{c_n\}$ and $\{d_n\}$ be solutions of the original Fibonacci recurrence, where c stands for cats and d for dogs. Cats and dogs accidentally fight, but the competition for space and food is the main factor of their relations.

If the birth rate is greater or (time to) maturity is smaller for dogs, then they will eventually oust cats (or the other way around), which is obvious without any recurrences. So we assume that $c_n = c_{n-1} + c_{n-2}$, $d_n = d_{n-1} + d_{n-2}$. Then $f_n = d_n - c_n$ determines the "winner." The unit of time can be a 1 hundred or even 1 thousand years for such a slow process.

The key is that this f_n solves the same recurrence. Indeed, $f_n = d_{n-1} + d_{n-2} - c_{n-1} - c_{n-2} = (d_{n-1} - c_{n-1}) + (d_{n-2} - c_{n-2})$. Generally, *any linear combination of solutions of a recurrence* $f_n = \sum_{i=1}^{k} a_i(n) f_{n-i}$ *is again a solution.* As above, *one solution* is an infinite sequence of numbers f_n. The coefficients $a_i(n)$ here can depend on n; constant $\{a_i\}$ are addressed in Theorem 4.2.2.

Thus the sequence in (4.2.5) can be interpreted as "the balance" between dogs and cats. The negative values of f_n correspond to centuries or millennia when cats dominated, however slowly but surely they became outnumbered by dogs for $n > 0$ if the balance is given by this sequence.

4.2.2 Binet's formula

Jacques Binet obtained a formula for f_n in 1750, i.e. 550(!) years after Fibonacci's book. This formula was also known to Euler (1707-1783) and other mathematicians. The technical tools needed here were perfectly available to many mathematicians:

$$\text{the roots of } x^2+ux+v=0 \text{ are } x_{1,2} = -\frac{u}{2} \pm \sqrt{\left(\frac{u}{2}\right)^2 - v}\,. \tag{4.2.6}$$

Expecting the exponential growth of the Fibonacci sequence f_n, let us try to find a "pure exponential" solution of the form $f_n = \lambda^n$. One has: $\lambda^n = \lambda^{n-1} + \lambda^{n-2}$ for any n. Obviously, $\lambda = 0$ is a solution: $f_n = 0$ for all $n > 0$. So let us assume that $\lambda \neq 0$. Then we divide by λ^{n-2} and obtain a *quadratic* equation, which we will call the ***master equation***:

$$\lambda^2 = \lambda + 1, \quad \text{with the roots} \quad \lambda_{1,2} = \frac{1 \pm \sqrt{5}}{2}. \tag{4.2.7}$$

Numerically, $\lambda_1 \approx 1.6, \lambda_2 = 1 - \lambda_1 \approx -0.6$. Thus, the sequences $(\lambda_1)^n$ and $(\lambda_2)^n$ are two different solutions. Here n can be an arbitrary integer, even negative.

The roots $\lambda_{1,2}$ are irrational, but *any* solution f_n of this recurrence can be represented as their linear combination: $f_n = C_1(\lambda_1)^n + C_2(\lambda_2)^n$, which is a particular case of Theorem 4.2.2 below. This includes the Fibonacci rabbits, so the problem is to find C_1 and C_2 such that $f_1 = 1, f_2 = 1$.

It is almost always convenient to switch here to f_0, f_1; any consecutive values will do, but these are the simplest. One has: $f_0 = 0, f_1 = 1$, and

$$f_0 = C_1(\lambda_1)^0 + C_2(\lambda_2)^0 = C_1 + C_2 = 0, \;\; f_1 = C_1(\lambda_1)^1 + C_2(\lambda_2)^1 = 1.$$

Using $C_2 = -C_1$, the second equation gives that $C_1(\lambda_1 - \lambda_2) = 1$ and $C_1 = 1/(\lambda_1 - \lambda_2) = 1/\sqrt{5}$.

Finally, Fibonacci's f_n are given by the formula

$$f_n = \frac{\left(\frac{1+\sqrt{5}}{2}\right)^n - \left(\frac{1-\sqrt{5}}{2}\right)^n}{\sqrt{5}}, \quad n = 1, 2, \cdots . \tag{4.2.8}$$

It is easy to see directly from this formula that $f_0 = 0, f_1 = 1$; always check that the resulting formula gives correct initial conditions! If even f_2 is what it is supposed to be, this formula will make sense.

4.2.3 General linear recurrences

This method can be readily extended to arbitrary *linear homogeneous recurrence relations with constant coefficients* assuming that the *master equation* has simple roots, which can be complex. This class of recurrences is from (i) in the following theorem.

Theorem 4.2.2 *(i) Given a recurrence relation*

$$f_n = a_1 f_{n-1} + a_2 f_{n-2} + \cdots + a_k f_{n-k}, \;\; n = k+1, k+2, \ldots, \tag{4.2.9}$$

with constant a_m let us assume that the corresponding master equation $\lambda^k = a_1\lambda^{k-1} + \cdots + a_{k-1}\lambda + a_k$ has k different roots $\lambda_1, \ldots, \lambda_k$. Then an arbitrary solution of this recurrence relation has the form $f_n^C = C_1\lambda_1^n + \cdots + C_k\lambda_k^n$ for some constants $C_1, \ldots, C_k$.

(ii) Continuing, given any g_n for $n = k+1, k+2, \ldots$, the solutions of the corresponding **inhomogeneous recurrence** *$\widetilde{f}_n = a_1\widetilde{f}_{n-1} + a_2\widetilde{f}_{n-2} + \cdots + a_k\widetilde{f}_{n-k} + g_n$ are as follows. Let $\widetilde{f}_n^{\bullet}$ be one of its solutions. Then an arbitrary solution can be uniquely represented in the form $\widetilde{f}_n = \widetilde{f}_n^{\bullet} + f_n^C$, where f_n^C is from (i) for some constants $C_1, \ldots, C_k$.*

Proof *. The fact that $f_n^{(i)} = \lambda_i^n$ is a solution for any λ_i is absolutely parallel to the case of the Fibonacci relation. So are any f^C, their linear combinations.

Any solution of (4.2.9) depends on k free values $f_1, \cdots f_k$. On the other hand, we have k free constants $C_1, \ldots, C_k$ in the formula for f_n^C. Thus the C-coefficients can be expected to be one-to-one with all solutions. This is not a justification of course. We need some algebra here.

First of all, it is convenient to switch to $f_0, \ldots, f_{k-1}$; any k consecutive f will do, but these are the simplest. Finding C_i for any given $f_0, \ldots, f_{k-1}$ requires solving the system of linear equations:

$$\begin{pmatrix} f_0 \\ f_1 \\ \vdots \\ f_{k-1} \end{pmatrix} = \begin{pmatrix} 1 & \cdots & 1 \\ \lambda_1 & \cdots & \lambda_k \\ \vdots & \ddots & \vdots \\ \lambda_1^{k-1} & \cdots & \lambda_k^{k-1} \end{pmatrix} \begin{pmatrix} C_1 \\ C_2 \\ \vdots \\ C_k \end{pmatrix}. \qquad (4.2.10)$$

So we need to check that the matrix (λ_j^{i-1}) here is invertible. If it is not, then there is a row that is a linear combination of the other rows, and we conclude that *all* λ_i are roots of a polynomial of degree smaller than k, which cannot have k different roots. Recall that all λ_i are assumed pairwise distinct.

This construction will be practiced a lot for $k = 2$, which is mostly sufficient in this book. Then $(\lambda_j^{i-1}) = \begin{pmatrix} 1 & 1 \\ \lambda_1 & \lambda_2 \end{pmatrix}$, and its determinant is $\lambda_2 - \lambda_1$ from Binet's formula. Generally, $Det(\lambda_j^{i-1})$ is the ***Vandermonde determinant*** $\prod_{1 \le i < j \le k} (\lambda_j - \lambda_i)$, which is nonzero for pairwise distinct λ_i. ❑

Theorem 4.2.3 * Systems of recurrence relations. *Let us assume that the vector-valued sequence $F_n = \{f_n^{(1)}, \ldots, f_n^{(k)}\}$, satisfies a* matrix *recurrence relation $F_n = MF_{n-1}$ for a constant $k \times k$ matrix M and $n = 1, 2, \ldots$. Here F_n are considered as column vectors, and we use matrix multiplication. The characteristic polynomial $P(x) = Det(M - x\mathbf{1}_k) = (-1)^k (x^k - \sum_{i=1}^k a_i x^{k-i})$ will be used, where $\mathbf{1}_k$ is the $k \times k$ identity matrix. Then for any given i from 1 to k, the sequence $f_n^{(i)}$ satisfies the* scalar *recurrence relation $f_n^{(i)} = a_1 f_{n-1}^{(i)} + a_2 f_{n-2}^{(i)} + \cdots + a_k f_{n-k}^{(i)}$ for $n = k, k+1, \ldots$.*

Proof. In contrast to counterparts of this theorem for systems of differential equations, the proof is very constructive. Let λ_i be the roots of $P(x)$, the *eigenvalues* of M, which are, for now, assumed to be pairwise distinct. We will need the following linear algebra fact. Due to $F_n = M^n F_0$ one has: $f_n^{(i)} = \sum_{j=1}^k C_j^{(i)} \lambda_j^n$ for $n = 0, 1, \ldots$ and some constants $C_j^{(i)}$ for $1 \le i, j \le k$. This readily results in the recurrence relations $f_n^{(i)} = \sum_{j=1}^k a_k f_{n-j}^{(i)}$: multiply $P(\lambda_i) = 0$ by powers of λ_i. The relations here are actually for any n provided that $a_k \neq 0$, but we need this relation for $n \geq k$. We obtain the required for simple roots λ_i.

The passage to any eigenvalues λ_i, simple or not, is as follows. The claim is some fact from abstract algebra. Namely, it states that if the entries of M and $f_n^{(i)}$ (all of them) satisfy certain algebraic quadratic-linear relations, which are $F_n = MF_{n-1}$, then some polynomial relations hold for the entries of M and $f_n^{(i)}$. Recall, that a_i are polynomials in terms of the entries of M.

Any matrix M is a limit of some matrices $\widetilde{M}(t)$ *with simple eigenvalues*: $M = \lim_{t\to 0} \widetilde{M}(t)$. Accordingly, any solution F_n of the recurrence $F_n = MF_{n-1}$ is a limit of certain $\widetilde{F}_n(t)$ satisfying $\widetilde{F}_n(t) = \widetilde{M}(t)\widetilde{F}_{n-1}(t)$ as $t \to 0$. To see this, construct $\widetilde{F}_n(t)$ for the initial conditions $\widetilde{F}_0(t) = F_0$ (the same for any t near 0). Its limit as $t \to 0$ will then be F_n for any $n \geq 0$.

We have already proved the claim of the theorem for $\widetilde{M}(t)$ using the simplicity of its eigenvalues. Hence, it holds if the eigenvalues begin to coincide under any limiting procedure. The key here is that this claim does not involve λ_i: it is in terms of the entries of M and $f_n^{(i)}$. See also Exercise 4.6.41. ❑

4.2.4 Generating functions II

In full generality, the next proposition must be given a *; however (ii) in Proposition 4.2.4 and other examples we consider are sufficiently straightforward (though somewhat computational). Generating functions are not among our core topics, but they provide a fundamental link between combinatorics and the theory of special functions. We calculate many of them, but do not really use them to solve related problems.

Proposition 4.2.4 *Let f_n be any solution of the recurrence $f_n = \sum_{i=1}^k a_i f_{n-i}$ for $k \geq 1$, where $n \geq k$. Recall that k is called the depth of this recurrence if $a_k \neq 0$; we will assume this.*

(i) Prove that the generating function *$G(t) = \sum_{n=0}^{\infty} f_n t^n$ is as follows: $G(t) = \frac{\sum_{i=0}^{k-1} u_i t^i}{1-\sum_{m=1}^{k} a_m t^m}$, where $u_m = f_m - \sum_{i=1}^{m} a_i f_{m-i}$ for $m = 0, 1, \ldots, k-1$. The matrix expressing $\{u_m\}$ in terms of $\{f_m\}$ is of determinant 1, so the latter can be obtained as linear combinations of the former with the coefficients polynomially depending on $\{a_i\}$.*

(ii) For $k = 2$: $G(t) = \frac{f_0 + f_1 t - a_1 f_0 t}{1 - a_1 t - a_2 t^2}$. Reprove this relation using Binet-type formulas. In particular, $G(t) = \frac{t}{1-t-t^2}$ for the Fibonacci numbers f_n.

Proof. **(i)** An immediate reformulation of the recurrence for f_n is the following relation: $G(t) = (\sum_{i=1}^k a_i t^i) G(t) + \sum_{i=0}^{k-1} u_i t^i$ for some u_i, which reflects that the recurrence begins with $n = k$. Comparing the coefficients of $1, t, t^2, \ldots,$

one has: $f_0 = u_0$, $f_1 = u_1 + a_1 f_0$, $f_2 = u_2 + a_1 f_1 + a_2 f_0$ and so on. Thus $u_m = f_m - \sum_{i=1}^{m} a_i f_{m-i}$ for $m = 0, 1, \ldots, k-1$. The determinant of the matrix expressing $\{u_m\}$ in terms of $\{f_m\}$ is 1. For instance, $f_0 = u_0$, $f_1 = u_1 + a_1 u_0$, $f_2 = u_2 + a_1(u_1 + a_1 u_0) + a_2 u_0 = u_2 + a_1 u_1 + (a_2 + a_1^2) u_0$, and so on.

(ii) For $k = 2$: $G(t) = \frac{u_0 + u_1 t}{1 - a_1 t - a_2 t^2} = \frac{f_0 + (f_1 - a_1 f_0)t}{1 - a_1 t - a_2 t^2}$. Assume that the master equation $\lambda^2 - a_1\lambda - a_2 = 0$ has 2 distinct solutions λ_1, λ_2. Then $f_n = C_1\lambda_1^n + C_2\lambda_2^n$, where $C_1 + C_2 = f_0, C_1\lambda_1 + C_2\lambda_2 = f_1$. Using that $\lambda_1 + \lambda_2 = a_1$, we obtain that $C_1\lambda_2 + C_2\lambda_1 = (\lambda_1 + \lambda_2)(C_1 + C_2) - (C_1\lambda_1 + C_2\lambda_2) = a_1 f_0 - f_1$. Applying these relations:

$$G(t) = \sum_{n=0}^{\infty} C_1(\lambda_1 t)^n + C_2(\lambda_2 t)^n = \frac{C_1}{1 - \lambda_1 t} + \frac{C_2}{1 - \lambda_2 t}$$
$$= \frac{(C_1 + C_2) - (C_1\lambda_2 + C_2\lambda_1)t}{1 - a_1 t - a_2 t^2} = \frac{f_0 - (a_1 f_0 - f_1)t}{1 - a_1 t - a_2 t^2}.$$

You can check directly that the same formula serves the case $\lambda_2 = \lambda_1$. Use that the fundamental solutions are λ_1^n and $n\lambda_1^n$; see Theorem 4.4.2 below. However, this is not necessary: the case $\lambda_2 = \lambda_1$ is *automatically* included in this formula because $G(t)$ obtained above does not involve $\lambda_{1,2}$ explicitly. So one can use it for $\lambda_1 \neq \lambda_2$, and tend $\lambda_2 \to \lambda_1$ in the formula for $G(t)$. ❑

As an application, we obtain that *rational functions are in one-to-one correspondence with solutions of linear homogeneous recurrences with constant coefficients.* The map is $\{f_n\} \mapsto G(t) = \sum_{m=0}^{\infty} f_n t^n$. In the opposite direction, let $G(t) = P(t)/Q(t)$, where P, Q are any polynomials and $Q(t) = t^k +$ lower terms for $k \geq 0$. Then the coefficients of $G(t) = \sum_{m=0}^{\infty} f_n t^n$ satisfy the recurrence of depth k associated with $Q(t)$ as above. One can assume that $Q(t)$ is of minimal possible degree k. Then k is the minimal possible depth of the corresponding recurrence for f_n. We note that the numerator $P(t)$ is of degree no greater than $p - 1$ if the recurrence is considered for $n \geq p$, where $p \geq k$. Any such polynomial $P(t)$ results in some solution of the recurrence provided by $Q(t)$ with $n \geq p$. This correspondence is not used in full generality in this book, but we will consider its various instances.

Problem 4.2.5 * *For the rational function $G(t)$ from Proposition 4.2.4 such that $a_k \neq 0$, assume that $G(b/t) = \epsilon\, t^\ell G(t)$ for $\epsilon = \pm 1$ and a positive integer ℓ. Its t-expansion will be denoted by $[G]_+ = \sum_{n=0}^{\infty} f_n t^n$. Extend f_n to negative n using the recurrence $f_n = \sum_{i=1}^{k} a_i f_{n-i}$ and prove that the symmetry $f_{-n-\ell} = -\epsilon\, b^n f_n$ holds for any integral n (not only positive).*

Solution. One can check directly that $\widehat{f_n} = -\epsilon\, b^{-n} f_{-n-\ell}$ satisfies the recurrence for f_n and has the same initial conditions, which gives that $\widehat{f_n} = f_n$

for any n. It is straightforward but somewhat technical. We provide a more transparent justification involving some general properties of the *expansions* of rational functions in terms of t and in terms of t^{-1}.

For any rational function $F(t)$, let $[F]_+$ and $[F]_-$ be its expansions in terms of t and t^{-1}, respectively. They are defined as follows. First, $[F]_+ = F = [F]_-$ for any sum of *polynomials* in terms of t or t^{-1}. Second, we have: $[\frac{t^p}{1-tP(t)}]_+ = t^p\sum_{n=0}^{\infty}(tP(t))^n$ and $[\frac{t^p}{1-t^{-1}P(t^{-1})}]_- = t^p\sum_{n=0}^{\infty}(t^{-1}P(t^{-1}))^n$ for any polynomials $P(t)$ in terms of t and any integers p, not only positive. For $G(t)$ from Proposition 4.2.4, these formulas give that $[G]_+$ coincides with $[G]_+$ there, and $[G]_- = [\frac{\sum_{i=0}^{k-1}u_i t^{i-k}}{t^{-k}-\sum_{m=1}^{k}a_m t^{m-k}}]_- = -\frac{1}{a_k}[\frac{\sum_{i=0}^{k-1}u_i t^{i-k}}{1+\sum_{m=1}^{k-1}(a_m/a_k)t^{m-k}-t^{-k}}]_-$.

The relations $(1-\sum_{i=1}^{k}a_i t^i)[G]_\pm = \sum_{i=0}^{k-1}u_i t^i$ hold, which result in the identity $(1-\sum_{i=1}^{k}a_i t^i)([G]_+ - [G]_-) = 0$. Let us obtain the latter identity directly from the definitions when $G = g_\alpha = \frac{1}{1-\alpha t}$ for any constant α. One has: $\frac{1}{1-\alpha t} = -\frac{(\alpha t)^{-1}}{1-(\alpha t)^{-1}}$, and $[g_\alpha]_+ - [g_\alpha]_- = \sum_{m=0}^{\infty}(\alpha t)^m + (\alpha t)^{-1-m} = \sum_{m=-\infty}^{\infty}(\alpha t)^m$. The multiplication by (αt) does not change the latter sum, so $([g_\alpha]_+ - [g_\alpha]_-)(1-\alpha t) = 0$.

This calculation is actually sufficient to prove the required identity for *any* G. It obviously holds for "generic" G: linear combinations of g_α, where α and the coefficients are generally complex numbers. An arbitrary G is a linear combination of some powers of g_α. However, such G can be obtained as proper limits of "generic" ones by making some α coincide in the limit, which gives the required for any G. We omit the details. This kind of argument was used when proving Proposition 4.2.4 in the case of some coinciding eigenvalues. It is not strictly necessary here, but we find it clarifying.

We will need the identity $([G]_+ - [G]_-)(1-\sum_{i=1}^{k}a_i t^i) = 0$ in the following equivalent form: $[G]_+ - [G]_- = \sum_{n=-\infty}^{\infty}f_n t^n$, where f_n is extended to negative n using the recurrence. Let us combine it with the general identity $t^{-\ell}[F]_\pm(\frac{b}{t}) = \left[t^{-\ell}F(\frac{b}{t})\right]_\mp$, which holds for any rational $F(t)$. The expansions "$\pm$" are transposed here due to the substitution $t \mapsto \frac{b}{t}$.

We obtain that $[G]_\pm(b/t) = [\epsilon\, t^{-\ell}G(b/t)]_\mp$, and $\epsilon\, t^{-\ell}([G]_+ - [G]_-)(b/t) = [G]_- - [G]_+$. Therefore: $\epsilon\, t^{-\ell}([G]_+ - [G]_-)(b/t) = \epsilon\, t^{-\ell}\sum_{n=-\infty}^{\infty}f_n(b/t)^n = \epsilon\, t^{-\ell}\sum_{n=-\infty}^{\infty}f_n b^n t^{-\ell-n} = [G]_- - [G]_+ = -\sum_{m=-\infty}^{\infty}f_m t^m$. Changing the summation index m in the latter expression to $n = -\ell - m$, we obtain that $\epsilon\, t^{-\ell}\sum_{n=-\infty}^{\infty}f_n b^n t^{-\ell-n} = -\sum_{n=-\infty}^{\infty}f_{-n-\ell}t^{-n-\ell}$. Thus, $f_{-n-\ell} = -\epsilon\, b^n f_n$.

❑

Let us demonstrate this procedure for $G(t) = \frac{1}{1-t-t^2}$, which satisfies the relation $G(b/t) = \epsilon t^\ell G(t)$ for $\epsilon = -1, b = -1$ and $\ell = 2$. One has: $[G]_+ = \sum_{m=0}^{\infty}(t+t^2)^m = 1 + t + 2t^2 + 3t^3 + \ldots$. Thus, $[G]_+ = \sum_{m=0}^{\infty}s_m t^m$, where

$s_n = d_n = f_{n+1}$ in terms of $d_n = \mathcal{N}_{2\times n}$ and the Fibonacci numbers f_n. Indeed, the denominator of G gives the recurrence $s_n = s_{n-1} + s_{n-2}$, so it suffices to see that $s_0 = d_0, s_1 = d_1$, or that $s_n = d_n$ for any 2 consecutive n.

Let us now calculate $[G]_-$. From the definition: $-[G]_- = [\frac{t^{-2}}{1+t^{-1}-t^{-2}}]_- = t^{-2}\sum_{m=0}^{\infty}(t^{-2}-t^{-1})^m = t^{-2}+(t^{-4}-t^{-3})+(t^{-6}-2t^{-5}+t^{-4})+\ldots = t^{-2}-t^{-3}+2t^{-4}+\ldots$. Thus, $[G]_+-[G]_- = \ldots+2t^{-4}-t^{-3}+t^{-2}-0\cdot t^{-1}+1+t+2t^2+\ldots = \sum_{n=-\infty}^{\infty} d_n t^n = \sum_{n=-\infty}^{\infty} f_{n+1}t^n$. The formula from the problem states that $d_{-n-2} = -\epsilon\, b^n d_n$. Indeed, $d_{-4} = 2 = (-\epsilon)(-1)^2 d_2 = d_2 = 2$, $d_{-3} = -d_1$. For $n = -1$: $d_{-1} = -\epsilon\,(-1)^{-1}d_{-1} = -d_{-1}$, which results in $d_{-1} = 0$.

4.3 Various examples

4.3.1 Tricolor problem

Let us calculate r_n, the number of colorings of $\overbrace{\square\square\cdots\square}^{n}$ with 3 colors, black, white, and gray such that there are no adjacent black boxes. *Composing a recurrence* must (always) begin with finding initial values. Similar to k_n (2 colors), $r_1 = 3$ and $r_2 = 3^2 - 1 = 8$. Let us use the symbol $\boxminus$ for the box that can be colored gray or white. For instance, $\boxminus\blacksquare = \blacksquare\blacksquare$ or $\square\blacksquare$, i.e. this pair can be "gray, black" or "white, black." The recurrence relation is as follows:

$$r_n = |\{\boxminus\overbrace{\boxast\cdots\boxast}^{n-1}\}| + |\{\blacksquare\overbrace{\boxminus\boxast\cdots\boxast}^{n-1}\}| = 2r_{n-1} + 2r_{n-2},\ n \geq 3.$$

The final stage is *solving the recurrence*. It begins with calculating r_0, which is 1 in this case. Indeed, $r_0 = 1 = \frac{r_2-2r_1}{2} = \frac{8-6}{2}$. Note that $r_{-1} = \frac{r_1-2r_0}{2} = \frac{1}{2}$, so rationals can occur for negative n even if all r_n are integers for positive n.

The master equation is $\lambda^2 = 2(\lambda+1)$, and its roots are $\lambda_{1,2} = 1\pm\sqrt{3}$. The last step is to find $C_{1,2}$ such that $r_n = C_1(\lambda_1)^n + C_2(\lambda_2)^n$ and $r_0 = 1, r_1 = 3$. One has: $C_1+C_2=1$, $C_2=1-C_1$,

$$C_1\lambda_1 + C_2\lambda_2 = 3,\ C_1(\lambda_1-\lambda_2)+\lambda_2 = 3,\ C_1 = \frac{3-\lambda_2}{2\sqrt{3}}.$$

Finally, $C_1 = \frac{1}{2} + \frac{1}{\sqrt{3}}$, $C_2 = \frac{1}{2} - \frac{1}{\sqrt{3}}$. ❑

If the calculation is correct then λ_2, C_2 must be obtained from λ_1, C_1 by the substitution $\sqrt{3} \mapsto -\sqrt{3}$, which follows from Galois theory (a topic for much later). This is convenient to double-check the formulas for λ_2, C_2 if λ_1 is irrational. Generally, it can be used for any master equations.

In this example $\lambda_1 \approx 2.73, \lambda_2 \approx -0.73$. Importantly, $|\lambda_2| < 1$, as it was in Binet's formula. We obtain that $r_n = \text{Round}[C_1(\lambda_1)^n]$ for sufficiently large n. It is easy to see that this holds for all $n = 0, 1, \ldots$.

Irrational numbers as processes. Similarly to the Round-formula for r_n, one has: $f_n = \text{Round}[\varphi^n/\sqrt{5}]$ for all $n = 0, 1, \cdots$, where $\varphi = \frac{1+\sqrt{5}}{2}$ is the celebrated ***golden ratio***, Leonardo da Vinci's "divine proportion."

Presumably, φ was used when constructing the Egyptian pyramids, though it is disputed: the Egyptians' "applied mathematics" was based on rational numbers. The usage of φ in the Parthenon and several other "famous instances of φ" also appeared far from certain. However, nautilus shells, solar system periods and spiral galaxies have some relation to φ and Fibonacci numbers; this is more solid. The nine successive orbital period ratios are close to fractions of Fibonacci numbers. See [28], which also contains a lot on the history of Fibonacci numbers.

Fibonacci numbers and φ are certainly present in music. It is a simple fact of psycho-acoustics that any two tones that one might play produce additional combination tones: difference tone and sum tone, with the sum and the difference of the frequencies of the two tones. The golden ratio occurs in music when φ is used as a music interval. It has interesting properties that the Western equal-tempered scale could scarcely hope to reproduce.

For the sake of completeness, let us mention the Elliott waves and that the number of petals (on certain species of flower) is frequently a Fibonacci number. However, many flowers have 4,6 petals. In 1935 Ralf Elliott made an uncanny prediction of a stock market bottom, but the Elliott Wave Oscillator based on Fibonacci numbers appeared far from being reliable: as they say, "Even a stopped clock is right twice a day." Anyway, φ frequently occurs in the physical world: playing cards, postcards, light switch plates, and widescreen televisions are all rectangular objects with the dimensions bearing roughly "golden" proportions.

In geometry, φ is the key for regular pentagrams and pentagons. Interestingly, Fibonacci worked with it, but never connected φ with his $\{f_n\}$. Simon Jacob (around 1564) noted that the ratio of consecutive Fibonacci numbers converges to the golden ratio: $f_{n+1}/f_n \to \varphi$ as $n \to \infty$. This is immediate from Binet's formula, since $\lambda_2^n \to 0$ as $n \to \infty$. A more precise version is the formula $f_n = \text{Round}[\varphi^n/\sqrt{5}]$, which works for any $n \geq 0$.

A fundamental fact is that the sequence $\frac{f_{n+1}}{f_n}$ provides the most accurate rational approximation for φ in the following sense. For f_n as the denomi-

[28] A. Drozdyuk, D. Drozdyuk, *Fibonacci, his numbers and his rabbits*, Choven Publishing Corp. Toronto, 2010.

nator, the accuracy here is about $\frac{1}{f_n f_{n+1}}$, much better than for the standard decimal fractions for φ. See Problem 5.5.4. More exactly, the deviation from φ is approximately $\frac{\alpha}{f_n f_{n+1}}$ for some α between 1 and $\frac{1}{\varphi}$ due to Exercise 4.6.4. Generally, the rational approximations $\frac{M}{N}$ of irrational numbers with "random" N provide the accuracy $\approx \frac{1}{2N}$. Let us comment on it.

The accuracy $\approx \frac{1}{2N}$ is granted because one can represent $N\varphi = M + \epsilon$ for some integer M and where ϵ satisfies $-\frac{1}{2} < \epsilon < \frac{1}{2}$. Therefore $|N\varphi - M| < \frac{1}{2}$ and $|\varphi - \frac{M}{N}| < \frac{1}{2N}$. This is actually a variant of division with remainder. Given N, the numerator M is unique here. If we take $N = f_n$, then $M = f_{n+1}$. The accuracy for such special N is $|\varphi - \frac{M}{N}| \approx \frac{const}{N^2}$, which is much better than the general $\frac{1}{2N}$.

For instance, $f_{35} = 9227465, f_{34} = 5702887$ and the ratio f_{35}/f_{34} is $1.618033988749909\ldots$. It deviates by 1.4×10^{-14} from the actual decimal fraction of φ with 16 digits, which is 1.618033988749895. The theoretical upper bound for this deviation is $\frac{1}{f_{34}f_{35}} \approx 2 \times 10^{-14}$ and the lower bound is $\frac{1}{\varphi f_{34} f_{35}} \approx 1.25 \times 10^{-14}$, which are sufficiently close to the actual deviation: 1.4×10^{-14}. Also, see Problem 7.2.5 and Exercise 5.6.37 (the cubic case).

We have arrived at the fundamental understanding of irrational numbers as processes. This is for any algebraic numbers λ, not just quadratic irrationalities. By definition, $\lambda^k = \sum_{i=1}^k a_i \lambda^{k-i}$ for rational a_i. Following (i) from Theorem 4.2.2, we consider the corresponding recurrence $f_n = \sum_{i=1}^k a_i f_{n-i}$, which is undoubtedly a "process." To go back to λ: $\lambda = \lim_{n\to\infty} f_{n+1}/f_n$, provided that f_n is a sufficiently general solution and λ has the greatest absolute value among all roots of the equation $x^k - \sum_{i=1}^k a_i x^{k-i}$. If f_n are integers or rational numbers, then we obtain a sequence of rational approximations of such dominant λ. The other roots λ can be obtained via some special sequences f_n too, but the latter are generally not "combinatorial."

For instance, we have the equivalence of φ and the recurrence $f_n = f_{n-1} + f_{n-2}$. Similarly, the sequence r_n/r_{n-1} gives the value of $1 + \sqrt{3}$ with the best possible accuracy. To give an example, the deviation from λ_1 is about 1.7×10^{-14} for $n = 24$. Here $\gcd(r_{2m+2}, r_{2m+1}) = 2^m = \gcd(r_{2m+1}, r_{2m})$ (check by induction). So the estimates are slightly different here from those for the Fibonacci numbers, where $\gcd(f_n, f_{n+1}) = 1$. The theoretical estimate for this deviation is $\approx 2^{n-1}/(r_n r_{n-1})$. For instance, it is approximately 2×10^{-14} for $n = 24$, which is close enough to the actual deviation.

Interestingly, $\sqrt{2}$ in this approach is not the "simplest" quadratic irrationality. The golden ratio φ is. It satisfies the simplest possible recurrence of depth 2, the Fibonacci one.

A "secret discovery" of the Pythagorean school was that $\sqrt{2}$ is not a rational number. Indeed, if $\sqrt{2} = p/q$ with $\gcd(p,q) = 1$, then $p^2 = 2q^2$, p is divisible by 2, p^2 is divisible by 4, and q is divisible by 2, which contradicts that $\gcd(p,q) = 1$. As we argued, irrationalities are rather *processes* than *numbers*. Let us obtain the recurrence for $1+\sqrt{2}$.

4.3.2 Growing stalagmites

Let us consider $2\times n$ *stalagmites*, represented horizontally (sorry!) by 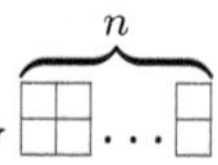. There are two kinds of boxes: some are *gaps*, which are marked by $\circ$ and some are filled, without any mark in the presence of $\circ$ in the diagram or marked by dots otherwise:  . The assumption is that *no two neighboring boxes (vertically or horizontally) are allowed to be gaps (marked by* $\circ$*)*; otherwise, the stalagmites break. Let the number of such configurations be h_n for $2\times n$. Obviously, $h_1 = 3$, but h_2 requires some consideration.

One admissible configuration is "all filled", for instance, . If there are two gaps ($\circ$), then it can only be , with the remaining boxes filled (we omit $\cdot$), or its rotation. So we have 2 such. If there is only one gap, we have 4 cases. They are: , and its 3 rotations. Finally, $h_2 = 7$. This approach becomes quite involved for $n > 2$; we need a better method.

The standard reduction of the size of the problem using some marked boxes (mostly corners) is possible here, but it does not work very well. Let us do the opposite. We begin with h_{n-1} and then *grow* $(n-1)$-stalagmites into n-stalagmites by adding (on the right) some  for $* = \circ$ or $* = \cdot$. We have at least 2 continuations: 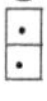 (always), and at least one by adding a column with $\circ$ somewhere.

There are (exactly) 3 ways to grow the following *special* $(n-1)$-stalagmites to size n. These ones have 2 nonempty boxes in the last column, i.e. $\cdots$ . Two of these three continuations were already counted above, so we need to add to $h_n = 2h_{n-1}$ the number of *special* $(n-1)$-stalagmites. Their number is exactly h_{n-2}: the column 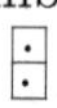 imposes no restrictions on the previous column. Thus we obtain that $h_n = 2h_{n-1} + h_{n-2}$, and $h_0 = 1 = h_2 - 2h_1 = 7-6$. Compare with the *Pell numbers* from Exercise 4.6.27, which provides all integer solutions of the *Pell equation*: $x^2 - 2y^2 = \pm 1$.

The next step is the master equation and finding its roots:

$$\lambda^2 = 2\lambda + 1,\ \lambda_1 = 1+\sqrt{2},\ \lambda_2 = 1-\sqrt{2}. \tag{4.3.11}$$

Then $h_n = C_1(\lambda_1)^n + C_2(\lambda_2)^n$, and: $C_1+C_2=1$ and $C_1\lambda_1+C_2\lambda_2=3$. Finally, we obtain that $C_1 = \frac{3-\lambda_2}{\lambda_1-\lambda_2} = \frac{2+\sqrt{2}}{2\sqrt{2}} = \frac{1+\sqrt{2}}{2}$ and $C_2 = \frac{1-\sqrt{2}}{2}$.

Generally, we do not always try to simplify the formulas in terms of square roots, but it makes sense in this case; it is hardly accidental that $C_1 = \lambda_1/2$ and $C_2 = \lambda_2/2$. We can present the formula as follows:

$$h_n = \frac{(1+\sqrt{2})^{n+1} + (1-\sqrt{2})^{n+1}}{2} \quad \text{for} \quad n = 0, 1, 2, \dots \,. \tag{4.3.12}$$

Applying Newton's binomial. Generally,

$$\frac{(1+x)^{n+1} + (1-x)^{n+1}}{2} = \sum_{m=0}^{2m \le n+1} \binom{n+1}{2m} x^{2m}, \tag{4.3.13}$$

which readily results from Newton's binomial (check). This formula is frequently used for the problems involving even and odd binomial coefficients. Making $x^2 = 2$, we arrive at:

$$h_n = \sum_{m=0}^{2m \le n+1} \binom{n+1}{2m} 2^m. \tag{4.3.14}$$

The simplicity of this formula is a clear invitation to try to find its proof via "tossing a coin," i.e. by some 1-1 correspondence.

Let us interpret this formula combinatorially. Let *plus-ℓ chain* be the following alternating sequence: 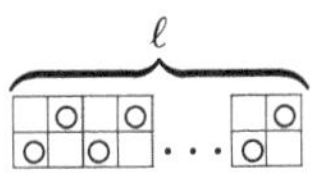. If it begins with , then we call it a *minus-ℓ-chain*. Here ℓ can be even or odd.

A stalagmite is a sequence of (nonempty) chains separated by some number of completely filled columns 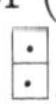, where at least one such filled column must occur between any consecutive chains.

A chain is completely described if we know its start, its end and its "sign". The starts/ends are some positions ***between*** the columns, which do not coincide (the chains are nonempty). The start of a chain is naturally before its end. The former can be before the 1st column; the latter can be after the last one. Any sequence of $2m$ such places can be uniquely interpreted as a sequence of m chains. We have $\binom{n+1}{2m}$ of choices for them. Finally, every chain has the "sign", "plus" or "minus"; so $\binom{n+1}{2m}$ must be multiplied by 2^m. This gives the required identity: a great example, connecting quite a few topics.

4.3.3 Predator-prey model

Going back to rabbits, the exponential growth of their population can be only for a short period of time. Such growth is unsustainable long term. This is a serious limitation for the usage of linear recurrence relations for modeling natural processes. There are essentially 3 ways to avoid the exponential growth: (1) *resonances*, to be considered next, (2) making the birth rate inversely proportional to time (later), and (3) predators, infections etc.

The following is a basic *predator-prey model*. Another type of this model, *rabbits-grass*, is suggested in Exercise 4.6.35. Let r_n be the number of rabbits at the nth month (or for any time-unit) and f_n the number of foxes in the area. Disregarding birth rate, maturity, and the life expectancy for r_n and for f_n, we obtain the following system of recurrence relations:

$$f_n = f_{n-1} + \alpha r_{n-1},\ r_n = r_{n-1} - \beta f_{n-1} \text{ for some } \alpha, \beta > 0. \tag{4.3.15}$$

The first equation means that foxes arrive in the area at a rate proportional to the number of rabbits there. The second equation means that one fox kills β rabbits during 1 month (one time-unit). Also, it includes the "emigration" of the rabbits from the area if f_n becomes "too big."

Generally speaking, β is supposed to be multiplied by $f_{n-1}r_{n-1}$, which is the number of "contacts" between foxes and rabbits: if $r_n \approx 0$, then there will be ≈ 0 contacts. This modification significantly influences the solutions as $n \to \infty$ and creates some algebraic challenges, which we prefer to avoid. Also, it is not always reasonable practically: if the number of rabbits drops below some threshold, then foxes will leave the area. The usage of (4.3.15) until r_n becomes negative describes many practical situations. Mathematically, the "saturation" can be incorporated via "linear models"; see Section 4.5.

We obtain that $r_{n-1} = \frac{1}{\alpha}(f_n - f_{n-1})$ from the first equation and then use the second:

$$\frac{1}{\alpha}(f_{n+1} - f_n) = \frac{1}{\alpha}(f_n - f_{n-1}) - \beta f_{n-1}.$$

Finally, $f_{n+1} = 2f_n - (1+\alpha\beta)f_{n-1}$, $\lambda^2 - 2\lambda + (1+\alpha\beta) = 0$ and $\lambda_{1,2} = 1 \pm \sqrt{1-(1+\alpha\beta)} = 1 \pm \sqrt{-\alpha\beta} = 1 \pm \sqrt{\alpha\beta}\,\imath$ for the ***imaginary unit*** $\imath = \sqrt{-1}$.

Dealing with complex numbers, which are $a + b\imath$ for real a and b, is no different from dealing with quadratic irrationalities, i.e. numbers like $x + y\sqrt{2}$ for rational x and y. One has: $(a + b\imath) + (c + d\imath) = (a + c) + (b + d)\imath$ for real a, b, c, and d, and the multiplication of complex numbers is:

$$(a + b\imath)(c + d\imath) = ac - bd + (ad + bc)\imath,\ 1/(a + b\imath) = (a - b\imath)/(a^2 + b^2).$$

Also, $|a + b\imath| = \sqrt{a^2 + b^2}$ extends the usual absolute value. Note that $|\lambda_{1,2}| = \sqrt{1+\alpha\beta} > 1$ for $\lambda_{1,2}$ we obtained above.

To simplify the formulas, let $\alpha = \beta = 1$. Then $\lambda_{1,2} = 1 \pm \imath$. Let $f_0 = 0, f_1 = 1$. Accordingly, $r_0 = f_1 - f_0 = 1$, and $r_1 = r_0 - f_0 = 1$. One has: $f_n = C_1(1+\imath)^n + C_2(1-\imath)^n$. Thus $C_1 = \frac{1}{2\imath} = -\frac{\imath}{2} = -C_2$, and

$$f_n = \frac{(1+\imath)^n - (1-\imath)^n}{2\imath} \text{ for } n = 0, 1, \ldots. \tag{4.3.16}$$

Of course, $\imath$ disappears after we use Newton's binomial here (check):

$$f_n = \binom{n}{1} - \binom{n}{3} + \binom{n}{5} + \ldots, n = 1, 2, \ldots .$$

We obtain that $\{f_1, f_2, \ldots\} = \{1, 2, 2, 0, -4, -8, -8, 0, 16, \cdots\}$ and (for rabbits) $\{r_1, r_2, \ldots\} = \{1, 1, 0, -2, -4, 0, 8, \cdots\}$.

In the calculus limit $n \to \infty$, we take a root of unity ζ of order m instead of $\imath$ and assume that $m \sim \sqrt{n}$. This results in the normal distribution times cos. Basically, our summation upon the division by 2^n becomes a counterpart of $\int_{-\infty}^{\infty} e^{-x^2/2} cos(\nu x) dx$ for proper ν. We leave this to motivated readers.

One needs to stop using this model after the 3rd month, when the number of rabbits becomes 0: foxes fully destroy the population of rabbits in 3 months. In reality, rabbits begin avoiding the area when the number of foxes increases beyond some threshold.

Negative solutions. It is possible to make the whole solution meaningful, even when r_n, f_n become negative. This is when we consider the difference between the number of rabbits and foxes in two areas subject to the same recurrence in (4.3.15) (with the same coefficients). In the context of epidemics in Section 4.5, we consider the difference of the number of infections in 2 areas, for instance in 2 hemispheres. Then the negative values make sense.

However, the magnitude of the oscillations will be approximately $(\sqrt{2})^n$, which is obviously unsustainable long term. One way to obtain more realistic models is to allow foxes and rabbits to leave or enter the area(s). Mathematically, we switch to $f_n = \epsilon f_{n-1} + \ldots$ and $r_n = \delta r_{n-1}$ from $f_n = f_{n-1} + \ldots$ and $r_n = r_{n-1} + \ldots$ for some $\epsilon, \delta \geq 0$.

Problem 4.3.1* *Assume the following relations for $n \geq 1$:*

$$f_n = \epsilon f_{n-1} + \alpha r_{n-1},\ r_n = \delta r_{n-1} - \beta f_{n-1} \text{ for some } \alpha, \beta > 0, \epsilon, \delta \geq 0. \quad (4.3.17)$$

Then the corresponding roots of the master equation are non-real and of absolute value 1 if and only if $\epsilon + \delta < 2$ and $\epsilon\delta + \alpha\beta = 1$. In this case, any sequences f_n, r_n are bounded (for all n).

Solution. As above, we obtain that $f_{n+1} - \epsilon f_n = \delta f_n - (\epsilon\delta + \alpha\beta) f_{n-1}$. Then $\lambda_{1,2} = \frac{\epsilon+\delta}{2} \pm \sqrt{\frac{(\epsilon+\delta)^2}{4} - (\epsilon\delta + \alpha\beta)}$. Also, $\lambda_1\lambda_2 = (\epsilon\delta + \alpha\beta)$. Thus, if λ_1 is not real, then λ_2 is its complex conjugation (sending $\imath$ to $-\imath$), $|\lambda_1|^2 = \epsilon\delta + \alpha\beta = |\lambda_2|^2$, and the boundness of any solution is equivalent to $\epsilon\delta + \alpha\beta = 1$. Imposing the latter, the roots $\lambda_{1,2}$ are non-real if and only if $(\epsilon + \delta) < 2$. ❑

Problem 4.3.2 *For $\epsilon = 0, \delta = 1, \alpha\beta = 1$ in (4.3.17), check that $f_{n+3} = -f_n$. In particular, let $f_1 = 1, f_2 = 1$. Check that $f_n = \frac{1}{2^{n-1}} \sum_{r=0}^{\infty} \binom{n}{2r+1}(-3)^r = \{0, 1, 1, 0, -1, -1\}$ for $n = 0, 1, 2, 3, 4, 5$ mod 6.*

Solution. Using that $f_n = \alpha r_{n-1}$, the recurrence for f_n becomes $f_{n+1} = f_n - f_{n-1}$, and the corresponding $\lambda_{1,2}$ solve the equation $\lambda^2 - \lambda + 1 = 0$. Therefore $\lambda^3 + 1 = (\lambda+1)(\lambda^2 - \lambda + 1) = 0$ for $\lambda = \lambda_{1,2} = \frac{1 \pm \sqrt{3}\imath}{2}$. So $f_{n+3} = -f_n$ for any $f_n = C_1\lambda_1^n + C_2\lambda_2^n$. Directly: $f_{n+3} = f_{n+2} - f_{n+1} = (f_{n+1} - f_n) - f_{n+1} = -f_n$. The solution $f_n = (\lambda_1^n - \lambda_2^n)/(\sqrt{3}\imath)$ results in the required identity. ❑

4.3.4 Back to tiling

A 3-mino problem. Complex numbers can appear without any "predators." Let us discuss the formula for $t_n = \mathcal{N}_{3\times n}^{\square\square\square}$, the number of perfect coverings of the $3 \times n$ shape with 3-minos.

Obviously, $t_1 = 1, t_2 = 1, t_3 = 2$. The same approach as for $d_n = \mathcal{N}_{2\times n}^{\square\square}$ (examine how the first box can be covered) gives that

$$t_n = t_{n-1} + t_{n-3}\,(n \geq 4),\ t_1{=}t_2{=}1,\ t_3{=}2,\ t_4{=}3,\ t_5{=}4,\ t_6{=}6. \qquad (4.3.18)$$

Note that this recurrence is for the Fibonacci rabbits with maturity at 2 months instead of 1 (starting with one 1-month-old pair).

The corresponding master equation is *cubic* now: $\lambda^3 = \lambda^2 + 1$. We use again (and again) Theorem 4.2.2. Two of its three roots, $\lambda_{2,3}$, are *complex* with the coinciding absolute values $|\lambda_{2,3}|$ smaller than 1; they are conjugate to each other. Recall that $|a + b\imath|^2 = a^2 + b^2$ for real a, b; the complex conjugation of $a + b\imath$ is $a - b\imath$ and it does not change the absolute value. The real root is the so-called *supergolden ratio*: $\lambda_1 \approx 1.465571231876768$. One has $\lambda_1\lambda_2\lambda_3 = 1 = |\lambda_2|^2\lambda_1$. So, as it was claimed, $|\lambda_2|^2 = |\lambda_3|^2 = 1/\lambda_1 < 1$ and $|\lambda_{1,2}| < 1$. The existence of a positive λ_1 with $|\lambda_1| > |\lambda_{2,3}|$ is some general property: see Exercise 4.6.14. This provides that $t_n \approx const\, \lambda_1^n$ for sufficiently large n. More exactly, $t_n = \text{Round}[C_1\lambda_1^n]$ for $C_1 \approx 0.611492$ and *all* $n \geq 0$. One has $t_n = C_1\lambda_1^n + C_2\lambda_2^n + C_3\lambda_3^n$. The exact algebraic formulas for λ_i, C_i are very involved; for instance, $\lambda_1 = \big(1 + (\frac{29+3\sqrt{93}}{2})^{1/3} + (\frac{29-3\sqrt{93}}{2})^{1/3}\big)/3$.

Problem 4.3.3 * *(i) Consider the system for g_n, s_n from Section 1.1.1. Show that $g_{2n} = 4g_{2n-2} - g_{2n-4}$, $g_0 = 1, g_2 = 3$, and the master equation is $\lambda^2 - 4\lambda + 1 = 0$. Find the algebraic formula for g_{2n}. Alternatively, you can use the characteristic polynomial for the matrix $\begin{pmatrix} 3 & 2 \\ 1 & 1 \end{pmatrix}$.*

(ii) Check that $g_{-2n-2} = g_{2n}$ for $n \geq 1$ for the extension of g_{2n} to negative n using the recurrence. Alternatively, deduce this from the relation $G(1/t) =$

$-tG(t)$ for the generating function $G(t)=\sum_{m=0}^{\infty} g_{2m}t^m=\frac{1-t}{1-4t+t^2}$; verify this formula. Use Proposition 4.2.4 and Problem 4.2.5.

(iii) Check $h_{n+1}-h_n-5h_{n-1}-h_{n-2}+h_{n-3}=0$ for the number $h_n=\mathcal{N}^{\square\square}_{4\times n}$ of perfect covering of "$4\times n$". Solve the corresponding master equation in radicals: divide the master equation by λ^2. Use that $h_0=1, h_1=1, h_2=5, h_3=11, h_4=36$ to check that $G(t)=\sum_{n=0}^{\infty} h_n t^n=\frac{1-t^2}{1-t-5t^2-t^3+t^4}$, and that $h_{-n-2}=h_n$ for any n. ❑

We refer to Problem 7.2.9 concerning the recurrence from Part (iii). Parts (i,ii) follow from what was discussed before.

Problem 4.3.4 *For any $m\geq 1$, prove that $N_n^m=\mathcal{N}^{\square\square}_{m\times n}$ satisfy some linear recurrence relations in terms of $n\geq 1$ with constant integer coefficients.*

Solution. We claim that there is a *system* of recurrence relations for N_n^m in terms of the numbers of perfect coverings by dominos of the following (smaller) shapes. Remove any even subset of boxes S from the *first two* columns of "$m\times n$" such that they have at least 1 box in the first column. These sets give the *types* of shapes.

The removal of some S makes the rest non-coverable; they can be discarded. Among the remaining ones, not all are really needed. For instance, it suffices to take one shape from each pair of those symmetric with respect to the reflection in the middle horizontal line (if they do not coincide).

Let N_n^S be the number of perfect coverings upon the removal of S; here n is the number of columns, complete or not. This is a straightforward generalization of g_n and s_n from Section 1.1.1.

When producing the recurrence relations for N_n^S, pick a "probe box" in the *first* column and analyze all possible ways to cover it by a domino. There will be no greater than 3 such ways. This provides an expression of N_n^S as a linear combination of $N_{n'}^{S'}$ for other types S', where $n'\leq n$ and the total number of boxes for the resulting shapes is smaller than that initially. Eventually a *finite* system of recurrence relations will be obtained with non-negative integer coefficients for the family $\{N_n^S\}$. Theorem 4.2.3 gives that the depth of the recurrence for N_n^m is no greater than the number of types.

For even m, it is more convenient to take subsets s of $[n]$ with an even number of elements as the types. They will be the sets of boxes removed from the *first* columns of "$m\times n$". For instance, $N_n^{\emptyset}=N_n^m$, and $N_n^s=N_{n-1}^m$ if s is the whole $[n]$ (the whole 1st column). Not all s are really needed for N_n^m. We can take only one of them from each pair of those related by the reflection symmetry of $[n]$. Also, if $[n]\setminus s$ contains any i such that $i-1$ and $i+1$ are (both) *not* in this set, then s can be omitted. This is because the box in the 1st column corresponding to such "isolated" i can be covered only horizontally. We obtain the estimate $(2^{m-1}+2^{m/2})/2$ for the depth of the recurrence for N_n^m, but the minimal depth is actually $\le 2^{m/2}$ (Kasteleyn). Combine formula (1.1.4) with Exercise 4.6.41(i). See also Problems 7.2.12, 7.2.13. ❑

Using computers. * Let us calculate the recurrence relations and generating functions for the $6\times n$ shape. We will employ formulas (1.1.2), (1.1.4) to produce $N_n^6 = \mathcal{N}_{M\times N}^{\square}$ for $M=6, N=n$. For $n=0,1,\ldots,12$:
$N_n^6 = \{1, 1, 13, 41, 281, 1183, 6728, 31529, 167089, 817991, 4213133, 21001799, 106912793\}$.
Finding the recurrence for N_n^6 is a good exercise. We suggest: write a program to find it, relying on the fact that it exists. Try the depths 1,2,.... Eventually your computer will find that it is 8. Also, see Problem 7.2.12. One has:
$N_n^6 = N_{n-1}^6 + 20N_{n-2}^6 + 10N_{n-3}^6 - 38N_{n-4}^6 - 10N_{n-5}^6 + 20N_{n-6}^6 - N_{n-7}^6 - N_{n-8}^6$.
This gives the denominator of the generating function $G_6(t)=\sum_{n=0}^{\infty} N_n^6 t^n$; it is $Den_6(t)) = 1-t-20t^2-10t^3+38t^4+10t^5-20t^6+t^7+t^8$. The numerator is $Num_6(t) = G_6(t)\cdot Den_6(t) = 1-8t^2-2t^3+8t^4-t^6$. Finally:

$$G_6(t) = \frac{1-8t^2-2t^3+8t^4-t^6}{1-t-20t^2-10t^3+38t^4+10t^5-20t^6+t^7+t^8}.$$

One has $G_6(-\frac{1}{t}) = -t^2 G_6(t)$. So $b=-1, \epsilon=-1, \ell=2$ in Problem 4.2.5, which gives that $N_{-n-2}^6 = -\epsilon\,(-1)^n N_n^6 = (-1)^n N_n^6$ for the extension of N_n^6 to negative n based on the recurrence. Generally, $N_{-n-2}^m = (-1)^{\frac{nm}{2}} N_n^m$, $N_{-1}^m = 0$ for any even m, which is due to Stanley.

4.4 Resonances etc.

4.4.1 Fission reactors

Fission nuclear reactors are based on *nuclear fission*. For instance, ^{235}U, Uranium-235 (with mass 235) can be used. Upon the absorption of one neutron (they are present in any mass), the nucleus becomes the unstable isotope ^{236}U. Then it splits into two lighter fission products, releasing energy and generating 2-3 fast neutrons; the average is about 2.64.

Unless they leave the mass (fissile material), the neutrons will then slow down and can be captured by other atoms. This is called a *fission chain reaction*. Its possibility was predicted by Leó Szilárd in 1933. The supercritical mass occurs when the number of neutrons grow exponentially, otherwise it is subcritical. The critical mass is the transition point: when the explosion begins.

In Fig. 4.2, the arrows represent neutron paths in the sphere of enriched uranium or plutonium. The stars are elementary fission reactions (scattering

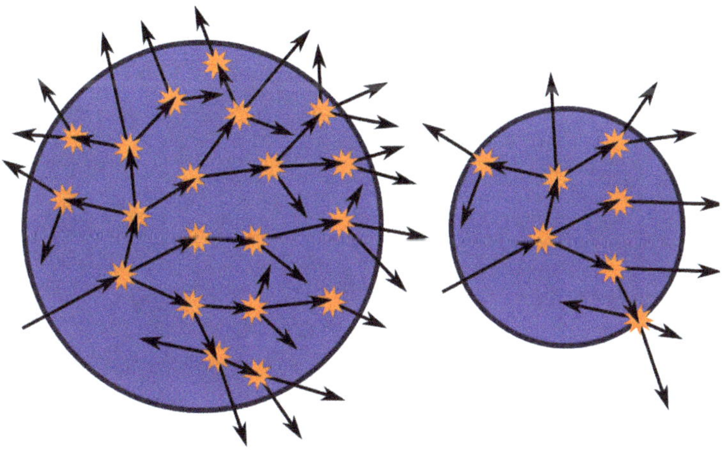

Figure 4.2: Supercritical and subcritical mass.

events); the average number of created neutrons is indeed about 2.5 in this figure, which matches ^{235}U (about 3.17 for Plutonium-239). There is too little active material in the right frame, so the fission reaction quickly ends. In the left frame, more material allows for more fission reactions, opening up the possibility of a self-perpetuating reaction.

Not too much was known about the danger of radiation and about the critical masses in the 1940s. Many physicists died due to the radiation, which becomes extreme when the mass approaches the critical one.

Fig. 4.3 shows a re-creation of the August 21, 1945 criticality accident at Los Alamos that occurred during the experiment aimed at the determination of critical mass of Plutonium. The mass became supercritical for a short period releasing deadly radiation. The byproducts of fission are extremely radioactive, much greater than plutonium itself.

Power reactors require constant protective measures to keep the number of neutrons in the mass constant and to make the release of nuclear energy linear. The mass must be somewhat below critical at any time, but sufficiently close to it. This process is self-regulating (mostly) in nuclear reactors with automated systems of several types and levels. Generally, this is involved because the chain reaction is exponential: the explosion or shutdown can quickly happen due to mismanagement or external factors.

The shutdown is of course much less dangerous than the explosion. Only

Figure 4.3: Plutonium critical mass.

about $1kg$ of the $64kg$ of uranium in "Little Boy" (the first nuclear weapon to be used during wartime) converted into energy, and yet the explosion came with devastating consequences, including radiation poisoning (from the same exploded $1kg$). Restarting a reactor after its failure or malfunction is very complicated, if possible at all.

Let us model *one-dimensional fission*. This is of course just a model; the real ones are three-dimensional, and fission is a very much stochastic process. Geometrically, the mass is then an interval of length L. We assume that 2 neutrons are generated per one fission instance and that a neutron at x "generates" (upon its absorption by the atom) a pair of neutrons at $x+R, x-R$. They can stay in the interval $[-L/2, L/2]$ or leave it. Let us comment on Fig. 4.4, demonstrating that the "critical mass" is $2R$.

For real fission, the angle correlations of the neutrons are complicated. For instance, the real probability that this angle is not far from 0 is high, which means that they move essentially in the same direction. Much depends on the energies and momenta of the fission products. However, our simplistic 1-dimensional model (when the newly-created neutrons move in opposite directions) captures some features of real fission, including the phenomenon of critical mass. See Problem 4.4.1.

By u_n, we denote the number of neutrons after n scattering events (fissions). We begin with one (red) neutron in this figure and show its "offspring" (the top diagrams) and then their "offspring" (the bottom diagrams). Its left half describes the "supercritical mass", which is the interval $[-L/2, L/2]$ under the inequalities $4R > L > 2R$ for L. Indeed, the initial neutron n from the interval $[-L/2+R, L/2-R]$ in this case becomes $2n, 2n, 4n, 4n, 8n$, etc., so the growth is exponential.

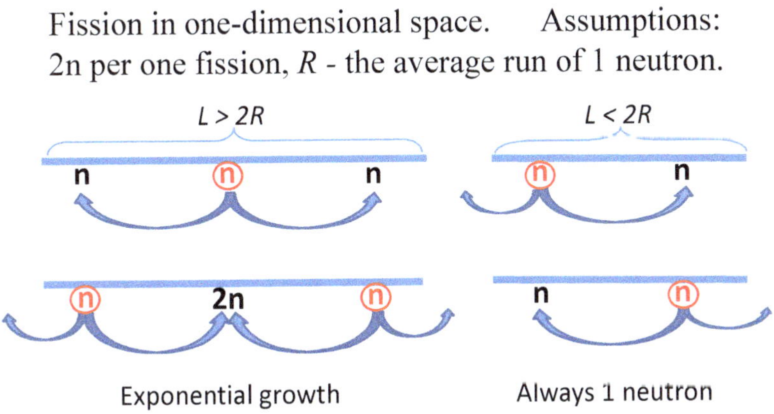

Figure 4.4: Supercritical and subcritical mass in dimension one.

Problem 4.4.1 *Let $L/2 = R+\epsilon$ for small $\epsilon > 0$. Check that $u_n = u_{n-1} + 2u_{n-2} - 2u_{n-3}$, and $u_n = C_1(\sqrt{2})^n + C_2(-\sqrt{2})^n + C_3$ for some constants C_1, C_2, C_3 for the number of neutrons u_n after n scattering events.*

Solution. Let x_n be the number of neutrons inside the open interval $(-\epsilon, \epsilon)$, y_n the number of those outside $(-R+\epsilon, R-\epsilon)$, and z_n the number of the remaining neutrons, those in $(-R+\epsilon, R-\epsilon) \setminus (-\epsilon, \epsilon)$. Then $y_n = 2x_{n-1}$, $x_n = y_{n-1}$ and $z_n = z_{n-1}$. Thus, $v_n = x_n + y_n$ satisfies the recurrence $v_n = 2v_{n-2}$ with the roots $\lambda_{1,2} = \pm\sqrt{2}$. Upon adding z_n, the recurrence for $u_n = v_n + z_n$ becomes $u_n = u_{n-1} + 2u_{n-2} - 2u_{n-3}$. One can check this by a direct calculation, but it is simpler to proceed as follows. Obviously, any v_n solves the recurrence for u_n (when $z_n = 0$), and any z_n is its solution (when $v_n = 0$). Thus the corresponding λ-roots are $\pm\sqrt{2}, 1$, and the master polynomial is $(\lambda^2-2)(\lambda-1) = \lambda^3 - \lambda^2 - 2\lambda + 2$. This gives the required recurrence and the general formula for u_n. ❑

The right half of this figure is for $R < L < 2R$. This is the case of "subcritical mass": the number of neutrons inside $[-L/2, L/2]$ remains 1 if the initial one is near the endpoints, and it becomes 0 inside this interval if the initial (red) one in the diagram at the top is near the center.

Let us modify this model to make the growth of u_n linear, which is a sort of one-dimensional fission reactor. See also Exercise 4.6.30.

Let $L = 2R$. We will count all neutrons in the open interval $(-2R, 2R)$. If the neutron is within this interval but outside of $(-R, R)$, then it will not be involved in further scattering. The whole point of nuclear reactors is to ensure the linear growth of the ***total*** number of produced neutrons and to keep the number of those ***inside*** the critical mass constant. Making the mass of fissile material reasonably close to the critical mass significantly increases the production of energy, which is the condition $L \approx 2R$ in our case.

We mention that the energy released by a reactor is mostly due to the kinetic energy of the fission fragments (about 85%); the total number of produced neutrons is proportional to it.

Let y_n be the number of neutrons outside $(-R, R)$, and x_n the number of those inside $(-R, R)$. Disregarding the neutron at the origin, we obtain that $x_n = x_{n-1}, y_n = x_{n-1} + y_{n-1}$. Thus, $x_{n-1} = y_n - y_{n-1} = y_{n-1} - y_{n-2}$, $y_n = 2y_{n-1} - y_{n-2}$, and $u_n = 2u_{n-1} - u_{n-2}$ for $u_n = x_n + y_n$.

Qualitatively, changing the geometry of the mass (its radius) from supercritical to subcritical is a passage from the recurrence $u_n = 2u_{n-1} - u_{n-3}$ to $u_n = 2u_{n-1} - u_{n-1} = u_{n-1}$. The first has solutions $u_n = C_1\lambda_1^n + C_2\lambda_2^n + C_3$ for $\lambda_{1,2} = \frac{1\pm\sqrt{5}}{2}$ (those for the Fibonacci recurrence; check this) and is of exponential growth if $C_1 \neq 0$; the second recurrence is constant. The recurrence above is a natural transition: the case of critical mass. Similarly, we can assume that $u_n = \nu u_{n-1} - u_{n-2}$, where ν changes from $2 - \epsilon$ (subcritical) to $2 + \epsilon$ (supercritical), which vaguely corresponds to moving ***cadmium rods*** inserted in nuclear reactors to absorb the extra neutrons, a common way to control their stability.

We arrive at a simple instance of ***resonance recurrence relations***, which are defined as those with repeated roots of the master equation:

$$u_n = 2u_{n-1} - u_{n-2} \Rightarrow \lambda^2 = 2\lambda - 1 \Rightarrow \lambda_1 = 1 \text{ is of multiplicity 2.} \quad (4.4.19)$$

The fundamental solutions are $u_n = n$ and $u_n = 1$ in this case, which is easy to see: $n = 2(n-1) - (n-2), 1 = 2 - 1$. They are linearly independent, and an arbitrary solution of this recurrence is $u_n = C_1 n + C_2$ for some constants $C_{1,2}$. So the growth is generally linear, which is the key for us. Informally, (4.4.19) describes neutrons with a "birth rate" of 2 and with minimal possible "life expectancy"; that is, staying within the critical mass.

Generalizing, let us allow multiple roots λ_i in Theorem 4.2.2.

Theorem 4.4.2 *If the master equation from Theorem 4.2.2 has a root λ_1 of multiplicity m_1, λ_2 of multiplicity m_2, and so on, then an arbitrary solution* *of the recurrence relation there has the form*

$$\begin{aligned} f_n^C = & C_1^{(0)}\lambda_1^n + C_1^{(1)} n\lambda_1^n + \cdots + C_1^{(m_1-1)} n^{m_1-1}\lambda_1^n \\ & + C_2^{(0)}\lambda_2^n + \cdots + C_2^{(m_2-1)} n^{m_2-1}\lambda_2^n + \ldots \end{aligned} \tag{4.4.20}$$

for some constants $C_i^{(j)}$. If $\widetilde{f}_n^{\bullet}$ is a solution of the corresponding inhomogeneous equations (with some g_n in the right-hand side), then its arbitrary solutions are in the form $\widetilde{f}_n^{\bullet} + f_n^C$ for some C.

Proof. Let us consider the most frequent particular case of this theorem when λ_1 is of multiplicity 2 and all other roots $\lambda_i (i = 2, \ldots)$ are simple. We will use Theorem 4.2.2, which covers the case of pairwise distinct roots of the master equation. We apply it to the sequence $\widetilde{\lambda} = \{\lambda_1, \widetilde{\lambda}_1 = (1+\varepsilon)\lambda_1, \lambda_2, \ldots\}$, where $\varepsilon > 0$ is sufficiently small. The coefficients of the corresponding recurrence will then polynomially (actually linearly) depend on ε. Indeed, they are the coefficients of the master equation for $\widetilde{\lambda}$, which is $(\lambda - \lambda_1)(\lambda - \widetilde{\lambda}_1)(\lambda - \lambda_2)\cdots = 0$. It has the following solution:

$$\widetilde{f}_n = \frac{(\widetilde{\lambda}_1)^n - (\lambda_1)^n}{\varepsilon} = \frac{(1 + n\varepsilon + O(\varepsilon^2))\lambda_1^n - \lambda_1^n}{\varepsilon} = n\lambda_1^n + O(\varepsilon),$$

where by $O(\varepsilon^k)$, we denote some polynomial (generally, a series) in terms of ε^ℓ for $\ell \geq k$, where the coefficients of ε^ℓ polynomially depend on λ_i. We use here Newton's binomial.

We see that the limit of $\widetilde{f}_n$ exists when $\varepsilon \to 0$ and it *automatically* satisfies the initial recurrence relation, which is when we set $\varepsilon = 0$. This limit is explicit: it equals $n\lambda_1^n$, so the latter sequence solves the recurrence in the case of the double root λ_1. This is the claim of the theorem for double λ_1.

If the multiplicity of λ_1 is 3, we set $\widetilde{\lambda} = \{\lambda_1, \lambda_1, \widetilde{\lambda}_1 = (1+\varepsilon)\lambda_1, \lambda_2, \ldots\}$. The solution with (sufficiently small) $\varepsilon > 0$ to be considered will now be $\widetilde{f}_n = \frac{n(\widetilde{\lambda}_1)^n - n(\lambda_1)^n}{\varepsilon} = n^2\lambda_1^n + O(\varepsilon)$. We obtain that $n^2\lambda_1^n$ is a solution upon $\varepsilon = 0$, as it is claimed in the theorem for the triple λ_1. Then we continue in the same way. Each time we use the previous step. This inductive procedure is applied as many times as necessary, namely $(m_1 - 1) + (m_2 - 1) + \ldots$ times.

We obtain the system of *fundamental solutions* of the recurrence for any root multiplicities $m_1, m_2, \ldots$. Their number is $m_1 + m_2 + \ldots$, which is the depth of the recurrence. As in Theorem 4.2.2, the last step is to justify that they are linearly independent. This requires some linear algebra.

In the case of λ_1 of multiplicity 2, let us check that the solutions $f_n = \lambda_1^n$ and $f'_n = n\lambda_1^n$ are linearly independent, i.e. if $af_n + bf'_n = 0$ for constant a, b

and for *any* n, then $a = 0 = b$. This is immediate since the matrix of the values of f_n, f'_n at $n = 0, 1$ is invertible. This matrix is $\begin{pmatrix} f_0 & f'_0 \\ f_1 & f'_1 \end{pmatrix} = \begin{pmatrix} 1 & 0 \\ \lambda_1 & \lambda_1 \end{pmatrix}$.

For an arbitrary multiplicity m_1, consider $f_n^{(k)} = n(n-1)\cdots(n-k+1)\lambda_1^n$ for $k, n = 0, 1, \ldots, m_1 - 1$; here $f_n^{(0)} = \lambda_1^n$. The linear independence of these solutions is equivalent to that for $n^k\lambda_1^n$, where $0 \le k \le m_1 - 1$. Indeed, the latter can be linearly expressed in terms of $f_n^{(k)}$, and vice versa.

The solutions $f_n^{(k)}$ are convenient because the corner above the diagonal of the matrix $(f_n^{(k)})$ for $k, n = 0, 1, \ldots, m_1 - 1$ will be all zeros; this corner is when $k > n$. The determinant of this matrix is the product of the diagonal entries. It is $\lambda_1^{m_1-1}$ with some nonzero coefficient of proportionality.

Generally, for any multiplicities, use the *Vandermonde determinant* provided after formula (4.2.10) for the sequence

$$\{\lambda_1, \lambda_1 + \varepsilon, \ldots, \lambda_1 + (m_1 - 1)\varepsilon, \lambda_2, \lambda_2 + \varepsilon, \ldots, \lambda_2 + (m_2 - 1)\varepsilon, \ldots\}.$$

Then assume that ε approaches 0, check that the smallest degree of ε in the resulting formula is $\mu = (m_1 - 1) + (m_2 - 1) + \ldots$, and find that the corresponding coefficient of ε^μ is nonzero. This proves that arbitrary solutions f_n are those described by (4.4.20). We will omit the details. ❑

The multiple λ-roots for homogeneous recurrence relations with constant coefficients generally occur because of external factors. This is called *resonance phenomenon*. There are mechanical resonances, acoustic resonances, electromagnetic resonances and so on. They generally result in linear or polynomial growth, which can be very useful or very dangerous. For fission reactors, the resonance phenomenon is a thin line between the subcritical mass and supercritical mass; the released energy must grow linearly.

Real fission. The real fission for uranium and plutonium is 3-dimensional. See e.g. [29] for an elementary calculation of the critical radius. The authors use that the average distance from the origin for d-dimensional random walk is proportional to $\lambda\sqrt{n}$ for some λ, where n is the number of the steps taken. Here d is arbitrary; we need this for $d = 3$. One random walk is a "history" of a particular neutron: the corresponding sequence of scatterings. The constant λ, *the mean free path between scattering events*, is about 4.57cm for highly enriched uranium and 5.79cm for ^{239}Pu. Here the following must be addressed. Quite a few neutrons generated near the border of the mass will leave the mass without scattering; the closer this event is to the border the greater the number of "escaped" neutrons.

---

[29] C. Chyba, C. Milne, *Simple calculation of the critical mass for highly enriched uranium and plutonium-239*, American Journal of Physics 82:10 (2014), 977–979.

4.4.2 Stopping times

The analysis of the fission is closely related to the random walks in domains, finite or infinite. The random walk will be our standard one: with the probability p for $x \mapsto x+1$ and q for $x \mapsto x-1$. If x is the amount of money one has in the process of *betting* when p is the probability to win \$1 then $q = 1-p$ is the probability to lose \$1. When playing roulette (American wheel): $p = \frac{9}{19}, q = \frac{10}{19}$.

Problem 4.4.3 *Assume that you begin with $x = k$ for an* integer *k belonging to the segment $[0, n]$, and that you lose if $x = 0$ and stop playing when $x = n$. What is the probability P_k to lose under p, q as above?*

Solution. One has $P_k = pP_{k+1} + qP_{k-1}$ for $1 \le k < n$, where $P_0 = 1$ and $P_n = 0$. Indeed, the 1st step from k can be to $k+1$ or $k-1$ with the probabilities p or q. For $p = 0$: $P_k = 1$ for $k < n$ and $P_n = 0$. If $p > 0$, then $P_k = \frac{1}{p}P_{k-1} - \frac{q}{p}P_{k-2}$. The master equation is $\lambda^2 - \frac{1}{p}\lambda + \frac{q}{p} = 0$, and

$$\lambda_{1,2} = \frac{1}{2p} \pm \sqrt{\frac{1}{4p^2} - \frac{q}{p}} = \frac{1}{2p} \pm \sqrt{\frac{(p+q)^2 - 4qp}{4p^2}} = \frac{1 \pm |q-p|}{2p} = \{\frac{q}{p}, 1\},$$

where we use that $p+q = 1$. Thus, the roots are $\lambda_1 = q/p$, $\lambda_2 = 1$; technically, $\lambda_1 = q/p$ if $q > p$ and 1 otherwise. Let us assume that $q \neq p$.

We obtain that $P_k = C_1\,(q/p)^k + C_2$. Then, $C_2 = 1 - C_1$ and $C_1\,(q/p)^n + 1 - C_1 = 0$. The latter relation gives that $C_1 = (1-(q/p)^n)^{-1}$. Finally:

$$P_k = \frac{(q/p)^k - (q/p)^n}{1-(q/p)^n} \quad \text{for } \ p \neq 0,\, q \neq p. \tag{4.4.21}$$

When $q/p \to 1$, *L'Hôpital's rule* gives that $P_k = \frac{n-k}{n}$; differentiate the numerator and denominator with respect to q/p considered as a variable. This formula can be readily obtained directly. The corresponding recurrence is that in (4.4.19), a simplest resonance one. ❑

Note that P_k becomes P_{n-k} if we transpose p and q. We transpose the roles of 0 and n: the reflection $k \leftrightarrow n-k$ and the corresponding switch $p \leftrightarrow q$. See also Problem 6.3.4 and Exercise 6.4.38. In the latter, we obtain via the Catalan numbers that $P_k \to (q/p)^k$ if $q < p$ and $n \to \infty$. This can be readily seen from (4.4.21).

This problem is clearly related to the fission (in dimension one), but now 1 neutron generates only 1, not 2 as in Fig. 4.4. The following problem is generally of obvious importance, theoretically and practically (including fission): for how long can a random walk stay in a bounded domain?

Problem 4.4.4 *Let D_k be the average number of steps needed to go from k to n for $0 \le k < n$ and under the conditions as above. Check that $D_k = pD_{k+1} + qD_{k-1} + 1$ for $1 \le k < n$ and find the formula for D_n.*

Solution. For $p = 0$: $D_k = k$ for $k < n$ and $D_n = 0$. Let $p \neq 0$. One has: $D_k = p(D_{k+1} + 1) + q(D_{k-1} + 1) = pD_{k+1} + qD_{k-1} + 1$. This is an inhomogeneous extension of the recurrence for P_n. According to Theorem 4.2.2(ii), if we can find one its solution $D_n^\bullet$ (any), then its arbitrary solution can be uniquely represented in the form $D_n = D_n^\bullet + P_n^C$, where P_n^C solves the recurrence for P_n with some $C_{1,2}$.

We can take here $D_n^\bullet = k/(q-p)$. Indeed: $\big(p(k+1)+q(k-1)\big)/(q-p)+1 = k/(q-p)$. We use that $p+q=1$. Thus, $D_k = k/(q-p) + C_1(q/p)^k + C_2$ for some $C_{1,2}$. The boundary conditions are $D_0 = 0, D_n = 0$. Accordingly:

$$D_k = \frac{k\big(1-(q/p)^n\big) - n\big(1-(q/p)^k\big)}{(q-p)\big(1-(q/p)^n\big)} \quad \text{for } p \neq 0,\, q \neq p. \tag{4.4.22}$$

Indeed, it is in the form $k/(q-p)+C_1(q/p)^k+C_2$ and satisfies $D_0 = D_n = 0$, which is sufficient for the justification. Its limit as $q/p \to 1$ can be readily calculated: differentiate the numerator and denominator twice in terms of q/p, considered as a variable. We obtain that $D_k = k(n-k)$ for $p = \frac{1}{2} = q$. Directly: solve $2D_k = D_{k+1} + D_{k-1} + 2$ following Exercise 4.6.31. ❑

As above, $0, n$ are on equal footing: we stop if any of them is reached. Thus, D_k remains unchanged if we switch $q \leftrightarrow p$ and $k \leftrightarrow n-k$. Multiply the numerator and denominator in (4.4.22) by p^n/q^n to see this algebraically.

Reflection at 0. An interesting twist is when we automatically increment the step to $x = 1$ after $x = 0$ is reached, i.e. add a reflection here. For the interpretation via betting, a player borrows \$1 at $x = 0$ and then continues playing. This does not change the probability P_k to reach 0 from k, but influences D_k: the boundary conditions will now be $D_0 = D_1+1$, $D_n = 0$. The corresponding formula is obtained in Exercise 4.6.23.

4.4.3 Human population

Let us try to model the growth of human population. We assume that the maturity period is about 20 years, which will be the time-unit, and that the reproduction period is until age 40. This corresponds to $k = 2$ in Problem 4.1.2. Let the birth rate be 1 for 20 years. So one pair gives birth to another pair during the period from 20+ to 40. This is of course a simplification, but a reasonable one. Life expectancy can be arbitrary beyond 40, say 60; it does not influence the recurrence but impacts the C-constants. The duration of the reproduction period is the key: see Problem 4.1.2.

Strictly speaking, we have *simple reproduction* for $k = 2$ and $b = 1$, i.e. the population is bounded. Let us demonstrate this assuming that the life expectancy is 60 when we begin with one young pair. Then (y_n, m_n, o_n), the numbers of young pairs (until 20), mature ones (from 20+ to 40), and the old ones (from 40+ to 60), will evolve as follows: $(1,0,0) \mapsto (0,1,0) \mapsto (1,0,1) \mapsto (0,1,0) \mapsto (1,0,1) \mapsto (0,1,0) \mapsto \ldots$. We see that the total number of pairs oscillates between 1 and 2 after some starting period.

The corresponding recurrence for the total number f_n of pairs at time n is (4.1.2) for $b = 1, k = 2$. We can count only those of age from 0 to 40 here: the recurrence will remain the same. See Section 4.1.2. The extension from (4.1.3), obtained there by adding constant solutions (for any b), is more relevant to address real population processes. This recurrence and the corresponding master polynomial are:

$$f_n = f_{n-1} + f_{n-2} - f_{n-3}, \quad \lambda^3 - \lambda^2 - \lambda + 1 = (\lambda - 1)^2(\lambda + 1). \tag{4.4.23}$$

Its arbitrary solution is in the form $f_n = C_1^{(0)} + C_1^{(1)} n + C_2(-1)^n$, which is of linear growth unless $C_1^{(1)} = 0$. If the maturity is disregarded here, then $f_{n-1} + f_{n-2} - f_{n-3}$ becomes $2f_{n-1} - f_{n-2}$, which is exactly (4.4.19).

One can adjust (4.4.23) to the actual maturity, reproduction period, life expectancy, and the birth rate. The depth of the recurrence will change accordingly. We note that a straightforward combinatorial model for the population growth results in (4.1.2), where all solutions are bounded for $b = 1, k = 2$. Its extension to (4.1.3) is somewhat similar to the passage from subcritical to critical mass we did for modeling nuclear reactors. Obtaining the resonance phenomenon generally requires such extensions.

A generalization of (4.4.23) is the recurrence $f_n = \alpha f_{n-1} + f_{n-2} - f_{n-3}$. It addresses the fact that no country is fully isolated: $\alpha > 1$ models the immigration to this country, $\alpha < 1$ describes the emigration. The classical example of $\alpha > 1$ is a Gold Rush (North Carolina from 1799, California ~ 1849, Alaska by 1896). Simplifying, people send letters to their friends, "there is gold here!" Not all recipients will come, but the number of newcomers will be basically proportional to the number of the letters, i.e. to the current number of gold seekers. Even small α here, say $\alpha = 1.1$, can have a huge impact long term: solve the corresponding master equation (numerically).

Another extension of (4.4.23) is the following *inhomogeneous* recurrence relation: $f_n = f_{n-1} + f_{n-2} - f_{n-3} + u$. Here u is the number of immigrants (counted in pairs) accepted by the country per one unit of time. It is assumed to be constant in contrast to the usage of α above due to administrative or other reasons. We justify that the growth of the population will then become *quadratic* in Problem 4.4.5.

Problem 4.4.5 *The population f_n in some country is expected to be described by the recurrence $f_n = f_{n-1} + f_{n-2} - f_{n-3}$. This country decides to allow immigration restricting it by 1 per unit of time; for instance, 1 can be 20M for the time-unit taken to be 20 years.*

Prove that any solution of the corresponding recurrence is in the form $f_n = \frac{n^2}{4} + C_1 n + C_2 + C_3(-1)^n$, so the potential linear growth of the population will change to quadratic. Solve it for $f_1 = f_2 = f_3 = 0$, i.e. when there is no population before the immigration begins.

Solution. The new recurrence is $f_n = f_{n-1} + f_{n-2} - f_{n-3} + 1$. First of all, $n^2 + (n-3)^2 = (n-1)^2 + (n-2)^2 + 4$, and $n^2/4$ is really a solution. Then, $f_n - n^2/4$ satisfies the initial *homogeneous* recurrence, so $f_n - n^2/4 = C_1 n + C_2 + C_3(-1)^n$ as we know. We have 3 equations for $C_{1,2,3}$:

$$\frac{1}{4} + C_1 + C_2 - C_3 = 0,\ 1 + 2C_1 + C_2 + C_3 = 0,\ \frac{9}{4} + 3C_1 + C_2 - C_3 = 0.$$

Thus, $\frac{5}{4} + 3C_1 + 2C_2 = 0$, $\frac{13}{4} + 5C_1 + 2C_2 = 0$. We obtain:

$$C_1 = -1,\ C_2 = \frac{7}{8},\ C_3 = \frac{1}{4} + C_1 + C_2 = \frac{1}{8}.$$

Finally, $f_n = n^2/4 - n + 7/8 + (-1)^n/8$. A direct justification of the fact that f_n is an integer is as follows: $n^2/4$ is an integer for even n or when $n^2 + 3 = 0 \bmod 4$ for odd n. ❑

As with any model, ours must be adjusted to the real processes. They are basic but seem flexible enough to address many features and regimes of population growth, including humans. Mathematically, the aim is to incorporate the linear and (later) power-type growth using linear recurrences. An alternative approach is by making their coefficients depend on time, to be discussed next.

4.4.4 Tree growth

To model the height of trees during their active growth period, we will need several simplifications. First, we assume that the radius r of a tree crown approximately coincides with the radius of its root system, which is typical for trees (with exceptions). Second, let the radius grow linearly in time. This reflects that the tree rings are basically evenly spaced. Third, we assume that "maturity" is present for trees: one-year cycles for many of them. So n will be the number of cycles (years).

Due to the first assumption, the root system, essentially a planar domain of area A proportional to r^2, provides nutrients for the whole tree, which is of volume V proportional to r^3.

Thus, the nutrients per one 1 cubic unit of the tree can be assumed proportional to A/V which is basically $1/r$, Finally, it can be taken proportional to $1/(time)$. The nutrients delivered to the top of the tree results in the corresponding increase of its height, so $1/(time)$ can be naturally considered the "birth rate" in this approach.

Assuming that a tree is exactly a ball and the root system is exactly a circle: $V = \frac{4}{3}\pi r^3$ and $A = \pi r^2$. So $A/V = 3/(4r)$, which is a bit smaller than $1/r$. However, the trees are not "spheres", and this is just a model anyway.

Under these simplifications and conditions, the recurrence relation for the *tree height* h_n at the end of the nth cycle (year n) becomes

$$h_n = h_{n-1} + \frac{1}{n-2} h_{n-2} \quad \text{for } n = 3, 4, \ldots . \tag{4.4.24}$$

Let us solve it "algebraically." One solution is obvious: $h_n^{(1)} = n$; indeed, $n = (n-1) + 1$. The second is $h_n^{(2)} = \delta_n = D_n/(n-1)!$, where D_n is the number of *derangements*; see relation (3.6.14). One has: $h_1^{(1)} = 1, h_2^{(1)} = 2$, and $h_1^{(2)} = 0, h_2^{(2)} = 1$. A natural challenge is to find the rate of tree growth in terms of the initial conditions $h_1 = a, h_2 = b$.

Generally, $h_n = C_1 h_n^{(1)} + C_2 h_n^{(2)}$ for this model. One has: $h_1 = a = C_1 h_1^{(1)} + C_2 h_1^{(2)} = C_1$, and $h_2 = b = C_1 h_2^{(1)} + C_2 h_2^{(2)} = 2C_1 + C_2$. Thus $C_1 = a$, and $C_2 = b - 2a$. Finally, $h_n = an + (b-2a)\delta_n$; it is of linear growth and the rate of growth $h_n - h_{n-1}$ (quickly) approaches $a + (b-2a)/e$ for the Euler number e. We use that $\delta_n \to n/e$ as $n \to \infty$. Check these calculations.

Thus we have an alternative "mechanism" for processes of linear growth. It can be due to resonance conditions (with some life expectancy) as above, or by "making" the birth rate inversely proportional to time as for trees.

Problem 4.4.6 *Assume that a tree is in its late period when the active area of the root system is proportional to the radius r, not to r^2. Prove that any solutions of the corresponding recurrence $h_n = h_{n-1} + \frac{1}{(n-2)^2} h_{n-2}$ reach the saturation as $n \to \infty$, i.e. $\lim_{n\to\infty} h_n = const$. When $h_1 = a, h_2 = b$, find* numerically *the constants A, B such that $\lim_{n\to\infty} h_n = Aa + Bb$.* ❑

Exercise 4.6.42 provides exact formulas for A, B after the following modification of the recurrence relation from Problem 4.4.6:

$$f_n = f_{n-1} + \frac{(n-1)}{(n-2)n(n+1)} f_{n-2}.$$

The simplest variant is by omitting the maturity:

Problem 4.4.7 * *Solve explicitly the recurrence* $f_n = (1+\frac{1}{(n-1)^2})f_{n-1}$, *which models the plant growth at late stages without the maturity delay. Deduce from the product expansion for* $\sinh(\pi x)$ *that* $\lim_{n\to\infty} f_n = f_1\frac{\sinh(\pi)}{\pi}$, *where* $sinh(x)=\frac{e^x-e^{-x}}{2}$.

Solution. The Weierstrass factorization for sinh is $\sinh(\pi x) = \pi x\prod_{n=1}^{\infty}(1+\frac{x^2}{n^2})$. Making here $x=1$, we obtain that $f_n = f_1\prod_{m=1}^{n-1}(1+\frac{1}{m^2})$ for $n\geq 1$, which gives the required limit. ❑

Let us now add life expectancy; the lifespan is finite for any tree. We will set it to be ≈ 10 years in the following problem.

Problem 4.4.8 *Assume that the after* 10 *years the tree growth satisfies the recurrence* $f_n = f_{n-1} + \frac{f_{n-2}}{(n-2)} - f_{n-10}$, *and* $f_n = n$ *for* $0\leq n\leq 9$. *Check numerically that this tree will live about* 15 *years:* $f_{14}>0, f_{15}<0$. ❑

4.5 Epidemic spread *

We begin with some basics. *Let* u_n *be the total number of infections at the* nth *moment from the beginning of some epidemic.* Infected individuals transmit the disease mostly during some initial period, which we make the unit of time, for example 1-2 weeks for Covid-19; so n will be in terms of such a time-unit.

The recurrence relation and the master equation are:

$$u_n = u_{n-1} + R_0\,(u_{n-1}-u_{n-2}),\ \lambda^2 = (1+R_0)\lambda - R_0.$$

Here R_0 is the average number of virus transmissions by one infected individual, the so-called ***initial reproduction number***. This happens during the initial period, so we multiply it by $u_{n-1}-u_{n-2}$, the number of newly-infected individuals at $(n-1)$.

We obtain: $\lambda_{1,2} = \frac{(1+R_0)}{2} \pm \sqrt{(1+R_0)^2/4 - R_0} = \{1, R_0\}$. Thus, $u_n = C_1 + C_2(R_0)^n$ for some $C_{1,2}$. If $C_2\neq 0$ and $R_0>1$, the growth of the total number of infections will be exponential.

This is unless the *herd immunity* is expected to be reached, when the so-called *SIR-model* is needed (in a discrete variant). Here *SIR* stands for "Susceptible-Infectious-Recovered." The herd immunity and the related saturation of the number of infections are when the total number of infections becomes comparable with the susceptible population. We omit the equations for the *SIR*-model; our approach will be different.

4.5.1 Power-type spread

The exponential growth happens only during short periods. The number of infections mostly grows as a power function in terms of time during the middle stages of epidemics. This was such for the waves of Covid-19. For instance, in Fig. 4.6 (below) the growth of the total number of detected infections was $\approx t^{3.65}$ for almost 5 months with high accuracy. So we need to significantly change the model above to address the power growth. Basically, there are 3 main mathematical possibilities to ensure that the spread is non-exponential:

(1) when R approaches 1, which results in "resonances" and can potentially provide some linear growth,

(2) the presence of *predators*, generally speaking, any forces restricting the growth (the spread for epidemics),

(3) when the "birth rate", the transmission rate in this context, becomes inversely proportional to time.

The resonances and the corresponding linear growth of the spread when $R \approx 1$ is a possibility, but this regime is unstable mathematically and not really compatible with the definition of R (the reproduction number). The periods of linear growth near the turning point are quite stable for epidemics, which makes (1) not too relevant.

Obviously, (2) is applicable: we do fight epidemics! However, this alone is insufficient: it does not address the power-type growth of the spread. A combination of (2) and (3), which provides the power-type growth, seems the key for modeling epidemics. Let us comment on it.

The spread of any disease is the growth of the "circles" of those infected; these are *combinatorial circles*. They are not geometric ones, in a map of the affected area. The contacts are mostly with your neighbors, but of course not only with them: people work, study, go shopping, travel.

As with basic waves in physics, our main assumption is that the rate of growth of the radii of these combinatorial circles is expected to be constant. The distance here is essentially the length of the chain of contacts between you and someone: a direct contact, a contact of a contact, and so on. For instance, your co-worker or your neighbor are of distance 1 from you. A family member of your co-worker or a co-worker of your neighbor are of distance 2, and so on. This is actually how the origins of infections are traced practically: those in charge try to obtain the tree of prior contacts of infected individuals and then follow this tree to identify further infections.

The individuals at the frontier of such a circle transmit the disease the most because of the following major factors:

(a) they are the "latest" and therefore in their most infectious stages,

(b) people inside the circle are "surrounded" by those with immunity.

We disregard here *reinfections* and assume that the contacts are mostly limited to your *combinatorial neighbors* in this circle, not only to your "geographic neighbors." So infected people inside the circle contribute to the transmission of the disease significantly smaller than those at its boundary.

The circle of infected people can be assumed a ball of dimension c; here we embed the graph of contacts in a space of sufficient dimension. Accordingly, u_n will be proportional to n^c. Here c can be actually any positive number, not only an integer. For instance, $c \approx 2$ if our contacts are mostly with those who live near us; however $c \approx 3.65$ in Fig. 4.6. It was always the case that $c > 2$ for the waves of Covid-19. We call c the ***transmission rate***; it reflects the virus strength and the basic average number of contacts in the area. In contrast to the *reproduction number* R, it is assumed unchanged for a given wave until it approaches the saturation. Knowing it and the intensity of protective measures (below) is essentially sufficient for modeling.

Disregarding the latent period when people are already infected but not infectious, the number of newly-infected people is basically the area of the *boundary* of this ball, i.e. the surface area of the corresponding c-dimensional sphere. Presenting this number as $c\,u_{n-1}/(n-1)$, we arrive at the following recurrence: $u_n - u_{n-1} = c\,\frac{u_{n-1}}{(n-1)}$.

Here the passage to difference or differential equations is standard for adding external "forces" (our next step). Let us justify that asymptotically (for sufficiently large n), n^c is really a solution to the recurrence above.

We have $u_n = (1+\frac{c}{n-1})(1+\frac{c}{n-2})\cdots(1+c)u_1$. Therefore, $\log(u_n) \approx \log(u_1) + c\sum_{m=1}^{n-1}\frac{1}{m} \approx \log(u_1) + c\log(n)$ (we use basic calculus). Thus $u_n \approx n^c u_1$. More exactly, $u_n \approx \frac{n^c}{\Gamma(c+1)}u_1$ for the Gamma function Γ. When $c = r$ for a positive integer r: $u_n = u_1\frac{1+r}{1}\frac{2+r}{2}\cdots\frac{1+2r}{1+r}\cdots\frac{n-1}{n-r+1}\cdots\frac{n+r-1}{n-1} = \frac{n(n+1)\cdots(n+r-1)}{r!}u_1 = \binom{n+r-1}{r}u_1$ (check). So u_n for $u_1 = 1$ extends $R(n,r) = \binom{n+r-1}{r}$ to arbitrary real r.

4.5.2 Adding protective measures

Any power growth is unsustainable long-term. If waiting for *herd immunity* is not an option, then *protective measures* are the main way to diminish the epidemic spread; of course, these also include self-imposed measures. The main methods are and always were the detection of infected people, quarantine periods, closing the places where the spread of infections is the most likely, and so on. These are hard measures; vaccination is one of the most radical. Without such measures, though they are almost always present in some forms, any particular wave will inevitably slow down when the circles of infections reach some limits, but this can take a lot of time.

For the most aggressive epidemic management, the recurrence is as follows

$$u_n - u_{n-1} = c\,\frac{u_{n-1}}{n-1} - p_{n-1},\;\; p_n - p_{n-1} = a\,u_{n-1}, \tag{4.5.25}$$

where a is the intensity of "hard measures", p_n is essentially the number of people protected from the virus due to the measures until time n.

Basically, we subtract the number of "preventions" from the total number of infections. In the second relation, the new protection measures are assumed proportional to the current number of cases, which is a very aggressive approach. The main point here is that the "isolation" of one infected individual prevents the number of virus transmissions roughly proportional to the time passed from this isolation. See [30].

It is not difficult to eliminate p_n from these relations. From the first one: $p_{n-1} = (1+\frac{c}{n-1})u_{n-1} - u_n$ and $p_n = (1+\frac{c}{n})u_n - u_{n+1}$. Utilizing the second relation:

$$\begin{aligned} p_n - p_{n-1} &= \Big(1+\frac{c}{n}\Big)u_n - u_{n+1} - \Big(1+\frac{c}{n-1}\Big)u_{n-1} + u_n \\ &= \Big(2+\frac{c}{n}\Big)u_n - \Big(1+\frac{c}{n-1}\Big)u_{n-1} - u_{n+1} = au_{n-1}. \end{aligned}$$

Finally, we obtain the recurrence relation:

$$u_{n+1} = \Big(2+\frac{c}{n}\Big)u_n - \Big(1+a+\frac{c}{n-1}\Big)u_{n-1}. \tag{4.5.26}$$

When $a = 0$, we have the solution $u_n = n$ for any c. This one is for the initial conditions $u_1 = 1, u_2 = 2$. For $a = 0$ and generic u_1, u_2, let us check that u_n becomes asymptotically proportional to n^c as $n \to \infty$.

Problem 4.5.1 *For $a = 0$ and $c > 1$ (the case of constant protective measures), check that the limit $\lim_{n\to\infty} u_n/n^c$ exists for any solution u_n. It is 0 only if $u_1 = p_1/(c-1)$; then $u_n = u_1 n$.*

Solution. Due to $a = 0$, p_n is some constant $d \geq 0$ and we obtain that $u_{n+1} = (1+\frac{c}{n})u_n - d$. By rescaling u_n (multiplying it by some proportionality coefficient), d can be assumed 0 or $c-1$. Let u_n^C and $\widetilde{u}_n$ be solutions of the recurrences for $d = 0$ and $d = c-1$, respectively, where u_n^C is normalized by the condition $\lim_{n\to\infty} u_n^C/n^c = C$. Here C determines u_n uniquely for $d = 0$; it is 0 only if $u_n = 0$ for all n in this case.

Given any solution $\widetilde{u}_n^\bullet$ of the recurrence $\widetilde{u}_{n+1} = (1+\frac{c}{n})\widetilde{u}_n - c + 1$ (for $d = c-1$), an arbitrary $\widetilde{u}_n$ is $\widetilde{u}_n^\bullet + u_n^C$ for some C. Due to a special choice of

---

[30] I. Cherednik, *Modeling the waves of Covid-19*, Acta Biotheoretica 70, 8 (2022), `https://doi.org/10.1007/s10441-021-09428-w`.

d, one can take $\widetilde{u}_n^\bullet = n$. Thus $\lim_{n\to\infty} \widetilde{u}_n/n^c = C$ for the corresponding C. Accordingly, this limit is zero only if $u_1 = \frac{d}{c-1}$ in terms of the u_n before their rescaling to $d = 0, c-1$. ❑

We note that when $c = 0$, this recurrence is not applicable to model u_n, the total number of infections. Indeed, u_n can only increase and the protective measures p_n cannot be negative. However, we obtain at $c = 0$ that $p_n = u_n - u_{n+1} < 0$.

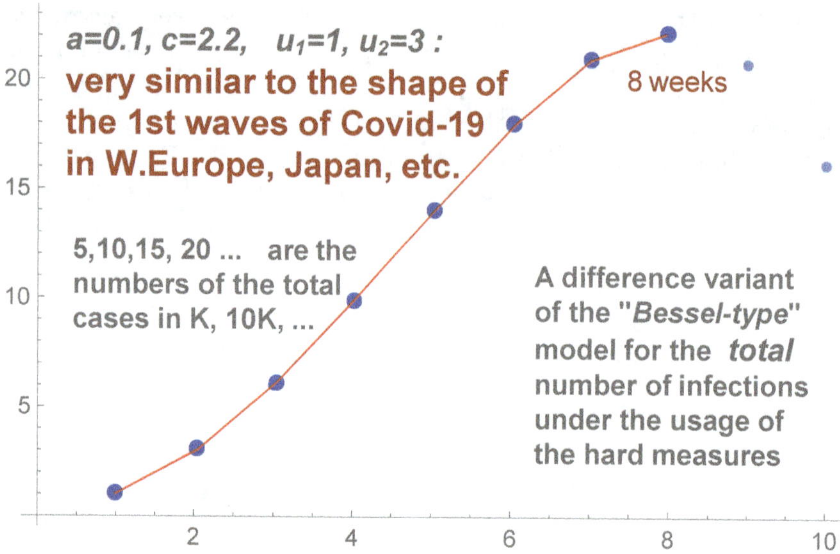

Figure 4.5: Epidemic spread under aggressive management.

Fig. 4.5 is an example of the practical calculations with (4.5.26). We adjust the parameters to make the graph similar to real curves for the total number of detected infections for Covid-19. Near and after the maximum value of u, a different recurrence must be used, which we omit. See the author's papers [30] and [31] quoted above and below. We take here $a = 0.1, c = 2.2$ and begin with $u_1 = 1, u_2 = 3$. This simulation must end at $n = 8$ (week 8): the *total* number of detected cases cannot decrease and the end of this graph is irrelevant. However, this restriction can be omitted if u_n is defined as the difference of the total numbers of infections in 2 areas.

4.5.3 Modeling Covid-19

A general problem is that basic linear recurrences mostly exhibit exponential growth. However *linear and power-type processes dominate in nature.* The first wave of Covid-19 in India is a typical example. Fig. 4.6 provides the graph of the ***total number of detected infections*** until October 7, 2020, and its power-type and Bessel-type approximations from [31].

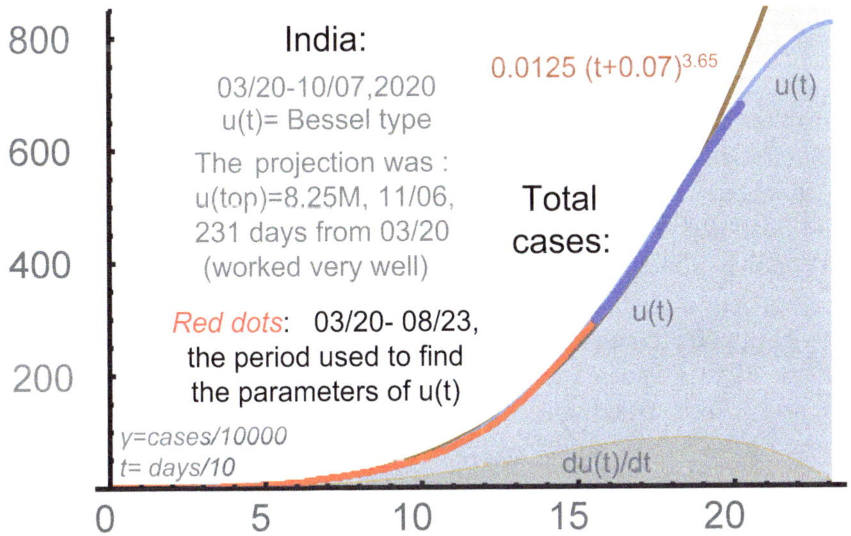

Figure 4.6: The growth of Covid-19 cases in India.

The x-axis in Fig. 4.6 is in units of time t measured as the number of days divided by 10, the y-axis is for the ***total*** number of detected infections divided by 10000. The curve is steep but certainly not exponential. The chart shows that the power-function approximation $0.0125(t+0.07)^{3.65}$ (brown) is quite good for the middle part of this curve: the thick one with visible dots inside. Its second (blue) part is a control period, the first (red) one was used to find power-type and Bessel-type approximations $u(t)$. The latter worked well until November 6, 2021 and continued to be accurate until the saturation.

The number of infections can grow like $1 \to R \to R^2 \to \dots$ for the ***reproduction number*** $R > 1$ only if the virus is new, the number of infected people is still relatively small and the protective measures are weak.

Discrete models are quite relevant here, since the data are not continuous curves. However, the following ***continuous*** variant of system (4.5.25) gave very good results for quite a few waves of Covid-19. It served well the periods almost until the saturation of the waves; see papers [30] and [31].

Let $u(t)$ be the total number of infections from the beginning of the wave (of Covid-19 or any epidemic with power-type spread). We also need the ***protection function*** $p(t)$. We will argue below that it can be assumed

[31] I. Cherednik, *Momentum managing epidemic spread and Bessel functions*, Chaos, Solitons & Fractals, 139 (October 2020), 110234; doi.org/10.1016/j.chaos.2020.110234.

$p(t) = \sum_i (t - t_i)$, where t is the current time, $\{t_i\}$ are the moments of all prior isolations. Then the basic system of differential equation is

$$\frac{du(t)}{dt} = c\,\frac{u(t)}{t} - p(t), \quad \frac{dp(t)}{dt} = a\,u(t)\,. \tag{4.5.27}$$

Here t is the time from the beginning the wave, though not always its very beginning, c is the *transmission rate*, a is the intensity of protective measures (or mechanisms). System (4.5.27) describes the impact of *hard measures* under the most aggressive response to the spread: the number of isolations is proportional to the total number of infections. When $a = 0$, we obtain the power growth: $u(t) \sim Ct^c$.

The parameter c can be measured experimentally during early stages of the waves. Mostly it was in the range $2.2 \le c \le 2.8$ for wave 1, but reached $c = 4.5$ and 5.75 in Brazil and India, the greatest values we saw.

There are two basic solutions, $u^1(t)$ and $u^2(t)$, of system (4.5.27):

$$u^{1,2}(t) = t^{(c+1)/2} J_{\pm\frac{c-1}{2}}(\sqrt{a}\,t), \quad J_\alpha(x) = \sum_{m=0}^{\infty} \frac{(-1)^m (x/2)^{2m+\alpha}}{m!\Gamma(m+\alpha+1)}\,; \tag{4.5.28}$$

$J_\alpha(x)$ are ***Bessel functions*** of the first kind; see [32] (Ch.3, S 3.1). Here $c \ge 0, c \ne 1$. The solution $u^1(t)$ (the sign is $+$) is the main; the second solution $u^2(t)$ (for $-$) is generally smaller and less important for modeling.

In Fig. 4.6, $c = 5.75$, $a = 0.035$ and

$$u(t) = 0.55\, t^{c/2+0.5}\big(J_{c/2-0.5}(\sqrt{a}\,t) + 0.2 J_{0.5-c/2}(\sqrt{a}\,t)\big).$$

Recall that $u(t)$ is the number of cases divided by 10000, and t is the number of days from August 23, 2020, the beginning of the wave, divided by 10.

For the *second wave in India*, the formula became

$$u(t) = 15\, t^{c/2+0.5}\big(J_{c/2-0.5}(\sqrt{a}\,t) + 0.4\, J_{0.5-c/2}(\sqrt{a}\,t)\big),$$
$$\text{where } c = 2.85, \ a = 0.14.$$

It was mainly from 03/25 to 05/25, 2021, followed by a period with relatively modest constant (and later decreasing) daily numbers of newly-detected cases.

We note that c dropped significantly vs. the 1st wave in India and the intensity of the protective measures increased. However, the scaling coefficient 15, which determines the magnitude of the wave, became much greater.

[32] G.N. Watson, *A treatise on the theory of Bessel functions*, 2nd Edition, Cambridge University Press, Cambridge, 1944.

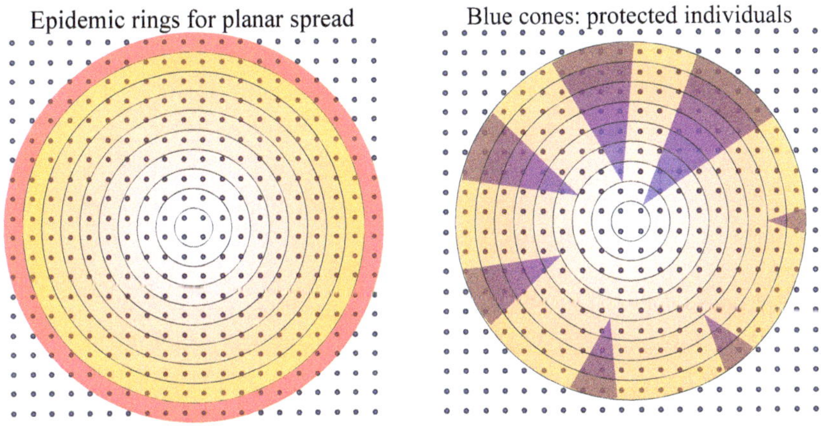

(a) Linear-type spread: $u(t) \sim t^2$. (b) Lasting effects of isolations.

Figure 4.7: Protection cones due to constant isolations.

The key argument in favor of power-type growth of $u(t)$ is that the wave of infections grows linearly in the *graph of contacts*. In this graph, individuals are represented as vertices and they are connected by edges if they are neighbors, co-workers or other *close* contacts. The dimension of this graph is the *transmission rate* c. The smallest c we observed was $c = 2.2$. It can be any number, not only an integer.

If only neighbors matter and they are assumed to be uniformly distributed, then $c = 2$. Fig. 4.7(a) shows the consecutive circles of infections with time in this case. If the current circle of infections is of radius $r - 1$ then those at the distance r from the origin, i.e. the closest to this circle, are the most likely to be infected next. For instance, $4, 10, 20$ people are infected after $1, 2, 3$ time-units: those in the 1st, 2nd and 3rd circles. Generally, this is with respect to the distance from the origin in the graph of contacts; the distance between two vertices there is the length of the shortest path between them.

The impact of isolations. Prompt isolations of infected individuals are the most efficient protective measures during epidemics. They are frequently preemptive; the transmissions are the most intensive during the first days-weeks after the infection.

Fig. 4.7(b) shows graphically what happens if 1 isolation per 1 time-unit is performed. Generally, isolations of infected people are supposed to protect their direct contacts, the contacts of their contacts and so on. The whole *cone* of protected people will result from 1 isolation with time.

For instance, if we isolate all 4 people in the first circle in this figure, the wave of infections stops. Even one isolation at a time will stop it eventually, i.e. will result in the saturation of the total number of those infected. Follow this figure to check that the time before the saturation is basically $\alpha \sum_{i=1}^{r-1} i \sim \frac{\alpha r^2}{2}$, where α is the angle of the cones (in radians) and $r \sim t$ is the radius. Thus, the whole r-circle will be "protected" as $\frac{\alpha r^2}{2} \sim 2\pi r$, i.e. for $t \sim \frac{4\pi}{\alpha}$, which is the duration of the epidemic wave in this setting.

To give examples beyond epidemics, if a branch of a tree is pruned, then this creates a hole in the tree crown growing with time. Generally, new branches are likely to use this space to grow and this hole may eventually vanish. However, periodic pruning will certainly diminish the crown.

Similarly, Fig. 4.7(a) can be interpreted as a simplified wave in the *neural network* in our brain triggered by a particular task. The growth of the number of neurons as in this figure is unsustainable long-term; "pruning" some of the directions (branches) of this process as in Fig. 4.7(b) can be the optimal way to limit the brain resources. Presumably, our brain reduces seemingly secondary or resource-consuming directions, though this can worsen the result.

Epidemic management. The graph of contacts is not a *tree* generally, you can be potentially infected by several individuals from the "previous circle" of infections, not by one. To address this we assume that the number of protected people due to 1 isolation grows *linearly* with time. So the lasting effect of isolations is assumed minimal. This is built in the definition of the *protection function* $p(t) = \sum_i (t - t_i)$, where t is the current time and t_i are the moments of time of *all* prior isolations of infected individuals. Then $dp(t)/dt$ is the *total* number of prior isolations. Generally, $dp(t)/dt$ is up to us to control, people and those in charge of *epidemic management.*

This derivative is a linear function of t in Fig. 4.7(b). Generally, we arrive at the following system:

$$\frac{du(t)}{dt} = c\,\frac{u(t)}{t} - p(t), \quad \frac{dp(t)}{dt} = a\,\frac{u(t)}{t}. \tag{4.5.29}$$

Note that it is different from (4.5.27): $a\frac{u(t)}{t}$ replaces $a\,u(t)$ there.

System (4.5.29) can be solved in terms of Bessel functions. The leading solution is $u(t) = t^{(c+1)/2} J_{c-1}(2\sqrt{at}\,)$ for $c > 1$. The saturation will occur significantly faster for (4.5.27) than for (4.5.29). The duration of the wave becomes $\sim \frac{1}{a}$ for the (4.5.29) versus $\sim \frac{1}{\sqrt{a}}$ for (4.5.27); recall that $a < 1$ and $\sqrt{a} > a$. Obviously, the saturation is supposed to be faster for (4.5.27).

The term $a\,u(t)$ in system (4.5.27) versus $a\frac{u(t)}{t}$ results in the following modification of Fig. 4.7(b). The total number of isolations until t must be assumed proportional to the *total* number of cases at t. Accordingly, the number of *new* isolations becomes $\sim r$, where r is the distance r from the origin; $c = 2$ in this figure. Qualitatively, the number of new isolations would be $1, 2, 3, \ldots$ instead of $1, 1, 1, \ldots$ in the figure above.

The formula for the duration t of the wave will now results from the relation $\alpha \sum_{i=1}^{t-1} i(t-i) \sim 2\pi t$. Thus, $t^2 \sim \frac{12\pi}{\alpha}$ and $t \sim \sqrt{\frac{12\pi}{\alpha}}$. See Exercise 3.7.11(i). This matches the formula $\sim \frac{1}{\sqrt{a}}$ obtained via Bessel functions.

4.5.4 Invasion ecology

System (4.5.27) describes well the size $u(t)$ of the prey population in the presence of $p(t)$ predators, where $u(t)$ is of power growth without predators. Then $dp(t)/dt$ can be assumed to be proportional to $u(t)$ (prey attracts predators from other areas), and $du(t)/dt$ must be diminished by $p(t)$ times some constant (predators feed on their prey). Actually the same approach (and Bessel-type functions) can be used to model ***tree growth***; see Fig. 4.7.

Let us consider the following predator-pray situation. We assume that there are two areas A_1 and A_2 with the plants (prey) necessary for the survival of some animals (predators). Let $u_i(t)$ be the number of plants in area A_i, which is supposed to be of linear growth in terms of time t. We have: $du_i(t)/dt = u_i(t)/t$ for $i = 1, 2$ in the absence of predators.

Now let the number of predators in A_i be $p_i(t)$. We assume that the main process in this ecosystem is the migration of the predators from one area to another depending on the ***prey gradient*** $u(t) = u_1(t) - u_2(t)$, though u_i and p_i can grow with time. The purpose of this model is to capture the oscillations of $u(t)$ and $p(t) = p_1(t) - p_2(t)$ in time. It is natural to assume that $dp(t)/dt$ is proportional to $u(t)$. We obtain that $dp(t)/dt = a\, u(t)$ for some constant a: *the greater availability of food the greater the number of animals moving to the area from the neighboring ones.*

The ***predator-prey relation*** is described by $du_i(t)/dt = u_i(t)/t - b\, p_i(t)$ for some parameter b, which is basically the amount of food (plants) one animal consumes during some time-unit. Here we need to assume the positivity of u_i, p_i. However, if we switch to the difference of the 2 differential equations above, i.e. to $du(t)/dt = u(t)/t - b\, p(t)$, then negative functions are perfectly meaningful. By rescaling the count of animals, we can make $b = 1$ here. Finally, we obtain system (4.5.27) for $c = 1$:

$$\frac{du(t)}{dt} = \frac{u(t)}{t} - p(t), \quad \frac{dp(t)}{dt} = a\, u(t). \tag{4.5.30}$$

It has the following solution:

$$u(t) = tJ_0(\sqrt{a}\, t), \quad \text{where } J_0(x) = \sum_{m=0}^{\infty} \frac{(-1)^m (x/2)^{2m}}{(m!)^2}. \tag{4.5.31}$$

We will omit the formula for the second *fundamental solution* of this system. It is not difficult to check directly that $tJ_0(\sqrt{a}\,t)$ solves the equation $\frac{du(t)}{dt} = \frac{u(t)}{t} - a\int_0^t u(\tau)d\tau$, equivalent to system (4.5.30) as $p(0) = 0$. One has:

$$u(t) = \frac{2}{\sqrt{a}}\sum_{m=0}^{\infty}\frac{(-1)^m(\frac{\sqrt{a}\,t}{2})^{2m+1}}{(m!)^2},\quad \frac{du}{dt}-\frac{u}{t} = \sum_{m=0}^{\infty}(-1)^m\frac{2m(\frac{\sqrt{a}\,t}{2})^{2m}}{m!\,m!}$$
$$= \sum_{m=1}^{\infty}(-1)^m\frac{2(\frac{\sqrt{a}\,t}{2})^{2m}}{(m-1)!\,m!} = -2\sum_{m=0}^{\infty}(-1)^m\frac{(\frac{\sqrt{a}\,t}{2})^{2m+2}}{(m+1)!\,m!}.$$

The last expression is, indeed, the negative of

$$a\int_0^t u(\tau)d\tau = 4\sum_{m=0}^{\infty}(-1)^m\frac{(\frac{\sqrt{a}\,t}{2})^{2m+2}}{2(m+1)m!\,m!} = 2\sum_{m=0}^{\infty}(-1)^m\frac{(\frac{\sqrt{a}\,t}{2})^{2m+2}}{(m+1)!\,m!}.$$

The following *asymptotic formula* is fundamental in the theory of Bessel functions J_α defined in (4.5.28):

$$J_\alpha(x) \sim \sqrt{\frac{2}{\pi x}}\cos\bigl(x-\frac{\pi\alpha}{2}-\frac{\pi}{4}\bigr)\ \text{ for sufficiently large } x.$$

See the reference [32] to "Watson" above: S 7.21. Thus, $u(t)$ can be approximated by $\mathcal{C}\,t^{1/2}\cos(\sqrt{a}\,t-\pi/4)$ for a constant $\mathcal{C}$ when t is sufficiently.

We obtain that the *asymptotic* period of $t^{-1/2}u(t)$ is $2\pi/\sqrt{a}$. So $u(t)$ will first increase starting with $t = 0$ and then go down approximately after $t = \pi/(2\sqrt{a})$. However, this estimate is too qualitative. One needs to plot the graph of $u(t)$ to find the actual point of its first maximum, the saturation of the first wave. The formula generally provides other waves and "predicts" that the magnitude of oscillations of $u(t)$ will grow as $t^{1/2}$ long-term, but this can be difficult to see in real populations.

Bison in Yellowstone. Bison were introduced in Yellowstone National Park in 1902. Their numbers for the years 1902-1931 are red dots in Fig. 4.8; the black dots are for the years when this number was unknown.

It perfectly matches our Bessel-type curves: a power growth with slowly diminishing exponent and the saturation. One can try to approximate this curve by the *sigmoid* function, which is basically $\frac{1}{1+\exp(-t)}$. However, its beginning is clearly not of this type and the end is far from that in the sigmoid. System (4.5.27) describes well the number $u(t)$ of bison; the solid curve is:

$$u(t) = 58\cdot\bigl(\frac{t}{5}\bigr)^{1/2+c/2}J_{c/2-1/2}(\sqrt{a}(t/5)),\ c=3,\ a=0.15,$$

where the year is $1900+t$ (t is the x-axis). The number of bison is $y+44$ for the corresponding y, so y is the change from year 1902, when the number of bison was 44. These numbers were:

1902 : 44, 1903 : 47, 1904 : 51, 1905 : 74, **1906** : –, 1907 : 84, 1908 : 95, **1909** : 118, **1910** : 149, 1911 : 168, 1912 : 192, 1913 : 215, **1914** : –, 1915 : 270, 1916 : 348, **1917** : 397, **1918** : –, 1919 : 504, 1920 : 501, 1921 : 602, **1922** : 647, 1923 : 748, 1924 : –, **1925** : 830, **1926** : 931, 1927 : 1008, 1928 : 1057, **1929** : 1109, 1930 : 1124, **1931** : 1192.

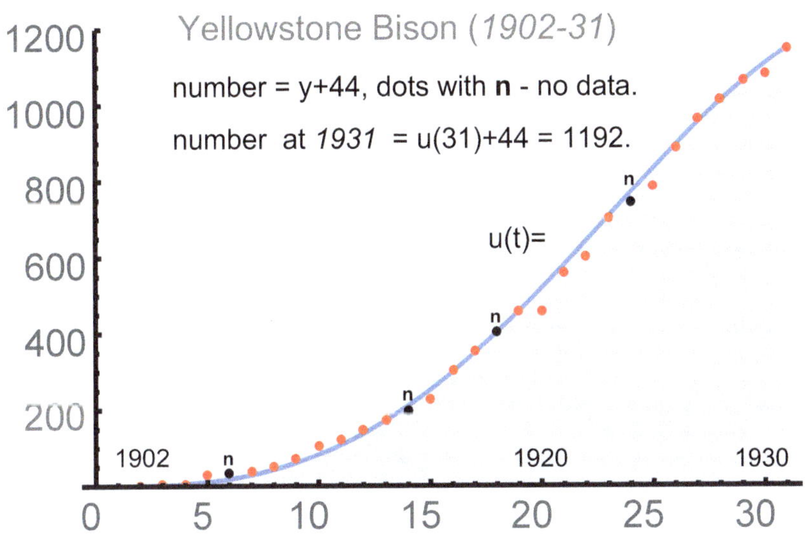

Figure 4.8: Bison herd in Yellowstone.

There were many factors that restricted the growth of bison population. The role of wolves and elks must be mentioned. However, disease, including various viruses, parasites and brucellosis, have a greater effect on the bison population. So $p(t)$ is a combination of various "prevention mechanisms".

In conclusion, we mention that Bessel-type curves as in Fig. 4.8 model very well *tree growth* through all phases of their life, *juvenile, mature* (essentially linear) and *senescent*. The corresponding $p(t)$ is due to tree ageing, disease, tree-killing insects and other factors. System (4.5.30) is basically sufficient.

4.6 Exercises

Exercise 4.6.1 $^{\heartsuit}$ *For the Fibonacci numbers f_n, use* mathematical induction *discussed in Section 3.3.3 to establish the following identities:*

(i) $f_1 + f_2 + \cdots + f_n = f_{n+2} - 1$ *for* $n \geq 1$,

(ii) $f_1 + f_3 + \cdots + f_{2n-1} = f_{2n}$, $f_2 + f_4 + \cdots + f_{2n} = f_{2n+1} - 1$,
(iii) $\sum_{i=0}^{n-1} f_{\epsilon+3i} = (f_{3n+\epsilon-1} - f_{\epsilon-1})/2$, *where* $n \geq 1$, $\epsilon = 0, 1, 2$.

Exercise 4.6.2$^\heartsuit$ *(i) Check that the Fibonacci numbers* f_n *satisfy the relations* $f_n = f_{n-1} + f_{n-2} = 2f_{n-2} + f_{n-3} = 3f_{n-3} + 2f_{n-4} = 5f_{n-4} + 3f_{n-5} = 8f_{n-5} + 5f_{n-6} = 13f_{n-6} + 8f_{n-7}$ *and* $f_n = 4f_{n-3} + f_{n-6}$.

(ii) Use these relations to prove that f_n *is divisible by* $2, 3, 5, 8, 13$ *and* 4 *if and only if* n *is correspondingly divisible by* $3, 4, 5, 6, 7$ *and* 6.

Exercise 4.6.3$^\diamondsuit$ *(i) Extending Exercise 4.6.2, prove by induction that* $f_n = f_{m+1}f_{n-m} + f_m f_{n-m-1}$ *for any* m *such that* $n - m - 1 \geq 0$.

(ii) For $N > 1$, *let* $m_N > 1$ *be minimal such that* N *divides* f_m, *i.e.* f_m *is zero modulo* N. *Use (i) to justify that* $f_n - (f_{m_N+1})^q f_r$ *is divisible by* N, *where* $n = qm_N + r$ *for* $0 \leq r < m_N$ *and* $n > m = m_N$ *as in (i).*

(iii) Check that $\gcd(f_n, f_{n+1}) = \gcd(f_{n-1}, f_n) = \cdots = 1$. *Then use* $\gcd(f_{m_N+1}, f_{m_N}) = 1$ *and (ii) to prove that* f_n *for* $n = jm_N$ *and only such* n *are divisible by* $N > 1$, *where* $j = 0, 1, 2, \ldots$ *and* m_N *is from (ii).*

(iv) Obtain from (iii) that f_n *is divisible by* f_m *if and only if* $n = jm$. *Namely, make* $N = f_m$ *there and use that* $m_N = m$. *Deduce from it that* $\gcd(f_n, f_m) = f_\ell$ *for* $\ell = \gcd(m, n)$.

Hint for (iv): Using (iii), f_k divides $\gcd(f_m, f_n)$ if and only if k divides $\ell = \gcd(m, n)$. Setting $N = \gcd(f_m, f_n)$, k is such if and only if it divides m_N. This formally gives that $\ell = m_N$. So f_ℓ divides N and is divisible by N.

Exercise 4.6.4$^\diamondsuit$ *(i) Check the formula* $f_{-n} = (-1)^{n+1} f_n$ *for the Fibonacci numbers extended to negative* n *via the Fibonacci recurrence; here* n *is arbitrary, positive or non-positive. Extend the relation* $f_n = f_{m+1}f_{n-m} + f_m f_{n-m-1}$ *from Exercise 4.6.3 to any integer* n, m. *For* $n = 1$, *obtain that* $f_{m+1}f_{1-m} + f_m f_{-m} = 1 = (-1)^m (f_{m+1}f_{m-1} - f_m f_m)$.

(ii) Let $n > 1$. *Using the latter relation,* $|\lambda_1 - \frac{f_n}{f_{n-1}}| < \frac{1}{f_n f_{n-1}}$ *for* λ_1 *from Binet's formula, and* $\frac{f_n}{f_{n-1}} < \lambda_1 < \frac{f_{n+1}}{f_n}$ *for even* n. *Also,* $|\lambda_1 - \frac{f_n}{f_{n-1}}| > \frac{1}{f_{n-1}f_n} - \frac{1}{f_{n+1}f_n} = \frac{1}{f_{n-1}f_{n+1}}$ *for any* n, *and* $|\lambda_1 - \frac{f_n}{f_{n-1}}| > \frac{1}{\lambda_1 f_n f_{n-1}}$ *for odd* n.

Exercise 4.6.5$^\heartsuit$ *The* ***Lucas numbers*** ℓ_n *solve the Fibonacci recurrence* $\ell_n = \ell_{n-1} + \ell_{n-2}$ *subject to* $\ell_0 = 2, \ell_1 = 1$.

(i) Check that $\ell_n = f_{n+1} + f_{n-1} = \lambda_1^n + \lambda_2^n$ *for* $\lambda_{1,2} = \frac{1 \pm \sqrt{5}}{2}$ *and any* n.

(ii) Use (i) and Binet's formula to check that $f_n f_m = \frac{1}{5}(\ell_{n+m} - (-1)^m \ell_{n-m})$ *and* $\ell_n \ell_m = \ell_{n+m} + (-1)^m \ell_{n-m}$ *for any* n, m.

Exercise 4.6.6 $^\diamondsuit$ *(i) Analyzing how the middle column can be covered by dominos in the shape* $2\times(2n+1)$, *check that* $d_{2n+1} = d_n^2 + 2d_nd_{n-1}$ *for* $n \geq 1$. *Similarly,* $d_{2n} = d_n^2 + d_{n-1}^2$. *See Section 1.1.1:* $d_n = \mathcal{N}^{\square\square}_{2\times n} = f_{n+1}$ *for the Fibonacci numbers* f_n. *Use the formula for* d_{2n+1} *to check that* $f_{2n} = f_n(f_n + 2f_{n-1}) = f_n\ell_n$ *for the* Lucas numbers ℓ_n *from Exercise 4.6.5.*

(ii) Use Binet's formula to generalize (i). For $m \geq 2$, *check that* $f_{mn}/f_n = \ell_{(m-1)n} + (-1)^n\ell_{(m-3)n} + (-1)^{2n}\ell_{(m-5)n} + \ldots$, *where the last term in the latter sum is* $(-1)^{nk}$ *for* $m = 2k+1$ *and* $(-1)^{n(k-1)}\ell_n$ *for* $m = 2k$.

Exercise 4.6.7 $^\heartsuit$ *For the* Lucas numbers *from Exercise 4.6.5, use Proposition 4.2.4 to calculate the* generating function $G(t) = \sum_{n=0}^\infty \ell_n t^n$: $G(t) = \frac{2-t}{1-t-t^2}$. *Using the generating function* $F(t) = \frac{t}{1-t-t^2}$ *for the Fibonacci numbers* $\{f_n\}$, *check that* $F^{(+1)}(t) = \frac{1}{1-t-t^2}$ *and* $F^{(-1)}(t) = 1 + \frac{t^2}{1-t-t^2}$ *are the generating functions for the sequences* $\{f_{n+1}\}$ *and* $\{f_{n-1}\}$. *Then restate Exercise 4.6.5(i) as the identity* $G(t) = F^{(+1)}(t) + F^{(-1)}(t)$.

Exercise 4.6.8 $^\heartsuit$ *Solve the recurrence relation* $a_m = 4^{2m-1}a_{m-1}$ *for* $a_1 = 4$. *See Exercise 1.4.39.*

Answer: 4^{m^2}.

Exercise 4.6.9 $^\heartsuit$ *Solve the recurrence relation* $g_n = 3g_{n-2}(n \geq 2)$ *with initial values* $g_0 = 0$ *and* $g_1 = 1$ *directly, and then via the master equation* $\lambda^2 = 3$.

Answer: $h_n = \frac{1}{2\sqrt{3}}\left((\sqrt{3})^n - (-\sqrt{3})^n\right) = 0, 1, 0, 3, 0, 9, \ldots$.

Exercise 4.6.10 $^\heartsuit$ *Compose and solve algebraically the recurrence for the number* m_n *of coloring of* n *consecutive boxes with white and* 6 *non-white colors such that any non-white color must be followed by at least one white.*

Exercise 4.6.11 $^\heartsuit$ *Compose and solve algebraically the recurrence relation for the* 5*-color problem: find the number* h_n *of colorings of "*$1\times n$*" using red and* 4 *other colors provided that there are no adjacent red boxes. Notice that* $\lambda_{1,2}$ *here are* 2 *times* $\lambda_{1,2}$ *from Section 4.3.2 on "growing stalagmites."*

Answer: $h_n = 4h_{n-1} + 4h_{n-2}$, $\lambda_{1,2} = 2(1 \pm \sqrt{2})$, $h_1 = 5, h_2 = 24, h_0 = 1$, $h_n = 2^n\left((\frac{1}{2} + \frac{3\sqrt{2}}{8})(1+\sqrt{2})^n + (\frac{1}{2} - \frac{3\sqrt{2}}{8})(1-\sqrt{2})^n\right)$.

Exercise 4.6.12$^{\heartsuit}$ *Let h_n denote the number of ways to perfectly cover the shape $1 \times n$, the line of n boxes, with monominos and dominos in such a way that no two dominos are consecutive.*

(i) Check combinatorially that $h_n = g_{n+1}$, where g_n is the number of perfect coverings of "$1\times n$" by monominos and 3-minos.

(ii) Obtain the recurrence relation satisfied by h_n and g_n, including finding h_i, g_i for $1 \le i \le 3$, and check that $g_n = \binom{n}{0} + \binom{n-2}{1} + \binom{n-4}{2} + \dots$.

(iii) Also, $G(t) = \sum_{m=0}^{\infty} g_m t^m = \frac{1}{1-t-t^3}$; compare with Proposition 4.2.4.

Hint for (iii): Expand $G(t) = \frac{1}{1-t-t^3} = 1 + (t+t^3) + (t+t^3)^2 \dots$.

Exercise 4.6.13$^{\heartsuit}$ *(i) Find but do not solve algebraically the recurrence relation for the number x_n of perfect coverings of the shape $1 \times n$ by monominos and dominos such that no two monominos are consecutive. Find $x_1, \dots, x_4$.*

(ii) Justify that $y_{n+2} = x_n$ for the number y_n of perfect coverings of "$1\times n$" by dominos and 3-minos. Check that $y_n = \sum_{2a+3b=n} \binom{a+b}{a}$, where $a, b \ge 0$, and $G(t) = \sum_{m=0}^{\infty} y_m t^m = \frac{1}{1-t^2-t^3}$.

Exercise 4.6.14$^{\clubsuit}$ *Given a recurrence relation $g_n = Ag_{n-1} + Bg_{n-2} + Cg_{n-3}$ with $A, B \ge 0$ and $C > 0$, let $P(\lambda) = \lambda^3 - A\lambda^2 - B\lambda - C$.*

(i) Then check that the corresponding master equation $P(\lambda) = 0$ has exactly one positive real solution λ_1; actually, it satisfies the inequality $\lambda_1 > A$. If the remaining 2 solutions λ_2 and λ_3 are real (they can coincide), then both are negative and $|\lambda_2| + |\lambda_3| = \lambda_1 - A$. In particular, $|\lambda_{2,3}| < \lambda_1$.

(ii) If λ_2 and λ_3 are complex (not real), then check that $\lambda_3 = \overline{\lambda_2}$ for the complex conjugation sending $a + b\imath$ to $a - b\imath$ for real a, b. Also, verify that $|\lambda_2|^2 = |\lambda_3|^2 = \lambda_2\lambda_3 < \lambda_1^2$ unless $A = 0 = B$, where $|a + b\imath|^2 = a^2 + b^2$ is the norm of $a + b\imath$. Here $\imath$ is the imaginary unit.

(iii) For any λ_i from (i) or (ii), check that $\lim_{n\to\infty} g_n/\lambda_1^n \to c$, where c is as follows: $g_n - c\lambda_1^n$ is a linear combination of λ_2^n and λ_3^n, or it is a linear combination of λ_2^n and $n\lambda_2^n$ if λ_2 is of multiplicity 2. In either case, check that g_{n+1}/g_n converges to λ_1 as $n \to \infty$ if $c \ne 0$. See Exercise 5.6.38(iii) concerning the condition $c \ne 0$.

Hints: (i) The existence of a positive root $\lambda_1 > A$ is due to $P(\lambda) > 0$ for sufficiently big $\lambda > 0$ and $P(C) < 0$. We have: $\lambda_1 + \lambda_2 + \lambda_3 = A \ge 0$, $\lambda_1\lambda_2 + \lambda_2\lambda_3 + \lambda_1\lambda_3 = -B \le 0$, $\lambda_1\lambda_2\lambda_3 = C > 0$. If all 3 roots are real, then the latter gives that either $\lambda_2, \lambda_3 < 0$ or all 3 are positive, which contradicts the relation for B. Thus $|\lambda_2| + |\lambda_3| = \lambda_1 - A$. Here A can be 0, but $|\lambda_{2,3}| > 0$; so $|\lambda_{2,3}| < \lambda_1$.

(ii) For complex (non-real) $\lambda_{2,3}$, one has: $\lambda_3 = \overline{\lambda_2}$, $|\lambda_2|^2 + \lambda_1(\lambda_2 + \lambda_3) = -B = |\lambda_2|^2 + A\lambda_1 - \lambda_1^2$, and $|\lambda_2|^2 = \lambda_1^2 - A\lambda_1 - B < \lambda_1^2$ unless $A = 0 = B$.

Exercise 4.6.15 ♣ *(i) Let H_n denote the number of ways to perfectly cover the shape $1 \times n$, the line of n boxes, with monominos, dominos and 3-minos, where the 3-minos are available in 2 colors. Check that $H_1 = 1, H_2 = 2, H_3 = 5$ and $H_n = H_{n-1} + H_{n-2} + 2H_{n-3}$ for $n \geq 4$; for $n = 3$, obtain that $H_0 = 1$.*

(ii) Find the master equation for H_n, and verify that $H_n = A2^n + B\rho^n + C\rho^{\ \ n}$ for $\rho^{\pm 1} = \exp(\pm 2\pi\imath/3)$, which are 2 roots of the equation $\lambda^2 + \lambda + 1 = 0$, and some constants A, B, C.

(iii) Obtain that $A = \frac{4}{7}$ and then deduce from (ii) that $H_n = Round[\frac{2^{n+2}}{7}]$ for any $n \geq 0$.

(iv) Use (iii) and the relation $\rho + \rho^{-1} = -1$ to find that $B = \frac{4-\rho}{21}$, $C = \frac{5+\rho}{21}$. It suffices to check that H_0, H_1, H_2 for H_n in (ii) are those in (i) (why?). Also, check directly that H_3 obtained via (ii) is 5 indeed; use that $\rho^3 = 1$.

Exercise 4.6.16 ♢ *Check the following relations for the number of perfect coverings $\mathcal{N}^{\square\square\square}_{6\times n}$ of the shape $6 \times n$ by 3-minos: $\mathcal{N}^{\square\square\square}_{6\times 4} = \mathcal{N}^{\square\square\square}_{6\times 3} + \mathcal{N}^{\square\square\square}_{3\times 4} + 2\mathcal{N}^{\square\square\square}_{3\times 3} = t_6 + t_4 + 2t_3 = 13$, $\mathcal{N}^{\square\square\square}_{6\times 5} = \mathcal{N}^{\square\square\square}_{6\times 4} + 3\mathcal{N}^{\square\square\square}_{3\times 4} = 13 + 3t_4 = 21$. We use here $t_3 = 2, t_4 = 3, t_6 = 6$ from (4.3.18).*

Hint: Use that there are 4 different ways to cover the 2 corners of the first 6-column. For $\mathcal{N}^{\square\square\square}_{6\times 4}$, one can also use that the middle vertical 6-line can intersect only 13 different configurations of 3-minos there.

Exercise 4.6.17 ♡ *Let us consider a sequence of signals "$\pm$" and pauses "p", where any two signals must always be separated by at least one "p". For instance, "$+p+pp-$" is admissible, but "$++$" and "$+-$" are not allowed anywhere. Denoting the number of such sequences of length n by u_n, check that $u_1 = 3$, $u_2 = 5$, $u_3 = 11$, and prove that $u_n = u_{n-1} + 2u_{n-2}, (n \geq 2)$. Find an algebraic formula for u_n.*

Answer: $u_2 = 9 - 4 = 5$, $u_3 = 27 - 12 - 12 + 8 = 11$, $u_n = \frac{4}{3}2^n - \frac{1}{3}(-1)^n$.

Exercise 4.6.18 ♡ *There are v vowels and c consonants in some alphabet. The words are "nice" if consonants are always separated by vowels. Here using different letters is not required and words can begin with any letter.*

(i) Justify the following recurrence relation for the number g_n of "nice" words of length n: $g_n = vg_{n-1} + vcg_{n-2}$. Calculate g_1 and g_2 for arbitrary c, v.

(ii) Assuming that $v = 2, c = 4$ in this alphabet, use (i) to find the algebraic formula for g_n.

Exercise 4.6.19 ♡ *Let s_n be the number of coverings of "$1\times n$" by dominos and monominos, where the former are available in 5 colors and the latter in 4 colors; $s_1=4, s_2=21$. Find the recurrence for s_n and solve it algebraically.*

Answer: $s_n = \frac{5}{6}5^n + \frac{1}{6}(-1)^n$.

Exercise 4.6.20 ♡ *Check that any solution $\widetilde{f}_n$ of the classical Fibonacci recurrence is a linear combination $C_1 f_n + C_2 d_n$ of the Fibonacci numbers $f_n = 1,1,2,3\cdots$ and $d_n = f_{n+1} = 1,2,3,5,\cdots$. Justify that if $\widetilde{f}_n$ are integers, then so are $C_{1,2}$.*

Exercise 4.6.21 ♣ *(i) Solve the "3-term relation" from the theory of orthogonal polynomials $(x+x^{-1})f_n = f_{n+1}+f_{n-1}$, where $x \neq 0$ (a parameter). The following are* **Schur functions** *and* **formal p-adic spherical functions** *for SL_2.*

(ii) Use (i) to justify that $f_n(x) = \frac{x^{n+1}-x^{-n-1}}{x-x^{-1}}$ is a unique solution of this recurrence with $f_0 = 1, f_{-1} = 0$ for $x \neq 1$. It becomes $f_n = n+1$ for $x=1$; check this by sending $x \to 1$. Also, $f_n = (-1)^n(n+1)$ for $x=-1$ (check).

(iii) Use (i) to check that $f_n(x;q) = x^{n+1}-x^{-n-1} - q^{-1}(x^{n-1}-x^{1-n})$ solves this recurrence for any n, positive and negative, under the condition $f_1 = qf_{-1}$ for $q \neq 0$. All such solutions for $x \neq \pm 1$ are $cf_n(x;q)$ for constant c. Also, $f_n = (-1)^{xn}(C_1 n + C_2)$ for $C_2 = \frac{q+1}{q-1}C_1$ for constant $C_{1,2}$ when $x=\pm 1$, $q \neq 1$, and $f_n = (-1)^{xn}C$ if $x = \pm 1, q = 1$ for constant C.

Hint: The master equation is $(x+x^{-1})\lambda = \lambda^2+1$, $\lambda_{1,2} = \frac{(x+x^{-1})\pm(x-x^{-1})}{2} = \{x, x^{-1}\}$. Thus, $f_n = C_1x^n + C_2x^{-n}$ unless $x=\pm 1$. Then make $x \to \pm 1$.

Exercise 4.6.22 ♣ *Consider an (infinite) tree of degree $q+1$, i.e. a connected graph without any cycles where every vertex is connected by edges with $q+1$ ones. A function $h(v)$ on its vertices v is called* **spherical** *if for any v: $\frac{1}{q+1}\sum_{v'} h(v') = \frac{x+x^{-1}}{q^{1/2}+q^{-1/2}}h(v)$, where the sum is over all vertices v' connected with v. The operation on the left-hand side is the* **Laplace operator for this tree**, *so $h(v)$ is its eigenfunction for the eigenvalue $\frac{x+x^{-1}}{q^{1/2}+q^{-1/2}}$. It is called a* **harmonic function** *if $x = q^{\pm 1/2}$. Let us pick one vertex to be the origin 0, and assume that $h(v) = h_n$, where n is the distance in this tree from v to 0, i.e. we assume that it depends only on n. Let $h_0 = 1$. It is* **p-adic spherical function for SL_2** *when $q = p^k$ and for a prime p. For such q or any $q \neq 0, -1$, we arrive at: $(1+q)\frac{x+x^{-1}}{q^{1/2}+q^{-1/2}}h_n = q^{1/2}(x+x^{-1})h_n = qh_{n+1}+h_{n-1}$, where $h_{-1} = h_1$, $h_0 = 1$. Check that its unique solution for $x \neq \pm 1$ is*

$$h_n = q^{-n/2}f_n(x;q)/f_0(x;q) = q^{-n/2}\frac{(x^{n+1}-x^{-n-1}) - q^{-1}(x^{n-1}-x^{1-n})}{(x-x^{-1})(1+q^{-1})}$$

for $f_n(x;q)$ from Exercise 4.6.21 and $n \geq 0$. Note that $h_n = 1$ for $x = q^{\pm 1/2}$.

Hint: Setting $f_n = h_n q^{n/2}$, one has: $f_0 = 1, f_1 = qf_{-1}$ and f_n satisfies $(x + x^{-1})f_n = f_{n+1} + f_{n-1}$.

Exercise 4.6.23 ♢ *Follow Problem 4.4.4 to check that the solution of the recurrence $D_k = pD_{k+1} + qD_{k-1} + 1$ for $p + q = 1$ under the boundary conditions $D_0 = D_1 + 1, D_n = 0$ is as follows:*

$$D_k = \frac{k-n}{q-p} - 2\frac{(q/p)^{k+1} - (q/p)^{n+1}}{(1-(q/p))^2} \quad \text{for } p \neq 0, q \neq p. \tag{4.6.32}$$

Obtain that $D_k = (n-k)(n+k)$ as $p = \frac{1}{2} = q$. Here D_k describes the average number of steps for a random walk with probabilities p and q of the steps $x \to x+1$ and $x \to x-1$ needed to reach n from k, where $0 \le k \le n$ and the reflection is performed at 0 (we make the step $0 \to 1$ if 0 is reached).

Hint: You need only to see that $D_k = k/(q-p) + C_1(q/p)^k + C_2$ for some $C_{1,2}$, and to check directly the boundary conditions. Then consider the limit $p/q \to 1$ or solve the corresponding inhomogeneous resonance recurrence.

Exercise 4.6.24 ♢ *(i) Let $S_n = \sum_{r=0}^{n} r\binom{n}{r}$ for $n \ge 0$. Then check combinatorially the recurrence $S_n = 2S_{n-1} + 2^{n-1}$ for $n \ge 1$ using the interpretation of $r\binom{n}{r}$ as the number of r-teams with a captain.*

(ii) Let $T_n = \sum_{r=1}^{n} r(r-1)\binom{n}{r}$ for $n \ge 1$. Then check combinatorially the recurrence $T_n = 2T_{n-1} + 2S_{n-1}$ for $n \ge 0$ using the interpretation of $r(r-1)\binom{n}{r}$ as the number of r-teams with a captain and a vice-captain.

Hint for (i): Any new player (a) can be included in an existing team (with the captain), (b) not used in any teams, and (c) added to any subset of players as the captain. For (ii), (c) reads: can be added to any team with the captain as its vice-captain, with a possible switch of roles with the existing captain.

Exercise 4.6.25 ♢ *Let b_n be the number of sequences of length n made from 0s and 1s that are assumed to contain 00 somewhere.*

(i) Check that $b_1 = 0$, $b_2 = 1$, $b_3 = 3$, $b_n = b_{n-1} + b_{n-2} + 2^{n-2}$ for $n \ge 2$.

(ii) Then prove that $b_n = 2^n - f_{n+2}$ for the Fibonacci numbers f_n.

Hint for (ii): $b_n = 2^n + g_n$, where $g_n = g_{n-1} + g_{n-2}$, $g_1 = -2$, $g_2 = -3$.

Exercise 4.6.26 ♡ *The Towers of Hanoi puzzle has three rods with n rings of different sizes stacked in decreasing order of size on one rod. A move consists of moving a ring from one rod to another so that it does not land on top of a smaller ring. Let t_n be the minimal number of moves required to move all the rings from the initial rod to another rod (any). Check that $t_n = 2t_{n-1} + 1$ with $t_1 = 1$ and $t_n = 2^n - 1$.*

Exercise 4.6.27 ♣ *The **Pell numbers** are defined from the following recurrence:* $P_n = 2P_{n-1} + P_{n-2}$, $P_0 = 0, P_1 = 1$.

(i) Find the algebraic formula for P_n*; check that* $\lim_{n\to\infty} \frac{P_n+P_{n-1}}{P_n} = \sqrt{2}$.

(ii) Prove by induction that $P_{m+n} = P_m P_{n+1} + P_{m-1} P_n$ *for any* $m \geq 1, n \geq 0$. *Also, verify* $P_{n+1}P_{n-1} - P_n^2 = (\pm 1)^n$ *and* $P_n = (P_{n+1}/2)^2 + (P_{n-1}/2)^2$ *for* $n \geq 1$. *Try to prove this:* $\sum_{i=0}^{4n-1} P_i = (P_{2n} + P_{2n+1})^2$ *(Santana, Diaz-Barrero, 2006).*

(iii) Deduce from the 2nd relation in (ii) that $x = P_n + P_{n-1}$ *and* $y = P_n$ *solve* Pell's equation $x^2 - 2y^2 = \pm 1$ *for* $n \geq 1$. *All positive integer solutions* $\{x, y\}$ *of this equation are such: Theorem 5.5.1.*

(iv) Use (ii) to prove that if m *divides* n *then* P_m *divides* P_n*; thus,* P_n *can be a prime number, called a **Pell prime**, only if* n *is prime.*

Solution for (i): $P_n = \frac{(1+\sqrt{2})^n - (1-\sqrt{2})^n}{2\sqrt{2}}$.
Hint for (iv): $P_{(k+1)m} = P_m P_{km+1} + P_{m-1} P_{km}$.

Exercise 4.6.28 ♠ *(i) Define the **Tchebychev polynomials** of the 2nd type by* $U_n(x) = 2^n \prod_{j=1}^n (x - \cos(\frac{j\pi}{n+1})) = 2^n \prod_{j=1}^n (x - \frac{e^{\pi\imath j/(n+1)} + e^{-\pi\imath j/(n+1)}}{2})$. *Check the following relation:* $U_n(\frac{u+u^{-1}}{2}) = \prod_{j=1}^n (u + u^{-1} - (e^{\pi\imath j/(n+1)} + e^{-\pi\imath j/(n+1)})) = \prod_{j=1}^n (u - e^{\pi\imath j/(n+1)})(1 - e^{-\pi\imath j/(n+1)} u^{-1}) = (u^{n+1} - u^{-n-1})(u - u^{-1})$.

(ii) Use Exercise 4.6.21 to prove the recurrence $U_n = 2xU_{n-1} - U_{n-2}$ *for* $n \geq 1$, *where* $U_{-1} = 0$, $U_0 = 1$, $U_1 = 2x$, $U_2 = 4x^2 - 1$.

(iii) Deduce from (ii) that $P_{n+1} = \imath^{-n} U_n(\imath)$ *for the* Pell numbers *from Exercise 4.6.27, where* $\imath^2 = -1$.

(iv) Obtain from (i) and (iii) that $2^M \Delta_L = P_{M+1}$ *for* Δ_L *from Kasteleyn's formula (1.1.4) (this is due to Kasteleyn-Stanley).*

Exercise 4.6.29 ♢ Mergesort algorithm. *When sorting* n *numbers, the algorithm divides them into several arrays, sorts them independently (what is called "divide and conquer"), and finally merges the resulting arrays, sorting in the process of merging. The time required for the last stage is approximately* n, *the worst-case scenario. Let it be* $T[n]$ *for the whole algorithm.*

(i) Let the two arrays be the first $p =$ *Ceiling*$[n/2]$ *numbers and the remaining* $n{-}p$ *numbers. Then* $T[n] = T[p] + T[n-p] + n$. *For* $n = 2^m$, *this recurrence becomes* $T[n] = 2T[n/2] + n$. *Check by induction that* $T[n] = 2^k T[n/2^k] + kn$ *for* $0 \leq k \leq m$, *and* $T[n] = n\log_2 n + n$, *where we set* $T[1] = 1$, $\log_2 x = \frac{\log x}{\log 2}$. *The latter satisfies the relation* $T[n] = 2T[n/2] + n$ *for any* n *(check), which gives the estimate* $\Theta(n \log n)$ *for* $T[n]$ *in really working algorithms.*

(ii) Split now the initial sequence of n *numbers into Floor*$[\sqrt{n}]$ *consecutive groups with about* $\sqrt{n}$ *numbers in each. The recurrence becomes* $T[n] =$

$\sqrt{n}\,T[\sqrt{n}]+n$. *Check that* $T[n]=n\log_2(\log_2 n)$ *solves it, which gives the estimate* $\Theta(n\log(\log n))$ *for "real"* $T[n]$ *in this regime under the worst-case scenario (and our assumptions): better than the estimate in (i) for large* n.

Hint for (ii): $n^{1/2}T[n^{1/2}] = n^{1/2}(n^{1/2}(\log_2\log_2 n - 1)) = n\log_2\log_2 n - n = T[n] - n$, where $T[n] = n\log_2\log_2 n$.

Exercise 4.6.30 $^{\diamondsuit}$ *In the notation of Problem 4.4.1, assume that the probability of one neutron scattering is* p $(0\le p\le 1)$*; otherwise, the neutron remains unchanged (with the probability* $q=1-p$*). Check that* $u_n = C_1(q+p\sqrt{2})^n + C_2(q-p\sqrt{2})^n + C_3$ *for some constants* C_1, C_2, C_3*, and that the recurrence for* u_n *with* $C_3 = 0$ *becomes* $u_n = 2u_{n-1} - u_{n-2}$ *in the limit* $p\to 0$.

Hint: In the notation from the solution of Problem 4.4.1: $y_n = 2px_{n-1} + qy_{n-1}, x_n = py_{n-1} + qx_{n-1}$, and $z_n = z_{n-1}$. The recurrence for $v_n = x_n + y_n$ is $v_n = 2qv_{n-1} + (2p^2 - q^2)v_{n-2}$.

Exercise 4.6.31 $^{\diamondsuit}$ *A straightforward way to cool a nuclear reactor is by separating some constant amount* c *of neutrons from the mass per one unit of time. Assume that the corresponding recurrence for the number of neutrons after* n *scatterings is* $u_n = 2u_{n-1} - u_{n-2} - c$.

(i) Check that $-\frac{c}{2}n^2$ *solves it, and therefore a general solution is given by the formula* $u_n = C_1 + C_2 n - \frac{c}{2}n^2$.

(ii) Calculate the time until the shutdown of the reactor, i.e. the greatest n *when* u_n *is still non-negative, assuming that* $c = 2$, $u_1 = 1$, $u_2 = 4$.

Answer: $u_n = 6n - 4 - n^2, u_5 = 1, u_6 < 0$.

Exercise 4.6.32 $^{\heartsuit}$ *Solve the following "predator-prey" system of recurrence relations with* $a_0 = 0, b_0 = 1$*:* $a_n = a_{n-1} + b_{n-1}$, $b_n = b_{n-1} - \frac{1}{4}a_{n-1}$. *What is the last* n *when* b_n *(for the prey) is positive?*

Hint: Express b in terms of a using the first relation and substitute.
Answer: $b_n = \frac{1}{2}\left((1+\frac{i}{2})^n + (1-\frac{i}{2})^n\right)$, $b_3 = \frac{1}{4}$ is the last positive.

Exercise 4.6.33 $^{\heartsuit}$ *Solve algebraically the recurrence* $a_n = \frac{10}{3}a_{n-1} - a_{n-2}$ *for* $a_0 = 1$, $a_1 = 1/3$. *It describes (small) creatures, where* 1 *is for* 1000 *of them or so, that multiply with the coefficient* 10/3 *during some time-unit and have life expectancy* 2 *units; the time to maturity is disregarded (very small).*

Answer: $a_n = (1/3)^n$.

Exercise 4.6.34 ◇ *Consider an arbitrary solution of the recurrence equation* $F_n+\sum_{i=1}^{k} a_i F_{n-i}=f_n$, *where* f_n *satisfies its own recurrence* $f_n+\sum_{i=1}^{\ell} b_i f_{n-i}=0$. *Let* $P(x)=x^k+\sum_{i=1}^{k} a_i x^{k-i}=0$, *and* $Q(x)=x^\ell+\sum_{i=1}^{\ell} b_i x^{\ell-i}=0$. *Prove that* F_n *is given by the formula* F^C *for homogeneous solutions of the recurrence relation for the master equation* $P(x)Q(x)=0$ *from Theorem 4.2.2 if all its roots are simple, or by that from Theorem 4.4.2 if there are multiple roots. For this, check that* $f_n = F_n^C+\sum_{i=1}^{k} a_i F_{n-i}^C$ *satisfies* $f_n+\sum_{i=1}^{\ell} b_i f_{n-i}=0$ *and then use that the number of free* C*-constants is* $k\ell$, *which equals the number of degrees of freedom for the recurrence under consideration.*

Exercise 4.6.35 ◇ *Consider an amount of grass* g_n *subject to the duplication* $g_n = 2g_{n-1}$ *for some unit of time, and rabbits* f_n *such that* $f_n = f_{n-1} + 2f_{n-1}$. *If such rabbits feed on this grass, the recurrence for grass becomes* $g_n = 2g_{n-1} - f_n$; *the recurrence for* f_n *remains unchanged. Solve it assuming that* $g_0 = 1, g_1 = 2, g_2 = 3$ *and obtain the corresponding formula for* f_n; *use Exercise 4.6.34. Check that* $g_n < 0$ *for* $n > 6$, *so this model (for these data) cannot be used as such after* $n = 6$.

Hint: The roots of the $PQ = (x^2 - x - 2)(x - 2)$ in Exercise 4.6.34 are 2 with multiplicity 2 and (-1) (simple). Thus $g_n = C_1 2^n + C_2 n 2^n + C_3(-1)^n = \frac{10}{9}2^n - \frac{1}{6}n2^n - \frac{1}{9}(-1)^n$; $f_n = 2g_{n-1} - g_n = \frac{1}{6}2^n + \frac{2}{9}(-1)^n$. One has $g_0, g_1, \ldots, g_6, g_7 = 1, 2, 3, 5, 7, 9, 7, -7$.

Exercise 4.6.36 ♡ *An* extraordinary subset X *of* $[n] = \{1, 2, \cdots, n\}$ *is such that its* smallest number *equals* $|X|$, *its size. For instance,* $\{1\}, \{2, 3\}, \{2, 4\}$ *are all such* X *for* $n = 4$. *Let* e_n *be their number for* $[n]$. *Check that* $e_n = e_{n-1}+e_{n-2}$ *and that* $e_n = f_n$ *for the Fibonacci numbers* f_n.

Hint: Use the map $X \mapsto (X + 1) \cup n$ for extraordinary subsets X in $[n - 2]$, where $\{x_1, x_2, \ldots\} + 1 = \{x_1 + 1, x_2 + 1, \ldots\}$.

Exercise 4.6.37 ◇ *A* 2-extraordinary subset X *in* $[n] = \{1, 2, \cdots, n\}$ *must have* $|X|$ *as its* second smallest number[33]. *For instance,* $X = \{12\}, \{134\}, \{2456\}$ *are such. Let* E_n *be their number in* $[n]$: $E_1 = 0, E_2 = E_3 = 1, E_4 = 3, \ldots$.

(i) Follow Exercise 4.6.36 to prove that $E_n = E_{n-1} + E_{n-2} + f_{n-3}$ *for the Fibonacci numbers* f_n, *where* $n \geq 3$.

(ii) Use Exercise 4.6.34 to check that $E_n = (C_1 n\lambda_1^n + C_2 n\lambda_2^n) + \frac{f_n}{5}$ *for* $\lambda_{1,2} = \frac{1\pm\sqrt{5}}{2}$ *and* $C_{1,2} = \frac{3\mp\sqrt{5}}{10}$. *Also,* $E_n = \frac{n\ell_{n-2}+f_n}{5}$ *for the* Lucas numbers ℓ_n.

[33] R. Grimaldi, *Extraordinary subsets: a generalization*, Fibonacci Quart. 55:2 (2017), 114–122.

Exercise 4.6.38 $^\heartsuit$ *Let h_n =Ceiling$[(2+\sqrt{2})^n]$; Ceiling$[x]$ is the closest integer no smaller than x. Check that $h_n = 4h_{n-1} - 2h_{n-2}$ starting with $n = 3$.*

Hint: Consider $(2+\sqrt{2})^n + (2-\sqrt{2})^n$.

Exercise 4.6.39 $^\heartsuit$ *Consider the sequence h_n =Ceiling$[(5+\sqrt{21})^n]$, Ceiling$[x]$ is the closest integer no smaller than x.*

(i) Check that $h_n = 10h_{n-1} - 4h_{n-2}$ starting with $n = 2$.

(ii) Prove that h_n is divisible by 10 for odd n and only for them.

Exercise 4.6.40 $^\heartsuit$ *Find a depth 2 recurrence relation for $x_n = 3^{n+1} - 5^n$. What is the non-homogeneous recurrence for $3^{n+1} - 5^n + n^2$?*

Exercise 4.6.41 $^\diamondsuit$ *Let $\lambda_1, \ldots, \lambda_n$ be pairwise distinct numbers, which can be generally complex, and $m = m_1 + \cdots + m_n$ for some $m_i \geq 1$.*

(i) Assume that S_n is a linear combination with constant coefficients of $n^j \lambda_i^n$ for $1 \leq i \leq n$, $0 \leq j < m_i$. Check that $S_n = aS_{n-1} + bS_{n-2} + \ldots$, where $(X-\lambda_1)^{m_1}(X-\lambda_2)^{m_2}\cdots = X^m - aX^{m-1} - bX^{m-2} + \ldots$.

(ii) Let $S_n = \sum_{r=0}^{n} r\binom{n}{r}$. Use formula (3.2.7) from Section 3.2.3 to check that $S_n = 4S_{n-1} - 4S_{n-2}$ for $n \geq 2$. You can also deduce it from the relation $S_n = 2S_{n-1} + 2^{n-1}$ from Exercise 4.6.24(i).

(iii) Setting $T_n = \sum_{r=0}^{n} \binom{4n}{4r}$, check that Exercise 3.7.41 results in the relation $T_n = 12T_{n-1} + 64T_{n-2}$ for $n \geq 2$.

Exercise 4.6.42 $^\spadesuit$ *Given a sequence of nonzero numbers $p_0 = 1, p_n \neq 0$ for $n \geq 1$, let $g_n = 1 + \sum_{m=1}^{n} (-1)^m \frac{1}{p_1 p_2 \cdots p_m}$ and let $\beta = \lim_{n\to\infty} g_n$ provided that $|p_1 p_2 \cdots p_m| \to \infty$ as $m \to \infty$. Define u_n for $n \geq 1$ by the formula $u_n = \frac{p_{n+1}}{(p_{n+2}-1)(p_{n+1}-1)}$ assuming that $p_n \neq 1$ for $n \geq 2$. We set $f_1 = g_1$, $f_n = g_n \prod_{m=2}^{n} \frac{p_m}{(p_m-1)}$ for $n \geq 2$.*

(i) Verify that $f_n = f_{n-1} + u_{n-2} f_{n-2}$ for $n \geq 3$. In particular, $f_n = f_{n-1} + \frac{1}{n-2} f_{n-2}$ for $p_m = m$, and $f_n = \delta_n = D_n/(n-1)!$ from (3.6.14). Check that $g_n = \frac{2^{n+1}+(-1)^n}{3\cdot 2^n}$ and $f_n = 2^{n-1} g_n = \frac{2^{n+1}+(-1)^n}{6}$ for $p_m = 2\,(m \geq 1)$.

(ii) Tree growth near saturation. For $p_m = m^2$, $g_n = 1 - 1 + \frac{1}{(2!)^2} - \frac{1}{(3!)^2} + \cdots + (-1)^n \frac{1}{(n!)^2}$ and the corresponding $f_n = g_n \frac{2n}{n+1}$, obtain the recurrence:

$$f_n = f_{n-1} + \frac{(n-1)}{(n-2)n(n+1)} f_{n-2}, \text{ where } n \geq 3,\ f_1 = 0, f_2 = 1/3 = f_3.$$

Accordingly, $\beta = 1 - 1 + \frac{1}{(2!)^2} - \frac{1}{(3!)^2} + \ldots$ *equals* $J_0(2) \approx 0.22389$ *in terms of the Bessel function* $J_0(x)$ *from (4.5.28), and* $\lim_{n\to\infty} f_n = 2\beta \approx 0.447782$. *See Fig. 4.9.*

(iii) Check that $\widetilde{f}_n = 1 - \frac{1}{n+1}$ *is another solution of the recurrence relation from (ii), and* $\phi_n = (3b-4a)f_n + 2a\widetilde{f}_n$ *solves this recurrence with* $\phi_1 = a, \phi_2 = b$. *This gives that* $\lim_{n\to\infty}\phi_n = 2\beta(3b-4a)+2a$. *See Exercise 5.6.41(ii) (the usage of generalized continued fractions).*

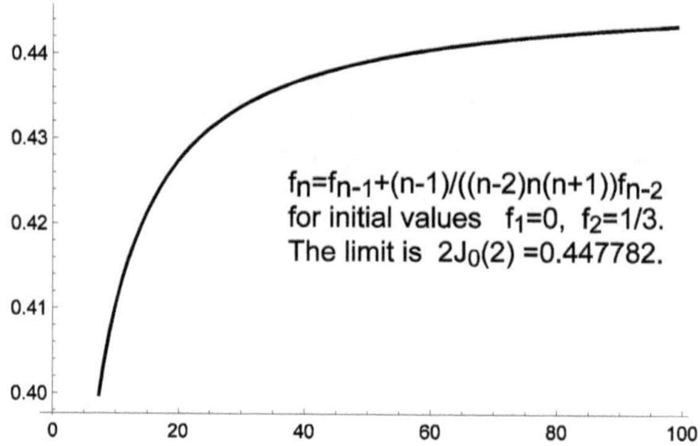

Figure 4.9: Modeling the saturation of tree growth.

Exercise 4.6.43 ♣ *Deduce from the recurrence relation* $\mathcal{D}_n^{(d)} = (n-1)\mathcal{D}_{n-1}^{(d)} + (n-1)(n-2)\cdots(n-d)\mathcal{D}_{n-d-1}^{(d)}$ *for the* **derangements** *of depth* $d \geq 1$ *in Proposition 3.6.2 the following recurrence:* $\kappa_n = \frac{n-1}{n}\kappa_{n-1} + \frac{1}{n}\kappa_{n-d-1}$ *for* $\kappa_n = \mathcal{D}_n^{(d)}/n!$, *where* $n \geq d+1$. *Then check by a direct substitution that this recurrence follows from the recurrence* $\kappa_n = \frac{1}{n}(1 + \sum_{i=d+1}^{n-d-1}\kappa_i)$, *where* $n \geq d+1$ *and we set* $\kappa_0 = 1, \kappa_i = 0$ *for* $1 \leq i \leq d$. *Under these initial conditions (for the derangements), both recurrences give the same* κ_n.

Exercise 4.6.44 ♠ *Check that the* z*-expansion of* $\varpi_d(z) = \frac{e^{-H_d}}{1-z}\exp(\sum_{j=1}^{d}\frac{1-z^j}{j})$ *is the* **generating function** *for* κ_n *from Exercise 4.6.43:* $\varpi_d(z) = \sum_{n=0}^{\infty}\kappa_n z^n$. *Here* $H_d = \sum_{j=1}^{d}\frac{1}{j}$, *which are the Harmonic numbers. This can be deduced from the definition of* $\mathcal{D}_n$, *but we suggest proving that the coefficients of* ϖ_d *satisfy the recurrence relations for* κ_n *and finding that the coefficients of* z^i *are* 0 *for* $1 \leq i \leq d$. *Let us provide a computer-generated calculation of the first* 11 *terms of the expansion of* $\varpi_d(z)$ *for* $d = 2$:

$$1 + \frac{1}{3}z^3 + \frac{1}{4}z^4 + \frac{1}{5}z^5 + \frac{2}{9}z^6 + \frac{19}{84}z^7 + \frac{107}{480}z^8 + \frac{361}{1620}z^9 + \frac{703}{3150}z^{10} + O\left(z^{11}\right).$$

Chapter 5

Modular Arithmetic

5.1 Congruences modulo n

The *modular split* of magic squares we discussed in Section 1.2.3 is a clear example of the potential of modular arithmetic for combinatorics. Conceptually, congruences and finite fields are "natural" sources of "randomness." It is not surprising that number theory plays a growing role in modern theoretical physics, though the connections here are always challenging. Metaphysically, modular arithmetic is a study of "alternative universes" where numbers, geometry, and physics are all finite. Here p-adic numbers are important, but we do not discuss them in this book.

5.1.1 Remainders-residues

The starting point is the definition of the *remainder* or *residue* of an integer m modulo n. Here m can be positive or negative; n is always assumed positive: $1, 2, 3, \ldots$. We set $m = qn + r$, where ***the residue*** r must satisfy the inequalities $0 \le r \le n-1$. The residue is obviously unique: if $r_1 > r_2$ are two residues for the same m, then $r_1 - r_2$ must be divisible by n, which is impossible because $n > r_1 - r_2 > 0$.

We write $r = m \bmod n$. In computer languages: $r = \mathrm{Mod}[m, n]$. We mostly stick to this definition in this book: mod *is the operation of taking the residue.* However, we write $a = b \bmod n$ in various modular calculations if these integers are ***congruent modulo*** n, i.e. $a - b$ is divisible by n. This is of course equivalent to $a \bmod n = b \bmod n$: if the remainders of a and b coincide then n divides $a - b$. This is the standard usage of mod.

Finding the residues is very constructive: we subtract from $m \ge 0$ the greatest possible multiple of n not exceeding m. The difference is r: it is in the required range. This procedure works for $m < 0$ too, but then we will subtract a negative multiple of n, i.e. add a proper positive multiple to obtain $0 \le r \le n-1$. Let us practice with the residues.

Some examples. If $r = 0$, then this means that m is divisible by n. If $r = m \bmod n$, then $-m \bmod n = n-r$ unless $r = 0$. For instance, $17 \bmod 5 = 17 - 15 = 2$, and $-17 \bmod 5 = 20 - 17 = 5 - 2 = 3$. Adding or subtracting any multiples of n does not change r. Practically, it is not always necessary to work out the division to find the residue r. We can calculate r in stages by subtracting from m some "convenient" multiples of n. This will diminish m, and we can proceed by induction until reaching the *range*: $0 \le r \le n-1$.

For instance, $250 \bmod 7 = 250 - 210 \bmod 7 = 40 \bmod 7 = 40 - 35 = 5$. *From now on, we will mostly omit intermediate instances of " mod" in similar calculations.* Putting " mod n" once will be sufficient; then we assume that all calculations are modulo n. We will write $250 \bmod 7 = 250 - 210 = 40 = 5$.

The key is that *taking the residue* is compatible with addition and multiplication. It is almost true that the sum/product of residues is the residue of the sum/product (which is the "ideal compatibility"), but some correction is necessary here. The following holds:

$$(a+b) \bmod n = \big((a \bmod n) + (b \bmod n)\big) \bmod n, \qquad (5.1.1)$$
$$(ab) \bmod n = \big((a \bmod n)(b \bmod n)\big) \bmod n.$$

We need to address here that the sum or the product of two residues modulo n can be possibly outside of the range, i.e. greater than $n-1$.

Let us justify the second formula (the first is simpler). The following notation for the residues is (and will be) convenient: $\overline{a} = a \bmod n$, $\overline{b} = b \bmod n$. Then $a = pn + \overline{a}$, $b = qn + \overline{b}$, and

$$(ab) \bmod n = (pn + \overline{a})(qn + \overline{b}) = (\overline{a}\overline{b}) + (\cdots)n = (\overline{a}\overline{b}) \bmod n.$$

Here $(\cdots) = p\overline{b} + \overline{a}q + \overline{a}\overline{b}n$, but we do not need to know this.

More generally, if we need to calculate the residue modulo n of any polynomial combination $P(x, y, z, \ldots)$ of some numbers $x, y, z, \ldots$, then the residue will remain the same if we replace these numbers by their residues $\overline{x}, \overline{y}, \overline{z}, \ldots$, or by any numbers $x', y', z', \ldots$ that are *congruent* to $x, y, z, \ldots$ modulo n.

So the claim is that $P(x, y, z, \ldots) = P(x', y', z', \ldots) \bmod n$ provided that $x = x' \bmod n$, $y = y' \bmod n$, $z = z' \bmod n, \ldots$. The purpose is of course to calculate $P(x, y, z, \ldots) \bmod n$. Taking here $\overline{x}, \overline{y}, \overline{z}$, and so on is natural, but not always the most "convenient." For instance, replacing 99 by -1 modulo 100 (which is not in the range) works very well in the following calculation: $99^{9999} \bmod 100 = (-1)^{9999} = -1 = 100 - 1 = 99$. Finding $m^k \bmod n$ is a classical demonstration of the power of modular methods. Important: *do not replace* 9999 *here by its remainder modulo* 100. These replacements can be done for m in m^k but not for the exponent k.

Problem 5.1.1 *Find* 999^{999} mod 7.

Solution. First of all, 999 mod 7 = 299 = 299 − 280 = 19 = 5. So $999^{999} = (-2)^{999} = -2^{999}$ mod 7. Here 2^{999} is still extremely large. Next: 2^0 mod 7 = 1, 2^1 mod 7 = 2, 2^2 mod 7 = 4, 2^3 mod 7 = 1. Then we stop because 2^{999} mod 7 = $(2^3)^{333} = 1^{333} = 1$, and 999^{999} mod 7 = −1 = 6. Note that the periodicity of the exponents here is modulo 3 (not modulo 7). Such periods are of fundamental importance in number theory. ❑

5.1.2 Modular numbers

We come to a fundamental discovery: there exist universes with only finitely many numbers, which are residues modulo n. The operations here are regular algebraic ones followed by the "truncation" modulo n. Everything is "finite" in such universes: geometry, physics, ... , anything at all.

Generally, any recursive process with a finite total number of possible outputs (for all steps) eventually becomes periodic. We assume that the outputs are used as inputs for the next steps, so the number of inputs is finite too. The periodic behavior begins after the first repetition of the inputs. An important example is Exercise 2.6.11 on repeating decimals, which we are going to solve in detail now.

Problem 5.1.2 *Prove that the decimal expansion of any rational number m/n becomes periodic starting from some decimal place. It is finite if and only if n is a product of some powers of* 2 *and* 5. *Otherwise, the length ℓ of the period of this expansion is no greater than $n-1$ and can become $n-1$. Check that it is exactly minimal $p > 0$ such that $10^p = 1$ mod n if* $\gcd(n, 10) = 1$.

Solution. Finding the decimals is a finite algorithm (a finite state machine): with finitely many inputs and outputs for any given n. The input is an integer x for the 1st step and any x from 0 to $n-1$ for other steps. The first step is exceptional: the output is the pair $\{q, r\}$ for $r = m$ mod n and $q = (m-r)/n$. Here q gives the integer part of the decimal fraction of m/n, and $x = r$ will be used as the input of the 2nd step.

For the 2nd and other steps, the ***output*** is a pair $\{q, r\} = \{$*the partial quotient q of $10x$ divided by n, and the corresponding residue* $r = 10x$ mod $n\}$. Then we set $x = r$ and continue. Thus, r from the previous step serves as the ***input*** x for the next one. The number q gives the corresponding digits.

For instance, $3/14 = 0.2\underbrace{142857}14\ldots$, where we underline the period. Namely, $3 = 0\cdot 14+3$, $3 \mapsto r$, $10r = 30 = 2\cdot 14+2$, $2 \mapsto r$, $10r = 20 = 1\cdot 14+6$, $6 \mapsto r$, $10r = 60 = 4\cdot 14+4$, which gives $3/14 \approx 0.214$, and so on.

If some input here is 0, then the algorithm terminates at this step and the decimal expansion of m/n is finite. This happens only if n is a product of powers of 2 and 5. Otherwise, the number of possible inputs is $(n-1)$ (no 0), and therefore ℓ can be no greater than $(n-1)$. Indeed, after the first repetition of the inputs, the process will obviously repeat itself.

Let us provide a formula for the pth digit d_p after dot of m/n assuming that $m < n$ and $\gcd(m, n) = 1$:

$$d_p = \text{Floor}[10r/n] \text{ for } r = 10^{p-1}m \bmod n.$$

So ℓ does *not* depend on m. If $\gcd(n, 10) = 1$, the period begins with the first digit after dot and ℓ is minimal $p > 0$ such that $10^p = 1 \bmod n$. For instance, it is $(n-1)$ for $1/7 = 0.\underbrace{142857}14\ldots$. Check these claims. ❑

Definition 5.1.3 *Let $\mathbb{Z}_n$, "zeē-en", be the set $\{0, 1, 2, \ldots, n-1\}$ with the addition and multiplication given by formulas (5.1.1). It has subtraction, $a - b \bmod n$, but generally no division. It is called the* ***ring of integers modulo*** *n.*

Actually, we are supposed to introduce new symbols for these operations in $\mathbb{Z}_n$, which are the classical ones followed by the truncation modulo n. We will use $\bar{+}, \bar{\times}$ a few times, but then will completely switch to the standard $+, \times$. We will put $\bmod n$ or $a, b \in \mathbb{Z}_n$ if some confusion is possible.

Geometrically, we now have a circle modulo n with 0 after $n-1$, a ring indeed instead of the classical line of integers. The inequalities "$a > b$" are meaningless in $\mathbb{Z}_n$, but anything else survives. However, there are deviations.

For instance, let us solve the equations $x^2 = x\bar{\times}x = 1$ modulo 12 by inspection. Its roots are $x_1 = 1, x_2 = -1 = 11, x_3 = 5, x_4 = -5 = 7 \bmod 12$. So we have 4 roots, which "contradicts" one of the main theorems of algebra: any equation of degree m has no greater than m roots. Conclude that this theorem generally fails modulo n. Also, $4\bar{\times}3=0 \bmod 12$, which never happens with our usual numbers (natural, rational, real, complex), where the product of two nonzero numbers is always nonzero.

5.1.3 Modular invertibility

The ***inverse*** (the reciprocal) of a in any algebraic theory is defined as u such that $ua = 1$. We will need some basics from arithmetic. Let us prove that $d = \gcd(m, n)$, ***the greatest common divisor*** of two positive integers m, n, coincides with $d' = \min\{am + bn > 0\}$ for any integers a, b (provided $am + bn > 0$). *First*, $d' \le m$ $(m = 1 \cdot m + 0 \cdot n)$, and $d' \le n$. *Second*, d' divides m and n. For this, let $m = qd' + r$, where $0 \le r \le d' - 1$: the division of m by

d' with the remainder r. Then $r = m - q(am+bn) = (1-qa)m + (-qb)n = a'm+b'n$ and can be only 0 since d' is minimal among all $am+bn > 0$. *Third*, d divides m and n, so it divides d' and $d' \geq d$. This is possible only if $d = d'$. Indeed, d' divides m, n, but d is the greatest such.

Generally, d' is more convenient to find than d. The optimal way is the famous ***Euclid's algorithm***. Assume that $m \geq n$ and set $\{m_0 = m, n_0 = n\}$, $\{m_1 = m \bmod n, n_1 = n_0\}$, $\{m_2 = m_1, n_2 = n_1 \bmod m_1\}$, and continue until $m_\ell = 0$ or $n_\ell = 0$. Since we always subtract some multiples of m or n in this construction, one has: $\gcd(m,n) = \gcd(m_1,n_1) = \gcd(m_2,n_2) = \cdots = \gcd(m_{\ell-1}, n_{\ell-1})$. Thus $d = \min\{m_{\ell-1}, n_{\ell-1}\}$.

For instance, $\gcd(119,91) = \gcd(119-91,91) = \gcd(28,91) = \gcd(28,91-3\cdot 28) = \gcd(28,7) = 7$, and $\gcd(n+1,n-1) = \gcd(2,n-1) = 2$ for odd $n > 1$.

If $\gcd(m,n) = 1$ then the integers $m, n > 0$ are called ***relatively prime*** or co-prime.

Let p be a ***prime number***: $p > 1$ divisible only by p and 1. Then $\gcd(p,n)$ is either 1 or p, where the latter means that n is divisible by p. So primes p are numbers $p > 1$ such that all numbers $n \geq 1$ with $n < p$ are relatively prime to p. A key property of relatively prime numbers is as follows.

Lemma 5.1.4 *If n divides $m\ell$ and $\gcd(n,m) = 1$, then n divides ℓ. For instance, if $\gcd(n,m) = 1$ and $\gcd(n,\ell) = 1$, then $\gcd(n, m\ell) = 1$.*

Proof. The relation $am+bn = 1$ gives that $am \bmod n = 1$. Thus $m\ell \bmod n = 0$ implies that $am\ell \bmod n = \ell \bmod n = 0$, which means the divisibility of ℓ by n. The second claim formally follows from the first. ❑

The ***Prime Factorization Theorem*** states that any integer $n > 1$ can be uniquely represented as $n = p_1^{a_1} \cdots p_m^{a_m}$ for prime p_i. The existence is obvious by induction. What is not obvious is that the set of pairs $\{p_i, a_i\}$ is uniquely determined by n. Let us assume that $n = p_1^{m_1} \cdots p_a^{m_a} = q_1^{\ell_1} \cdots q_\ell^{b_\ell}$ for a different set of pairs $\{q_i, b_i\}$. Reducing similar factors, we obtain that there exists some prime q that divides a product of powers of p_i for prime numbers p_i different from q. This is impossible due to Lemma 5.1.4 because $\gcd(q, p_i) = 1$.

Proposition 5.1.5 *The number $a \in \mathbb{Z}_n$ has an inverse u if and only if $\gcd(a,n) = 1$. If u exists, then it is unique, and any equation $ax = b$ for $b \in \mathbb{Z}_n$ then has a unique solution, which is $x = bu$. We call such a number a* ***invertible*** *(modulo n) and put $x = b/a$ correspondingly.*

Proof. This is direct from Lemma 5.1.4. Let us provide a constructive proof similar to that of the Chinese Remainder Theorem from Section 2.1.3. Consider the *multiplication line* for a, which is $\{0, a, 2a, 3a, \ldots, (n-1)a\}$, where

$2a = a + a \bmod n = 2\bar{\times}a$ and so on. If $ia = ja$ here for $i < j$, then $(j-i)a = 0 \bmod n$, which means that $(j-i)a$ is divisible by n. Since a is relatively prime to n, this can be only if $(j-i)$ is divisible, which is impossible because $0 < (j-i) < n-1$. So all ia here are different (modulo n). Thus the multiplication line is a permutation of $\{0, 1, \ldots, n-1\}$ (Pigeonhole Principle (II)), and 1 occurs somewhere, which proves the existence of u. Similarly, any b from $\mathbb{Z}_n$ occurs exactly once in this multiplication line, which proves the existence and uniqueness of x such that $xa = b$ (including the uniqueness of u for $b = 1$). ❑

The multiplication lines are among the key instances of *modular randomness*: they provide a simple and universal way to produce *random* permutations (of the set of residues). They are really random when a is somewhere in the middle of the range of invertibles.

For relatively small n, solving the equations $ax = b$ in $\mathbb{Z}_n$ *by inspection* is quite reasonable: examine $x = 0, 1, 2, \ldots, n-1$. A more systematic way is the usage of Euclid's algorithm explained in the next problem.

Problem 5.1.6 *Use Euclid's algorithm to find all integer solutions of the equation $99x + 133y = 1$, and then calculate $\frac{1}{99} \bmod 133$ and $\frac{1}{133} \bmod 99$.*

Solution. We begin with $\{m_0 = m = 133, n_0 = n = 99\}$ and obtain: $\{m_1 = m - n = 133 - 99 = 34, n_1 = n\}$, $\{m_2 = m_1, n_2 = n - 2m_1 = 99 - 2 \cdot 34 = 31\}$, $\{m_3 = m_1 - n_2 = 34 - 31 = 3, n_3 = n_2\}$, $\{m_4 = m_3, n_4 = n_2 - 10m_3 = 31 - 10 \cdot 3 = 1\}$. Substituting: $n_2 = n - 2m_1 = n - 2(m-n) = 3n - 2m$, $m_3 = m_1 - n_2 = (m-n) - (3n-2m) = 3m - 4n$, and $n_4 = n_2 - 10m_3 = (3n-2m) - 10(3m-4n) = 43n - 32m = 1$. Thus, $\{x_0 = 43, y_0 = -32\}$ is a solution. Arbitrary solutions are then $\{x = 43 + 133k, y = -32 - 99k\}$ for integer k (use the lemma). Accordingly, $1/99 \bmod 133 = x_0 = 43$ and $1/133 \bmod 99 = y_0 = -32 = 99 - 32 = 67$. ❑

As an example for Proposition 5.1.5, the list of all invertibles in $\mathbb{Z}_{12}$ is $\{1, 5, 7, 11\}$. All of them except 1 are prime numbers, so $\gcd(a, 12) = 1$ is granted for them. Any other numbers mod 12 are either divisible by 2 or by 3. Note that if a and b are invertibles, then $\frac{1}{a}$ and ab are also invertible; check that $\frac{1}{a} = a$ for the four listed, so the equation $X^2 = 1$ has 4 solutions in $\mathbb{Z}_{12}$.

Let us mention here *Euler's formula* for the number of invertibles modulo n: $\varphi(n) = (p_1^{a_1} - p_1^{a_1-1}) \cdots (p_\ell^{a_\ell} - p_\ell^{a_\ell-1})$, where $n = p_1^{a_1} \cdots p_\ell^{a_\ell}$ is the prime factorization. See Exercise 5.6.13. It indeed gives: $\varphi(12) = (3-1)(4-2) = 4$.

Definition 5.1.7 *For a prime p, the* **prime field** mod p *is $\mathbb{Z}_p$; another notation is $\mathbb{F}_p$. Any $a \neq 0$ is invertible in $\mathbb{F}_p$, which follows from Proposition 5.1.5, and a/b are well-defined for any $b \neq 0$ (as for rationals).*

Thanks to this "definition-proposition," $\mathbb{F}_p$ is very much similar to rational numbers. For instance, any polynomial $P(X) = X^n + \ldots$ of degree n can have no more than n roots in $\mathbb{F}_p$ (not true in $\mathbb{Z}_{12}$: see above). Indeed, if x_1 is a root, then we divide: $Q(X) = P(X)/(X - x_1) = X^{n-1} + \ldots$, and then proceed by induction on degree. In particular, formula (4.2.6) can be used for quadratic $P(X)$ if $p \neq 2$ and the square root there belongs to $\mathbb{F}_p$.

As an example of calculations with modular rationals, let us "transfer" the formula $\frac{1}{2} + \frac{1}{3} = \frac{5}{6}$ to $\mathbb{F}_7$. We need two multiplication lines:

$$\left\{ \begin{array}{c|cccccc} 2\times: & 1 & 2 & 3 & 4 & 5 & 6 \\ = & 2 & 4 & 6 & 1 & 3 & 5 \end{array} \right\}, \quad \left\{ \begin{array}{c|cccccc} 3\times: & 1 & 2 & 3 & 4 & 5 & 6 \\ = & 3 & 6 & 2 & 5 & 1 & 4 \end{array} \right\},$$

where we omit 0. We actually do *not* multiply to obtain these lines: the procedure is simply to *add* 2 or 3 modulo 7 to the previous entry. By finding the preimages of 1 in these lines, i.e. we obtain: $\frac{1}{2} = 4$, $\frac{1}{3} = 5$, and $\frac{1}{2} + \frac{1}{3} = 4 + 5 = 2$ in $\mathbb{F}_7$. On the other hand, $\frac{5}{6} = \frac{-2}{-1} = 2$: it is of course the same!

Note that such multiplication lines allow us to calculate directly any b/a, not only $1/a$. For small n, it makes sense to calculate once and forever the addition and multiplication tables and then use them as *lookup tables*. The most famous such tables are for $\mathbb{Z}_2$: $\begin{array}{c|cc} + & 0 & 1 \\ \hline 0 & 0 & 1 \\ 1 & 1 & 0 \end{array}, \quad \begin{array}{c|cc} \times & 0 & 1 \\ \hline 0 & 0 & 0 \\ 1 & 0 & 1 \end{array}$. They are the keys for computers. Notice that the multiplication table is "simpler" than the addition one! Let us provide the tables for $\mathbb{Z}_4$:

+	0	1	2	3
0	0	1	2	3
1	1	2	3	0
2	2	3	0	1
3	3	0	1	2

×	0	1	2	3
0	0	0	0	0
1	0	1	2	3
2	0	2	0	2
3	0	3	2	1

We mention here that Exercise 5.6.45 states that the first one, which is a Latin square, has no orthogonal mate; this problem is a combinatorial *puzzle*: no arithmetic is necessary. Let us discuss the connections with Latin squares.

5.2 Applications

5.2.1 Latin squares and MOLS

We fix $n > 1$ and pick any $r \in \mathbb{Z}_n$ such that $\gcd(r, n) = 1$, i.e. r is an invertible element in $\mathbb{Z}_n$. Recall that the latter means that r is relatively prime to n: $\gcd(n, r) = 1$.

From now on, the row numbers and the column numbers in the considered $n \times n$ squares will be from 0 to $n - 1$. Namely, $0 \leq i \leq n - 1$ will be the row number, and $0 \leq j \leq n-1$ the column number. In the formulas involving i, j, they will be treated as residues modulo n.

In this section, the notation $\{a, b, c, \ldots\}$ will be used for *sequences*, not for *sets*. The sequences will be mostly columns or rows of some squares. We add "sequence" if some confusion with the set-theoretical notion of union of objects $\{a, b, c, \cdots\}$ is possible.

Let the entry a_{ij} at the cross of the ith row and jth column be $a_{ij} = i + jr \bmod n$, treated as an element in $\mathbb{Z}_n$. We obtain an $n \times n$ square with the entries in $\mathbb{Z}_n$.

There is no repetition in its columns and rows, so this is a Latin square of size $n \times n$. Indeed:

$$\begin{aligned} &\text{if } a_{ij} = a_{i'j}, \text{ then } \ i + jr = i' + jr, \ i' - i = 0, \text{ and } i' = i; \\ &\text{if } a_{ij} = a_{ij'}, \text{ then } \ i + jr = i + j'r, \ (j' - j)r = 0, \ j' = j. \end{aligned} \tag{5.2.2}$$

In the 2nd line, we simply divide by r using its invertibility, and obtain that $j' - j = 0$, which means that j and j' must coincide. *This Latin square will be denoted by LS_r.*

Proposition 5.2.1 *Let us assume that r_1 and r_2 are invertibles in $\mathbb{Z}_n$, and that $r_1 - r_2$ is invertible too. For $\mathbb{F}_p$, where $n = p$ is prime, this simply means that r_1, r_2, and $r_1 - r_2$ are nonzero in $\mathbb{F}_p$; so we have $(p-1)(p-2)$ choices for r_1 and r_2 here. Then $L_1 = LS_{r_1}$ and $L_2 = LS_{r_2}$ form* MOLS, *i.e. the corresponding $L_1 \vee L_2$, the bi-square obtained by combining the corresponding entries from L_1 and L_2, contains all pairs $\langle u, v \rangle$ for $u, v \in \mathbb{Z}_n$, and each pair occurs once.* ❑

We prove a somewhat more general version below. Now let us provide an example. The calculations of LS_r are quite simple if the multiplication line for r is known. Let $n = 5$, and $r_1 = 2, r_2 = 3$. We need the following multiplication lines modulo 5:

$$\left\{ \begin{array}{c|ccccc} 2\times: & 0 & 1 & 2 & 3 & 4 \\ = & 0 & 2 & 4 & 1 & 3 \end{array} \right\}, \ \left\{ \begin{array}{c|ccccc} 3\times: & 0 & 1 & 2 & 3 & 4 \\ = & 0 & 3 & 1 & 4 & 2 \end{array} \right\},$$

Due to the formula $a_{ij} = i + jr$, these multiplication lines will be the first rows in the corresponding L_1 and L_2; their first columns will be simply $\{0, 1, 2, 3, 4\}$. *No further calculations are needed if the 1st row and the 1st column are known.* One cyclically reads the entries in the first row (the multiplication line) starting with $k = 0, 1, 2, 3, 4$. The corresponding line will then be the row for $i = k$ in LS_r. This can be easily done well beyond $n = 5$. For $n = 5$, and $r_1 = 2, r_2 = 3$:

$$L_1 = \left\{ \begin{array}{c|ccccc} i\backslash j & 0 & 1 & 2 & 3 & 4 \\ \hline 0 & \mathbf{0} & \mathbf{2} & \mathbf{4} & \mathbf{1} & \mathbf{3} \\ 1 & 1 & 3 & 0 & 2 & 4 \\ 2 & 2 & 4 & 1 & 3 & 0 \\ 3 & 3 & 0 & 2 & 4 & 1 \\ 4 & 4 & 1 & 3 & 0 & 2 \end{array} \right\}, \quad L_2 = \left\{ \begin{array}{c|ccccc} i\backslash j & 0 & 1 & 2 & 3 & 4 \\ \hline 0 & \mathbf{0} & \mathbf{3} & \mathbf{1} & \mathbf{4} & \mathbf{2} \\ 1 & 1 & 4 & 2 & 0 & 3 \\ 2 & 2 & 0 & 3 & 1 & 4 \\ 3 & 3 & 1 & 4 & 2 & 0 \\ 4 & 4 & 2 & 0 & 3 & 1 \end{array} \right\}.$$

Then we form $\widetilde{L} = L_1 \vee L_2$ by combining the corresponding entries; see below. All of its entries are different, indeed. Finally, if $\langle u, v\rangle$ is an entry in $\widetilde{L}$, we employ the formula $\langle u, v\rangle \mapsto 1+u+nv$ from Section 1.3.3 to construct the corresponding magic square, denoted by MS. The numbers u, v are treated here as normal integers, *not* as modular numbers.

$$\widetilde{L} = \left\{ \begin{array}{c|ccccc} i\backslash j & 0 & 1 & 2 & 3 & 4 \\ \hline 0 & 00 & 23 & 41 & 14 & 32 \\ 1 & 11 & 34 & 02 & 20 & 43 \\ 2 & 22 & 40 & 13 & 31 & 04 \\ 3 & 33 & 01 & 24 & 42 & 10 \\ 4 & 44 & 12 & 30 & 03 & 21 \end{array} \right\}, \quad MS = \left\{ \begin{array}{ccccc} 0 & 1 & 2 & 3 & 4 \\ \hline 1 & 18 & 10 & 22 & 14 \\ 7 & 24 & 11 & 3 & 20 \\ 13 & 5 & 17 & 9 & 21 \\ 19 & 6 & 23 & 15 & 2 \\ 25 & 12 & 4 & 16 & 8 \end{array} \right\}.$$

The resulting MS is not only magic but also *pandiagonal magic* (check).

Let us provide *justifications.* We will begin with a particular case of Proposition 5.2.1.

Corollary 5.2.2 *Let* $\gcd(r_1 \pm 1, n) = 1$ *and* $\gcd(r_2 \pm 1, n) = 1$. *Then there is no repetition in the corresponding pandiagonals and anti-pandiagonals. In* $\mathbb{F}_p$, *these conditions simply mean that* $r_{1,2} \neq \pm 1$. *Accordingly, the procedure* $\langle u, v\rangle \mapsto 1 + u + nv$ *for the bi-entries* $\langle u, v\rangle$ *of* $\widetilde{L} = L_1 \vee L_2$ *results in a pandiagonal magic square.*

Proof. The absence of repetitions in columns and rows of L_1 and L_2 is because they are Latin due to $\gcd(r_{1,2}, n) = 1$. So the vertical and horizontal sums are really *magic*: all are equal to $n(n^2+1)/2$.

Due to Section 1.2, pandiagonals are given by the formulas $i-j = c \bmod n$, and anti-pandiagonals by $i + j = c \bmod n$ for some c.

Now, if there is a repetition in the pandiagonal corresponding to c in the Latin square LS_r, then $i - j = i' - j' \bmod n$ and $i + rj = i' + rj' \bmod n$ for some pairs i, j and i', j'. Hence $i-i' = j-j' \bmod n$ and $i-i' = r(j'-j) \bmod n$. Thus, $(i-i')(1+r) = 0 \bmod n$, and $i = i'$ upon the division by $(1+r)$ (we used its invertibility). This gives that $j = j'$. So all entries are different in any pandiagonal. Similarly, use $(1-r)$ for the anti-pandiagonals. ❑

The following is a generalization of Proposition 5.2.1.

Proposition 5.2.3 *Let $a, b, c, d \in \mathbb{Z}_n$ be such that (i) a, b, c, and d are relatively prime with n, and (ii) $ad - bc$ is also relatively prime with n. In $\mathbb{F}_p$ for prime p, these conditions become: (i) $a, b, c, d \neq 0$, and (ii) $ad - bc \neq 0$. As above, $i, j \in \mathbb{Z}_n$ will be used as row and column numbers.*

(A) Setting $L_1 = \big(ai+bj\big)$, $L_2 = \big(ci+dj\big)$ modulo n, they are Latin squares due to (i) and form MOLS due to (ii).

(B) The MS constructed as in Corollary 5.2.2 from $L_1 \vee L_2$ is pandiagonal magic if all of $a \pm b$ and $c \pm d$ are invertible modulo n.

Proof. The justification for (B) is parallel to that of Corollary 5.2.2. The corresponding L_1 and L_2 have no repetition in any pandiagonal and any anti-pandiagonal. For (A), the coincidence of the entries in $L_1 \vee L_2$ for some i, j and i', j' means that $ai + bj = ai' + bj'$ and $ci + dj = ci' + dj'$ in $\mathbb{Z}_n$. Equivalently, $a(i - i') + b(j - j') = 0$ and $c(i - i') + d(j - j') = 0$. However, this system of equations for $(i - i')$ and $(j - j')$ can only have the zero solution if $ad - bc$ is invertible. This is standard for 2×2 linear systems. Thus $i = i'$, $j = j'$. ❑

5.2.2 Hill ciphers

Cryptograms were the beginning of cryptography. They are simple substitution ciphers, which were considered unbreakable for many centuries. The "key" for them is the ***substitution table***. Now we have many programs online that will break any *sufficiently long* cryptograms for you. For example, see the book by Simon Singh [34] about the history of "secrecy."

The magazines devoted to "solving cryptograms" were very popular in the 19th century. Without computers, such an activity can be entertaining. The usage of computers made this kind of intellectual game outdated; computers can check a lot of variants in no time versus the dictionaries.

The frequencies of the letters are very helpful for deciphering (not only for cryptograms). For instance, "e" is the most frequent character in English. Actually, "space" is the most frequent. Omitting spaces is not actually an option for real messages: the text can become very difficult to read. However, providing them makes breaking cryptograms relatively simple.

In the 20th century, a method of choice for secret communications was the usage of books, especially those that are commonly available everywhere. Then one letter can be encoded in a great variety of ways: the page number, the line number and its number in the line. This is actually a substitution cipher but with multiple "ciphertexts": a polyalphabetic one. When used

[34] S. Singh, *The code book: The science of secrecy from ancient Egypt to quantum cryptography*, Anchor (2000), `https://www.math.uci.edu/~brusso/freshman6.pdf`.

really randomly it is safe, ... unless your adversaries find the book (in your apartment or luggage). Later, digital files replaced books.

Enigma was a great "polyalphabetic" cipher machine; it was designed around the 1920s. It has an electromechanical rotor mechanism that scrambles the 26 letters of the alphabet. If plain text is entered, the illuminated letters are the ciphertext. However, it appeared vulnerable when it was used massively by the Germans during World War II. Frequent phrases like "nothing to report" and some stereotypes in its usage appeared sufficient to "solve" it. The operators were supposed to strictly follow the instructions and use Enigma (any such device) "in the most random way," which is not realistic to expect when a system is used so extensively.

Hill ciphers. Lester Hill suggested in 1929 a simple approach, which is quite useful even now in combination with other tools. Let us take sufficiently large n. Generally, $n \approx 40$ is sufficient for all letters and symbols. However, the greater n is, the more difficult it is to break the cipher. Also, large n can be used to address the frequencies of letters. The probability is about 11% for "e", 8.5% for "a" and 7% for "t" (Concise Oxford Dictionary, 1995). The average length of an English word is about 5, which makes the probability of space about $\frac{1}{6} \approx 17\%$. Accordingly, the probablities of e, a, t become $\frac{5}{6}$ times $11, 8.5, 7\%$. So we can assign 17 symbols to "space", 9 to "e", 7 to "a", 6 to "t" and so on, just to give you some idea. Special consideration is needed when the messages contain many numbers.

Then the procedure is basically as follows. First, replace the letters and symbols in the message by the corresponding residues modulo n, and split them into evenly-sized groups: with m symbols in each. Add some random symbols at the end to make the length of the message divisible by m. Second, take an invertible matrix A of size $m \times m$ with the entries in $\mathbb{Z}_n$; this will be *the key*. Apply A to each of these m-groups, considered as m-columns, and send the resulting m-groups to the recipient. To decode the message, apply A^{-1} and then replace the resulting modular numbers by the letters.

Hill himself did this for $m = 3$. In the case $m = 2$, the message is presented as a sequence of pairs of elements in $\mathbb{Z}_n$. The *key* is then an invertible matrix $A=\begin{pmatrix} a & b \\ c & d \end{pmatrix}$, which will be applied to pairs of elements from $\mathbb{Z}_n$ considered as columns. Thus, we impose condition (ii) in Proposition 5.2.3, which is $\gcd(ad - bc, n) = 1$, but do not assume that $a, b, c,$ or d are relatively prime to n here. This condition is irrelevant for Hill ciphers.

An example. For instance, let $m = 2, n = 41,\ A = \begin{pmatrix} a & b \\ c & d \end{pmatrix} = \begin{pmatrix} 3 & 12 \\ 2 & 17 \end{pmatrix}$, then $Det(A) = ad - bc = 27$, and $B = A^{-1} = \begin{pmatrix} q & r \\ s & t \end{pmatrix} = \begin{pmatrix} 31 & 36 \\ 6 & 32 \end{pmatrix}$ mod n.

Check that $AB = \begin{pmatrix}10\\01\end{pmatrix}$ mod n. The encryption of the "message" (x, y, z, f) is $(u, v, w, g) = (ax + by, cx + dy, az + bf, cz + df)$ mod n. To decipher (decode): $(x, y, z, f) = (qu + rv, sw + tg, qu + rv, sw + tg)$ mod n. For instance, the encryption of $(18, 31, 10, 8)$ is $(16, 30, 26, 8)$. Use your computer to check this and verify that applying B to the latter is, indeed, the former. In *Mathematica*, $\{u, v\} = A.\{x, y\}$, and $B = Mod[Det[A] * B * ModularInverse[Det[A], n], n]$.

With $m = 4$, it is getting really difficult to "solve" the corresponding cipher due to the unknown n and a lot of possible "keys" A. Also, you can do some transformations of the text before applying A and employ similar "tricks" to make this even more difficult.

The problem with many ciphers is when the attacker can send some (many) texts and see their encryptions, which are called "known-plaintext attacks." This is quite a concern for any cipher. However, the ones based entirely on linear algebra are especially vulnerable for such attacks. If an encryption system is available to *all*, then anyone can run it as many times as desired and then do "reverse engineering."

The attacker must of course know the main principles of the cipher and assume that there will be some tricks and random choices, which are always present in real systems. However, "linear encryption systems" always involve risks. Generally, ciphers become much more difficult to "solve" if some "non-linearity" is added to modular linear methods.

Upon some modifications, modular matrix multiplication is heavily used in cryptography, say in "AES" (Advanced Encryption Standard). The usage of finite fields and relevant algebraic geometry over such fields provides almost unlimited possibilities here.

5.2.3 Error correcting codes

Here we generally need systems $L_1, L_2, \ldots, L_k$ of Latin squares of size $n \times n$ that are *pairwise MOLS*, which means that every bi-square $L_i \vee L_j$ for $i \neq j$ contains all (ordered) pairs $\langle u, v \rangle$ for $u, v \in \mathbb{Z}_n$. The length k must be such that kn is no smaller than the number of symbols in our messages. Part (i) of the following problem is entirely combinatorial.

Problem 5.2.4 *(i) Prove that $k < n$ for any set $L_1, L_2, \ldots, L_k$ of pairwise orthogonal Latin squares of size $n \times n$.*

(ii) Use the modular method to construct at least one such set of Latin squares with $k = n-1$ assuming that n is a prime number.

Solution. To prove (i), let us begin with k (independent) *renamings* in every square L_m to make the first rows $\{0, 1, \ldots, n-1\}$ in all of them. The MOLS property is preserved upon such renamings.

This gives that for any given $0 \le i, j \le n-1$, the entries at (i, j) in $L_1, \ldots, L_k$ (the same position for all) must be all different. Indeed, if we have s and s for L_l and L_m at this position, where $l \neq m$, then the corresponding bi-entry in $L_l \vee L_m$ will be $\langle s, s \rangle$. However, *all* such pairs (of coinciding type) already occur in the first row. This proves that $k \le n$: otherwise, at least one $\langle s, s \rangle$ is granted at any (i, j) for some l, m due to Pigeonhole Principle (I).

Now let r be the entry in the *second* row of the *first* column of L_1: $\begin{pmatrix} 0 & 1 & 2 & \cdots \\ r & \cdots & \cdots & \cdots \end{pmatrix}$. Obviously, $r \neq 0$: the first column of L_1 already has 0 as the entry. The entries must be different in all $\{L_m\}$ at this position. This follows from the same logic as in the previous paragraph: if they coincide in L_l and L_m for some $l \neq m$, then the corresponding entry is $\langle s, s \rangle$ in $L_l \vee L_m$, which is impossible.

Assuming now that $k = n$, we obtain that *all* numbers $\{0, 1, \ldots, n-1\}$ must occur at this position in some L_m due to Pigeonhole Principle (II). As we demonstrated above for $m = 1$, this is impossible since 0 is already in the first column of L_m (at $(0, 0)$ in all of them). This proves (i).

For the existence in (ii), we take $LS_{r_1}, \ldots, LS_{r_{n-1}}$ for $r_1 = 1, \ldots, r_{n-1} = n-1$ assuming that n is a prime number. Then Proposition 5.2.1 gives that such LS are pairwise MOLS. ❑

We state without explanation that the usage of such systems of LS generally provides the best possible protection against the noise of almost any kind; this is proven (at least) if n is a power of a prime number. Let us take $n = 5$ and $k = 3$; so we will use 5 frequencies, $0, 1, 2, 3, 4$, to transmit messages over powerlines or somehow with an alphabet of $15 = 5 \cdot 3$ letters. We take $r_{1,2,3} = 1, 2, 3$ in (ii) of Problem 5.2.4. The table of *codes*, the corresponding combinations of consecutive frequencies, for these letters will be as follows:

	0	1	2	3	4
A	**0**	**1**	**2**	**3**	**4**
B	1	2	3	4	0
C	2	3	4	0	1
D	3	4	0	1	2
E	4	0	1	2	3

,

	0	1	2	3	4
F	**0**	**2**	**4**	**1**	**3**
G	1	3	0	2	4
H	2	4	1	3	0
I	3	0	2	4	1
J	4	1	3	0	2

,

	0	1	2	3	4
K	**0**	**3**	**1**	**4**	**2**
L	1	4	2	0	3
M	2	0	3	1	4
N	3	1	4	2	0
O	4	2	0	3	1

.

Here L_1 is the addition table; $LS_{2,3}$ were considered after Proposition 5.2.1.

For instance, "A" is encoded by the consecutive signals at frequencies $0, 1, 2, 3, 4$, "B" by $1, 2, 3, 4, 0$ and so on. The number of different frequencies is obviously greater than we need for 15 letters. However, if any 3 columns are deleted here in all squares (3 signals are lost, we know which), the message can be recovered. Similarly, if any 3 (known to us) frequencies of the 5 are not readable, i.e. if we delete their symbols from the tables, the recovery is possible. This kind of noise is not unusual when using powerlines or similar signal carrying methods. Say, if the frequencies 0, 1, 2 are not readable, then the letters will be sent as the following sequences:

	0	1	2	3	4
A	–	–	–	**3**	**4**
B	–	–	3	4	–
C	–	3	4	–	–
D	3	4	–	–	–
E	4	–	–	–	3

,

	0	1	2	3	4
F	–	–	**4**	–	**3**
G	–	3	–	–	4
H	–	4	–	3	–
I	3	–	–	4	–
J	4	–	3	–	–

,

	0	1	2	3	4
K	–	**3**	–	**4**	–
L	–	4	–	–	3
M	–	–	3	–	4
N	3	–	4	–	–
O	4	–	–	3	–

.

The sequences with "–" are still sufficient to recover the letters (if we know which frequencies are unreadable). We can extend the alphabet by 5 more letters by adding LS_{r_4} for $r_4 = 4$, which is the subtraction table mod 5:

	0	1	2	3	4
P	**0**	**4**	**3**	**2**	**1**
Q	1	0	4	3	2
R	2	1	0	4	3
S	3	2	1	0	4
T	4	3	2	1	0

omit 012 $\Longrightarrow$

	0	1	2	3	4
P	–	**4**	**3**	–	–
Q	–	–	4	3	–
R	–	–	–	4	3
S	3	–	–	–	4
T	4	3	–	–	–

.

5.2.4 Nonattacking queens

Let us begin with placing n *nonattacking rooks* on an $n \times n$ chessboard. Rooks can attack only vertically or horizontally. The origin of this name is from the Persian "rukh", a (war) chariot. There can be only 1 rook in any row. So given any i $(0 \le i \le n-1)$, we must provide the position p_i of the corresponding rook (only one) in this row, and p_i must be all different. This is simply a reformulation of the definition of the n-permutation $\{i \mapsto p_i\}$. So the number of the configurations of n nonattacking rooks is $n!$.

The problem about placing n nonattacking queens on an $n \times n$ chessboard is from 1848. Queens attack vertically, horizontally, and diagonally (including anti-diagonals). This is doable for any $n \ge 4$ (a theorem). There are only 12 solutions up to rotations and reflections for $n=8$; the total is 92. This is about 8-permutations with some special properties; $8! = 40320$ is not infinity, so computers can be used. The "n-queen problem" is a standard benchmark for "back-tracking algorithms" in computer science; such algorithms basically mean "redo the previous step if no continuation appears possible."

An interesting twist is to go to the torus. The conditions are that no 2 queens are allowed to be in the same *pandiagonal* or *anti-pandiagonal*. In other words, we replace the plane chessboard by the toroidal one, the one on a doughnut. We call such queens *pan-nonattacking*. This is by connecting the top and the bottom edges of the columns and the left and right edges of the rows correspondingly.

It is not always possible to put n pan-nonattacking queens on the $n \times n$ board. The maximal number is n if and only if $\gcd(n, 6) = 1$ (Pólya); we will call these maximal configurations *full*. As Paul Monsky proved, this number is $(n-1)$ if and only if n is not divisible by 3 or by 4; otherwise, $(n-2)$ pan-nonattacking queens can always be found. The exact (algebraic) formula for the total number of such configurations is due to Timothy Chow and Mario Velucchi. So no greater than 6 pan-nonattacking queens can be placed on the 8×8 board. See Exercises 5.6.50 and 5.6.51. modular arithmetic appears to be the key to obtain the results mentioned above.

Let us employ linear modular formulas to construct *full configurations* of pan-nonattacking queens. For the sake of uniformity, we continue to number rows (i) and columns (j) starting with 0, as above. They will be treated as modular numbers (elements in $\mathbb{Z}_n$) in the formulas.

Proposition 5.2.5 *For $n > 0$, let a be an invertible number in $\mathbb{Z}_n$ such that $a \pm 1$ are invertible too. They exist if and only if $\gcd(n, 6) = 1$. In particular, any $a \neq 0, 1, p-1$ can be taken in $\mathbb{Z}_p = \mathbb{F}_p$ for prime p. Then the queens on the $n \times n$ board placed at positions (i, j) where $j = ai + b \bmod n$ for $i = 0, 1, \ldots, n-1$ form a pan-nonattacking configuration for any $b \in \mathbb{Z}_n$.*

Proof. Since a is invertible, such queens are nonattacking vertically and horizontally. Indeed, the multiplication line $\{ai, 0 \le i < n-1\}$ is then a permutation for such, as we justified and used above.

If the queens at $(i, ai+b)$ and $(i', ai'+b)$ belong to the same pandiagonal, then $i - ai - b = i' - ai' - b \bmod n$ and $(i - i')(1-a) = 0$ in $\mathbb{Z}_n$, which gives that $i = i'$ due to the invertibility of $(1-a)$. For anti-pandiagonals, the coincidence relation is $i + ai + b = i' + ai' + b \bmod n$, and we use the invertibility of $(1+a)$.

Let us check that $\gcd(n, 6)$ must be 1 for such a to exist. Any triple $a-1, a, a+1$ contains at least one even number, and exactly one number here is divisible by 3. Thus, n cannot be divisible by 2 or by 3, if $a-1, a, a+1$ are all relatively prime to n. The condition $\gcd(n, 6) = 1$ means that the smallest prime factor of n is 5 or greater. However, then $1, 2, 3$, and 4 are invertible modulo n, all of them. Thus $a = 2$ and $a = 3$ satisfy the required gcd-conditions for $a-1, a, a+1$. We obtain that n pan-nonattacking queens always exist for $\gcd(n, 6) = 1$. ❑

By no means have we proved that it is impossible to find a (full) pan-nonattacking configuration of n queens if $\gcd(n,6) \neq 1$. We just found that our modular method is too weak to construct a full set of such queens for n divisible by 2 or by 3. See Exercise 5.6.50 for a proof that such configurations really do not exist if $\gcd(n,6) \neq 1$; this is a combinatorial fact, but the justifications rely on modular calculations.

Note that our construction always produces *pan-nonattacking* queens (if any). Modular methods describe the geometry of the torus square boards. The "torus diagonals" are pandiagonals, so they inevitably occur in modular considerations (not just diagonals).

This is similar to the modular LS constructed above. Recall that the formula $a_{ij} = ai + bj$ in $\mathbb{Z}_n$ for invertible a and b produces modular LS, where $0 \leq i, j \leq n-1$ are the indices of rows and columns. If $a+b$ and $a-b$ are invertible in $\mathbb{Z}_n$, then we obtain *pandiagonal Latin squares.* By definition, they are LS that have no repetitions of the entries in all pandiagonals and all anti-pandiagonals. See Proposition 5.2.3.

There is an immediate passage from the *pandiagonal* Latin squares to *pan-nonattacking* queens. Namely, let us fix any u $(0 \leq u \leq n-1)$ and place the queens at the boxes of such a LS with the entry u. They will be pan-nonattacking queens then. This is a simple combinatorial fact.

Let us establish the relation of the corresponding modular constructions. We will do this for Latin squares LS'_r given by the formula $a'_{ij} = ri + j$ provided the conditions $\gcd(r,n) = 1$ and $\gcd(r \pm 1, n) = 1$. Such LS'_r is the *matrix transposition* of LS_r defined by the formula $a_{ij} = i + rj$; we will use them instead of LS_r here and below.

The relation $a'_{ij} = u$ means that $u = ri + j \bmod n$. We obtain that $j = (-r)i + u \bmod n$, i.e. this is the formula from Proposition 5.2.5 for $a = -r, b = u$.

Note that the conditions $\gcd(r \pm 1, n) = 1$ are necessary and sufficient for the corresponding LS to have no repetitions at the main diagonal and the main anti-diagonal (no "pan"). If they hold, then *all* pandiagonals and anti-pandiagonals will automatically have the same property.

Fig. 5.1 is an example of the passage from pandiagonal Latin squares to pan-nonattacking queens for $n = 7$. We take $r = 3$ and construct the corresponding LS'_r. The configuration of queens will then be for $a = -r = 4 \bmod 7$. Bold-red is used in this figure to mark the squares with 0 as their entry in the resulting Latin square (on the left). Considered as queens, these boxes form a pan-nonattacking configuration.

Recall that we do not use the formula $a'_{ij} = ri + j \bmod n$ for all entries of modular LS. The multiplication line for 3 modulo 7 is sufficient to know for ob-

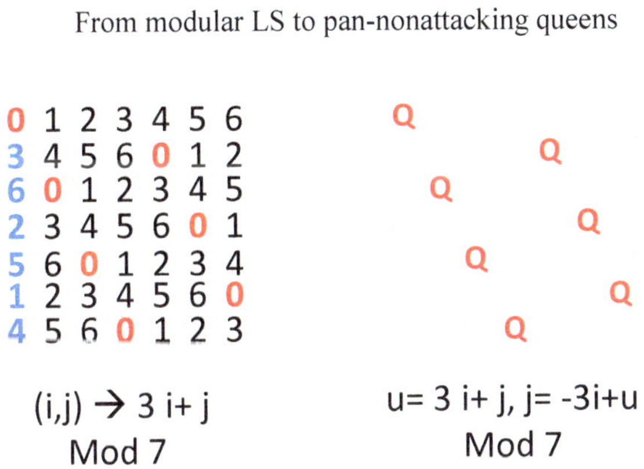

Figure 5.1: From LS to nonattacking queens.

taining LS'_3 (for $n = 7, r = 3$); it was calculated after Definition 5.1.7. It gives the first column of LS'_3. The other columns are obtained then by "reading" the first column cyclically (from top to bottom) starting with $0, 1, 2, 3, 4, 5, 6$, which constitute the first row of LS'_3. In the variant with $a_{ij} = i + rj$, the first row is the multiplication line for r.

Note that the construction from Proposition 5.2.5 with $b = 0$ readily gives a configuration of nonattacking queens in the $(n-1) \times (n-1)$ board, where we omit the first row and the first column. Indeed, the first queen is always at $(0, 0)$ for $b = 0$, so the first column and the first row are free of queens and can be "deleted." The remaining $(n-1)$ queens are obviously nonattacking in this reduced board (with $i > 0$ and $j > 0$).

These queens are not *pan-nonattacking* though. The pandiagonals and anti-pandiagonals must be recalculated modulo $(n-1)$. The following holds for the resulting $(n$–$1)$-configuration: there are no queens on the main diagonal and anti-diagonal and any queen at (i, j) that does not belong to this "cross" can be extended to a full $(n-1)$-configuration for $a = j/i \bmod n$ (with $b = 0$).

5.3 Finite fields

5.3.1 The field with 4 elements

A fundamental discovery was that the modular construction is not the only one to obtain *finite fields*. Generally, ***fields*** $\mathbb{F}$ are defined as sets with addition, subtraction, multiplication, and where the division by any nonzero elements is possible. They must have 0 and 1, and satisfy all standard algebraic properties. For instance, $b - a$ for any a and b/a for $a \neq 0$ are well defined: these values are the (unique) solutions of the equations $x+a=b,\ xa=b$ in $\mathbb{F}$.

In particular, a field $\mathbb{F}$ is a *group* with respect to its addition. A group is a set with the operation $(a,b)\mapsto c$, where the symbols $+$, $\times$, or no symbols between a,b are used for this operation: $c=a+b, c=a\times b, c=ab$.

The symbol $+$ is always used for addition (as for usual numbers); the standard properties must be satisfied: $(a+b)+c=a+(b+c)$, $a+b=b+a$, $0+a=a+0=a$. The existence of $-a$ is imposed for any a.

Similarly, the set $\mathbb{F}^*=\{a\in\mathbb{F}, a\neq 0\}$ is a group with respect to the multiplication for a field $\mathbb{F}$. Then $(ab)c=a(bc)$, $1a=a1=a$ for any a, and the existence of $1/a$ holds for $a\neq 0$. We mostly use no symbol between a and b in ab for the multiplication, though $a\times b$ and $a\cdot b$ are used from time to time too. Finally, the connection between the addition and the multiplication is the distributive law: $a(b+c)=ab+bc$ for any a,b,c.

Note that generally, $ab=ba$ is not required in groups. The group of permutations $\mathbf{S}_n$ is an example when this does not always hold. So general defining properties are $a(bc)=(ab)c$, $ae=a=ea$ for the unit element e and the existence of a^{-1} such that $aa^{-1}=e=a^{-1}a$. The unit element e is unique (if it exists), and so is a^{-1} for a given a.

All these properties are a direct generalization of those for the fields of *rational numbers* or *real numbers*; the standard notation is $\mathbb{Q},\mathbb{R}$.

Characteristic of $\mathbb{F}$. Let $\mathbb{F}$ be a finite field; $q=|\mathbb{F}|$ be the number of its elements. Its ***characteristic*** is defined as the order of 1 for the addition: the smallest $m>0$ such that $m\cdot 1=\overbrace{1+\cdots+1}^{m}=0$. We claim that it is a prime number p. Indeed, if $m=ab$ for $1<a,b<m$ then $m\cdot 1=(\overbrace{1+\cdots+1}^{a})\overbrace{1+\cdots+1}^{b})=0$, which gives that either $a\cdot 1=0$ or $b\cdot 1=0$ because $\mathbb{F}$ is a field (and we can divide by nonzero elements). However, this contradicts the minimality of m and it must be a prime number.

Let us check that p divides q. The Lagrange Theorem readily gives this. It is not difficult to obtain that actually $q=p^k$ for some k; see Section 7.3.1.

Problem 5.3.1 *Lagrange Theorem. The order of any element x in a finite group G, which is the smallest $m>0$ such that $x^m=1$, divides the order $|G|$ of G, which is its number of elements. More generally, the order of any subgroup H of G divides its order. The subsets $H\subset G$ are called subgroups if $h'/h=h'h^{-1}\in H$ for any $h,h'\in H$, so this implies that H contains the unit element e of G, which is $0,1$ or id for $\mathbb{F},\mathbb{F}^*$ or $\mathbf{S}_n$.*

Solution. Consider $H=\{1,x,x^2,\cdots,x^{u-1}\}$, where u is the order of x in G. This is a subgroup of G; indeed, h'/h are in the form x^v for some $0\le v<u$, which is for any $h',h\in H$.

Let $aH = \{ah, h \in H\}$; any such set contains u elements. Then G can be represented as the union of subsets $H, a_1H, a_2H, \cdots, a_kH$ for some $a_i \in G$, where any two such subsets either totally coincide or do not intersect. Indeed, if they intersect, then $a_ih = a_jh'$ for some $h, h' \in H$, and $a_j^{-1}a_i = h'h^{-1} \in H$. We obtain that $a_iH = a_jH$, since H is a subgroup. Thus, $|G|/u$ is the number of different subsets a_iH in this presentation of G, which gives the required divisibility. The same argument works for any subgroup $H \subset G$. ❑

In particular, we obtain that $p = 2$ for any field $\mathbb{F}_4$ with 4 elements (if it exists). Assuming that $\mathbb{F}_4$ exists, let us use the notation $0, 1, \widehat{2}, \widehat{3}$ for its elements; 0 and 1 are actually 0 and 1, but $\widehat{2}$ and $\widehat{3}$ are just symbols, which have nothing to do with the residues 2 and 3. The corresponding addition and multiplication tables are

$+$	0	1	$\widehat{2}$	$\widehat{3}$
0	0	1	$\widehat{2}$	$\widehat{3}$
1	1	0	x	y
$\widehat{2}$	$\widehat{2}$	x	0	z
$\widehat{3}$	$\widehat{3}$	y	z	0

,

$\times$	0	1	$\widehat{2}$	$\widehat{3}$
0	0	0	0	0
1	0	1	$\widehat{2}$	$\widehat{3}$
$\widehat{2}$	0	$\widehat{2}$	u	v
$\widehat{3}$	0	$\widehat{3}$	v	w

,

where x, y, z, u, v, w are some elements from $\mathbb{F}_4$, which we are going to find. The zeros on the diagonal of the addition table are because $p = 2$. Indeed $a + a = a(1 + 1) = 0$ for any a in $\mathbb{F}_4$.

Let us begin with the analysis of the multiplication table. Due to the existence of division, its subsquare $\begin{bmatrix} 1 & \widehat{2} & \widehat{3} \\ \widehat{2} & u & v \\ \widehat{3} & v & w \end{bmatrix}$ must be Latin. Indeed, the columns and rows are "multiplication lines" and must be permutations of $\{1, \widehat{2}, \widehat{3}\}$ because division exists! Therefore v cannot be $\widehat{2}$ or $\widehat{3}$: look at the row and the column through v. So v can be only 1, and therefore $u = \widehat{3}, w = \widehat{2}$.

The addition table must be Latin too (subtraction exists!). There is only one possibility here: $x = \widehat{3}, y = \widehat{2}, z = 1$. We conclude that if $\mathbb{F}_4$ exists, then the tables must be as follows:

$+$	0	1	$\widehat{2}$	$\widehat{3}$
0	0	1	$\widehat{2}$	$\widehat{3}$
1	1	0	$\widehat{3}$	$\widehat{2}$
$\widehat{2}$	$\widehat{2}$	$\widehat{3}$	0	1
$\widehat{3}$	$\widehat{3}$	$\widehat{2}$	1	0

,

$\times$	0	1	$\widehat{2}$	$\widehat{3}$
0	0	0	0	0
1	0	1	$\widehat{2}$	$\widehat{3}$
$\widehat{2}$	0	$\widehat{2}$	$\widehat{3}$	1
$\widehat{3}$	0	$\widehat{3}$	1	$\widehat{2}$

. (5.3.3)

It looks like we are done with the construction of a field with 4 elements, but this is not the end. We proved that if such a field exists, then the corresponding addition and multiplication tables can be only as in (5.3.3). The properties of these operations were used for some elements, but not for all. We still need to check *all* instances of the additive and multiplicative associativity, and the distributive law (very many!). This can be done: are all satisfied indeed. A systematic theory of fields and Galois theory makes all these checks unnecessary, but they are a must in the combinatorial approach we have employed above.

5.3.2 4×4 Magic squares

At last, we have tools for constructing 4×4 MOLS and magic squares using number theory! The usage of $\mathbb{Z}_4$ is no good for this. Indeed, modular MOLS require r_1, r_2 and $r_1 - r_2$ to be relatively prime to 4, i.e. all odd, which is obviously impossible.

Proposition 5.2.1 in the variant for $\mathbb{F}_p$ requires *only* the existence of the division, so the same procedure *readily* holds for $\mathbb{F}_4$ and for any finite field.

Given $r \in \mathbb{F}_4 \setminus 0$, i.e. any nonzero element in $\mathbb{F}_4$, the Latin square LS'_r is the matrix (a'_{ij}), where $a'_{ij} = ri + j \in \mathbb{F}_4$; the indices i, j and the entries are now from $\mathbb{F}_4$. The calculation of a'_{ij} requires the usage of the tables in (5.3.3). We obtain that for any two nonzero $r_1 \neq r_2$ in $\mathbb{F}_4$ the corresponding $L_1 = LS'_{r_1}$ and $L_2 = LS'_{r_2}$ form MOLS. There are $(4-1)(4-2) = 6$ choices for r_1 and r_2.

Let us calculate these LS for $r_1 = \widehat{2}, r_2 = \widehat{3}$:

$\widehat{\mathbf{2}}$	0	1	$\widehat{2}$	$\widehat{3}$
0	0	1	$\widehat{2}$	$\widehat{3}$
1	$\widehat{2}$	$\widehat{3}$	0	1
$\widehat{2}$	$\widehat{3}$	$\widehat{2}$	1	0
$\widehat{3}$	1	0	$\widehat{3}$	$\widehat{2}$

,

$\widehat{\mathbf{3}}$	0	1	$\widehat{2}$	$\widehat{3}$
0	0	1	$\widehat{2}$	$\widehat{3}$
1	$\widehat{3}$	$\widehat{2}$	1	0
$\widehat{2}$	1	0	$\widehat{3}$	$\widehat{2}$
$\widehat{3}$	$\widehat{2}$	$\widehat{3}$	0	1

. (5.3.4)

We are going to now omit hats in $\widehat{2}$ and $\widehat{3}$ in the $L_1 \vee L_2$ below. Treating $0, 1, 2,$ and 3 as numbers in this bi-square, we can employ the mapping $\langle u, v\rangle \mapsto 1 + u + 4v$. The result will be the corresponding MS:

$\vee$	0	1	2	3
0	00	11	22	33
1	23	32	01	10
2	31	20	13	02
3	12	03	30	12

$\mapsto$

MS	0	1	2	3
0	1	6	11	16
1	15	12	5	2
2	8	3	14	9
3	10	13	4	7

. (5.3.5)

The diagonal sums must be checked directly. Since there are no repetitions in the (main) diagonals and anti-diagonals of L_1 and L_2, they equal 34, the magic sum for $n=4$. Note that this square is not *pandiagonal magic*.

Finally, let us replace $\{0, 1, 2, 3\}$ by $\{\spadesuit, \heartsuit, \diamondsuit, \clubsuit, \}$ for u, and $\{0, 1, 2, 3\}$ by $\{A, K, Q, J\}$ for v in every entry $\langle u, v\rangle$ of the bi-square. We arrive at the solution of the 16-card problem from Fig. 2.2: $\begin{pmatrix} \spadesuit A & \heartsuit K & \diamondsuit Q & \clubsuit J \\ \diamondsuit J & \clubsuit Q & \spadesuit K & \heartsuit A \\ \clubsuit K & \diamondsuit A & \heartsuit J & \spadesuit Q \\ \heartsuit Q & \spadesuit J & \clubsuit A & \diamondsuit K \end{pmatrix}$.

Exercise 1.4.20, a combinatorial one, states that the addition table for $\mathbb{F}_4$, coinciding with L there, has only two orthogonal mates $M_{1,2}$ with the first row $1, 2, 3, 4$. They were calculated there. We obtain: $M_1 = L_2$ and $M_2 = L_1$. It is of course possible to take $r = 1$ (any nonzero r is suitable); the corresponding $M_{r=1}$ will be orthogonal to $M_{1,2}$.

As at was for $\mathbb{Z}_p$, any nonzero r in an arbitrary finite field $\mathbb{F}$ can be taken; the addition table is for $r = 1$. Generally, we have $|\mathbb{F}| - 1$ *pairwise* orthogonal Latin squares of size $|\mathbb{F}|$; see Problem 5.2.4(i). We will describe all finite fields later; they are uniquely determined by the number of elements, which can be any power p^m of a prime number p. They provide error-correcting codes for $(p^m - 1)p^m$ symbols. The usage of finite fields for error-correcting codes, in cryptography and cybersecurity is quite natural.

5.3.3 Field extensions

Let us eliminate the necessity to check that the tables in (5.3.3) really define a field. The first step is to observe by inspection that there is a unique quadratic equation with the coefficients in $\mathbb{F}_2 = \mathbb{Z}_2$ that has no roots in $\mathbb{F}_2$. It is $x^2 + x + 1 = 0$ (check that $0, 1 \in \mathbb{F}_2$ are not its roots). The second will be to add its roots to $\mathbb{F}_2$, exactly as $\sqrt{2}$ is added to rational numbers. See Problem 5.3.2 below for a general construction for arbitrary $\mathbb{F}_p$.

We set: $\mathbb{F}_4 = \{a + b\lambda,\ a, b \in \mathbb{F}_2\}$, assuming that $\lambda^2 + \lambda + 1 = 0$. Equivalently, $\lambda^2 = \lambda + 1$, since $-1 = 1$ in $\mathbb{F}_2$. There are exactly 4 *new numbers*, $0, 1, \lambda$ and $1 + \lambda$, and so $|\mathbb{F}_4| = 4$. Here we need to check that all 4 are different, i.e. if $a + b\lambda = a' + b'\lambda$ then $a = a'$ and $b = b'$. Let $b \neq b'$ in this relation (otherwise a and a' must coincide). Then $(a - a') = (b' - b)\lambda$, and therefore $\lambda = \frac{a-a'}{b'-b}$ belongs to $\mathbb{F}_2$, which is impossible. This construction is absolutely parallel to adding $\sqrt{2}$ to the rationals!

The addition and multiplication are quite straightforward: $(a + b\lambda) + (c + d\lambda) = (a + b) + (c + d)\lambda$, and

$$\begin{aligned}(a + b\lambda)(c + d\lambda) &= ac + bd(\lambda + 1) + ad\lambda + bc\lambda \\ &= (ac + bd) + (bd + ad + bc)\lambda \ \text{ for any } \ a, b \in \mathbb{F}_2,\end{aligned}$$

where we multiplied the binomials in the usual way, using the distributive law, and then replaced λ^2 by $\lambda + 1$.

The last check is the existence of division, which is explicit: $1/\lambda = \lambda + 1$ and $1/(\lambda+1) = \lambda$. So we do have a field with 4 elements. Our combinatorial approach is not totally redundant: we proved that if such a field exists, then it is *unique*. The identification is $\lambda = \widehat{2}$ and $1 + \lambda = \widehat{3}$.

Note that the equation $x^2+x+1 = 0$ has its two solutions in $\mathbb{F}_4$: the second one is $1+\lambda$. This is a very simple calculation. Without any calculations, $1+\lambda$ is a root of some quadratic polynomial and does not belong to $\mathbb{F}_2$, but there is only one such polynomial with coefficients in $\mathbb{F}_2$.

It is not difficult to prove that there exists a field $\mathbb{F}_{p^m}$ with p^m elements for any prime p and $m \geq 1$. We will do this in Section 7.3.1. It suffices to find an ***irreducible*** polynomial of degree m with coefficients in $\mathbb{F}_p$. Reducible polynomials are by definition products of polynomials of nonzero degrees (with the coefficients in $\mathbb{F}_p$). The fact that all finite fields with the same number of elements (which is p^m) are *isomorphic* is not difficult too.

Let us discuss now arbitrary quadratic extensions, the fields obtained by adding "quadratic irrationalities" (we will give an exact definition).

Problem 5.3.2 *Let $\mathbb{F}$ be any field, for instance, $\mathbb{F}$ can be a finite field, the field of rational numbers $\{m/n\}$, or a field of all real numbers. Assuming that $P(X) = X^2 - aX - b$ has no roots in $\mathbb{F}$, let $\mathbb{F}[\lambda] = \{r + s\lambda\}$ for all r and s in $\mathbb{F}$ and for λ satisfying the relation $P(\lambda) = 0$. Check that (i) $\{r + s\lambda\}$ is 0 if and only if $r = 0$ and $s = 0$, (ii) sums and products of the elements $(r + s\lambda)$ can be represented in the same form, and (iii) that you can divide by nonzero elements. The division is explicit: present $1/(r + s\lambda)$ as $\widetilde{r} + \widetilde{s}\lambda$.*

Solution. Let us assume that $r + s\lambda = 0$. Then $r = 0$ if $s = 0$. If $s \neq 0$, then $\lambda = -r/s$, which is impossible since $P(X)$ has no roots in $\mathbb{F}$. Thus, we obtain that both, r and s, must be zero, which proves (i).

The addition of "new numbers" $r + s\lambda$ is component-wise as above. The products are straightforward too:

$$(r_1 + s_1\lambda)(r_2 + s_2\lambda) = (r_1r_2 + s_1s_2(a\lambda + b)) + (r_1s_2 + s_1r_2)\lambda$$
$$= (r_1r_2 + bs_1s_2) + (r_1s_2 + s_1r_2 + as_1s_2)\lambda.$$

For the division formula, we use that the second root of $P(X)$ is $a - \lambda$. Then, $\frac{1}{r+s\lambda} = \frac{r+s(a-\lambda)}{(r+s\lambda)(r+s(a-\lambda))} = \frac{r+sa-s\lambda}{r^2-s^2b-rsa}$. We obtained this formula assuming that at least one of r or s is nonzero.

It is instructional to check directly that $r^2 - s^2b - rsa \neq 0$ for such a if at least one $r, s \in \mathbb{F}$ is nonzero. If it is 0 and $s \neq 0$, then we divide $r^2 - s^2b - rsa \neq 0$ by s^2 and obtain $(\frac{r}{s})^2 - a(\frac{r}{s}) - b = 0$. This cannot be zero because $P(X)$ has no roots in $\mathbb{F}$. Thus, s must be zero. However, this implies that $r^2 = 0$, and that $r = 0$. ❑

5.3.4 Extensions of any degree

We will continue using X for the variable, and x for special values of X (in fields). Let $P(X) = X^m + \ldots$ be any polynomial of degree $deg(P) = m$ with coefficients in the *base field* $\mathbb{F}$, which will be $\mathbb{F}_p$ in the applications below. We take x satisfying $P(x) = 0$ and add to $\mathbb{F}$ any polynomials of x with the coefficients in $\mathbb{F}$; the numbers from $\mathbb{F}$ are included. This is called *the extension of* $\mathbb{F}$ *by* x and is denoted by $\mathbb{F}[x]$. This generalizes adding λ to $\mathbb{F}_2$ in Problem 5.3.2, or adding $\sqrt{2}$ to rational numbers. Obviously, we can add and multiply the elements of $\mathbb{F}[x]$, but division by any nonzero elements is not always possible. The terminology is as follows: $\mathbb{F}[x]$ is a *ring* (it has addition and multiplication), but it is not always a *field* (which requires the division).

More exactly, *new numbers* (elements of $\mathbb{F}[x]$) are linear combinations of $1, x, x^2, \ldots, x^{m-1}$ with the coefficients from $\mathbb{F}$. Indeed, the powers x^n for $n \geq m$ can be expressed this way using the relation $P(x) = 0$ as many times as necessary. Here x is just a symbol. Only the relation $P(x) = 0$ is supposed to be used in any algebraic manipulations involving x. More exactly, $R(x) = 0$ means that $R(X) = Q(X)P(X)$ for some polynomial $Q(X)$.

For instance, we obtain that there can be no nonzero polynomial $R(X)$ with degree $deg(R) < m$ such that $R(x) = 0$. Equivalently, $\sum_{i=0}^{m-1} c_i x^i$ is zero if and only if $c_i = 0$ for all i. Equivalently, 2 new numbers y and z coincide if and only if all their coefficients coincide. To prove the latter, consider $y - z$. This gives that the number of elements of $\mathbb{F}_p[x]$ is p^m for any $P(X)$ of degree m. We use that the coefficients c_i from $\mathbb{F}_p$ are arbitrary, and new numbers are zeros if and only if all c_i are zero.

The terminology "new numbers" can be somewhat misleading. For instance, if $P(X) = X^2$, then we extend the basic field by x such that $x^2 = 0$ and define new numbers as above. However, $x \neq 0$ and $x^2 = 0$, which never happens with "true numbers." This cannot happen if $P(X)$ is *irreducible*, a notion that is key for the theory of fields and their extensions.

Irreducibility. Similarly to composite (natural) numbers, *reducible polynomials* $P(X)$ can be written as products $P(X) = Q(X)R(X)$ for some polynomials Q and R of nonzero degrees. The coefficients of these polynomials are in $\mathbb{F}_p$, or in any base field $\mathbb{F}$. *Irreducible* $P(X)$ are those that are not such products.

Lemma 5.3.3 *A product of polynomials $Q(X)R(X)$ is divisible by an irreducible $P(X)$ only if Q or R is divisible by P. Let us prove it.*

Proof. We will use the long division of polynomials over $\mathbb{F}$. Given any two such polynomials R_1 and R_2, we have a unique presentation $R_1 = Q_1R_2 + R_3$, where

$deg(R_3) < deg(R_2)$ and Q_1 and R_3 have coefficients in $\mathbb{F}$. The justification is the same as for usual polynomials (with rational, real or complex coefficients).

Employing the long division to R_2 and R_3 we obtain that $R_2 = Q_2R_3 + R_4$ where $deg(R_4) < deg(R_3)$. Then we do the same for R_3, R_4 and continue until the end of this chain of long divisions, which will be the relation $R_i = Q_iR_{i+1}$ for some i. Here either $R_{i+1} = c$ for $c \neq 0$ in $\mathbb{F}$ or $deg(R_{i+1}) > 0$. In the latter case, R_{i+1} divides R_i, so it divides R_{i-1}, so it divides R_{i-2} and so on. Eventually we obtain that R_{i+1} divides R_2 and R_1 in this case. This is actually a counterpart of Euclid's algorithm; $D = \gcd(R_1, R_2)$ is R_{i+1} divided by its leading coefficient (to make it *monic*). Working backwards from R_{i+1} to R_2, R_1, we obtain that $AR + BP = D$ for some polynomials A and B.

Let us assume now that $R = R_1$ is not divisible by an irreducible $P = R_2$ and QR is divisible by P for some polynomial Q. Then $D = 1$ due to the irreducibility of P and $AR + BP = 1$ for some A, B. We obtain that $AQR + BQP = Q$ and Q is divisible by P. ❑

This lemma is the key in the following proposition.

Proposition 5.3.4 *Let $P(X)$ be an irreducible polynomial of degree $m > 0$ with coefficients in the base field $\mathbb{F}$, $\mathbb{F}[x]$ the extension of $\mathbb{F}$ by x satisfying $P(x) = 0$, and $y = \sum_{i=0}^{m-1} c_i x^i$ any nonzero "new number", an element of $\mathbb{F}[x]$. Then the powers $1, y, y^2, y^3, \ldots$ (all of them) are linearly dependent over $\mathbb{F}$, and there exists a polynomial $Q(X) \neq 0$ with coefficients in $\mathbb{F}$ such that $Q(y) = 0$ in $\mathbb{F}[x]$. Let $Q(X) = \sum_{i=0}^{\ell} a_i X^i$ be such a monic polynomial of minimal possible degree ℓ (normalized by the condition $a_\ell = 1$). Then $a_0 \neq 0$, which results in the following explicit inversion formula:*

$$\frac{1}{y} = \frac{a_1}{a_0} + \frac{a_2}{a_0} y + \cdots + \frac{a_\ell}{a_0} y^{\ell-1}.$$

Proof. Let $y = R(x)$ for $R(X) = \sum_{i=0}^{m-1} c_i X^i$. The powers $1, y, y^2, \ldots$ in $\mathbb{F}_p[x]$ will eventually become linearly dependent. Thus, $Q(y) = 0$ for some nonzero polynomial $Q(X)$ of degree $\ell > 0$. We assume that ℓ is minimal possible. Here $\ell \leq m$ because the greatest number of linearly independent elements in $\mathbb{F}_p[x]$ over $\mathbb{F}$ is m, but we will not need this in the proof.

Let us justify that $Q(0) \neq 0$. Otherwise, $Q(X) = \widetilde{Q}(X)X$ for some $\widetilde{Q}$ of degree $\ell - 1$. In terms of x, $Q(R(x)) = 0$, which means that $\widetilde{Q}(R(X))R(X)$ is divisible by $P(X)$ by the definition of the ring $\mathbb{F}[x]$. However, P is irreducible and $R(X)$ is not divisible by $P(X)$ because $deg(R) < m$ by construction. This gives that $\widetilde{Q}(R(X))$ must be divisible by $P(X)$. Therefore $\widetilde{Q}(y) = 0$ in $\mathbb{F}[x]$, and ℓ can be diminished, which is impossible. We conclude that $a_0 \neq 0$, and $\frac{1}{y} = \frac{a_1}{a_0} + \frac{a_2}{a_0} y + \cdots + \frac{a_\ell}{a_0} y^{\ell-1}$, where $Q(X) = \sum_{i=0}^{\ell} a_i X^i$. ❑

The invertibility of nonzero elements proves that $\mathbb{F}[x]$ is a field; it has p^m elements for $\mathbb{F} = \mathbb{F}_p$. Irreducible polynomials $P(X)$ over $\mathbb{F}_p$ exist for any degrees $m > 0$; we will address this in Section 7.3.1, and will prove that fields with the same number of elements are isomorphic to each other. This justifies the notation $\mathbb{F}_{p^m}$ for them.

Problem 5.3.5 * *Add to* $\mathbb{F}_3 = \mathbb{Z}_3$ *the element* $\lambda = \sqrt{-1}$ *and check directly that you will obtain* $\mathbb{F}_9$*, the field with* 9 *elements. Similar to the prime fields, use* $\mathbb{F}_9$ *to construct Latin squares* LS'_r *by the formula* $\{i,j\} \mapsto ri+j$ *for any* $r \neq 0$ *in* $\mathbb{F}_9$*. Let* $\mathbb{F}_9 = \{a_0, \ldots, a_8\}$*, providing that* $a_m + a_{8-m} =$ *const for all* m*. Check that* LS'_r *have all different entries on the main diagonal and anti-diagonal for this ordering if* $r \neq 1, 2$*. Calculate examples of MOLS for* $r_1 \neq r_2$ *and then the corresponding* 9×9 *magic squares using the formula* $\langle u, v\rangle \mapsto 1 + u + 9v$*. Investigate applications of such LS to Sudoku; are any* $r \neq 0$ *suitable?* ❑

5.4 Back to Fibonacci numbers

5.4.1 Modular Pascal's triangle

We will discuss in this section the divisibility of Fibonacci numbers f_n by prime numbers p and the corresponding Pisano periods. These two problems are closely related. Finding the periods and "patterns" can be very challenging even for simple recurrences over finite fields.

Let us begin with an example of Pascal's triangle modulo $p = 7$. This is a two-dimensional recurrence. Instead of finding the periods, we need to capture the corresponding repeated patterns. Fig. 5.2 demonstrates what can be expected here for binomial coefficients. The colors are used for their remainders modulo 7. Here $\binom{0}{0}$ is at the bottom-left corner. The white boxes are with $\binom{n}{r}$ divisible by 7; they are the simplest to analyze.

Problem 5.4.1 *(i) Use the formula for* $\binom{n}{r}$ *to check that* $\binom{p^m}{r}$ *is divisible by* p *for a prime* p *except for* $r = 0$ *or* $r = p^m$.

(ii) As an application, prove that the upper-right triangle in the rectangle 49×49 *containing the bottom row in Fig. 5.2 is really "all white."*

Solution. The divisibility of $\binom{p}{r} = \frac{p(p-1)\cdots(p-r+1)}{r!}$ by p for $0 < r < p$ is obvious because $r!$ is the product of numbers smaller than p and cannot be divisible by p. Thus, p in the numerator provides the required divisibility. A variant of this argument works for (pure) powers p^m, but it is simpler to use Newton's binomial. We have $(x+y)^p = \sum_{r=0}^{p^m} \binom{p}{r} x^r y^{p-r} = x^p + y^p \bmod p$, and therefore,

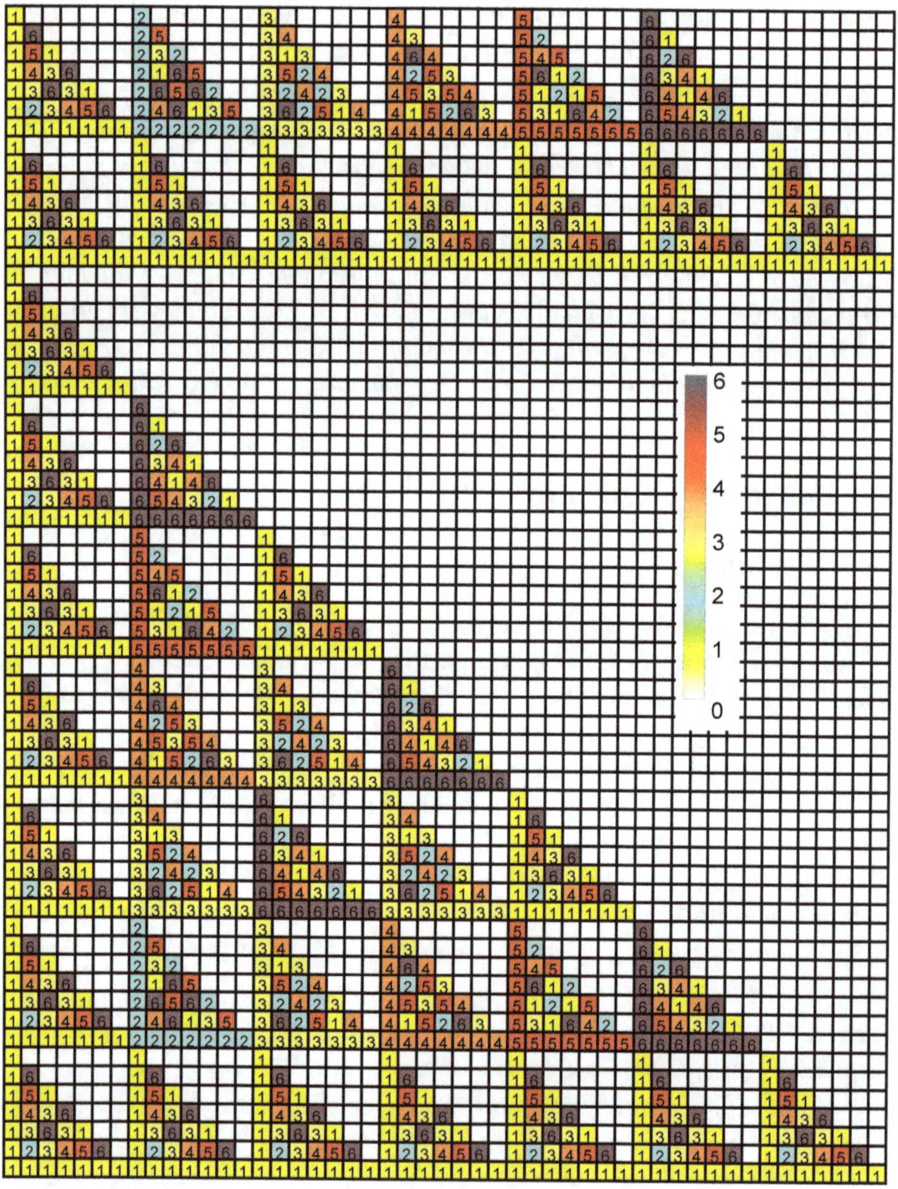

Figure 5.2: Pascal's triangle modulo 7.

upon the coefficient-wise reduction modulo p:

$$(x+y)^{p^m} = \sum_{r=0}^{p} \binom{p^m}{r} x^r y^{p^m-r} = (((\ldots(x+y)\overbrace{^p)^p)\ldots)^p}^{m} = x^{p^m} + y^{p^m}. \qquad (5.4.6)$$

As for (ii), the "diagonal" connecting $\binom{49}{0} = 1$ and $\binom{49}{49} = 1$, next to the triangle with the vertices $(0,0)$, $(48,0)$ and $(0,48)$, is "all white" (divisible by 7) except for boxes $(49,0)$ and $(0,49)$, which are green. Thus, the recurrence relation for binomial coefficients proves that all boxes in the triangle above this diagonal and inside "49×49", are, indeed, divisible by 7. ❑

5.4.2 Pisano periods

Fibonacci numbers are the sums of shallow diagonals in Pascal's triangle, but their divisibility by prime numbers p cannot be deduced from the periodicity of binomial coefficients (we will not even try). It requires different tools: Binet's formula, the analysis of the corresponding field extensions, and even the generalized Riemann hypothesis (to address this problem in full).

Let $\bar{f}_n$ for $n = 0, 1, 2\ldots$ be the sequence of Fibonacci numbers in $\mathbb{F}_p = \mathbb{Z}_p$. The divisibility simply means that $\bar{f}_n = 0$ in $\mathbb{F}_p$. The ***Pisano period*** is the minimal t_p such that $\bar{f}_n = \bar{f}_{n+t_p}$ for all integral n (positive or not).

Proposition 5.4.2 *Let m_p be the minimal $m > 0$ satisfying $\bar{f}_m = 0$ in $\mathbb{F}_p$. We claim that all such m are $m_p, 2m_p, 3m_p, \ldots$, and that $t_p = \alpha m_p$, where $\alpha > 0$ is the smallest exponent such that $u_p^\alpha = 1$ for $u_p = \bar{f}_{m_p+1}$ in $\mathbb{F}_p$, i.e. α is the order of u_p in the group $\mathbb{F}_p^*$.*

Proof. If $\bar{f}_m = 0$ for some $m > 0$, then the sequence starting with $\bar{f}_m = 0$ becomes $0, u\bar{f}_1, u\bar{f}_2, u\bar{f}_3 \bmod p, \ldots$, where $u = \bar{f}_{m+1}$. To see this let us compare the sequence $\bar{f}_{m+i}$ for $i = 0, 1, 2, \ldots$ with $\bar{f}_i$, which is $0, 1, 1, 2, \ldots \bmod p$. Due to 0 in the beginning of both sequences, the former will be proportional to the latter. Recall that if 2 consecutive terms coincide in 2 solutions of any linear recurrence of depth 2, then they coincide identically, i.e. for all n. The coefficient of proportionality is obviously $\bar{f}_{m+1}$.

Continuing $\bar{f}_n$ after m: $\bar{f}_{2m} = 0, \bar{f}_{2m+1} = u^2, \ldots, \bar{f}_{3m} = 0, \bar{f}_{3m+1} = u^3$, and so on. Let $m = m_p$. Then $\{\bar{f}_n\}$ is the union of consecutive sequences $u^i\{\bar{f}_0, \ldots, \bar{f}_{m_p-1}\}$ for $i = 0, 1, 2, 3, \ldots$. Only $u^i\bar{f}_0$ are zeros here.

We obtain that all m such that $\bar{f}_m = 0$ are exactly multiples of m_p. Also, the minimal t such that $\bar{f}_t = 0$ and $\bar{f}_{t+1} = 1$, which is the Pisano period t_p, is αm_p. Indeed, u^a becomes 1 for the first time (after $a = 0$) exactly for $a = \alpha$, which is the order of u.

Recall that we consider here the Fibonacci sequence for all n, positive and negative, so the period t_p can be counted starting from any place. It is convenient to begin with $\bar{f}_0 = 0$, but not really necessary. ❑

For the first 15 prime numbers $p = p_n$, let us provide the table of the corresponding m_p and the Pisano periods t_p. The numbers s_p will be described in Theorem 5.4.3. They are as follows: for any $p \neq 5$, either m_p divides $(p-1)$, and then $s_p = 1$, or it is divisible by $(p+1)$, and then $s_p = -1$. We note that m_p cannot divide both: then it would be 1 or 2, which is impossible. Also, we set $s_5 = 0$. The table is:

n	1	2	3	4	5	6	7	8	9	10	11	12	13	14	15
p_n	2	3	5	7	11	13	17	19	23	29	31	37	41	43	47
m_p	3	4	5	8	10	7	9	18	24	14	30	19	20	44	16
t_p	3	8	20	16	10	28	36	18	48	14	30	76	40	88	32
s_p	−1	−1	0	−1	1	−1	−1	1	−1	1	1	−1	1	−1	−1

The number of $s_p = -1$ in this table is greater than the number of $s_p = 1$. Generally, the values ± 1 are equally distributed for large p due to Tchebyshev's Theorem on the prime numbers in arithmetic sequences. We need here Quadratic Reciprocity to be discussed later.

5.4.3 Modular Binet's formula

We provide here a complete proof of the following theorem on m_p and t_p. The claim that m_p and t_p divide $p^2 - 1$ is relatively straightforward; the rest of the proof is actually with *. In particular, the justification of the relations from (iii) and the case of $s_p = -1$ are somewhat technical. Some additional information can be found, for instance, at [35].

The *order* of any element a in a finite group G, denoted by $ord(a)$, is the smallest $m > 0$ such that $a^m = 1$. Recall that the Lagrange Theorem gives that $ord(a)$ divides $|G|$, the order of G. See Problem 5.3.1. We write a^m, but it will be ma if the additive structure is considered.

Let us apply it to $\mathbb{F}^*_{p^k} = \{\mathbb{F}_{p^k} \setminus \{0\}\}$, which is the group of nonzero elements in $\mathbb{F}_{p^k}$; k will be $1, 2$ below. We obtain that the order of any a divides $p^k - 1$.

Theorem 5.4.3 *(i) If the equation $X^2 = X + 1$ has a solution $\lambda \in \mathbb{F}_p$, we set $s_p = 1$, except for $p = 5$, when we set $s_5 = 0$. Otherwise, we set $s_p = -1$. Due to Quadratic Reciprocity (below), $s_p = 1$ if and only if $p = \pm 1$ mod 10.*

[35] https://en.wikipedia.org/wiki/Pisano_period.

(ii) If $p=2$, then $m_p=t_p=3$. If $p=5$, then $m_5=5$, $t_p=20$. Otherwise t_p is always even, and there are 2 cases:

$$(\alpha)\ s_p=1:\ \ \textit{then } m_p \textit{ and } t_p \textit{ divide } (p-1)$$
$$(\beta)\ s_p=-1:\ \ \textit{then } m_p \textit{ divides } (p+1), \textit{ and}$$
$$t_p \textit{ is } 2(p+1) \textit{ divided by an odd factor.}$$

(iii) Let $p>5$. For $s_p=1$: if $p=1 \bmod 4$ then m_p divides $\frac{p-1}{2}$, and $p=3 \bmod 4$ implies $m_p=2 \bmod 4$. Also, if m_p is odd then $p=1 \bmod 4$ (so m_p divides $\frac{p-1}{4}$), and if 4 divides m_p then m_p divides $\frac{p-1}{2}$. Now for $s_p=-1$: if $\frac{p-1}{2}$ is odd then 4 divides m_p, and if $\frac{p-1}{2}$ is even then m_p is odd.

Proof. Binet's formula can be perfectly used modulo $p\neq 5$: $\bar{f}_n=\frac{\lambda_1^n-\lambda_2^n}{\lambda_1-\lambda_2}$ for $n=0,1,2,\ldots$. Here λ_1 is a solution of $\lambda^2=\lambda+1$ in $\mathbb{F}_p$ or $\mathbb{F}_{p^2}$. Then the second solution is $\lambda_2=1-\lambda_1=-1/\lambda_1$.

Here $p=5$ is exceptional because $\lambda^2-\lambda-1=(\lambda-3)^2$ in $\mathbb{F}_5$, and this is the case of a multiple root. Theorem 4.4.2, covering resonances, gives that $\bar{f}_n=n3^{n-1} \bmod 5$. This is easy to justify directly. Indeed, modulo 5:

$$(n+1)3^n=n3^{n-1}+(n-1)3^{n-2}\Leftrightarrow 9(n+1)=3n+(n-1).$$

We will exclude $p=2$ and $p=5$ from the considerations below. The relations $m_2=3=t_2$ and $m_5=5$ are simple to check. From the latter, $t_5=5k$ for some k. We obtain that $\bar{f}_{n+5k}=(n+5k)3^{n+5k-1}=n3^{n+k-1} \bmod 5$ must be equal to $n3^{n-1} \bmod 5$, which is possible only if k is divisible by 4, the order of $3 \bmod 5$. Thus $t_5=20$, which of course can be seen by inspection.

Let $p\neq 2,5$. Then $\lambda_{1,2}=\frac{1\pm\sqrt{5}}{2}$ and $s_p=1$ if and only if $\sqrt{5}\in\mathbb{F}_p$. Applying *Quadratic Reciprocity*, proven in Problem 7.3.17, the corresponding condition for odd $p\neq 5$ is that p is a square in $\mathbb{F}_5$; use that $\frac{p-1}{2}\frac{5-1}{2}$ is even (see there). This readily gives that $p=\pm 1 \bmod 5$. Since it is odd, we arrive at $p=\pm 1 \bmod 10$ claimed in the theorem.

For instance, $5=7^2 \bmod 11$, $5=9^2 \bmod 19$; however, 5 is not a square modulo 3, 7, 13 or 17. Respectively, $s_{11}=s_{19}=1$, but $s_3=s_7=s_{13}=s_{17}=-1$.

To analyze the divisibility of m_p and t_p we will employ Binet's formula. It gives that $\bar{f}_m=0$ if and only if $(\lambda_2/\lambda_1)^m=1$. More exactly, the necessary and sufficient condition for such m is $\lambda_1^{2m}=(-1)^m$, where we use that $\lambda_2=-\lambda_1^{-1}$. Hence $m_p=ord(-\lambda_1^2)$, the order of $-\lambda_1^2$ in $\mathbb{F}_p^*$ or $\mathbb{F}_{p^2}^*$.

Let us come to t_p. Obviously, m_p divides t_p. Assuming that $\bar{f}_m=0$, the next term is

$$\bar{f}_{m+1}=\lambda_1^m(\lambda_1-\lambda_2)/(\lambda_1-\lambda_2)=\lambda_1^m.$$

Thus, the *Pisano period* t_p is the minimal $t > 0$ such that $(\lambda_2/\lambda_1)^t = 1$ and $\lambda_1^m = 1$. Equivalently, $\lambda_1^t = 1 = \lambda_2^t$. These conditions are considered in $\mathbb{F}_p$ if $\lambda_{1,2}$ are there, or in $\mathbb{F}_{p^2}$ otherwise (both conditions must be satisfied). Using that $\lambda_2 = -1/\lambda_1$, we obtain that $\lambda_1^{2t} = (-1)^t$. Therefore $(-1)^t = 1$ in $\mathbb{F}_p$ and t must be even. Vice versa, if t is even then the condition $\lambda_1^t = 1$ implies that $\lambda_2^t = (-1/\lambda_1)^t = 1$. We obtain that t_p is the minimal *even* t such that $\lambda_1^t = 1$.

An alternative justification of the fact that t_p is the minimal t such that $\lambda_1^t = 1 = \lambda_2^t$ (equivalently, minimal even t such that $\lambda_1^t = 1$) is as follows. We have $\lambda^n = f_n\lambda + f_{n-1}$, which results from $\lambda^2 = \lambda + 1$ (check). If $\bar{f}_n = 0$ and $\bar{f}_{n+1} = 1$, then $\bar{f}_{n-1} = 1$, and $\lambda^t = 1$ must hold for $\lambda = \lambda_1$ and for $\lambda = \lambda_2$.

Let us now prove the divisibility claims for m_p and t_p.

The divisibility for $s_p = 1$. In this case, $\lambda_1 \in \mathbb{F}_p$ and $\lambda_1^{p-1} = 1$ due to Fermat's Little Theorem or the Lagrange Theorem for $\mathbb{F}_p^*$. This gives the divisibility of $p - 1$ by m_p and t_p; this is for $s_p = 1$.

Next, $\lambda_1^{2m} = (-1)^m$ holds for $m = \frac{p-1}{2}$ if $p = 4k + 1$ for some k, i.e. if $p = 1 \bmod 4$. Therefore, m_p divides $\frac{p-1}{2} = 2k$ in this case.

To obtain the remaining constraints for m_p, let x be the order of λ_1^2; it divides $\frac{p-1}{2}$. Then the order of $(-1)\lambda_1^2$ is $m_p = 2x$ for odd x, and the smallest m such that $x = 2k$ divides $(1 + k)m$ (for some k). We use that $(-1) = \lambda_1^k$, which follows from the cyclicity of the group generated by λ_1^2. Indeed, λ_1^k is not 1 due to $0 < k < 2k$, and its square is 1; so it can be only -1. Using that $\gcd(k, 1 + k) = 1$, we obtain that $m_p = k$ for odd k and $m_p = 2k$ for even k. In the latter case, $p - 1$ must be divisible by $2k$, i.e. in this case: $p = 1 \bmod 4$ and m_p is divisible by 4. For odd $m_p = k$, we obtain that $m_p = x/2$ divides $\frac{p-1}{4}$. This gives the required relations in (iii).

The case $s_p = -1$.* This case is somewhat more involved than the one considered above: we mark this part of the proof by * .

For such s_p, $\widetilde{\lambda}_1 = \lambda_1^p$ satisfies the same equation as λ_1. This is direct: $\widetilde{\lambda}_1^2 = (\lambda_1^2)^p = (\lambda_1 + 1)^p = \lambda_1^p + 1 = \widetilde{\lambda} + 1$. Generally, the map $x \mapsto x^p$ is called the ***Frobenius automorphism***. It obviously preserves the multiplication. Also, $(x + y)^p = x^p + y^p$ in any field of characteristic p (minimal $p > 1$ such that $p \cdot 1 = 0$). So it is compatible with addition too and is really an *automorphism* of the field $\mathbb{F}_{p^2}$. See also the solution of Problem 5.4.1.

It preserves the elements from $\mathbb{F}_p$ inside $\mathbb{F}_{p^2}$, so the coefficients of the polynomial $X^2 - X - 1$ remains fixed under its action and its roots in $\mathbb{F}_{p^2}$ are permuted by this automorphism. We obtain that $\widetilde{\lambda}_1$ can be either λ_2 or λ_1.

The latter is impossible. Due to Fermat's Little Theorem the equation $X^p - X = 0$ has p different solutions in $\mathbb{F}_p$. It cannot have more than p different roots, since the degree of $X^p - X$ is p. Note that this argument proves the identity $X^p - X = \prod_{i=0}^{p-1}(X-i)$ modulo p. Thus, $\lambda_1^p = \lambda_1$ implies that λ_1 must belong to $\mathbb{F}_p$ in this case, which contradicts $s_p = 1$.

Divisibility of m_p for $s_p = -1$. We see that $\lambda_1^p = \lambda_2$. Therefore $\lambda_2/\lambda_1 = \lambda_1^{p-1}$ and $m_p = ord(\lambda_1^{p-1})$ in this case. The orders of any elements in $\mathbb{F}_{p^2}^*$ divide p^2-1 because of the same Lagrange Theorem. Thus, $(\lambda_2/\lambda_1)^{p+1} = \lambda_1^{p^2-1} = 1$, and m_p divides $(p+1)$.

To obtain the relations from (iii), let x be the order of λ_1^2 in $\mathbb{F}_{p^2}^*$; then x divides $(p+1)$. Following the reasoning for $s_p = 1$: m_p is $2x$ for odd x, and minimal $m > 0$ such that $2k$ divides $(1+k)m$ for $x = 2k$. Let us combine these relations with $m_p = ord(\lambda_1^{p-1}) = ord((\lambda_1^2)^{\frac{p-1}{2}})$. The latter order is x/d for $d = \gcd(\frac{p-1}{2}, x)$. We obtain that x can be only even for $s_p = -1$, and there are 2 cases, which are exactly those from (iii):

(a) $m_p = x/2 = k$ for odd k, $d = \gcd(\frac{p-1}{2}, 2k) = 2$, which gives that $p = 1 \bmod 4$ and $m_p = k$ is odd;

(b) $m_p = x = 2k$ for even k, $d = \gcd(\frac{p-1}{2}, 2k) = 1$, which gives that $p = 3 \bmod 4$ and $m_p = 2k$ is divisible by 4.

Divisibility of t_p for $s_p = -1$.

One has: $\lambda_1^p = \lambda_2 = -1/\lambda_1$. It implies that $\lambda_1^{p+1} = -1$ and $\lambda_1^{2(p+1)} = 1$. Thus t_p, which is minimal even integer $t > 0$ such that $\lambda_1^t = 1$, equals $2(p+1)/\ell$ for some divisor ℓ of $2(p+1)$. If ℓ is even here, then t_p must be a divisor of $(p+1)$, which implies that $\lambda_1^{p+1} = 1$. This is impossible because $\lambda_1^{p+1} = -1$. We conclude that ℓ is odd. ❑

Note that in the list we provided: $t_p < p-1$ for $s_p = 1$ occurs only for $p = 29$, and $t_p < 2(p+1)$ for $s_p = -1$ occurs only for $p = 47$. For $p = 47$: $t_{47} = 32 = 2(47+1)/3$, smaller than $2(47+1)$.

Concerning m_p, the divisibilities in (iii) generally cannot be improved. For instance, $m_p = 20$ for $p = 41$ $(s_p = 1)$, and $\frac{p-1}{2} = m_p$; the statement was that m_p divides $\frac{p-1}{2}$ if $p = 1 \bmod 4$. Though we do not claim that we found all possible divisibility constraints for m_p.

We mention that the chances for s_p to be 1 or -1 are 50–50 due to the Dirichlet Theorem; though -1 is more frequent in our table: the odds there are 9 : 5. The questions about the *density* of p such that $t_p < p-1$ for $s_p = 1$ and $t_p < 2(p+1)$ for $s_p = -1$, and parallel questions for m_p require the (generalized) Riemann hypothesis, so they are all open problems.

The theorem can be extended to *master equations* of any degree d; see Exercise 5.6.35. The method we used gives that the sequence $\sum_{i=1}^{d} C_i\lambda_i^n$ is periodic modulo p with the period $p^m - 1$ or one of its divisors, where m is such that *all* roots of the corresponding master equation over $\mathbb{Z}_p = \mathbb{F}_p$ belong to $\mathbb{F}_{p^m}$. Note that m can be greater than d; see the next problem.

Problem 5.4.4 *Check that all roots of $P(X) = (X^2+2)(X^3+2)$ considered over $\mathbb{Z}_7$ belong to $\mathbb{F}_{7^6}$, a field with 7^6 elements, and not to any smaller field.*

Solution. First of all, X^2+2, X^3+2 are irreducible, i.e. cannot be non-trivially factored over $\mathbb{Z}_7$. Consider the squares and cubes of $\pm1, \pm2, \pm3$, which shows that these polynomials have no roots in $\mathbb{Z}_7$, and which is sufficient for their irreducibility. Then we add to $\mathbb{Z}_7$ a root x_1 of X^2+2. Generally, X^3+2 can be factored in $\mathbb{Z}_7[x_1]$, but this is not the case: it remains irreducible there. Indeed, if it has a root x_2 in $\mathbb{Z}_7[x_1]$, then $\mathbb{Z}_7[x_2]$, which is $\mathbb{F}_{7^3}$, must belong to $\mathbb{F}_{7^2}$. This is obviously impossible since it is bigger. Thus, adding x_2 to $\mathbb{Z}_7[x_1]$ generates a field $\{a + bx_2 + cx_2^2\}$ with any a, b, c in $\mathbb{F}_7[x_1]$, which has $(7^2)^3$ elements. The natural notation is $\mathbb{Z}_7[x_1, x_2]$. It is a minimal field containing x_1, x_2. Any other roots of X^2+2 and X^3+2 belong, correspondingly, to $\mathbb{Z}_7[x_1]$ and $\mathbb{Z}_7[x_2]$ (check). Generally, see (7.3.16). So they belong to $\mathbb{Z}_7[x_1, x_2]$. ❑

5.5 Pell's equations *

The star here is because this is not a core topic for us, though this section is connected to the *Pell numbers* and the usage of *continued fractions* for "optimal" rational approximations of irrationalities. Rational approximations of quadratic irrationalities based on recurrence relations are among the core topics in this book. Also, the periods of solutions of Pell's equations modulo prime numbers are directly related to the Pisano periods.

Generalized continued fractions are of importance for us, and not only for algebraic numbers. See Exercise 5.6.40 concerning $1/e$. A more exotic example is $J_0(2)$ in terms of the Bessel function J_0; see Exercise 5.6.41(ii).

5.5.1 Algebraic approach

Solving Pell's equation is equivalent to an explicit description of the groups of invertible elements, frequently called *units*, in the rings $x + y\sqrt{d}$ for integers or half-integers x and y. Here and below $d > 0$ is not a perfect square. This group has one generator up to multiplication by ±1, and this generator corresponds to the fundamental solution of the corresponding Pell equation. It is a particular case of the Dirichlet Theorem. This theorem will not be used; we will define and prove in this section everything we need.

Theorem 5.5.1 *We consider the class of positive integer solutions (x, y) of the* ***Pell equation*** *$x^2 - dy^2 = \epsilon$ for any ϵ from the set $\{1, -1\}$. Here d is a positive integer and not a perfect square.*

*(i) Prove that any such solution is as follows. Take a solution (x_1, y_1) with positive integers such that $u_1 = x_1 + y_1\sqrt{d}$ is the smallest possible, called a **fundamental solution**. The corresponding ϵ is allowed to be $+1$ or -1; let it be ϵ_1 for u_1. Represent $u_n = (u_1)^n = x_n + y_n\sqrt{d}$. Then any solution with ϵ from $\{1, -1\}$ is (x_n, y_n) for some $n \geq 1$; the corresponding ϵ_n for u_n is ϵ_1^n.*

(ii) For $d = 2$, check that $u_1 = 1+\sqrt{2}$. Represent: $u_n = P_{n-1}+P_n(1+\sqrt{2})$: $x_n = P_n + P_{n-1}$, $y_n = P_n$. In particular, $P_0 = 0, P_1 = 1, P_2 = 2$. Justify that $P_n = 2P_{n-1} + P_{n-2}$, which is sufficient to calculate all P_n and all x_n, y_n. The numbers P_n here are the Pell numbers *from Exercise 4.6.27.*

Proof. First of all, considering ± 1 in the right-hand side of Pell's equation is equivalent to the invertibility of $x+y\sqrt{d}$ in the ring $\{a+b\sqrt{d}\}$ for integers a, b. Indeed, if $x + y\sqrt{d}$ is invertible in this ring, then its conjugation $x - y\sqrt{d}$ is invertible too, and so is the product $(x+y\sqrt{d})(x-y\sqrt{d}) = x^2 - dy^2$. However, only ± 1 are invertible in the ring of (usual) integers, so the right-hand side can only be ± 1. Then $(x + y\sqrt{d})^{-1} = \pm(x - y\sqrt{d})$. We will use that the conjugation $u = x + y\sqrt{d} \mapsto \overline{u} = x - y\sqrt{d}$ is a ring automorphism: the conjugation of the product is the product of conjugations.

(i) If (x, y) solve the Pell equation, then $u = x + y\sqrt{d}$ will be called *u-solutions.* Let (x, y) and (x', y') be two solutions of the Pell equation with ϵ and ϵ' from $\{1, -1\}$, and u, u' the corresponding u-solutions. The key observation is that $\widetilde{u} = uu' = \widetilde{x} + \widetilde{y}\sqrt{d}$ is another u-solution, which is for $\widetilde{\epsilon} = \epsilon\epsilon'$.

We obtain that (x_n, y_n) for $u_n = u_1^n$ solves the Pell equation for $\epsilon_n = \epsilon_1^n$ in the class of its solutions with positive integers x, y. Similarly, $u_n^{-1} = \epsilon_n \overline{u_n}$ is an integral u-solution *of non-positive type.*

If a u-solution u' with positive x', y' is not in the form u_1^n for some $n \geq 1$, then one can find n such that $u_1^n < u' < u_1^{n+1}$ for some $n \geq 1$. We use that u_1 is minimal and that $u_1^n \to \infty$ as $n \to \infty$; this can be done for any positive real numbers $u' > u_1 > 1$. We set $\widetilde{u} = u'/u^n = \widetilde{x} + \widetilde{y}\sqrt{d}$ if $\epsilon_n = \epsilon'$ or $\widetilde{u} = u^{n+1}/u'$ otherwise. Thus, $1 < \widetilde{u} < u_1$.

Using the second of these inequalities, $1/\widetilde{u} = \widetilde{x} - \widetilde{y}\sqrt{d} > 1/u_1 > 0$. Thus $\widetilde{u} + 1/\widetilde{u} = 2\widetilde{x} > 0$. Let us check that $\widetilde{y} > 0$ follows from the first inequality. One has: $1 > 1/\widetilde{u} = \widetilde{x} - \widetilde{y}\sqrt{d}$. This is impossible if $\widetilde{x} > 0$ and $\widetilde{y} < 0$ because they are integers and $d > 1$. Finally, $\widetilde{u}$ is a u-solution with positive $\widetilde{x}, \widetilde{y}$ and smaller than u_1, a contradiction.

(ii) Obviously, $u_1 = 1 + \sqrt{2}$, which is for $\epsilon_1 = -1$. Setting $x_n = P_n + P_{n-1}$, $y_n = P_n$, we obtain that $u_1^n = P_n + P_{n-1} + P_n\sqrt{2} = u_1 \cdot u_1^{n-1} = (3P_{n-1} + P_{n-2}) + (2P_{n-1} + P_{n-2})\sqrt{2}$. The coefficient of $\sqrt{2}$ gives that $P_n = 2P_{n-1} + P_{n-2}$, which is sufficient to determine all P_n using $P_0 = 0, P_1 = 1$.

❑

Algebraic integers and units. This is to connect Pell's equations with the theory of algebraic numbers; the readers can omit this discussion and go from here to Problem 5.5.2. The definitions we provide below will not be used in the book (except in a couple of exercises). Though the discussion of the quadratic case right before this problem is quite clarifying: it explains why half-integers naturally occur in Pell's equations.

The general setting is as follows. ***Algebraic integers*** are defined as roots x of *irreducible* polynomials $P(X) = X^n + \sum_{i=0}^{n-1} a_i X^i$ with rational coefficients a_i that are assumed integers. Notice that the first coefficient is 1; such polynomials are called ***monic***. A ***unit*** is an algebraic integer u such that $v = 1/u$ is an algebraic integer too. The necessary and sufficient condition for this is simple: $a_0 = \pm 1$. Recall that $a_0 \neq 0$ due to the irreducibility of $P(X)$ (which is over any field of coefficients). Let us justify this.

One has: $Q(X) = X^n P(1/X) = 1 + \sum_{i=0}^{n-1} a_i X^{n-i}$, which is irreducible over rationals if and only if $P(X)$ is irreducible. Obviously, v is a root of $Q(X)$. Upon the division by the leading coefficient: $\frac{1}{a_0} Q(X) = X^n + \sum_{i=0}^{n-1} \frac{a_i}{a_0} X^{n-i} + \frac{1}{a_0}$. So v is an algebraic integer if and only if *all* coefficients here are integers. This includes $\frac{1}{a_0}$, so a_0 must be ± 1. However then $\frac{a_i}{a_0} = \pm a_i$ are all integers. We obtain that $a_0 = \pm 1$ is necessary and sufficient for u to be a unit.

It is not difficult to prove that the product of units is a unit, but we will not do this here; see Problem 7.3.10. Product of invertible elements are invertible elements in any theories.

Let us consider units v in the ring generated by some fixed algebraic integer x, a root of monic irreducible $P(X)$ of degree n with integer coefficients. So $v = \sum_{i=0}^{n-1} c_i x^i$. Its *Norm* is defined as $(-1)^n (b_0)^{n/m}$, where $Q(v) = 0$ and $Q(X) = X^m + \sum_{i=0}^{m-1} b_i X^i$ is assumed to be irreducible with integer coefficients b_i. It is called the ***minimal polynomial*** of v (over rational numbers). We use here that m divides n. For sufficiently general coefficients c_i, the degrees m and n coincide and $Norm(v) = (-1)^n b_0$.

The most general version of Pell's equations is the equation $Norm(v) = \pm 1$. If v is an algebraic integer, then this is necessary and sufficient for v to be a unit. In this form, we arrive at a Diophantine equation of degree n (generally) in terms of undetermined integers $\{c_i\}$, though not only integers can occur here; there can be units u with fractional c_i in terms of x^i.

Quadratic extensions. In this case, units are roots of $X^2 + bX + \epsilon$ for an integer b and ϵ from $\{1, -1\}$, assuming that this polynomial has no rational roots. Actually, it suffices to check that no integers are its roots; then there will be no rational roots too. This is a simple exercise on divisibility.

Explicitly, $u = \frac{-b \pm \sqrt{D}}{2}$ for $D = b^2 - 4\epsilon$. The discriminant D is positive here unless $\epsilon = +1$ and either $b = 0$ or $b = 1$. If it is negative, then the units are $u = \pm\imath$ or $u = \frac{1\pm\sqrt{3}\imath}{2}$ for $\imath = \sqrt{-1}$. Together with ± 1, the units constitute a cyclic group of order 4 in the first case, and a cyclic group with 6 elements in the second. This is the full answer in the "complex case." The cyclicity of any finite multiplicative group in a field is granted by Proposition 7.3.3(ii).

We need the "real case", when $D > 0$ and it is not a perfect square. Let $D = dc^2$ for some integer c and an integer $d > 0$, which is then not a perfect square. One has: $u = (-b \pm c\sqrt{d})/2$. We now fix d and find all integers b, c such that u is a unit.

The answer is that $(\pm b \pm c\sqrt{d})/2$ is a unit if and only if b and c are integers and $dc^2 = b^2 - 4\epsilon$, i.e. b and c are solutions of the Pell equation for ± 4, which is $b^2 - dc^2 = 4\epsilon$. Equivalently, we need to solve Pell's equation with half-integers $x = \frac{b}{2}, y = \frac{c}{2}$ for ± 1 to obtain all units. This is a natural generality here. The following problem is the corresponding extension of Theorem 5.5.1.

Problem 5.5.2 *Let $v_1 = x_1 + y_1\sqrt{d}$ be the minimal solution of the equation $v\overline{v} = x^2 - dy^2 = \epsilon$ for ϵ in $\{1, -1\}$ with positive half-integers or integers x and y, where $d > 0$ is not a perfect square.*

(i) Following Theorem 5.5.1, prove that any such solution is one of (x_n, y_n) for $v_1^n = x_n + y_n\sqrt{d}$, where $n = 1, 2, \ldots$. In particular, some power of v_1 is the fundamental solution u_1 from Theorem 5.5.1.

(ii) Check that $v_1 = \frac{1+\sqrt{5}}{2}$ is the minimal solution in (i) for $d = 5$. Then, $u_1 = v_1^3 = \frac{1+3\sqrt{5}+3\cdot 5+5\sqrt{5}}{8} = 2 + \sqrt{5}$. Both are for $\epsilon = -1$. For $\epsilon = 1$, the minimal (fundamental) solutions are $v_1^2 = \frac{3+\sqrt{5}}{2}$ and $u_1^2 = 9 + 4\sqrt{5}$.

Solution. **(i)** The justification is exactly as for (i) in Theorem 5.5.1. The only place where we used that x and y are integers was the conclusion that the inequality $1 > x - y\sqrt{d}$ is impossible for $x > 0$ and $y < 0$. This holds for half-integers: the minimal value of this expression is $\frac{1+\sqrt{d}}{2} > 1$.

(ii) For $d = 5$, we need only to check that v_1 is minimal. Then $u_1 = v_1^3$ (check). The passage to the minimal solutions with $\epsilon = +1$ is direct: they are v_1^2 and u_1^2. See Exercise 5.6.39, where $d = 5 \bmod 8$ is discussed; this condition is necessary for the existence of half-integer (not integral) solutions. ❑

The existence of u_1 and v_1 is not included in these statements, which potentially can make this construction "empty." To fix this, it suffices to find just one (any) solution of $x^2 - dy^2 = \epsilon$ for any choice of $\epsilon = \pm 1$ with nonzero integer or half-integer x and y. If such a solution exists, then $x^2 - dy^2 = 1$ is solvable in terms of positive integers: consider the powers of $v = |x| + |y|\sqrt{d}$ to see this. The following lemma provides a solution with positive integers x and y for $\epsilon = 1$, i.e. in the classical, the strongest possible, variant. The first of its proofs was due to Lagrange (1768).

Lemma 5.5.3 *The equation $x^2 - dy^2 = 1$ can be solved in terms of positive integers x and y if $d > 0$ is not a complete square.*

Proof. For any $m \geq 1$, let us consider $z_i = i\sqrt{d}$ and their fractional parts $\{z_i\} = z_i - \text{Floor}[z_i]$, where $0 \leq i \leq m$. Recall that Floor$[x]$ is the largest integer $\leq x$ (its integer part for $x \geq 0$). The numbers $\{z_i\}$ never coincide for different i because $\sqrt{d}$ is irrational. Therefore there exist some $i > i'$ such that $0 < |\{z_i\} - \{z_{i'}\}| < \frac{1}{m}$, which is due to the Pigeonhole Principle. Explicitly: $|(i-i')\sqrt{d} - j| < \frac{1}{m} \leq \frac{1}{i-i'}$ for some integer $j > 0$. Setting $x_m = j$, $y_m = (i - i')$ and $u_m = x_m + y_m\sqrt{d}$, we obtain that $|\overline{u_m}| < \frac{1}{m} \leq \frac{1}{y_m}$ for the conjugation $\overline{u_m} = x_m - y_m\sqrt{d}$ as above. Thus, $|\frac{x_m}{y_m} - \sqrt{d}| < \frac{1}{y_m^2}$.

The sequence of pairs (x_m, y_m) may have repetitions, but it will contain infinitely many different pairs as $m \to \infty$. One has: $|x_m^2 - dy_m^2| = |u_m\overline{u_m}| < \frac{x_m + y_n\sqrt{d}}{y_m} = \frac{x_m}{y_m} + \sqrt{d} < 2\sqrt{d}+1$. Since $x_m^2 - dy_m^2$ is an integer, we obtain that there exists some integer k between $-2\sqrt{d}-1$ and $+2\sqrt{d}+1$ such that there are infinitely many different pairs (x_m, y_m) solving $x_m^2 - dy_m^2 = k$ for this k. Within this sequence, we can find at least two different pairs, corresponding to some u_m and $u_{m'}$, satisfying the congruences $x_{m'} = x_m \bmod k$ and $y_{m'} = y_m \bmod k$. We used the Pigeonhole Principle again! There are infinitely many such pairs but we need only 2 of them.

Let $u = u_{m'}/u_m = \frac{1}{k}(x_{m'} + y_{m'}\sqrt{d})(x_m - y_m\sqrt{d}) = \frac{1}{k}(x_{m'}x_n - dy_{m'}y_m + (y_{m'}x_m - y_m x_{m'})\sqrt{d})$. Setting $u = x + y\sqrt{d}$, one has: $x^2 - dy^2 = 1$. We performed such divisions above. Importantly, x and y are integers, though possibly not positive. Indeed, $x_{m'}x_m - dy_{m'}y_m + (y_{m'}x_m - y_m x_{m'})\sqrt{d} = x_m x_m - dy_m y_m + (y_m x_m - y_m x_m)\sqrt{d} \bmod k = 0$, which provides the divisibility of u by k. Finally, replace x, y by $\pm x, \pm y$ if necessary to obtain positive x, y. ❑

The process of approximation by rationals we used is general: given any irrational $\alpha > 0$ and $m > 0$, there exist positive integers x_m, y_m such that $|\frac{x_m}{y_m} - \alpha| < \frac{1}{y_m^2}$, and y_m can be made here arbitrarily large. However, such an estimate can be obtained only for (very) special denominators y_n; otherwise the accuracy cannot be better than $\frac{1}{2y_m}$. The application of this construction to solving Pell's equation is specific to quadratic irrationalities.

5.5.2 Rational approximations

The approximation procedure used in Lemma 5.5.3 can be sharpened. We will begin with a general construction of "best possible rational approximations," and then will essentially redo it via continued fractions. An application of continued fractions to finding fundamental solutions of Pell's equations will be discussed at the end. The readers can finish this chapter here, though this topic is clarifying; Theorem 5.5.7 later is the best for obtaining the fundamental solutions (and its justification is really interesting).

For an irrational number $\alpha > 0$, we construct the following sequence of integer points: $(\widetilde{x}_{2m-1}, \widetilde{y}_{2m-1})$ and $(\widetilde{x}_{2m}, \widetilde{y}_{2m})$ such that

$$\frac{\widetilde{x}_{2m-1}}{\widetilde{y}_{2m-1}} < \alpha < \frac{\widetilde{x}_{2m}}{\widetilde{y}_{2m}} \quad \text{for} \quad m \geq 1,\ \widetilde{x}_n \geq 0,\ \widetilde{y}_n > 0. \tag{5.5.7}$$

We set $(\widetilde{x}_0, \widetilde{y}_0) = (\text{Ceiling}[\alpha], 1)$, $(\widetilde{x}_1, \widetilde{y}_1) = (\text{Floor}[\alpha], 1)$ and then proceed by induction as follows. Given $(\widetilde{x}_n, \widetilde{y}_n)$, let $(\widetilde{x}_{n+2}, \widetilde{y}_{n+2})$ be the integer pair with the minimal possible $\widetilde{y}_{n+2}$ such that $\widetilde{y}_{n+2} > \widetilde{y}_n$ and $dist_{n+2} < dist_n$. Here and below $dist_n$ is the distance from $(\widetilde{x}_n, \widetilde{y}_n)$ to the line $x = \alpha y$. So the rational approximation $\widetilde{x}_{n+2}/\widetilde{y}_{n+2}$ improves $\widetilde{x}_n/\widetilde{y}_n$ and y_{n+2} is the smallest new denominator after y_n when the improvement is possible.

Note that the distance from (x_n, y_n) to the line $x = \alpha y$ is $\frac{|x_n - \alpha y_n|}{\sqrt{1+\alpha^2}}$. Use the *dot-product* or simply calculate the *cosine* of the angle between the ray from $(0,0)$ to (x_n, y_n) and this line, so minimizing the distances to the line $x = \alpha y$ is equivalent to minimizing $|x_n - \alpha y_n|$.

We will now "merge" the odd and even sequences of pairs we constructed.

First, for any $m \geq 1$, let i_m be the greatest i such that $\widetilde{y}_{2i-1} < \widetilde{y}_{2m}$. Check that $i_m \leq i_{m+1}$; repetitions of i_m are possible. We obtain a set $I = \{i_m\}$, where $m = 1, 2, 3, \ldots$. For each $i \in I$, let $m(i)$ be the greatest m° such that $i_{m^\circ} = i_m$: this is the m-index for the last repeat of i in I. Check that $m(i) < m(i')$ and that $\widetilde{y}_{2i-1} < \widetilde{y}_{2m(i)} < \widetilde{y}_{2i'-1}$ when $i < i'$ for $i, i' \in I$.

Second, using $m(i)$ defined above for $i \in I$, we combine $y_{2i-1} = \widetilde{y}_{2i-1}$ for $i \in I$ and the corresponding $\widetilde{y}_{2m(i)}$ in one (strictly) increasing sequence $\{y_n\}$. Namely, we set $y_{2i} = \widetilde{y}_{2m(i)}$. Naturally, x_n will be the corresponding x-components. Then (5.5.7) holds and we will have the following relations.

Problem 5.5.4 *For the sequence of the pairs (x_n, y_n) with increasing y_n constructed above for irrational α and $r_n = \frac{x_n}{y_n}$, prove that*

$$0 < \alpha - r_{2m-1} < 1/(y_{2m} y_{2m-1}),\ 0 < r_{2m} - \alpha < 1/(y_{2m} y_{2m-1}) \ \text{for}\ m \geq 1.$$

Solution. Let us check that the triangle $\widetilde{\Delta}_m$ in the plane with the vertices at the points $(0,0)$, $(\widetilde{x}_{2i_m-1}, \widetilde{y}_{2i_m-1})$, $(\widetilde{x}_{2m}, \widetilde{y}_{2m})$ contains no other integer points. Indeed, any such point (x', y') with $\frac{x'}{y'} > \alpha$ would contradict the construction of the sequence $(\widetilde{x}_{2m}, \widetilde{y}_{2m})$: its distance to the line $x = \alpha y$ would be smaller than $dist_{2m}$. Generally, $y' > y_{2i_m-1}$ is not impossible because $\widetilde{\Delta}_m$ contains points with such y'. However, (x', y') would then be closer to the line $x = \alpha y$ than $(\widetilde{x}_{2i_m-1}, \widetilde{y}_{2i_m-1})$. Since it is an integer point, either $(x', y') =$

$(\widetilde{x}_{2i'-1}, \widetilde{y}_{2i'-1})$ for some $i' > i$ in I or there exists such $(\widetilde{x}_{2i'-1}, \widetilde{y}_{2i'-1})$ in $\widetilde{\Delta}_m$ (closer to the line $x = \alpha y$ than (x', y')). Then $\widetilde{y}_{2i'-1} < \widetilde{y}_{2m}$, a contradiction with the definition of i_m.

Similarly, the inequality $\widetilde{y}_{2i-1} < \widetilde{y}_{2m(i)} < \widetilde{y}_{2i'-1}$ is granted, because no $\widetilde{y}_{2i'-1}$ for $i' > i$ in I can occur before $\widetilde{y}_{2m(i)}$ due to the definition of i_m.

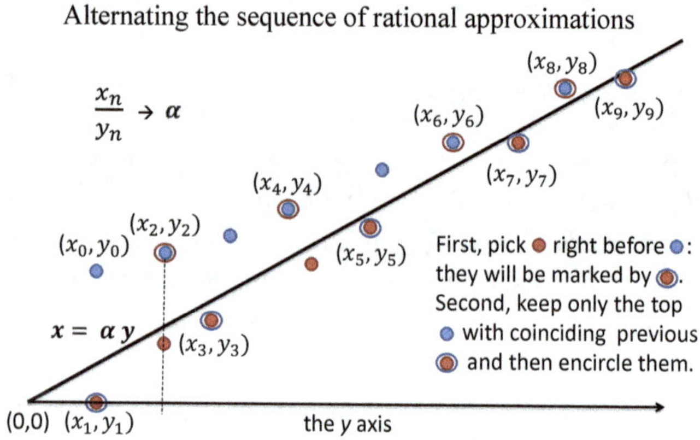

Figure 5.3: Optimal approximations by rationals.

Our construction is presented in Fig. 5.3. First, we reduce the set of points $(\widetilde{x}_{2m-1}, \widetilde{y}_{2m-1})$ picking only those "right before" $(\widetilde{x}_{2m}, \widetilde{y}_{2m})$, which is for $\widetilde{y}_{2i_m-1}$. Second, among $(\widetilde{x}_{2m}, \widetilde{y}_{2m})$ with coinciding i_m, we keep only the point with the greatest $\widetilde{y}_{2m(i)}$; here $i \in I = \{i_m\}$. So the resulting sequence of (x_n, y_n) is "alternating" (below-above-below- the line for α), and the inequalities $y_{n+1} > y_n$ hold by construction.

Let Δ_m be the triangle of type $\widetilde{\Delta}$ considered above with vertices at the points $(0,0)$, (x_{2m-1}, y_{2m-1}), (x_{2m}, y_{2m}). So it contains only 3 integer points at its boundary, and no integer points inside it. Thus its area is $\frac{1}{2}$ from Pick's Theorem; see Exercise 6.4.27.

Algebraically, this area is given by the formula: $\frac{1}{2}(x_{2m}y_{2m-1} - x_{2m-1}y_{2m})$, which is a standard property of the dot-product of 2 vectors. Thus:

$$x_{2m}y_{2m-1} - x_{2m-1}y_{2m} = 1, \text{ and } \frac{x_{2m}}{y_{2m}} - \frac{x_{2m-1}}{y_{2m-1}} = \frac{1}{y_{2m}y_{2m-1}}, \tag{5.5.8}$$

whose relations readily result in the required inequalities. ❑

The sequence of rational approximations of α we obtained is sharper than that used in the main step in the proof of Lemma 5.5.3, so it can be used for its proof. Namely, for $n = 2m-1$ we have the inequality $|\frac{x_n}{y_n} - \alpha| < \frac{1}{y_n\, y_{n+1}}$. The one we used when proving the lemma was $|\frac{x_n}{y_n}-\alpha|<\frac{1}{y_n^2}$, which was weaker since $y_{n+1}>y_n$, significantly weaker for $\alpha=\sqrt{d}$. Generally, the new sequence satisfies $\min\{\frac{x_{n+1}}{y_{n+1}}-\alpha,\alpha-\frac{x_n}{y_n}\}<\frac{1}{2y_n^2}$ (see below). Such improvements are of importance, but they are insignificant for the proof of the lemma.

5.5.3 Continued fractions

An alternative approach to rational approximations, the 3rd we employ, is the usage of ***continued fractions***. The generalized continued fractions are defined in Exercises 5.6.40 and 5.6.41; the usual ones are for $a_i=1$. We set $[b_0,b_1,b_2,\ldots]=b_0+\cfrac{1}{b_1+\cfrac{1}{b_2+\ldots}}$, provided the convergence (see below). Given α, b_i are determined as follows: $b_0=$Floor$[\alpha], r_0=\alpha-b_0$ (the zero step), $b_1=$Floor$[\frac{1}{r_0}]$, $r_1=\frac{1}{r_0}-b_1$, $b_2=$Floor$[\frac{1}{r_1}]$, $r_2=\frac{1}{r_1}-b_2$, and so on. Here $0\le r_i<1$ and b_i are *positive* integers for $i>0$. We terminate this process if $r_m=0$: at the first such m. This happens if and only if α is rational, which follows from *Euclid's algorithm* (check). Note that $b_m\ge 2$ if $m>0$. Indeed, $1/r_{m-1}=b_m+r_m$ for $r_m=0$ and $b_m=1$ would give that $r_{m-1}=1$, which is impossible. We will sometimes set $b_i=0$ for $i>m$ for rational α.

Generally, the convergence of infinite $[b_0,b_1,b_2,\ldots]$ is granted if $b_i\ge 1$ or $b_i=0$ for sufficiently large i (see below). It converges to α if b_i are as above, so we can write $\alpha=[b_0,b_1,b_2,\ldots]$.

The corresponding successive rational approximations are defined upon the truncations of the sequence of b_i. One has: $[b_0,b_1,b_2,\ldots,b_n]=P_n/Q_n$ for $P_n=b_nP_{n-1}+P_{n-2}$, $Q_n=b_nQ_{n-1}+Q_{n-2}$ for $n\ge 2$. When $n<2$: $P_0=b_0, P_1=b_0b_1+a_1$, $Q_0=1, Q_1=b_1$. For instance, $P_1/Q_1=[b_0,b_1]=b_0+1/b_1=\frac{b_1b_0+1}{b_1}$, $P_2/Q_2=[b_0,b_1,b_2]=\frac{b_2(b_1b_0+1)+b_0}{b_2b_1+1}$. The fractions P_n/Q_n are called the ***convergents of the continued fraction***.

Here P_n,Q_n are positive integers for $n>0$ satisfying $\gcd(P_n,Q_n)=1$ due to the relation $P_nQ_{n-1}-Q_nP_{n-1}=(-1)^{n-1}a_1\cdots a_{n-1}a_n=(-1)^{n-1}$ from Exercise 5.6.40(i). The proof is by induction on n. $P_nQ_{n-1}-Q_nP_{n-1}=(b_nP_{n-1}+P_{n-2})Q_{n-1}-(b_nQ_{n-1}+Q_{n-2})P_{n-1}=-(P_{n-1}Q_{n-2}-Q_{n-1}P_{n-2})=(-1)^{n-1}$. Thus, $\frac{P_n}{Q_n}$ are *reduced fractions* and knowing the ratio P_n/Q_n is sufficient to recover P_n and Q_n.

Also, we obtain that $\frac{P_n}{Q_n}-\frac{P_{n-1}}{Q_{n-1}}=(-1)^{n-1}\frac{1}{Q_nQ_{n-1}}$, which readily gives that $\alpha=\lim_{n\to\infty}\frac{P_n}{Q_n}$. We need only to check that this difference tends to 0 since $\frac{P_n}{Q_n}$ is an alternating sequence. More exactly, one can use that $Q_n\ge f_{n+1}$ for the Fibonacci numbers f_n (for irrational α). Note that $x_n=P_{n+1}, y_n=Q_{n+1}$

satisfy (5.5.8), which gives another proof of the key step of Lemma 5.5.3.

We will need below $\langle\alpha\rangle_n = [\,b_n, b_{n+1}, \ldots]$ and the relation:

$$\alpha = [\,b_0, b_1, \ldots, b_{n-2}, b_{n-1} + 1/\langle\alpha\rangle_n\,] = \frac{(b_{n-1} + 1/\langle\alpha\rangle_n)P_{n-2} + P_{n-3}}{(b_{n-1} + 1/\langle\alpha\rangle_n)Q_{n-2} + Q_{n-3}}$$
$$= \frac{\langle\alpha\rangle_n(b_{n-1}P_{n-2} + P_{n-3}) + P_{n-2}}{\langle\alpha\rangle_n(b_{n-1}Q_{n-2} + Q_{n-3}) + Q_{n-2}} = \frac{\langle\alpha\rangle_n P_{n-1} + P_{n-2}}{\langle\alpha\rangle_n Q_{n-1} + Q_{n-2}}.$$

The first equality is directly from the definition; here $\langle\alpha\rangle_n \neq 0$ is required.

Problem 5.5.5 *(i) Let* $\alpha = \frac{a}{b} = [\,b_0, b_1, \ldots, b_m\,]$ *for relatively prime integers* $a, b > 1$. *Check that the pair* $(x{=}Q_{m-1}, y{=}P_{m-1})$ *solves the linear Diophantine equation* $ax - by = (-1)^{m-1}$. *Moreover, this solution is the smallest with positive integers. More exactly,* $0 < x < b$ *and* $0 < y < a$*; such a solution always exists and is unique if* $\gcd(a, b) = 1$ *for* $a, b \geq 2$.

(ii) From the definition: $[\,b_0, b_1, \ldots, b_m\,] = [\,b_0, b_1, \ldots, b_m - 1, 1\,]$, *called* short *(canonical) and* long *continued fraction for* $\frac{a}{b}$*; recall, that* $b_m \geq 2$. *Setting* $[b_0, b_1, \ldots, b_m - 1] = P'_m/Q'_m$ *(the* m*th convergent of this continued fraction), check that the pair* $(x' = Q'_m, y' = P'_m)$ *is the smallest positive integer solution of* $ax' - by' = (-1)^m$, *and that* $x' = b - x, y' = a - y$.

Solution. The relation $[\,b_0, b_1, \ldots, b_m\,] = [\,b_0, b_1, \ldots, b_m - 1, 1\,]$ for $b_m \geq 2$ is the only one among finite continued fractions with integer $b_i \geq 1$ for $i > 0$ (check). The formulas $ax - by = (-1)^{m-1}$ and $ax' - by' = (-1)^m$ follow from the general one $P_nQ_{n-1} - Q_nP_{n-1} = (-1)^{n-1}$. The recurrence relations for P_m, Q_m give that $0 < x < b$ and $0 < y < a$. See also Problem 7.3.11.

The general existence and uniqueness of a solution of $ax - by = \pm 1$ in this range is as follows. We can always assume that $1 \leq x \leq b - 1$ by reducing it modulo b and adjusting y accordingly. Then y must be positive and $y = (xa \pm 1)/b \leq (b-1)a/b + 1/b \leq a - a/b + 1/b < a$. Any other positive solutions are with $x \mapsto x + kb$ for $k > 0$, which are $\geq b$.

Similarly, $0 < x' < b$, $0 < y' < a$ and (x', y') is a unique solution in this range of $ax' - by' = (-1)^m$. However, then $x' = b - x, y' = a - y$. This also follows from the recurrences $a = P'_{m+1} = P'_m + P'_{m-1}, b = Q'_{m+1} = Q'_m + Q'_{m-1}$ since $P'_{m-1} = P_{m-1}$ and $Q'_{m-1} = Q_{m-1}$. ❑

The following theorem essentially states that the *convergents* P_n/Q_n to irrational α, give its best rational approximations. Here and in Theorem 5.5.7 below, a convenient reference is [36].

[36] S.H. Yang, *Continued fractions and Pell's equation*, 2008, `https://www.math.uchicago.edu/~may/VIGRE/VIGRE2008/REUPapers/Yang.pdf`.

Theorem 5.5.6 *For an irrational number $\alpha > 0$, assume that $|\frac{P}{Q} - \alpha| < \frac{1}{2Q^2}$ for integers $P, Q > 0$ such that $\gcd(P,Q) = 1$. Then $\frac{P}{Q} = \frac{P_n}{Q_n}$, i.e. this fraction is the nth convergent to α for certain n.*

Proof. We have $|\alpha - \frac{P_n}{Q_n}| < \frac{1}{Q_nQ_{n+1}} < \frac{1}{2Q_n}$ provided here and below that $n \geq 2$ (then $Q_n \geq 3$). Let us assume $0 < Q \leq Q_n$ and $P/Q \neq P_n/Q_n$.

Step 1. We are going to prove that $|\alpha - \frac{P}{Q}| > |\alpha - \frac{P_n}{Q_n}|$ for such P, Q.

First, take $Q = Q_n$ and assume that $P \neq P_n$. Then

$$\left|\frac{P}{Q} - \frac{P_n}{Q_n}\right| \geq \frac{1}{Q_n} \text{ and } \left|\alpha - \frac{P_n}{Q_n}\right| < \frac{1}{2Q_n};$$

we use that $P \neq P_n$ (for $Q = Q_n$) and $Q_n \geq f_{n+1}$ for the Fibonacci numbers f_n, so $Q_n \geq 3$. By the *triangle inequality*:

$$\left|\alpha - \frac{P}{Q}\right| \geq \left|\frac{P}{Q} - \frac{P_n}{Q_n}\right| - \left|\alpha - \frac{P_n}{Q_n}\right| > \frac{1}{Q_n} - \frac{1}{2Q_n} = \frac{1}{2Q_n} > \left|\alpha - \frac{P_n}{Q_n}\right|.$$

We obtain that $P = P_n$ gives the best approximation to α for $Q = Q_n$.

Second, let $0 < Q < Q_n$. Expressing P, Q as $P = uP_n + vP_{n-1}$ and $Q = uQ_n + vQ_{n-1}$ for some u, v (to be found next), we obtain that

$$u = \frac{PQ_{n-1} - QP_{n-1}}{P_nQ_{n-1} - P_{n-1}Q_n} = \pm(PQ_{n-1} - QP_{n-1}),$$
$$v = \frac{PQ_n - QP_n}{P_nQ_{n-1} - P_{n-1}Q_n} = \pm(PQ_n - QP_n),$$

where we used (again!) that $P_nQ_{n-1} - Q_nP_{n-1} = (-1)^{n-1}$.

The numbers u, v are *nonzero* integers. Indeed, if $u = 0$ then $P/Q = P_{n-1}/Q_{n-1}$, which is impossible. If $v = 0$, then $P/Q = P_n/Q_n$, which is impossible too ($Q < Q_n$). The signs of u and v are opposite: otherwise $uQ_n + vQ_{n-} \geq Q_n + Q_{n-1} > Q_n$. The signs of $\alpha - \frac{P_{n-1}}{Q_{n-1}}$ and $\alpha - \frac{P_n}{Q_n}$ are opposite too. Therefore, $u(Q_n\alpha - P_n)$ and $v(Q_{n-1}\alpha - P_{n-1})$ have the same sign and we arrive at the relation

$$Q\alpha - P = (uQ_n + vQ_{n-1})\alpha - (uP_n + vP_{n-1}) = u(Q_n\alpha - P_n) + v(Q_{n-1}\alpha - P_{n-1}),$$

which gives the inequality:

$$|Q\alpha - P| = u \cdot |Q_n\alpha - P_n| + v \cdot |Q_{n-1}\alpha - P_{n-1}| > \text{Min}\{|Q_n\alpha - P_n|, |Q_{n-1}\alpha - P_{n-1}|\}.$$

For instance, the inequality $|Q\alpha - P| > |Q_{n-1}\alpha - P_{n-1}|$ gives that for $Q > Q_{n-1}$, the deviations $|Q\alpha - P|$ with any P become strictly greater than that for Q_{n-1} (with P_{n-1}) until $Q = Q_n$, when the corresponding P must be

P_n. We conclude that if P/Q gives the best approximation to α among any fractions x/y with $y \le Q$, then $Q = Q_n$ and $P = P_n$ for some n.

Step 2. As we already used, α is always strictly between P_n/Q_n and P_{n+1}/Q_{n+1} for any n. More generally, let $r_{2m-1} = \frac{x_{2m-1}}{y_{2m-1}}$ and $r_{2m} = \frac{x_{2m}}{y_{2m}}$ be any pair of consecutive approximations constructed in Problem 5.5.4. More precisely, we need the following to proceed: $\frac{x_{2m-1}}{y_{2m-1}} < \alpha < \frac{x_{2m}}{y_{2m}}$, $y_{2m} > y_{2m-1}$ and $x_{2m}y_{2m-1} - x_{2m-1}y_{2m} = 1$.

Assume that $|\frac{x}{y} - \alpha| \ge \frac{1}{2y^2}$ for (x_{2m-1}, y_{2m-1}) and (x_{2m}, y_{2m}). Then

$$\begin{aligned}\frac{1}{y_{2m}\, y_{2m-1}} &= \frac{x_{2m}}{y_{2m}} - \frac{x_{2m-1}}{y_{2m-1}} = \Big(\frac{x_{2m}}{y_{2m}} - \alpha\Big) + \Big(\alpha - \frac{x_{2m-1}}{y_{2m-1}}\Big)\\ &\ge \frac{1}{2y_{2m}^2} + \frac{1}{2y_{2m-1}^2} = \frac{y_{2m}^2 + y_{2m-1}^2}{2y_{2m}^2\, y_{2m-1}^2}.\end{aligned}$$

Upon the multiplication of the first and the last fractions by $2y_{2m}^2\, y_{2m-1}^2$, we obtain that $2y_{2m}y_{2m-1} \ge y_{2m}^2 + y_{2m-1}^2$. This is impossible because $y_{2m} > y_{2m-1}$. Thus, we conclude that either (x_{2m-1}, y_{2m-1}) or (x_{2m}, y_{2m}) satisfies the inequality $|\frac{x}{y} - \alpha| < \frac{1}{2y^2}$. For instance, one from any two consecutive P_n/Q_n must satisfy the latter inequality.

Step 3. We now assume that $|\frac{P}{Q} - \alpha| < \frac{1}{2Q^2}$ for $P, Q > 0$ such that $\gcd(P, Q) = 1$. Step 1 provides that it is a convergent to α if it is the best approximation to α among any fractions x/y with $y \le Q$.

If this is not true, then there exists x/y such that $x/y \ne P/Q$ and $|y\alpha - x| < |Q\alpha - P|$ for $y \le Q$. Then $|y\alpha - x| < 1/(2Q)$ and we obtain that

$$\frac{1}{yQ} \le \frac{|yP - xQ|}{yQ} = \Big|\frac{P}{Q} - \frac{x}{y}\Big| \le \Big|\frac{P}{Q} - \alpha\Big| + \Big|\alpha - \frac{x}{y}\Big| < \frac{1}{2Q^2} + \frac{1}{2yQ} = \frac{y+Q}{2yQ^2}.$$

Upon the multiplication by yQ:

$$1 < \frac{y+Q}{2Q}, \quad \text{which results in } \ 2Q < y + Q \ \text{ and } \ y > Q.$$

The latter contradicts to $y \le Q$, which concludes Step 3 and the proof. ❑

Theorem 5.5.7 *Assuming that the integer $d > 1$ is not a perfect square, let $\alpha = \sqrt{d} = [b_0, b_1, b_2, \ldots]$, and ℓ be the period of the sequence $\{b_i\}$ (it is periodic) for $i \ge 1$: the minimal $l > 0$ such that $b_{i+l} = b_i$ for such i.*

(i) Then $(P_{\ell-1}, Q_{\ell-1})$ is the fundamental solution for $x^2 - dy^2 = \epsilon$, where $\epsilon \in \{\pm 1\}$. Recall that it is the solution with positive integers x and y such that $x + \sqrt{d}y$ is minimal possible. The sign is $\epsilon = (-1)^\ell$. There are no integer solutions (x, y) with $\epsilon = -1$ if the period ℓ is even.

(ii) The sequence $(P_{n\ell-1}, Q_{n\ell-1})$ for $n = 1, 2, \ldots$ provides all positive integer solutions, where the corresponding ϵ_k is ϵ^k. Setting $x_n = P_{n\ell-1}, y_n = Q_{n\ell-1}$, $u_n = x_n + y_n\sqrt{d}$, one has $u_n = u_1^n$. In particular, if ℓ is odd, then $(P_{2\ell-1}, Q_{2\ell-1})$ is the fundamental solution of $x^2 - dy^2 = 1$: the one with minimal $x + \sqrt{d}y$ for positive integers x, y.

(iii) As an example, $\sqrt{21} = [4, \{1, 1, 2, 1, 1, 8\}]$; the period is $\{\cdots\}$. So $\ell = 6$, and $(x_1, y_1) = (P_5, Q_5) = (55, 12)$ is the fundamental solution of $x^2 - 21y^2 = 1$. Another example: $(x_1, y_1) = (P_4, Q_4) = (70, 13)$ is the fundamental solution for $x^2 - 29y^2 = -1$, where $\sqrt{29} = [5, \{2, 1, 1, 2, 10\}]$ and $\ell = 5$.

Proof. Let us outline the main steps.

(a) We set $x = P_{\ell-1}, y = Q_{\ell-1}$, Recall that $\langle\alpha\rangle_n = [b_n, b_{n+1}, \ldots]$ for $n = 1, 2, \ldots$. Then $\langle\alpha\rangle_1 = \langle\alpha\rangle_{\ell+1}$, and $\alpha = \frac{\langle\alpha\rangle_{\ell+1}P_\ell+P_{\ell-1}}{\langle\alpha\rangle_{\ell+1}Q_\ell+Q_{\ell-1}} = \frac{\langle\alpha\rangle_1 P_\ell+P_{\ell-1}}{\langle\alpha\rangle_1 Q_\ell+Q_{\ell-1}}$, where $\langle\alpha\rangle_1 = 1/(\alpha - b_0)$. Hence: $\alpha = \frac{P_\ell+P_{\ell-1}(\alpha-b_0)}{Q_\ell+Q_{\ell-1}(\alpha-b_0)}$. Upon the multiplication by the denominator and equating the rational and irrational parts in the resulting formula (recall that $\alpha = \sqrt{d}$):

$$Q_\ell = b_0 Q_{\ell-1} + P_{\ell-1},\ P_\ell = b_0 P_{\ell-1} + Q_{\ell-1}d.$$

From the 1st relation: $b_0 = \frac{Q_\ell - P_{\ell-1}}{Q_l}$. Using this, the 2nd relation becomes: $P_{\ell-1}^2 - dQ_{\ell-1}^2 = P_{\ell-1}Q_\ell \quad Q_{\ell-1}P_\ell = (-1)^\ell$.

(b) Let us assume that $x = P_{m-1}, y = Q_{m-1}$ satisfy $x^2 - dy^2 = \pm 1$ for some m; then $\pm 1 = (-1)^m$. We have: $\alpha = \frac{[\alpha]_m P_{m-1}+P_{m-2}}{[\alpha]_m Q_{m-1}+Q_{m-2}}$. Reversing the argument in (a), we deduce from $P_{m-1}^2 - dQ_{m-1}^2 = (-1)^m$ that $\langle\alpha\rangle_m$ is $\sqrt{d}$ plus some integer. We omit this calculation. This is sufficient to conclude that $\langle\alpha\rangle_1 = \langle\alpha\rangle_{m+1}$. Indeed, adding an integer to $\langle\alpha\rangle_m$ does not influence $\langle\alpha\rangle_{m+1}$. We obtain that ℓ divides m.

This divisibility and the reasoning from (a) show that $(P_{n\ell-1}, Q_{n\ell-1})$, where $n = 1, 2, \ldots$, constitute *all* solutions of $x^2 - dy^2 = \epsilon$ with positive integers x and y (for $\epsilon \in \{\pm 1\}$). For instance, $(P_{2\ell-1}, Q_{2\ell-1})$ is the fundamental solution for $x^2 - dy^2 = 1$ if ℓ is odd. For even ℓ, all integer solutions are with $\epsilon = +1$. The formula $u_n = u_1^n$ follows from Theorem 5.5.1.

(c) Finally, we rewrite $P^2 - dQ^2 = \pm 1$ for positive integers P, Q as $|\frac{P}{Q} - \sqrt{d}| = \frac{1}{(P+Q\sqrt{d})Q}$, where the latter is obviously smaller than $\frac{1}{2Q^2}$ for $d > 3$. Then Theorem 5.5.6 is applied, which gives that (P, Q) is (P_m, Q_m) for some m, and (b) is used. The cases $d = 2, 3$ can be examined directly. ❑

We mention that the period $\{a_1, \ldots, a_\ell\}$ for $\alpha = \sqrt{d}$ always ends with $a_\ell = 2b_0$, and $a_k = a_{\ell-k}$ for $k = 1, 2, \ldots, \ell - 1$ (Galois). This can be clearly seen in the examples from Part (ii). The following is of interest: there are infinitely many d for any given period $\ell \geq 1$ (Friesen).

Generally, we do not need here continued fractions to calculate x_n, y_n for $n > 1$ if $u_1 = x_1 + y_1\sqrt{d}$ is known. The formula $u_n = u_1^n$ and the following recurrences for x_n and y_n are the main tools.

Let $x_1 = a, y_1 = b, \lambda_1 = a + b\sqrt{d}, \epsilon = \epsilon_1$. Then λ_1 and $\lambda_2 = \overline{\lambda_1} = a - b\sqrt{d}$ solve the *master equation* $\lambda^2 = 2a\lambda - \epsilon$: $2a = \lambda_1 + \lambda_2$ and $\epsilon = \lambda_1\lambda_2$. We obtain that $x_n = (\lambda_1^n + \lambda_2^n)/2, y_n = (\lambda_1^n - \lambda_2^n)/(2\sqrt{d})$. Therefore they satisfy both sequences, the recurrence $f_n = 2af_{n-1} - \epsilon f_{n-2}$, and can be uniquely determined from the initial conditions: $x_0 = 1, y_0 = 0, x_1 = a, y_1 = b$.

For instance, let $d = 15$. Then the fundamental solution is $(x_1, y_1) = (4, 1)$ for $\epsilon_1 = 1$. Accordingly, $a = 4, b = 1, \epsilon = 1$ and the recurrence for x_n, y_n becomes $f_n = 8f_{n-1} - f_{n-2}$ with the initial conditions $x_0 = 1, x_1 = 4$, and $y_0 = 0, y_1 = 1$. For instance, the pairs (x_n, y_n) for $1 \le n \le 7$ are $(4,1), (31,8), (244,63), (1921,496), (15124,3905), (119071,30744), (937444,242047)$.

5.5.4 Generalizing Pisano periods

Let us discuss counterparts of *Pisano periods* for solutions of Pell's equations.

Problem 5.5.8 *Let $\tau_p(d)$ be the period of the sequence of pairs $\{(x_n, y_n)\}$ modulo p for any prime p, and $n = 1, 2, \ldots$. It is by definition the smallest $m > 0$ such that $x_m = 1 \bmod p$ and $y_m = 0 \bmod p$.*

Let $p \ne 2$. Then $\tau_p(d)$ divides $(p+1)$ for $\epsilon_1 = 1$ or $2(p+1)$ for $\epsilon_1 = -1$ if $d \bmod p$ is not a square in $\mathbb{Z}_p$. In the latter case, $2(p+1)/\tau_p(d)$ is always odd. If d is a nonzero square in $\mathbb{Z}_p$, then $\tau_p(d)$ divides $p - 1$ for any ϵ_1 and must be even for $\epsilon = -1$. If $d = 0 \bmod p$, then the period divides $2p$ for $\epsilon = +1$ in $x_1^2 - dy_1^2 = \epsilon$ and can be 4 or p if $\epsilon = -1$. For $p = 2$, $\tau_p(d)$ divides 2.

Solution. In the first case, we construct $\mathbb{F}_{p^2}$ by adding $\sqrt{d}$ to $\mathbb{F}_p = \mathbb{Z}_p$ and then use the Lagrange Theorem for $\mathbb{F}_{p^2}^* = \mathbb{F}_{p^2} \setminus \{0\}$. The order of $\lambda_1 = (x_1 + y_1\sqrt{d})$ considered in $\mathbb{F}_{p^2}^*$ (the period of the sequence $\{\lambda_1^m\}$ for integer m) coincides with $\tau_p(d)$. Following the case of $s_p = 1$ in the proof of Theorem 5.4.3, we have $\lambda_2 = \epsilon\lambda_1^{-1} = \lambda_1^p$. Hence, $\lambda_1^{p+1} = \epsilon$ and the order of λ_1 divides either $(p+1)$ for $\epsilon = 1$ or $2(p+1)$ for $\epsilon = -1$. Here $2(p+1)/\tau_p(d)$ for $\epsilon = -1$ cannot be even, because then λ_1^{p+1} would be 1.

Recall that $p \ne 2$. We note that it is possible that $x_1 \bmod p = 1$ and $y_1 \bmod p = 0$. For instance, this occurs for $d = 11$. Then $\sqrt{11} = [3, \{3, 6\}], \ell = 2, x_1 = 10, y_1 = 3$; so take $p = 3$.

Let us now assume that d is not divisible by p and $\delta = \sqrt{d}$ belongs to $\mathbb{Z}_p$ for prime $p > 2$. Generally, the multiplication of the pairs (x, y) corresponding to that for $x + y\sqrt{d}$ is as follows: $(x, y) \cdot (x', y') = (xx' + dyy', xy' + yx')$. The addition is component-wise. Considering x and y modulo p, the pairs (x, y) form a ring; the notation will be $\widetilde{\mathbb{Z}}_p(d)$. We claim that it can be identified with the ring $\{\mathbb{Z}_p, \mathbb{Z}_p\}$ of pairs $\{a, b\}$ with the component-wise addition and multiplication. The map is $f : (x, y) \mapsto \{a = x + y\delta,\, b = x - y\delta\}$. The inverse map is $g : \{a, b\} \mapsto (\frac{a+b}{2}, \frac{a-b}{2d})$. Both maps are compatible with the addition and multiplication. The period of the sequence $w, w^2, w^3, \ldots,$ for an arbitrary element w in $\{\mathbb{Z}_p, \mathbb{Z}_p\}$ is $(p-1)$ or its divisor; we use that the multiplication is component-wise. So $m = \tau_p(d)$ is the smallest $m > 0$ such that $\lambda_1^m = 1, (\epsilon/\lambda_1)^m = \epsilon^m = 1$ in $\mathbb{Z}_p$, which gives the required.

The identification of $\widetilde{\mathbb{Z}}_p(d)$ with $\{\mathbb{Z}_p, \mathbb{Z}_p\}$ fails for $d = 0 \bmod p$ or for $p = 2$; these cases require a different consideration.

If $d = 0 \bmod p$, then $(x, y)^2 = (x^2, 2xy), (x, y)^3 = (x^3, 3x^2y), \ldots$. Generally, $(x, y)^m = (x^m, mx^{m-1}y)$ for any $m > 0$, which gives that $(x, y)^p = (x, 0)$. For $\epsilon = +1$, $x = \pm 1 \bmod p$. Thus $(1, y)^p = (1, 0)$ for $x = 1$, and $(1, y)^{2p} = (1, 0)$ for $x = -1$; accordingly, $\tau_p(d)$ divides $p, 2p$. If $\epsilon = -1$, then x is of order 4 for $p \neq 2$ and $\tau_p(d)$ is either 4 for $y = 0 \bmod p$ or $4p$ otherwise.

When $p = 2$ and $d = 1 \bmod 2$, then $(x, y) \bmod 2$ can be $(1, 0)$ or $(0, 1)$. Then $(0, 1)^2 = (1, 0)$ and $\tau_p(d) = 2$ if (x, y) is $(0, 1)$ modulo 2. ❑

We see that this problem is a special instance of Theorem 5.4.3 extended to general (linear, constant) recurrences of depth 2. Let us provide the list of periods $\tau_p(d)$ for the first 15 prime numbers for $d = 5, 15$; $\epsilon_1 = -1$ for 5 and $+1$ for 15. One has: $\sqrt{5} = [2, \{4\}], \ell = 1$ and $\sqrt{15} = [3, \{1, 6\}], \ell = 2$. Here $\sigma_p(d)$ is 0 if p divides d, $+1$ if d is a square in $\mathbb{Z}_p$, and -1 otherwise, the ***Jacobi symbol*** of d modulo p. Indeed, $\tau_p(d)$ divides $p - \sigma_p(d)$ in this table for $p \neq 2, 5, 3$ for $d = 15$. It divides $p - 1$ for $d = 5$ for $\sigma_p(5) = 1$, and $2(p+1)$ for $\sigma_p(5) = -1$. Also, $\tau_5(5) = 4 \cdot 5, \tau_5(15) = 2 \cdot 5$, $\tau_3(15) = 3$.

In this table, $\tau_p(d)$ are *smaller* than the maximum values from the proposition when $p = 17, 19, 23, 29, 31, 47$ for $d = 5$ and $p = 3, 13, 31, 37, 41, 43$ for $d = 15$.

n	1	2	3	4	5	6	7	8	9	10	11	12	13	14	15
p_n	2	3	5	7	11	13	17	19	23	29	31	37	41	43	47
$\tau_p(5)$	2	8	20	16	10	28	12	6	16	14	10	76	40	88	32
$\sigma_p(5)$	1	−1	0	1	1	−1	−1	1	−1	1	1	−1	1	−1	−1
$\tau_p(15)$	2	3	10	6	10	7	16	20	24	30	8	19	21	7	48
$\sigma_p(15)$	1	0	0	1	1	−1	1	−1	−1	−1	−1	−1	−1	1	−1

We mention that the line for $\tau_p(5)$ has many coinciding entries with the line for the Pisano periods t_p (Fibonacci numbers) from the table right before Section 5.4.3. This is of course not incidental: t_p is the minimal even $t > 0$ such that $v_1^t = 1 \bmod p$ for $v_1 = \frac{1+\sqrt{5}}{2}$, and $\tau_p(5)$ is the minimal $\tau > 0$ such that $u_1^\tau = 1 \bmod p$ for $u_1 = 2 + \sqrt{5} = v_1^3$. See Problem 5.5.2(ii).

5.6 Exercises

Exercise 5.6.1 $^\heartsuit$ *Calculate without any special software* 10^{999} *modulo* 41.

Hint: Examine 10^5 modulo 41.

Exercise 5.6.2 $^\heartsuit$ *Assume that* $\gcd(m,n) = 1$ *for positive integers* n, m. *Prove that* $\gcd(\ell n{+}m, \ell(n{+}m){+}m) = 1$ *if and only if* $\gcd(\ell, m) = 1$. *For instance,* $\gcd(1{+}\ell n, 1{+}\ell(n{+}1)) = 1$ *for any* $n, \ell \geq 0$.

Hint: If a prime number p divides $\ell n + m, \ell(n+m)+m$, then it divides ℓm, and $\ell(\ell n + m) - \ell m = \ell^2 n$. Due to $\gcd(m,n) = 1$, p must divide ℓ. Therefore it divides $m = (\ell n + m) - \ell n$. Any common divisor of ℓ, m obviously divides $\gcd(\ell n + m, \ell(n+m)+m)$.

Exercise 5.6.3 $^\heartsuit$ *Check that* $\gcd(2^{2m} - 1, 2^{m+1} - 1)$ *for* $m \geq 1$ *is* 1 *or* 3, *where the latter occurs exactly for odd* m.

Hint: Justify and use that $\gcd(2^m - 1, 2^{m+1} - 1) = 1$.

Exercise 5.6.4 $^\heartsuit$ *(i) Use* Euclid's algorithm *to see that* $\gcd(204, 457){=}1$. *Check this using the prime factorization. Find* $\gcd(204, 458)$ *and* $\gcd(204, 459)$.

(ii) Calculate the continued fraction *of* 204/457. *Drop the last digit in its presentation* $[1, \ldots, *]$ *and calculate the corresponding rational number* $\frac{y}{x}$. *Check that* $204x - 457y = -1$.

Answer: $\frac{204}{457} = [0,2,4,6,8], [0,2,4,6] = \frac{25}{56}$.

Exercise 5.6.5 $^\heartsuit$ *(i) Employ the last step of* Euclid's algorithm *for* $\gcd(75, 28)$ *to calculate an integer solution of the equation* $75x - 28y = 1$ *and find* $\frac{1}{28}$ mod 75 *and* $\frac{1}{75}$ mod 28. *(ii) Similarly, find* $\frac{1}{157}$ mod 225 *and* $\frac{1}{225}$ mod 157.

Exercise 5.6.6 $^\diamondsuit$ *Employ* continued fractions *and Problem 5.5.5 to find the integer solutions of the equations* $225x - 157y = \pm 1$ *(both) in the range* $0 < x < 157$, $0 < y < 225$. *Use this to find* $\frac{1}{157}$ mod 225 *and* $\frac{1}{225}$ mod 157.

Hint: Check that $\frac{225}{157} = [1,2,3,4,5]$, where $b_0 = 1$ and $m = 4$ in the notation of Problem 5.5.5. Accordingly, $x = Q_3, y = P_3$ solve $225x - 157y = -1$; $y/x = 43/30 = [1,2,3,4]$. Then use $[1,2,3,4,5] = [1,2,3,4,4,1]$ to check that $x' = Q'_4 = 127, y' = P'_4 = 182$ for the latter continued fraction solve $225x - 157y = 1$. Actually, $x' = 157 - x, y' = 225 - y$, so no new calculation is needed and the knowledge of (x,y) is sufficient.

Exercise 5.6.7 $^{\diamondsuit}$ *The* McNugget numbers *are all possible numbers of McDonald's Chicken McNuggets available in the UK in the 1980s using the standard boxes of* $6, 9$ *and* 20 *nuggets. This definition is due to Henri Picciotto.*

(i) Prove that all numbers from 44 *onward are McNugget numbers. See Problem 7.5.7(i) in full generality and (ii) there for some special sequences.*

(ii) Find all solutions to the equation $3x + 20y = 43$ *for integers* x, y, *and then check that it has no positive integer solutions with* $x > 1$. *This gives us that* 43 *is not a McNugget number.*

(iii) Assuming that the 6-boxes are missing, check that 151 *cannot be represented as* $9x + 20y$ *with positive integer* x *and* y, *but any number starting with* $152=(9-1)(20-1)$ *are such. See Problem 7.5.5(i).*

Hint for (i): You need only to check $44, \ldots, 49$, which is direct.
Hint for (ii): Use that $\{x = 1, y = 2\}$ is such a solution; then find all of them.
Hint for (iii): Use that the pair $\{x = -1, y = 8\}$ gives 151; also, $1 = 20 \cdot 5 - 9 \cdot 11 = 9 \cdot 9 - 20 \cdot 4$, $2 - 20 - 9 \cdot 2$, and so on; or follow Problem 7.5.5(i).

Exercise 5.6.8 $^{\heartsuit}$ *Prove that* $p^m - 1$ *is divisible by* $p^k - 1$ *for a prime number* p *and* $m > k \geq 1$ *if and only if* k *divides* m.

Hint: Set $m = qk + r$ for $q > 0, 0 \leq r < k$. Then $p^m - 1 = p^m - p^r + p^r - 1 = (p^{qk} - 1)p^r + (p^r - 1)$, and $r - 0$ is necessary and sufficient for the divisibility.

Exercise 5.6.9 $^{\diamondsuit}$ *Let* ℓ_n *be the minimal* x *such that* $10^x = 1 \bmod n$, *where we assume that* $\gcd(n, 10) = 1$. *Equivalently, it is the length of the period of the decimal expansion of* m/n *if* $\gcd(m, n) = 1$; *use Problem 5.1.2. Check that* $\ell_7 = 6, \ell_n - k$ *for* $n = 11\ldots1$ *(k times), and that* $\ell_{7n} = 6k/\gcd(6, k)$ *unless* 6 *divides* k. *For* $k = 6$, *use a computer to find that the length of the period of the decimal fraction for* $1/(7n)$ *is* 42, *which is* ℓ_{777777}.

Hint: Use that $99\ldots9 = 10^k - 1$ is divisible by 7 if and only if $n = 11\ldots1$ is such (k units). Thus, $\ell_{7n} = lcm(\ell_7, \ell_n)$. Then find the decimal fraction for $1/11\ldots1$ and that for $1/777777$ (use a computer for the latter).

Exercise 5.6.10 $^{\heartsuit}$ *Compute the addition table in* $\mathbb{Z}_8$ *and the multiplication tables for invertible elements there. Why are they Latin squares?*

Exercise 5.6.11 $^{\heartsuit}$ *(i) Check that all nonzero elements of* $\mathbb{Z}_{11}$ *are powers of* 2. *Equivalently,* $2^{10} = 1$ *and* 10 *is the smallest such exponent. Using the standard terminology:* 10 *is the order of* 2 *in the group of* 10 *nonzero elements of* $\mathbb{Z}_{11}$, *so* 2 *is a* primitive root $\bmod 11$. *See also Exercise 5.6.17 below.*

(ii) Check without calculations that 10 *is the smallest such exponent (the order) for* $2, 2^3, 2^7, 2^9$ *and only for these elements in* $\mathbb{Z}_{11}$*; then find them. So there are* 4 primitive roots *there.*

(iii) Verify that there are no elements of order 4 *in the group of* 4 *invertible elements in* $\mathbb{Z}_8$*. So this group is not cyclic and has no primitive roots.*

Exercise 5.6.12$^\heartsuit$ *Check that* 2 *is a* primitive root *in* $\mathbb{Z}_{13}$*, i.e. its order is* 12 *in the group of its nonzero elements, the greatest possible. Use this to find that all primitive roots* mod 13 *are* $2, 6, 7, 11$*. See also Exercise 5.6.17 below.*

Hint: The primitive roots are $2, 2^5, 2^7, 2^{11} = 1/2 \bmod 13$.

Exercise 5.6.13$^\diamondsuit$ *(i) Check that the invertible numbers modulo* p^m *for a prime number* p *are exactly those with nonzero remainders* mod p*. Deduce from this that their number is* $(p-1)p^{m-1}$*.*

(ii) Use the Chinese Remainder Theorem to prove that $\varphi(m)\varphi(n)=\varphi(mn)$ *if* $\gcd(m,n) = 1$*, where* $\varphi(n)$ *is the number of invertible elements modulo* n*, known as the* ***Euler*** φ***-function****. Thus:* $\varphi(p_1^{m_1}p_2^{m_2}\cdots)=(p_1-1)p^{m_1-1}(p_2-1)p^{m_2-1}\cdots$*, where* p_i *are pairwise distinct prime numbers and* $m_i \geq 1$

(iii) Prove that $\sum_{d|n}\varphi(d) = n$ *for any* $n \geq 1$*, where the summation is over all divisors* d *of* n*. Then use the* Möbius inversion *from Exercise 3.7.45(ii) to prove that* $\varphi(n) = n(1 - \sum_i \frac{1}{p_i} + \sum_{i \neq j} \frac{1}{p_i p_j} + \ldots)$ *for* $n = p_1^{m_1}p_2^{m_2}\cdots$ *, which is the same formula as in (ii) (check).*

Exercise 5.6.14$^\diamondsuit$ *Check that the map* $F : x \mapsto x^p$ *is an automorphism of the cyclic group* $\mathbb{Z}_{p^m-1}$ *with* $p^m - 1$ *elements for prime* p *and* $m \geq 1$*. Its order is* m*:* $F^m =$ id *(the identity map), and* $F^k \neq$ id *for* $1 < k < m$*. Deduce from this that* m *divides* $\varphi(p^m - 1)$ *for the Euler function* φ *from Exercise 5.6.13.*

Hint: F^k =id if and only if $x^{p^k} = x$ for all $x \in \mathbb{Z}_{p^m-1}$, which is possible only if m divides k. The number of cyclic generators of $\mathbb{Z}_{p^m-1}$ is $\varphi(p^m - 1)$, where this set is invariant under the action of F (under any automorphism). The orbit $x, F(x), \ldots, F^{m-1}(x)$ has exactly m elements for any x in this set (why?), which gives the required divisibility.

Exercise 5.6.15$^\clubsuit$ *(i) Let* $P'(X) = dP(X)/dX$ *(the derivative) for a polynomial* $P(X) = X^n + \sum_{i=1}^n a_i X^{n-i}$ *with* a_i *in* $\mathbb{Z}_{p^m}$*, and let* $P_k = P$ mod p^k*,* $P'_k = P'$ mod p^k *for* $1 \leq k \leq m$*.*

Prove that P *has no greater than* n *roots in* $\mathbb{Z}_{p^m}$ *assuming that* P'_1 *is nonzero for* any *root* x_1 *of* P_1 *in* $\mathbb{Z}_p$*. For this, check that any root* x_k *of* P_k *in*

$\mathbb{Z}_{p^k}$ can be uniquely lifted to a root x_{k+1} of P_{k+1} in $\mathbb{Z}_{p^{k+1}}$ for $1 \le k < m$ if $P_k'(x_k) \neq 0 \bmod p$. The lift $\widetilde{x} \in \mathbb{Z}_{p^{k+1}}$ of $x \in \mathbb{Z}_{p^k}$ is with respect to the map $\widetilde{x} \bmod p^{k+1} \mapsto x = \widetilde{x} \bmod p^k$. This is Hensel's Lemma.

(ii) Check that X^2-1 has 4 roots in $\mathbb{Z}_{2^m}$ for $m \ge 3$; find them. Justify directly (without using Hensel's Lemma) that the roots of $X^2 - X$ in $\mathbb{Z}_{2^m}$ are only 0 and 1 for any $m \ge 1$.

Hints. Use that generally $P(x+v) = P(x) + vP'(x) + \frac{v^2}{2}(\cdots) + \ldots$. If x_k is a root of $P_k(X)$ in $\mathbb{Z}_{p^k}$, then $x_{k+1} = x_k - P(x_k)/P'(x_k) \bmod p^{k+1}$ is a unique root of P_{k+1} in $\mathbb{Z}_{p^{k+1}}$ such that $x_{k+1} = x_k \bmod p^k$. In this formula, x_k is lifted to $\mathbb{Z}_{p^{k+1}}$ in any manner. Use Exercise 5.6.13.
For (ii): The roots of $x^2 = 1$ are $\pm(1 \pm 2^{m-2}) \bmod 2^m$ for any choice of signs $\pm$: consider $(a_0(1+a_1 2+a_2 4+\cdots+a_{m-1}2^{m-1}))^2$ with undetermined coefficients $a_i = \pm 1$. The product $x(x-1)$ is divisible by 2^m only when either $x = 0 \bmod 2^m$ or $x-1$ is such because one of them is always odd.

Exercise 5.6.16 $^\diamond$ *Check that the multiplicative group $\mathbb{Z}_{p^m}^*$ of invertible elements in $\mathbb{Z}_{p^m}$ for prime p and $m \ge 1$ contains $\mathbb{Z}_{p^m}^1 = 1 + p\mathbb{Z}_{p^m}$, the subgroup formed by $1+px \bmod p^m$ for integers x. Prove that $|\mathbb{Z}_{p^m}^1| = p^{m-1}$, and that it is a cyclic group generated by $1+up$ for any u relatively prime to p, i.e. prove that the elements of $\mathbb{Z}_{p^m}^1$ are exactly $(1+up)^k$, where $0 \le k \le p^{m-1}-1$.*

Hint: The order of $\mathbb{Z}_{p^m}^1$ is the number of $x \bmod p^{m-1}$, which is p^{m-1}. Use Newton's binomial to check that $(1+up)^k = 1 \bmod p^m$ if and only if k is divisible by p^{m-1}, which gives the cyclicity of $\mathbb{Z}_{p^m}^1$.

Exercise 5.6.17 $^\diamond$ *(i) Use Exercise 5.6.16, Exercise 5.6.13, and the cyclicity of $\mathbb{Z}_p^*$ proven in Proposition 7.3.3 to verify that $\mathbb{Z}_{p^m}^*$ is cyclic for any odd prime p and $m \ge 1$, i.e. it has a* **primitive root**, *which is by definition an element α such that any β in $\mathbb{Z}_{p^m}^*$ can be uniquely represented as $\beta = \alpha^k$ for some $0 \le k < \varphi(p^m) = |\mathbb{Z}_{p^m}|$. Here $\varphi(p^m) = p^{m-1}(p-1)$.*

(ii) Check that the map $\nu : \mathbb{Z}_{2p^m}^* \to \mathbb{Z}_{p^m}^*$, which sends $x \mapsto x \bmod p^m$, identifies these groups for any odd prime p and $m \ge 1$. More exactly, ν is a bijection such that $\nu(ab) = \nu(a)\nu(b)$ for any $a, b \in \mathbb{Z}_{2p^m}^*$. For instance, $\mathbb{Z}_{50}$ is cyclic; find its primitive roots. Prove that the group $\mathbb{Z}_n^*$ is never cyclic unless $n = 2, 4, p^m$ or $2p^m$ for odd prime p ($m \ge 1$).

(iii) Justify that α^k is primitive in $\mathbb{Z}_n^*$ for a primitive α (when it exists) if and only if $\gcd(k, \varphi(n))=1$. Check that the number of primitive roots in $\mathbb{Z}_{p^m}$ for odd prime p and $m \ge 2$ is $\varphi(\varphi(p^m))=\varphi((p-1)p^{m-1})=\varphi(p-1)\varphi(p^{m-1})=(p-1)p^{m-2}\varphi(p-1)$ for the Euler function $\varphi(n)$. See Exercise 5.6.13 for the properties of the Euler function.

Hint for (i): Take any α' such that $\alpha \bmod p$ is a cyclic generator of $\mathbb{Z}_p^*$ (the case of $\mathbb{F}_p^*$ from Proposition 7.3.3). Then the order of α' is $(p-1)p^{m'}$ for some $m' \le m$ due to the Lagrange Theorem in Problem 5.3.1. If $m' = m$ this is the required primitive root. If $m' < m$ take $\alpha = \alpha'(1+p)$. Thus, primitive roots in $\mathbb{Z}_{50}$ are those in $\mathbb{Z}_{25}$, which are $\pm 2(1+5u) \bmod 25$ for invertible u in $\mathbb{Z}_5$: $\pm 2(1 \pm 5) \bmod 25$ for any choice of signs $\pm$. We use that the primitive roots in $\mathbb{Z}_5$ are 2 and $3 = -2$.
Hint for (ii): The natural map $\mathbb{Z}_{p^m}^* \to \mathbb{Z}_{2p^m}^*$ is an identification because $\varphi(2p^m) = \varphi(p^m)$. For general n, use that the product of 2 cyclic groups (with component-wise multiplication) is cyclic if and only if their orders, the numbers of their elements, are co-prime. Exercise 5.6.15(ii) states that $\mathbb{Z}_{2^m}^*$ is not cyclic for $m \ge 3$. Actually, it is isomorphic to the group $\mathbb{Z}_2 \times \mathbb{Z}_{2^{m-2}}$ with the component-wise multiplication due to Exercise 5.6.16.

Exercise 5.6.18$^\diamond$ *(i) Use the cyclicity of $\mathbb{F}_{p^m}^*$, proved in Proposition 7.3.3, to justify that x in this group is a square for an odd prime p if and only if $x^{\frac{p^m-1}{2}} = 1$. For $m = 1$, this is* Euler's criterion.

(ii) Apply (i) to see that -1 is a square in $\mathbb{F}_{p^m}^*$ if and only if $\frac{p^m-1}{2}$ is even, i.e. either $p = 1 \bmod 4$ for any m or m is even for $p = 3 \bmod 4$.

(iii) Check that $-1 = y^2$ for some y in $\mathbb{Z}_{p^m}^*$ for odd prime p if and only if $p = 1 \bmod 4$ (for any $m \ge 1$); use Exercise 5.6.17(i).

Hint for (i): Using the (cyclic) generator α of $\mathbb{F}_{p^m}^*$, $x^{\frac{p^m-1}{2}} = 1$ if and only if $x = \alpha^k$ for $k\frac{p^m-1}{2}$ divisible by $p^m - 1$, which means that k is even.
Hint for (ii): $(-1)^{\frac{p^m-1}{2}} = 1$ if and only if $\frac{p^m-1}{2}$ is even.

Exercise 5.6.19$^\diamond$ *(i) Use 6^2 and 7^2 to check that $-7, 6$ are squares in $\mathbb{Z}_{43}$. Consider $4 \cdot 11, 4 \cdot 13 \bmod 43$ to see that $11, 13$ are squares in $\mathbb{Z}_{43}$, which gives that $-32, -30$ are squares too. Then obtain that $-2, -3, -5$ are squares there. Using that (-1) is not a square due to Exercise 5.6.18(ii), conclude that $2, 3, 5, 7$ are not squares in $\mathbb{Z}_{43}$; so $11, 13$ are the first squares among the prime numbers $1 < q < 43$.*

(ii) Continuing (i), check that $17 = \frac{8}{3} \bmod 43$ is a square, $19 = -\frac{5}{2} \bmod 43$ is not, $23 = \frac{3}{2} \bmod 43$ is a square; do this for the remaining prime $23 < q < 43$.

(iii) To double-check (i), apply Problem 7.3.16(ii) and Quadratic Reciprocity from Problem 7.3.17(ii) to $2, 3, 5, 7, 11, 13$ modulo $p = 43$.

Exercise 5.6.20$^\diamond$ *(i) Let $p = 1 + 2q$ for odd* prime *numbers p and q. Prove that primitive roots modulo p (cyclic generators of $\mathbb{Z}_p^*$) are exactly $m \ne \pm 1 \bmod p$ that are not quadratic residues (not squares in $\mathbb{Z}_p^*$). For instance, either m or*

$-m$ is primitive for any $m \neq \pm 1$ in $\mathbb{Z}_p^*$ for such p and the number of primitive roots, denoted by $Prim_p$, is $q-1=\frac{p-3}{2}$.

(ii) Justify that for any prime $p > 3$ such that $p = 3 \bmod 4$ and not $1+2q$ as above, $Prim_p < \frac{p-3}{2}$. For $p = 23 = 1 + 2 \cdot 11$, check that -2 and 11 are primitive and find all $q - 1 = 10$ primitive roots modulo $p = 23$.

Hint: The group $\mathbb{Z}_p$ is isomorphic to the product $\mathbb{Z}_2 \times \mathbb{Z}_q$ of the groups $\mathbb{Z}_2$ and $\mathbb{Z}_q$. Its only non-trivial subgroups are $\mathbb{Z}_2 = \{\pm 1\}$ and $\mathbb{Z}_q$. Thus, the set of primitive roots is $-1 \times (\mathbb{Z}_q \setminus \{1\})$. For any $p = 4k + 3$, justify and use that $\mathrm{Prim}_p = \varphi(p-1)$ for the Euler function. For $p = 23$, use that $2 = 25 - 23$.

Exercise 5.6.21$^\heartsuit$ *(i) Check De Polignac's formula: if p^μ is the greatest prime power divisor of $n!$ for any prime p, then $\mu = \left[\frac{n}{p}\right] + \left[\frac{n}{p^2}\right] + \cdots + \left[\frac{n}{p^s}\right]$, where s satisfies $p^s \le n < p^{s+1}$, and $[x]$ =Floor$[x]$ is the integer part of x.*

(ii) Expanding $n = r_0 + r_1 p + r_2 p^2 + \cdots + r_s p^s$, where $0 \le r_i < p$, check that $\mu = (r_1 + r_2 p + r_3 p^2 \ldots) + (r_2 + r_3 p + \ldots) + \ldots = r_1 \frac{p-1}{p-1} + r_2 \frac{p^2-1}{p-1} + r_3 \frac{p^3-1}{p-1} + \ldots$, and that $\mu = \frac{n-(r_0+r_1+\cdots+r_s)}{p-1}$.

Exercise 5.6.22$^\heartsuit$ *Use Exercise 5.6.21 to count the number of zeros at the end of $50!$, $100!$ and $500!$. Check that this number is n for $(5n)!/n!$, where $n \ge 1$.*

Answer: 12, $\frac{100-4}{4}$ and $\frac{500-4}{4}$.

Exercise 5.6.23$^\diamondsuit$ *Show that any prime number $p > 5$ divides some number of the form $1111 \cdots 1$, and prove that there are infinitely many such numbers for each prime $p > 5$.*

Hint: Consider the sequence $10^n - 1 \bmod p$.

Exercise 5.6.24$^\heartsuit$ *Show that $n^7 - n$ is divisible by 42 for any $n \ge 1$.*

Hint: Find the m-periods of $a^m \bmod p$ for $p = 2, 3, 7$ and any a.

Exercise 5.6.25$^\heartsuit$ *Check that $P(X) = X^3 - X + 1$ is not a product $P(X) = Q(X)R(X)$ of any 2 polynomials Q and R of smaller degree with the coefficients in $\mathbb{Z}_3 = \{0, 1, 2\}$. Use this to justify that the field $\mathbb{F}$ obtained by adding to $\mathbb{Z}_3$ an element α satisfying $P(\alpha) = 0$ (and adding all α^m for $m \ge 0$) has 27 elements. Check that $P(\beta) = \beta^3 - \beta + 1 = 0$ for $\beta = 1 + \alpha$, i.e. that β is another root of $P(X) = 0$, and express β^{-1} in terms of α.*

Exercise 5.6.26 $^\diamondsuit$ *(i) Check that an* **irreducible** *polynomial* $P(X) = X^3+aX^2+bX+c$ *with the coefficients in* $\mathbb{Z}_2$ *can be either* $P_1(X) = X^3+X+1$ *or* $P_2(X) = X^3+X^2+1$. *Recall: the irreducibility of* $P(X)$ *means that it cannot be factored into polynomials of degree less than 3 with coefficients in* $\mathbb{Z}_2$.

(ii) Define $\mathbb{Z}_2[\lambda] = \{x = a+b\lambda+c\lambda^2\}$ *for* λ *satisfying the relation* $P_1(\lambda) = 0$ *and arbitrary* a, b, *and* c *in* $\mathbb{Z}_2$. *Use the irreducibility of* P_1 *to check that* x *in this form can be zero only if* $a = b = c = 0$, *and calculate the formulas for the addition and multiplication of such* x.

(iii) Calculate $\frac{1}{\lambda}$ *and check that it is a root of* $P_2(X)$. *Calculate* $\frac{1}{\lambda+1}$ *and check that it is a root of* $P_1(X)$. *Check that the "new numbers"* x *either satisfy* $P_1(x) = 0$ *or* $P_2(x) = 0$ *(not both!); the "old numbers" are those from* $\mathbb{Z}_2$.

Exercise 5.6.27 ♣ *(i) Check that if a finite field* $\mathbb{F}'$ *contains* $\mathbb{F}$, *then* $|\mathbb{F}'| = |\mathbb{F}|^k$ *for the numbers of elements,* $|\mathbb{F}'|$ *and* $|\mathbb{F}|$, *of these fields, where* k *is as follows. It is the greatest number of linearly independent elements in* $\mathbb{F}'$, *where linear combinations are considered with coefficients in* $\mathbb{F}$. *In particular, the number of elements in any finite field* $\mathbb{F}$ *is a power of its* **characteristic** p, *the smallest integer* $p > 1$ *such that* $p \cdot 1 = 0$ *in* $\mathbb{F}$, *which we know to be a prime number.*

(ii) Deduce from (i) that $1, x, x^2, \ldots, x^k$ *must be linearly dependent with the coefficients in* $\mathbb{F}$ *for any* x *in* $\mathbb{F}'$. *Equivalently,* $P(x) = 0$ *for some nonzero polynomial* P *of degree no greater than* k *with coefficients in* $\mathbb{F}$. *Check that* $P(0) \neq 0$ *if this* P *is of minimal possible degree (use that* $\mathbb{F}'$ *is a field), and employ* P *to obtain a formula for* $1/x$.

(iii) Any field $\mathbb{F}'$ *with 8 elements cannot contain* $\mathbb{F}_4$ *due to (i) (check). Therefore its elements not from* $\mathbb{Z}_2$ *must be roots of some irreducible cubic polynomials* $P(X)$ *over* $\mathbb{Z}_2$ *(justify). Use this and Exercise 5.6.26 to prove directly that any such* $\mathbb{F}'$ *is isomorphic to* $\mathbb{Z}_2[\lambda]$ *constructed there.*

Exercise 5.6.28 $^\diamondsuit$ *(i) Check that if* $P(X) = X^4 + a$ *has a root* λ *in some field* $\mathbb{F}$ *of characteristic 5 containing* a, *then all its roots there are in the form* $x\lambda$ *for* $x \in \mathbb{Z}_5^* = \mathbb{Z}_5 \setminus \{0\}$.

(ii) Use this to prove the irreducibility of $P(X)$ *if* a *is not a square in* $\mathbb{F}$. *Check that it is reducible otherwise, though it has roots in* $\mathbb{F}$ *if and only if* $a = b^4$ *for some* b *there.*

(iii) Let $\lambda^4 + 2 = 0$ *over* $\mathbb{Z}_5$. *Then* $\mathbb{Z}_5[\lambda] = \{a + b\lambda + c\lambda^2 + d\lambda^3\}$ *for* $a, b, c, d \in \mathbb{Z}_5$ *is a field with* 5^4 *elements. Calculate* $1/\lambda$ *and* $1/(\lambda+1)$ *as polynomials of* λ.

Hints: (ii) If a is not b^2 for any b in $\mathbb{F}$, then $\lambda \notin \mathbb{F}$ and the reducibility of $P(X)$ means that $P(X) = Q(X)R(X)$ for quadratic polynomials Q, R. Then

$\lambda(1+x), x\lambda^2$ must be from $\mathbb{F}$ for some x there, which gives that $x=-1$, $\lambda^2 \in \mathbb{F}$, and $a=-(\lambda^2)^2$; use that $-1=2^2$ in $\mathbb{Z}_5$.
(iii) Employ the formula $\lambda^4+1=(\lambda+1)(\lambda^3-\lambda^2+\lambda-1)=-1$.

Exercise 5.6.29$^\heartsuit$ *Count the number of irreducible* monic *polynomials in the form X^2+aX+b over $\mathbb{F}_q$. Here $a,b\in\mathbb{F}_q$ and $q=p^k$ for prime p $(k\geq 1)$.*

Hint: By subtracting the number of reducible polynomials from the total number of (monic) quadratic polynomials, which is q^2, we obtain $(q^2-q)/2$.

Exercise 5.6.30$^\diamondsuit$ *Use the* Möbius function *$\mu(n)$ from Exercise 3.7.45 and Exercise 5.6.27 to check that the number $N_q(n)$ of irreducible polynomials $X^n+a_1X^{n-1}+\ldots$ with the coefficients in $\mathbb{F}_q$ is $\frac{1}{n}\sum_{d|n}\mu(d)q^{n/d}$, where the summation is over all divisors d of n (indicated by $d|n$).*

Hint: The sum $\sum_{d|n} dN_q(d)$ is the total number of elements of $\mathbb{F}_{q^n}$; use Exercise 5.6.27. Then the inverse Möbius transform from Exercise 3.7.45 of the function $|\mathbb{F}_{q^n}|=q^n$ gives that $nN_q(d)$ equals $\sum_{d|n}\mu(d)q^{n/d}$.

Exercise 5.6.31$^\spadesuit$ *Rewrite the formula from Exercise 3.7.43 for q-binomial coefficients as follows*

$$\begin{bmatrix} n \\ r \end{bmatrix}_q = \frac{(1-q^n)(1-q^{n-1})\cdots(1-q^{n-r+1})}{(1-q)(1-q^2)\cdots(1-q^r)} = \frac{(q^n-1)(q^n-q)\cdots(q^n-q^{r-1})}{(q^r-1)(q^r-q)\cdots(q^r-q^{r-1})}.$$

Consider the field $\mathbb{F}_q$ with q elements. Prove by a direct calculation that the number of r-dimensional subspaces in the vector space $\mathbb{F}_q^n$ equals $\begin{bmatrix} n \\ r \end{bmatrix}_q$.

Hint: To construct an r-dimensional subspace, there are q^n-1 choices for v_1 there, then (q^n-q) choices for v_2, linearly independent of v_1, then (q^n-q^2) choices of v_3, linearly independent of v_1, v_2 and so on. However, we must divide the resulting number by the total number of r linearly independent vectors (bases) in $\mathbb{F}_q^r$, which is obtained in the same way.

Exercise 5.6.32$^\diamondsuit$ *Given a prime number p, we justified that the first $m>0$, denoted by m_p, such that $f_m=0 \bmod p$ for the Fibonacci numbers f_m is the period of the sequence of all m satisfying $f_m=0 \bmod p$. Check that this holds for m_n defined for any $n>1$ (not only for prime numbers $n=p$) as the minimal m such that $f_m=0 \bmod n$. Check that $m_{ab}=m_am_b$ if $\gcd(a,b)=1$.*

Exercise 5.6.33 ♢ *The number of invertibles modulo* p^k *for prime* p *and* $k \geq 1$ *is* $\varphi(p^k) = p^{k-1}(p-1)$ *(Exercise 5.6.13). Use this to justify that* m_{p^k}*, defined as the smallest number* m *such that* p^k *divides the Fibonacci number* f_m*, is a factor of* $(p-1)p^{k-1}$ *provided that the polynomial* $X^2 - X - 1$ *has a root in* $\mathbb{F}_p = \mathbb{Z}_p$*. [Otherwise, it divides* $(p+1)p^{k-1}$*, but this is beyond this problem.]*

Exercise 5.6.34 ♢ *A combination of Binet's formula and Newton's binomial results in the following formula for the Fibonacci numbers:*

$$f_n = \frac{\lambda_1^n - \lambda_2^n}{\sqrt{5}} = \frac{1}{2^{n-1}} \sum_{m=0}^{2m+1\leq n} \binom{n}{2m+1} 5^m.$$

Check without any reference to the recurrence for f_n *that the right-hand side is an integer, i.e. that there will be no powers of 2 in the denominator of the resulting fractions. Use that* λ_1, λ_2 *and* $\frac{\lambda_1^n-\lambda_2^n}{\sqrt{5}}$ *are well defined in* $\mathbb{F}_4$.

Exercise 5.6.35 ♣ *Exercise 4.6.15 provides that* $H_n = \frac{4}{7}2^n + \frac{4-\rho}{21}\rho^n + \frac{5+\rho}{21}\rho^{-n}$ *is an integer for any* $n \geq 0$*, where* $\rho = \exp(2\pi\imath/3) = -\frac{1}{2} + \frac{\sqrt{3}}{2}\imath$.

(i) Check that the complex conjugation*, preserving real numbers and sending* $\imath$ *to* $-\imath$*, transforms* ρ *to* ρ^{-1} *and, accordingly, does not change* H_n*. Use the relation* $\rho+\rho^{-1}=-1$*. This proves directly that* H_n *are real numbers.*

(ii) Due to the modular variant of Theorem 4.4.2, solutions of the recurrence relation $H_n = H_{n-1}+H_{n-2}+2H_{n-3}$ *in* $\mathbb{Z}_3$ *are* $A+B\,n+C\,(-1)^n$ *for any constant coefficients* A, B, C *from* $\mathbb{Z}_3$*. Justify this by considering the master equation modulo 3. Similarly, check that the solutions of this recurrence in* $\mathbb{Z}_7$ *are in the form* $A\,2^n + B\,n2^n + C\,4^n$ *for* A, B, C *in* $\mathbb{Z}_7$.

(iii) For the initial conditions $H_0 = 1 = H_1$ *and* $H_2 = 2$ *find the corresponding coefficients of these presentations by solving the corresponding linear equations for* A, B, C *in* $\mathbb{Z}_3$ *and* $\mathbb{Z}_7$*. Also, check that* $H_n \bmod 2 = f_{n+1} \bmod 2$ *for the Fibonacci numbers* f_n.

Answer for (iii): $H_n = 2n + (-1)^n \bmod 3$, $H_n = 4(2^n + 4^n) - n2^n \bmod 7$.

Exercise 5.6.36 ♣ *Use the formula* $H_n = \frac{4}{7}2^n + \frac{4-\rho}{21}\rho^n + \frac{5+\rho}{21}\rho^{-n}$ *from Exercise 5.6.35 to justify that* $H_n - 1$ *is divisible by a prime number* $p \neq 2, 3, 7$ *if* $p^2 - 1$ *divides* n*. Moreover,* p *divides* $(H_n - 1)$ *if* $p - 1$ *divides* n *provided that the polynomial* $X^2 + X + 1$ *has roots in* $\mathbb{Z}_p$ *for such* p.

Hint: The proof is the same as that in the corresponding part of Theorem 5.4.3. Assume that $p \neq 2, 3, 7$ and use that $\lambda^{p^2-1} = 1 \bmod p$ for all 3 (different) roots λ of the master equation (one of them is 2).

Exercise 5.6.37 ◇ *For the sequence t_n from formula (4.3.18) and $r_n = t_{n+1}/t_n$ for $n \geq 0$, prove that $\lim_{n\to\infty} r_n = \lambda_1$, where $\lambda_1 \approx 1.465571231876768$ is the unique positive root of the polynomial $P(X) = X^3 - X^2 - 1$. Use that $|\lambda_{2,3}|^2|\lambda_1| = 1$ to obtain the estimate $|r_n - \lambda_1| < Ct_n^{-3/2}$ for some constant $C > 0$. Actually, $C = 0.67322$ is sufficient for all $n \geq 0$; C_{max} is reached at $n = 53$. Thus, the accuracy is $\approx q^{-3/2}$ for the rational approximations $\frac{p}{q}$ of λ_1 with the denominators $q = t_n$, which is better than the accuracy $\frac{1}{2q}$ for the approximations with "random" denominators q. Recall that it is $\approx q^{-2}$, i.e. better than $\approx q^{-3/2}$, for* quadratic *irrationalities λ_1 and the denominators q obtained via the corresponding recurrences.*

Exercise 5.6.38 ♣ *Let $g_n = C_1\lambda_1^n + C_2\lambda_2^n$ for arbitrary real or complex constants $C_1, C_2 \neq 0$ and any (real or complex) $\lambda_1 \neq \lambda_2$.*

(i) If g_n in this sequence are rational numbers for all sufficiently large n, prove that $a = \lambda_1 + \lambda_2$ and $b = -\lambda_1\lambda_2$ must be rational numbers, i.e. $\lambda_{1,2}$ are roots of the polynomial $X^2 - aX - b$ with rational coefficients.

(ii) Assume that the master equation of some recurrence is $P(\lambda) = 0$ for a cubic polynomial $P(X)$ with rational coefficients that has no rational roots; let $\{\lambda_1 \neq \lambda_2\}$ be 2 its roots (from 3). Prove that nonzero $g_n = C_1\lambda_1^n + C_2\lambda_2^n$ cannot take rational values for all sufficiently large n for any $C_{1,2}$.

(iii) Using (ii), apply Exercise 4.6.14(iii). Impose the conditions for A, B, C there. Check that if g_n are rational numbers for sufficiently large n and $P(X)$ has no rational roots, then the condition $c \neq 0$ (there) is satisfied, and $\lim_{n\to\infty} g_{n+1}/g_n = \lambda_1$ for the unique positive root λ_1 of $P(X)$.

Hints: (i) One has $g_n = ag_{n-1} + bg_{n-2}$; $\widetilde{g}_n = g_{n+1}$ is another solution, which is non-proportional to g_n. This gives that a, b are rational.
(ii) If $C_i = 0$ then λ_{3-i} must be rational: a contradiction. When $C_1 \neq 0 \neq C_2$, $P(X)$ is divisible by a quadratic polynomial with rational coefficients due to (i). The quotient is linear, and $P(X)$ then has a rational root (impossible).

Exercise 5.6.39 ◇ *(i) Use Lemma 5.5.3 to see that the equation $x^2 - dy^2 = 4$ always has a solution with even integers x and y if $d > 0$ is not a perfect square. Justify that if it has a solution with odd integers, then $d = 5 \bmod 8$.*

(ii) Check that there are no odd solutions (both x and y) to this equation for $d = 37$ (actually, 37 is the smallest such). For this, find that the smallest $w = x + \sqrt{37}y$ with positive integers x and y such that $x^2 - 37y^2 = -4$ is $12 + 2\sqrt{37}$. Then use Problem 5.5.2. As an application, prove that $w = 146 + 24\sqrt{37}$ is the smallest w resulting in a solution of $x^2 - dy^2 = +4$.

Hints: (i) Consider $\widetilde{x} = x/2, \widetilde{y} = y/2$, which satisfy $\widetilde{x}^2 - d\widetilde{y}^2 = 1$ and are described by Problem 5.5.2. For $d = 5 \bmod 8$, $m^2 = 1 \bmod 8$ for any odd m. (ii) Any $(x + y\sqrt{37})/2$ for $x^2 - y^2\sqrt{37} = \pm 4$ is a power of $(12 + 2\sqrt{37})/2$.

Exercise 5.6.40 ♠ *The **generalized continued fraction** $b_0 + \overset{\infty}{\underset{i=1}{\mathrm{K}}} \frac{a_i}{b_i}$ is defined as* $x = b_0 + \cfrac{a_1}{b_1 + \cfrac{a_2}{b_2 + \dots}}$. *We set $x_n = b_0 + \overset{n}{\underset{i=1}{\mathrm{K}}} \frac{a_i}{b_i}$, i.e. we omit the a_i and b_i in the continued fraction for x for all $i > n$. The convergence, i.e. the equality $x = \lim_{n\to\infty} x_n$, is addressed in Exercise 5.6.41.*

(i) Define $P_n = b_n P_{n-1} + a_n P_{n-2}$, $Q_n = b_n Q_{n-1} + a_n Q_{n-2}$ for $n \ge 2$, where $P_0 = b_0, P_1 = b_0 b_1 + a_1$, $Q_0 = 1, Q_1 = b_1$. Check that $x_n = P_n/Q_n$ for $n \ge 1$ and $P_n Q_{n-1} - Q_n P_{n-1} = (-1)^{n-1} a_1 \cdots a_{n-1} a_n$. For instance, $x = \frac{1+\sqrt{5}}{2}$ for $b_0 = 1, a_i = b_i = 1 (i \ge 1)$; in this case: $P_n = f_{n+2}$, $Q_n = f_{n+1}$ for the Fibonacci numbers f_n $(n \ge 0)$.

(ii) Deduce the recurrence $\kappa_n = \frac{n-1}{n}\kappa_{n-1} + \frac{1}{n}\kappa_{n-2}$ for $\kappa_n = D_n/n! = \delta_n/n$ from (3.6.14), where $\kappa_0 = 1, \kappa_1 = 0, \kappa_2 = 1/2$. For instance, $\kappa_{3,4,5,6,7} = \{\frac{1}{3}, \frac{3}{8}, \frac{11}{30}, \frac{53}{144}, \frac{103}{280}\}$. Use that $\lim_{n\to\infty} \kappa_n = \frac{1}{e}$ to check that $\kappa_n = b_0 + \overset{n}{\underset{i=1}{\mathrm{K}}} \frac{a_i}{b_i}$ for $b_0 = 1, b_1 = 1, a_1 = -1$ and $a_i = \frac{1}{i}, b_i = \frac{i-1}{i}$ for $i \ge 2$. Accordingly, $x = 1/e$, and $|x_n - x| < \frac{a_{n+1}a_n \cdots a_2}{Q_n Q_{n+1}} = \frac{1}{(n+1)!}$ for $n \ge 0$. What is the expression for x_5?

Hint for (i): Express $x_n = b_0 + \overset{n-1}{\underset{i=1}{\mathrm{K}}} \frac{a'_i}{b'_i}$ for $a'_{n-1} = a_{n-1}b_n, b'_{n-1} = b_n b_{n-1} + a_n$ and a_i, b_i otherwise. By induction, $P_n = b'_{n-1}P_{n-2} + a'_{n-1}P_{n-3} = (b_n b_{n-1} + a_n)P_{n-2} + a_{n-1}b_n P_{n-3} = b_n(b_{n-1}P_{n-2} + a_{n-1}P_{n-2}) + a_n P_{n-2} = b_n P_{n-1} + a_n P_{n-2}$, where we used the induction statement twice.
In $P_n Q_{n-1} - Q_n P_{n-1}$, express P_n, Q_n using the recurrence and then proceed by induction. For $n = 1$: $P_1 Q_0 - Q_1 P_0 = (b_0 b_1 + a_1) - b_1 b_0 = a_1$.
Hint for (ii): Due to (i), $\{a_i\}$ and $\{b_i\}$ provide the recurrence for P_n, Q_n coinciding with that for κ_n for $n \ge 2$. For Q_n: $Q_0 = 1, Q_1 = b_1 = 1, Q_2 = b_2 Q_1 + a_2 Q_0 = 1$, and $Q_n = 1$ for any n. Thus, $x_n = P_n$ and we only need to identify 2 initial terms for P_n and κ_n. The estimate $|x_n - x| < \frac{1}{(n+1)!}$ follows from (i); use that $Q_n = 1$. By definition: $x_5 = 1 - \cfrac{1}{1 + \cfrac{1/2}{1/2 + \cfrac{1/3}{2/3 + \cfrac{1/4}{3/4 + 1/5}}}}$.

Exercise 5.6.41 ♣ *Use the formula $\frac{P_n}{Q_n} - \frac{P_{n-1}}{Q_{n-1}} = (-1)^{n-1}\frac{a_1 a_2 \cdots a_n}{Q_n Q_{n-1}}$ from Exercise 5.6.40(i) to obtain the following expression for $x = b_0 + \overset{\infty}{\underset{i=1}{\mathrm{K}}} \frac{a_i}{b_i}$. Let $Q_n =$*

$b_nQ_{n-1} + a_nQ_{n-2}$ for $Q_0 = 1, Q_1 = b_1$. Then $x = b_0 + \frac{a_1}{Q_1} - \frac{a_1a_2}{Q_1Q_2} + \cdots + (-1)^m\frac{a_1\cdots a_{m+1}}{Q_mQ_{m+1}} + \ldots$ provided the convergence of the latter sum. Assuming that $R_m = \frac{a_1\cdots a_{m+1}}{Q_mQ_{m+1}}$ has a constant sign for sufficiently large m, check that the convergence of R_m to 0 as $m \to \infty$ is necessary and sufficient for x to exist. Note that no explicit knowledge of P_n is necessary in this approach.

(i) Obtain from this that $\frac{1+\sqrt{5}}{2} = 1 + \frac{1}{f_1f_2} - \frac{1}{f_2f_3} - \frac{1}{f_3f_4} + \ldots$ for the Fibonacci numbers f_n. Also, check that we arrive at $1/e = \sum_{n=0}^{\infty}(-1)^n\frac{1}{n!}$ for the example in Exercise 5.6.40(ii).

(ii) Use Exercise 4.6.42 to prove that $J_0(2) = 1 - 1 + \frac{1}{(2!)^2} - \frac{1}{(3!)^2} + \ldots$ equals $b_0 + \operatorname*{K}_{i=1}^{\infty}\frac{a_i}{b_i}$ for $b_0 = 1$, $a_1 = -1/2$, $a_2 = 1/6$, $b_1 = 1/2$, $b_2 = 1$, and $a_n = \frac{n-1}{(n-2)n(n+1)}$, $b_n = 1$ for $n \geq 3$.

Hints: $\frac{P_n}{Q_n} = (-1)^{n-1}\frac{a_1a_2\cdots a_n}{Q_nQ_{n-1}} + (-1)^{n-2}\frac{a_1a_2\cdots a_{n-1}}{Q_{n-1}Q_{n-2}} + \ldots$.
(ii): Use that $Q_n = \frac{n}{n+1}$ is a solution of the recurrence relation $Q_n = Q_{n-1} + a_nQ_{n-2}$ for $n \geq 3$; see Exercise 4.6.42(ii,iii). Then adjust a_n for $n = 1, 2$ to obtain that $\frac{a_1\cdots a_n}{Q_{n-1}Q_n} = -\frac{1}{(n!)^2}$; here $n \geq 1$ and we set $Q_0 = 1$.

Exercise 5.6.42$^\heartsuit$ *(i) Use the formula $\{i, j\} \mapsto ri+j$ for i, j from $\mathbb{Z}_n$ with $n = 8$ and $r = 5$ to construct a Latin square of order 8. Then do this for the formula $\{i, j\} \mapsto ai + bj$ with $a = 3, b = 2$.*

(ii) Check that there exist no invertible a, b, c, d in any $\mathbb{Z}_{2m}$ such that $ad - bc$ is invertible. Check that they exist in $\mathbb{Z}_n$ for any odd $n > 1$, and the Latin squares given by the formulas $\{i, j\} \mapsto ai + bj$ and $\{i, j\} \mapsto ci + dj$ are mutually orthogonal.

(iii) Use the construction from (ii) to obtain MOLS from $\mathbb{F}_8$. Namely, take nonzero a, b, c, d in $\mathbb{F}_8$ such that $Det\begin{pmatrix} ab \\ cd \end{pmatrix} = ad - bc \neq 0$. Check that the total number of such a, b, c, d is $7^4 - 7^3$.

Hint for (iii): Calculate the number of nonzero a, b, c, and d satisfying $ad-bc = 0$ and then subtract it from 7^4.

Exercise 5.6.43$^\diamondsuit$ *(i) Let us assume that $r_1, r_2, r_1-r_2, r_1-1, r_2-1$ are relatively prime to n. Due to Corollary 5.2.2, the corresponding MOLS $\{LS'_{r_1}, LS'_{r_2}\}$ will have the following property: there will be no repetitions of entries in the anti-pandiagonals of the Latin squares LS'_{r_1} and LS'_{r_2} (the pandiagonals are not considered). Check that no r_1, r_2 satisfying these conditions exist in $\mathbb{Z}_9$.*

(ii) Let us relax these conditions for r_1, r_2 in $\mathbb{Z}_9$; we assume now that $r_1, r_2, r_1 - r_2$ are all not divisible by 3. The standard passage MOLS$\to$ MS is as follows. We construct $LS'_{r_{1,2}}$ for $\{i, j\} \mapsto r_{1,2}i + j$ in $\mathbb{Z}_9$ and then

apply the map $\langle u, v\rangle \mapsto 1 + u + 9v$ *for the entries* $\langle u, v\rangle$ *of the corresponding* $LS'_{r_1} \vee LS'_{r_2}$. *Use computers to check that the resulting* 9×9 *square with the entries* $1, 2, \ldots, 9^2$ *never has* $369 = \frac{9(9^2+1)}{2}$ *as its diagonal and anti-diagonal sum for any choice of such (relaxed)* r_1, r_2.

Hints: (i) If $r_2 = r_1 + 1 \bmod 3$, the triple $r_1 - 1, r_1, r_2$ contains 1 number divisible by 3. If $r_2 = r_1 + 2 \bmod 3$, then $r_1 - 1, r_1, r_2 - 1$ is such a triple.
(ii) Using computer, the diagonal sum is either 45, or 288, or 360 (not 369) for 9 admissible pairs $\{r_1 < r_2\}$ in $\mathbb{Z}_9$ (resulting in *MOLS*). The anti-diagonal sums for such $r_{1,2}$ are $378, 405, 450$, so never 369 too.

Exercise 5.6.44 ♣ *(i) For any finite field* $\mathbb{F}_q$, *where* $q = p^m$, *prove that its addition table L and the subtraction table M, given by* $(x, y) \mapsto x - y$, *are in the same isotopy class when considered as abstract Latin squares of size* $q \times q$. ***Isotopy equivalence*** *means: in the same class under permutations of the rows, columns, and the symbols used as the entries (here the elements from* $\mathbb{F}_q$*). See Exercise 1.4.18.*

(ii) The ***main equivalence class*** *of Latin squares, is defined when the isotopy equivalence is extended by the permutations of* i, j, a *(6 of them), where* i, j *are the row and column numbers, and a is the corresponding entry. Use that the L above is* symmetric, *i.e. invariant with respect to the transformation* $i, j, a \mapsto j, i, a$, *to check that the images of L under these 6 permutations are either L or M or* $-M$ *for M from (i).*

Exercise 5.6.45 ♢ *Prove "combinatorially" that the addition table of* $\mathbb{Z}_4$ *considered as a Latin square has no orthogonal mate. This is a puzzle, which is not related to the modular construction in this chapter. The case of any even order is Problem 7.3.15; see also Exercise 5.6.46.*

Hint: Use that if the orthogonal mate exists, then its entry x at the bottom-left corner cannot appear in the 1st column, in the last row, and in the diagonal through this box. Then use that x must occur 3 more times somewhere.

Exercise 5.6.46 ♣ *Use Exercise 1.4.18(iii) and Exercise 5.6.45 to check that up to* isotopy *of* 4×4 *Latin squares, only the addition table for* $\mathbb{Z}_4$ *has no orthogonal mates.*

Exercise 5.6.47 ♡ *(i) Find all nonattacking configurations of 4 queens in "4×4" without considering all* $4! = 24$ *permutations. (ii) Justify that no greater than 2 pan-nonattacking queens can be placed in the* 4×4 *chessboard.*

Hints: (i) Check that no queen in this configuration can occupy box $(0,0)$. (ii) You can assume that one queen is at $(0,0)$ using horizontal and vertical shifts of "4×4" modulo 4.

Exercise 5.6.48 ♡ *For prime p, the formula $i\mapsto j = ai \mod p$ for $1\le i\le p-1$ gives the positions (i,j) of $p-1$ nonattacking queens on a $(p-1)\times(p-1)$ chessboard provided that $a\neq 0,\pm 1 \mod p$. Here i,j are the row and column numbers; $i=0$ and $j=0$ are omitted.*
For $p=11$, find the parameter a such that there is a queen at $(i=5, j=7)$, and plot the remaining 9 nonattacking queens for such a in the 10×10 square as above (with $i,j>0$). Justify that such an extension exists in this 10×10 chessboard for any initial position instead of $(5,7)$ unless it is at the main diagonal there or the main anti-diagonal.

Exercise 5.6.49 ♡ *To practice the procedure from the previous problem, let $n=7$. Extend the following 2 pan-nonattacking queens in the 7×7 chessboard to a full set of 7 pan-nonattacking queens there:*

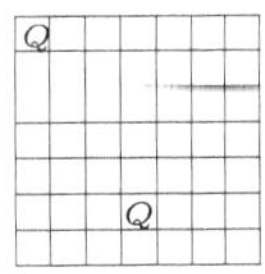

.

Check that the configurations of any nonattacking queens with Q at $(0,0)$ can be naturally restricted to some nonattacking configurations of $n-1$ queens in "$(n-1)\times(n-1)$". They are never pan-nonattacking: the main diagonal in "$(n-1)\times(n-1)$" contains no queens.

Exercise 5.6.50 ♠ *Let $M(n)$ be the maximal number of* pan-nonattacking queens *on an $n\times n$ chessboard.*

(i) Prove that $M(n)\le n-1$ for even n. It suffices to impose here only the anti-pandiagonal nonattacking conditions. To prove this assume that $M(n)=n$ and consider $\sum i_m$, $\sum j_m$, and $\sum(i_m+j_m)$ for the positions (i_m, j_m) of the queens. This will lead to a contradiction for even n.

(ii) Prove that $M(n)\le n-1$ if $\gcd(n,6)=3$. For this, assume that $M(n)=n$ and then apply the identity $\sum_{k=0}^{n-1}k^2 = \frac{(n-1)n(2n-1)}{6}$ from Exercise 3.7.10 to the sums of $\{i_m^2\}$, $\{j_m^2\}$ and $\{(i_m\pm j_m)^2\}$. Take $i_m = m-1$ here.

(iii) Prove that $M(n)<n-1$ if $\gcd(n,6)=3$. Assuming that there are $n-1$ pan-nonattacking queens, check that it is always possible to extend this configuration to that with n such queens (T. Klöve), which contradicts (ii).

Hints: (i) Assuming that there are n such queens, $\sum_{m=1}^{n} i_m = \sum_{m=1}^{n} j_m = \sum_{m=1}^{n}(i_m+j_m) = X = \frac{n(n-1)}{2} \mod n$; we use that i_m+j_m are all distinct. Recall that $0\le i_m, j_m\le n-1$. Thus, $X = X+X$ and $X = 0 \mod n$, which is impossible for even n.

(ii) We can take $i_m = m-1$; then j_m will be the position of the queen in row $m-1$. Let $Y = \frac{(n-1)n(2n-1)}{6} \bmod n$, $Z = \sum_{m=1}^{n-1} m j_m \bmod n$. One has: $\sum_{m=1}^{n} j_m^2 = \sum_{m=1}^{n}(m-j_m)^2 = \sum_{m=1}^{n}(m+j_m)^2 \bmod n$. The last two equalities give that $Y = 2Y+2Z = 2Y-2Z$. Thus, $2Y = 0$ and $\frac{(n-1)n(2n-1)}{3} = 0 \bmod n$. The latter is possible only if $n = \pm 1 \bmod 3$. We followed Klöve.
(iii) Assume that $M(n) = n-1$ when $\gcd(n,6) = 3$. Let $i = a, j = b, i-j = c, i+j = d$ be the indices of the missing row, column, pandiagonal and the anti-pandiagonal. Then $a+\sum_{m=1}^{n-1} i_m = b+\sum_{m=1}^{n-1} j_m = c+\sum_{m=1}^{n-1}(i_m-j_m) = d+\sum_{m=1}^{n-1}(i_m+j_m) = n(n-1)/2 \bmod n$. Thus $a-b = c, a+b = d$ in $\mathbb{Z}_n$. However, this gives that we can add a queen at (a,b) to this $(n-1)$-configuration and obtain a full n-configuration of nonattacking queens for $\gcd(n,6) = 3$, which contradicts (ii).

Exercise 5.6.51 ♠ *(O. Heden) For even n, assume that there exist $(n-1)$ pan-nonattacking queens on an $n\times n$ chessboard with the indices (i_m, j_m) for $1 \le m \le n-1$, and that no queen is in the first row and the first column. Check that modulo n: $\sum_{m=1}^{n-1} i_m = \sum_{m=1}^{n-1} j_m = \frac{n(n-1)}{2} = \frac{n}{2}$, $\sum_{m=1}^{n-1}(i_m \pm j_m) = 0$. Obtain from these relations that the set $\{i_m \pm j_m\}$ in $\mathbb{Z}_n$ for any choice of $\pm$ must be $\{0,1,\ldots,n-1\}\setminus\{\frac{n}{2}\}$ (i.e. with omitted). This readily gives that $\sum_{m=1}^{n-1}(i_m+j_m)^2+(i_m-j_m)^2 = 4\frac{(n-1)n(2n-1)}{6} \bmod (2n)$. Next, deduce from this relation that $4n^3-3n^2+2n = 0 \bmod (12n)$ and $4n^2-3n+2 = 0 \bmod 12$. Finally, obtain from the latter congruence that n cannot be divisible by 4 or 3. So we obtain that the maximal number of pan-nonattacking queens is $\le n-2$ when n is divisible by 4 or 3. [Actually, it is exactly $n-2$ for such n due to Monsky, but this is beyond this problem.]*

Exercise 5.6.52 ♣*Sum of Two Squares Theorem. If $p = a^2+b^2$ for* prime *$p > 2$ and integer $a,b > 0$, then one of them is even, $p = u^2+4v^2$ for integer $u,v > 0$ and $p = 4k+1$ for some $k > 0$. Prove that any such p can be represented in this form, following D. Zagier, R. Heath-Brown, J. Liouville. Introduce the set $X = \{(u,v,w) \mid u^2+4vw = p\}$ with integer $u,v,w > 0$ and its 2* involutions:

$$I_1(u,v,w) = (u,w,v), \quad I_2(u,v,w) = \begin{cases} (u+2w, w, v-u-w) & \text{if } v-w > u \\ (2v-u, v, u-v+w) & \text{if } v-w < u < 2v \\ (u-2v, u-v+w, v) & \text{if } u > 2v \end{cases}.$$

Check that, indeed, $(I_2)^2 =$id: A_3A_1, A_1A_3 and A_2A_2 are identities for the transformation A_i in the ith line of I_2. Also, I_2 preserves X, which is obvious for I_1. The inequalities here are necessary for the positivity of $I_2(u,v,w)$. One of them must hold since $u = v-w$ or $u = 2v$ are impossible in X (why?). Check that there is only one fixed point of I_2: for the 2nd line and with $u = v$, which results in $u^2+4uw = p$ and $u = 1 = v, w = k$. Then conclude that $|X|$ is odd and I_1 must have fixed points in X, which are with $v = w$.

Chapter 6

Graphs and Catalan Numbers

6.1 Introduction to graphs

6.1.1 Isomorphisms

Finite graphs are an important part of combinatorics. We provide only the basics, but the isomorphisms of graphs, Eulerian circuits and Hamiltonian cycles are a must in any introduction. Some topics are moved to the Exercises and the last chapter, including planar graphs, labeled trees, and line graphs.

A ***graph*** G is a collection V of ***vertices*** (nodes) connected by ***edges***, which will always be (in this chapter) without orientation. The edge collection will be denoted E. The ***order*** of G is $|V|$, the number of its vertices. The ***degree*** deg_v, or simply d_v, of a vertex $v \in V$ is the number of edges connected to it. The loop edges from v to v are counted twice in d_v. Obviously: $\sum_v d_v = 2|E|$ where $|E|$ is the total number of edges (order of the edge set) of G. Hence, the sum of the degrees over all the vertices in a graph is even, a simple but useful observation.

We allow multiple edges between the vertices. The loops from v to v are considered multiple edges too; we will frequently use the symbol $\circlearrowleft$ for them. ***Simple graphs*** are graphs without multiple edges. For Eulerian trails below, allowing multiple edges is natural. Hamiltonian paths are defined for simple graphs. The graphs will be assumed *connected* almost always, which means that any two vertices are connected by a ***walk***: a sequence of edges where any two neighboring ones share a vertex.

Definition 6.1.1 *Graphs G_1 and G_2 with their respective sets of vertices V_1 and V_2 are called isomorphic if there exists an identification $f : V_1 \to V_2$, a one-to-one correspondence between these sets, such that the number of edges between any v and v' in V_1 equals the number of edges between $f(v)$ and $f(v')$*

in V_2. We then call f an ***isomorphism*** *of graphs. An isomorphism f is called an* ***automorphism*** *of G if $G_1 = G = G_2$.*

More rigorously, we need to extend f to some $f^e : E_1 \to E_2$, an identification of the respective sets of edges, such that $f^e(e_1)$ connects $f(v_1)$ and $f(v_1')$ in G_2 if e_1 connects v_1 and v_1' in G_1. Such an extension always exists; it is unique for simple graphs G. *We will always call f (alone) an isomorphism, not the corresponding pair $\{f, f^e\}$.*

It is easy to construct isomorphic graphs: permute the vertices and plot the corresponding (number of) edges between the new vertices. It can be significantly more difficult to see that graphs are in the same isomorphism class, especially if the number of vertices is large enough.

Given a set of n vertices, the simplest G are when there are no edges at all, and when any two different vertices are connected with exactly one edge. The latter graphs are called ***complete graphs***; the usual notation is K_n. In both cases, the *automorphisms* are exactly the permutations of the vertices.

The number $|V|$, the number $|E|$, and the (unordered!) set of degrees $\{d_v\}$ for all $v \in V$ are *isomorphism invariants* of graphs. They must coincide for isomorphic graphs, but this is generally far from being sufficient.

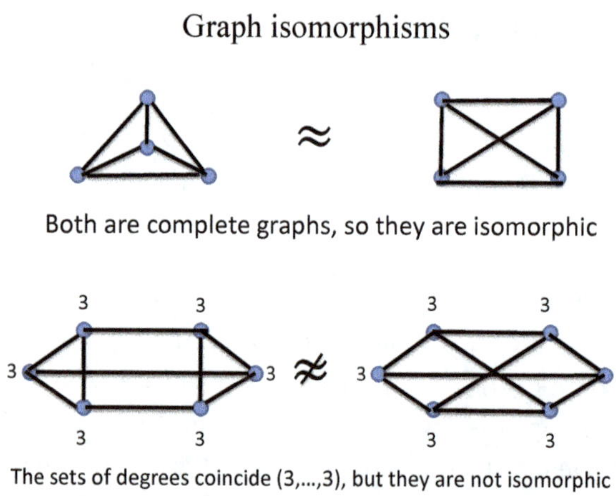

Figure 6.1: Isomorphic and non-isomorphic graphs.

In the upper half of Fig. 6.1, both graphs are *complete*; indeed, any two vertices are connected by one edge. Complete graphs with the same number of vertices are isomorphic. Actually, these graphs are two different projections of the tetrahedron on a plane. *Note that intersections of edges of planar images of graphs are generally not vertices!* We will encircle them (or mark somehow in the figures) if they are vertices.

There are several proofs that the other two graphs in Fig. 6.1 are non-isomorphic; we will use the notation $\not\approx$ for non-isomorphic graphs. Generally, you are supposed to check that for any permutation of the vertices they never become isomorphic. However, mostly we try to find some "special features" which are present or absent in the graphs under consideration.

The simplest proof that they are not isomorphic is that the first graph has "triangles" (3-cycles), when the second has none. One can also count the "squares" (4-cycles): 2 and 9, respectively. Another approach is to see that the first is planar and the second is not. See Fig. 6.2. By a planar graph G, we mean that it can be drawn in the plane with the intersections only at its vertices. Isomorphic graphs must be planar simultaneously. The second graph is not planar, which is Exercise 6.4.25(iii). Let us mention that any finite graph can be embedded into 3-dimensional space avoiding intersections except at the vertices: not a difficult theorem.

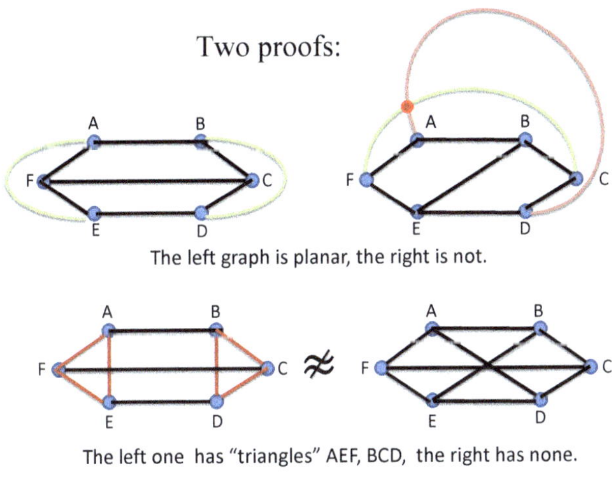

Figure 6.2: Two proofs of *non-isomorphism*.

6.1.2 Eulerian trails

The concept of the Eulerian trail is due to Euler; its origin was the *Königsberg bridges problem*, which Euler solved and generalized in 1736. It was not really a difficult problem (for him), but it is rightfully considered a beginning for graph theory. At that time, Königsberg's residents enjoyed strolling through their city, and wondered if every part of the city could be visited by walking across each of the seven bridges exactly once and returning to the starting location. Fig. 6.3 shows the reformulation of the problem using a graph with 4 vertices, where each landmass is represented by a vertex, and each bridge is represented by an edge. After viewing this depiction and making a cursory effort to complete such a walk, a negative answer is almost immediate, but Euler did not stop there, and was able to give a solution for any graph.

In the present day (at the time of this publication,) there are no bridges **b** and **d** in the city, so such a walk through the remaining 5 exists, but only from "B" to "D", or from "D" to "B".

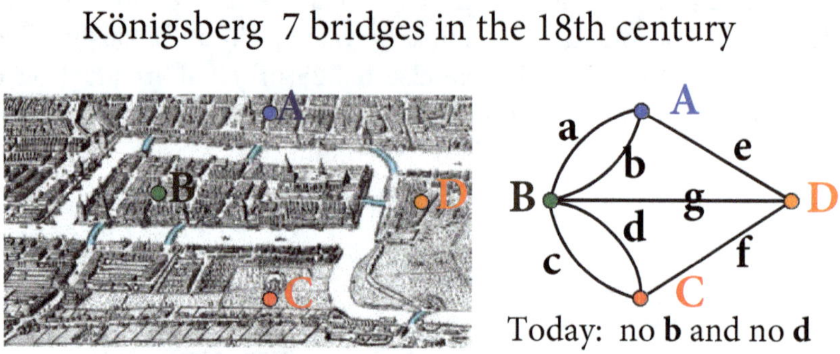

Figure 6.3: Königsberg bridges, past and present.

Several definitions. As above, a *walk* is a sequence of edges where any neighboring edges share a vertex. A ***path*** is a walk without repetitions of vertices. A ***cycle*** is a closed (reentrant) path. A ***trail*** (as in a park) is a walk where no edge can be taken twice. Visiting vertices repeatedly comes without any restrictions in a trail. A ***loop*** is a closed (reentrant) trail. Finally, ***Eulerian trails*** are trails through all edges of a connected graph. ***Eulerian circuits*** are reentrant Eulerian trails.

Theorem 6.1.2 *(I) Given a connected graph G, an Eulerian circuit exists if and only if the degrees d_v of all vertices $v \in V$ are even.*
(II) A non-reentrant Eulerian trail exists if and only if there are exactly two vertices, A and B, with odd degrees; then this trail must be from A to B.

Proof. The conditions of the theorem are almost obviously *necessary*. Indeed, if an Eulerian circuit exists, then it "enters and exits" at any vertex. Since it goes through all edges, it goes through every vertex A exactly $d_A/2$ times, so all d_A must be even. Similarly, it must begin at a vertex of odd degree and end at another vertex of odd degree in case (II); all other vertices must then have even degrees.

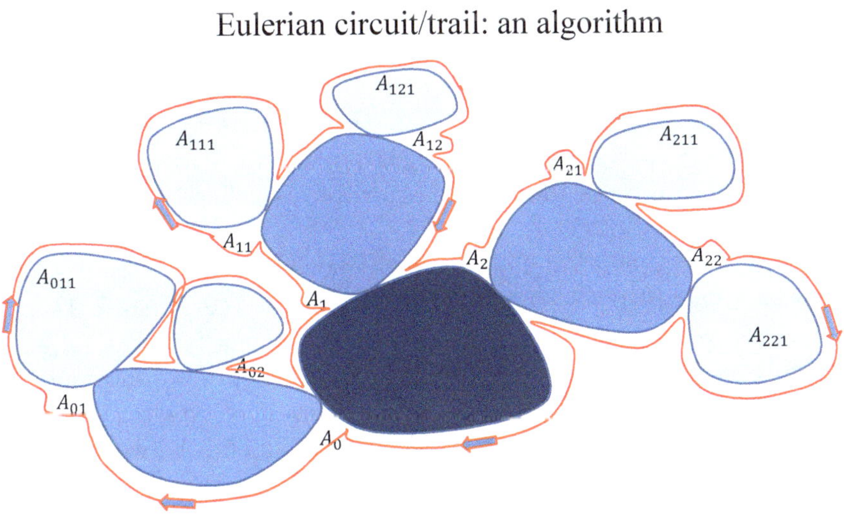

Figure 6.4: Constructing Eulerian circuits.

The proof in the opposite direction is not difficult too, and, importantly, it is very much constructive. We now assume that all degrees are even except for A and B in case (II). We present it as an algorithm (in case someone wishes to use it for a screensaver, see below); Fig. 6.4 is actually sufficient.

Level 0. We move arbitrarily from some vertex A avoiding repetition of edges until we reach a vertex where there are no "free" edges left to exit. In case (II), let A be one of the two vertices of odd degree.

Since the degrees are assumed all even (but for the two vertices A and B in (II)), this process can stop only at the initial A for (I), and at B for (II). Otherwise, any "in" can be followed by some "out", and we can continue walking. Let us denote this trail by a; it is a loop for (I).

Level 1. We set $A'_0 = A, a'_0 = a$. Then we go along a'_0, from A'_0, and employ the same procedure starting with every vertex A'_i on our way, *avoiding the edges already "taken"* (from a'_0).

Deleting, disregarding to be exact, the edges in a'_0 (but not vertices!) will result in a subgraph satisfying again the condition "all degrees are even." Indeed, every "in" in a'_0 was followed by an "out", so deleting a'_0 will diminish the degrees by even numbers. The resulting graph may be disconnected, but this does not matter for the proof.

If there are no "free edges" left at $A'_i \in a'_0$, we go to the next vertex along a'_0: continue on with A'_{i+1}. The corresponding *loop* will be denoted a'_i.

Level 2, etc. We set $A''_{i0} = A'_i, a''_{0i} = a'_i$, and go along this loop repeating the above procedure for every vertex A''_{ij} on our way, and producing loops a''_{ij} of the "second level."

Then we continue with higher levels until there will be no single "free edge" left. We use that if a vertex A belongs to some loop of level ℓ, then *all* edges from A that were not taken at the previous steps will occur in the loops of levels $> \ell$. We must begin such loops from A again and again until all edges from A are exhausted; the algorithm is defined this way. Then we use the connectivity of G.

Finally, the *envelope* is defined as the sequence of the vertices $A'''^{\cdots}_{ijk\ldots}$ *ordered lexicographically*: $\{0, 0, \cdots, 0, 0\}, \{0, 0, \cdots, 0, 1\}, \cdots, \{0, 0, \cdots, 1, 0\}, \ldots$. Let us provide a picture: Fig. 6.4. Recall that we identify: $A'_i = A''_{i0} = A'''_{i00}$, $A''_{ij} = A'''_{ij0}$, and so on; e.g. $A = A'_0 = A''_{00}$. In this figure, we show loops, loops "from" loops (loops of the 1st level), and so on. Then we "read" the vertices of G following the "envelope" (red). The only difference between (I) and (II), is that a is not a loop for (II): it is a trail from A to B. ❑

Eulerian circuits on closed surfaces. A simple but fruitful example is a coordinate net on a closed surface without a boundary. Actually, one can take any *closed* curves in any surface.

An example is the graph for a coordinate net on the torus in Fig. 6.5. The boxes are vertices, all degrees are 4. It is from a paper on optimization of computer nets; Eulerian circuits can be of use for such nets and in *networks*.

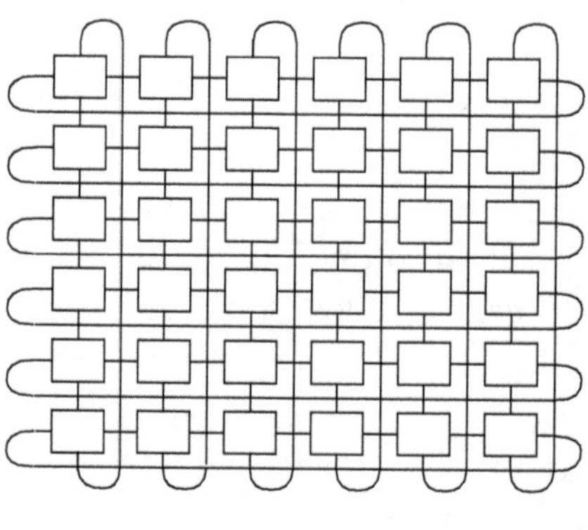

Figure 6.5: Torus chessboard as a network.

All intersections will then be considered vertices; all portions of these curves between the intersection (without such intersections inside them!) will be the edges. All degrees are even because adding a closed curve through a vertex increases its degree by 2. Thus, Eulerian circuits exist for such graphs, and we can employ Euler's algorithm for constructing them.

The algorithm in Fig. 6.4 can be used to design some advanced versions of the screensaver "3D pipes," a famous one, and one of the oldest. Take a picture of some 3D object, for example, a photo of a statue.

Supply the surface of this object with a coordinate net (a mesh). This is standard for making 2D images 3D-like, but you must use only closed curves. Then write a program producing a random Eulerian circuit for this net considered as a graph as above. This program will run before the screensaver starts (this is fast). Then, when the screensaver starts: it will plot on the computer screen the edges of this Eulerian circuit consecutively in "slow motion." The screen will look random for some time; eventually, the user will see your object!

Does this face have an Eulerian circuit/trail?

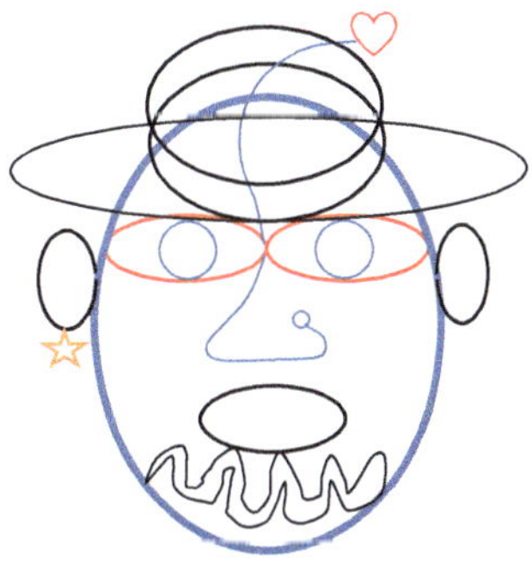

All intersections are considered vertices.

6.1.3 Hamiltonian cycles

A "dual problem" is to find paths in a graph G visiting *every* vertex exactly once, called ***Hamiltonian paths***. For instance, we want to drive through all cities in a country, but we desire to avoid revisiting any city. The usage of edges/roads is unlimited, so we replace multiple edges between vertices by single connections (and delete ↻). Recall that a path is $A - B - C - \ldots$, where the connections are those by the edges of G: A is connected to B by an edge, B is connected to C, and so on. No vertex must be used twice in a path. The classical problem is to find a ***Hamiltonian cycle*** in G: a reentrant Hamiltonian path, where the first and the last vertices are linked.

The theory of these cycles is harder than that of Eulerian trails. The following criterion due to Øystein Ore is about the best we *generally* have. The simplest example of a graph that has a Hamiltonian cycle is of course a "circle": $A - B - C - \cdots - Z - A$. This example is actually very far from those covered by the next theorem. Recall that a graph is called *simple* if it has no multiple edges, including loop edges ↻ (from A to A).

Theorem 6.1.3 *A simple connected graph G of order n (with n vertices) has a Hamiltonian cycle if $deg_A + deg_B \geq n$ for any two vertices $A \neq B$.*

Proof *. We do not require the connectivity of G because it *automatically* follows from these inequalities. Indeed, if G is a disjoint union of G' and G''

with n' and n'' vertices, respectively, then $deg_{A'}+deg_{A''} \le (n'-1)+(n''-1) = n-2 < n$ for any A' in G' and any A'' in G''.

Let $A_1 - A_2 - \cdots - A_{m-1} - A_m$ be the longest possible Hamiltonian path in G: with all different A_i by definition. There are several possibilities, which are the corresponding consecutive steps to go from here.

(a) Firstly, we assume that A_1 and A_m are linked by an edge. In this case and if $m = n$, adding this edge gives the required Hamiltonian *cycle*. If $m < n$, the edge between A_1 and A_m can be used to increase m; this goes as follows. Let $m < n$ and C be a vertex not in this path linked to some A_i by an edge; it exists since G is connected. If $i \neq m$, then $C - A_i - A_{i+1} - \cdots - A_m - A_1 - \cdots - A_{i-1}$ has length $m+1$: impossible due to the maximality of m. If $i = m$, we take $C - A_m - A_1 - \cdots - A_{m-1}$, also with length $m+1$.

(b) Thus, we can now assume that m is the maximal possible and that A_1 and A_m are not linked, i.e. the path is not reentrant. In this case, any vertex C linked to A_1 or linked to A_m must appear in the sequence $A_2, \cdots, A_{m-1}$. Otherwise, the path $C - A_1 - \cdots - A_m$ or the path $A_1 - \cdots - A_m - C$ would have the length $m+1$, which contradicts the maximality of m. Let us represent the vertices linked to A_1 by $\odot$, those linked to A_m by $\times$ and the vertices linked to both by $\otimes$. Finally, the symbol $\Box$ will be used for the vertices among the A_i $(1 < i < m)$ not linked to A_1 and to A_m.

(c) The key point of the proof is the following argument, under the assumptions from (b). If there exists one pair of neighboring vertices, A_iA_{i+1}, such that A_{i+1} is linked to A_1, i.e. of type $\odot$ or $\otimes$, and A_i is linked to A_m, i.e. of type $\times$ or $\otimes$, then we can form a path of the same length as follows:

$$A_{i+1} - A_1 - A_2 - \cdots - A_i - A_m - A_{m-1} - A_{m-2} \cdots - A_{i+2}.$$

Graphically,

$$\overset{\longleftarrow}{\ulcorner A_1} \to A_2 \to \cdots \to \underset{\longrightarrow}{A_i \urcorner \;\llcorner A_{i+1}}A_{i+2} \leftarrow \cdots \leftarrow \lrcorner A_m.$$

Since the endpoints, A_{i+1} and A_{i+2}, in this path are linked, it is reentrant. We arrive at (b), and obtain the required contradiction.

(d) Let *blocks* be connected components of the set of A_i $(1 < i < m)$ linked to A_1 or to A_m (or to both), i.e. those between squares $\Box$. We deduced in (c) from the maximality of m that the path $A_1 - A_2 - \cdots - A_m$ is the following sequence of *blocks*, shown by $\{\cdots\}$:

$$A_1\,\{\odot\cdots\odot\otimes\times\cdots\times\}\Box\cdots\Box\{\odot\cdots\odot\otimes\times\cdots\times\}\Box\cdots\,A_m.$$

Indeed, $\times$ or $\otimes$ cannot occur before $\odot$ (if they are present) due to (c), and, similarly, $\otimes$ cannot occur after $\times$. Here $\otimes$ can be present in a block or can be absent. If they are present then there can be only one such: if there are two $\otimes$ in a block, then they must be neighbors, which was excluded in (c).

(*e*) The end of the proof is "numerical." Given a block $\mathcal{B}_j$ for $1 \le j \le \ell$, let r_j, s_j, t_j be the numbers of $\odot, \otimes, \times$ in $\mathcal{B}_j$. We obtain that A_1 is linked exactly to $d_1 = \sum_{j=1}^{\ell}(r_i+s_i)$ vertices in G, and A_m is linked to $d_m = \sum_{j=1}^{\ell}(s_i+t_i)$ vertices. These numbers are the degrees of A_1 and A_m respectively. Therefore, $d_1 + d_m = \sum_{j=1}^{\ell}(r_j + t_j) + 2\sum_{j=1}^{\ell} s_j = (m-2) + \sum_{j=1}^{\ell} s_j - |\text{squares}|$. Here $|\text{squares}|$ is the number of $\square$ in this path, which is at least $\ell - 1$. The number of $\otimes$ is no greater than ℓ. Thus, $d_1 + d_m \le (m-2) + \ell - (\ell - 1) = m - 1$. We have $d_1 + d_m \le m - 1 \le n - 1$, which contradicts the inequality of the theorem. Note that here $m = n$ contradicts to (b): A_1 and A_m were *not* supposed to be linked in (b). ❑

The graphs at Fig. 6.1 are simple examples when the inequality of the theorem holds, which is for all of them. Thus the Hamiltonian cycles exist there. If you remove any edge from the second two graphs there, this inequality will not be satisfied. but the Hamiltonian cycles will still exist there (check this).

Generally, the claim of the theorem is actually too weak: "unreasonably many roads" are requested in this "country." There are many graphs with not many edges that have such cycles. The proof is nice, but the class of the graphs from the theorem is actually not too far from *complete graphs*, where $d_A + d_B = 2(n-1)$ for any two vertices A, B and Hamiltonian cycles are *any* permutations of the vertices.

6.1.4 Knight's tours and networks

There is quite a famous puzzle with $d_A + d_B$ smaller than the number of vertices: ***the knight's tour problem.*** It is old, presumably from India in the 9th century CE. The *knight* chess piece moves by two squares vertically and one horizontally or one vertically and two horizontally (that is, in an L shape.) We create a graph from the chessboard by letting the vertices be the squares and the edges be between the squares connected by 1 knight move. The "tour" means that all squares must be visited and exactly once; mostly we are interested in *reentrant* ones. There are from 2 to 8 edges from any vertex, so $4 \le d_A + d_B \le 16$, smaller than required in Theorem 6.1.3.

A reentrant solution obtained by the chess-playing "Mechanical Turk" is presented in Fig. 6.6. This was a famous *hoax* machine: it purported to consist of an animatronic chess-playing figure (dressed in late 18th century *Turquerie*) and controlled by unknown and mysterious machinery in the box

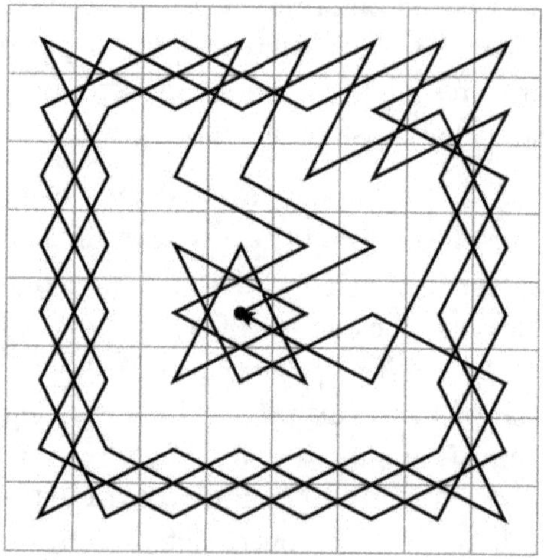

Figure 6.6: A solution of the knight's tour problem.

that supported it. In reality, a human was actually playing and creating solutions for the knight's tour problem (while hiding in the box.) A renewed interest in this problem is due to *neural networks*.

Problem 6.1.4 *There are no reentrant knight's tours for the* 3×4 *chessboard, and there are exactly* 8 *non-reentrant (open) solutions.*

Solution. It is not difficult to find a non-reentrant tour:

1	4	7	10
12	9	2	5
3	6	11	8

Here $1, 2, \ldots, 12$ show consecutive knight's moves. Let us use these numbers to label the corresponding squares. Then the graph of connections between the squares under knight's moves is as follows:

$$\begin{array}{ccccccc} 1 & - & 6 & - & 7 & - & 8 \\ | & & | & & | & & | \\ 2 & & 5 & & 12 & & 9 \\ | & & | & & | & & | \\ 3 & - & 4 & - & 11 & - & 10 \end{array}.$$

To double-check that the number of edges is 14 (so we did not miss any), use that it is $\frac{1}{2}\sum_{i=1}^{12} d_i$, where 4 degrees d_i are 3 and the remaining 8 are 2.

So we have only 2 *bridges* between the 6-cycles $1-2-3-4-5-6-1$ and $7-8-9-10-11-12-7$; they are $6-7$ and $4-11$. Bridges are, naturally, edges between connected subgraphs of a given graph G. If a graph G is a pair of 2 cycles, $\mathcal{A}$ and $\mathcal{B}$, with 2 bridges, $A_1 - B_1$ and $A_2 - B_2$ between them, a Hamiltonian cycle exists if and only if G contains the square $A_1 - B_1 - B_2 - A_2 - A_1$, where $A_1 - A_2$ and $B_1 - B_2$ are edges in $\mathcal{A}, \mathcal{B}$, correspondingly (check). This is not the case for the 6-cycles above.

One bridge is sufficient for non-reentrant solutions. There are 2 choices for this bridge to take, and 2 choices for going around each of the 6-cycles (clockwise and counterclockwise): $2 \times 2 \times 2$. Namely, if the bridge $6 - 7$ is taken, the endpoints can only be $1 \leftrightsquigarrow 8, 1 \leftrightsquigarrow 12, 5 \leftrightsquigarrow 8, 5 \leftrightsquigarrow 12$. For $4-11$, they are $3 \leftrightsquigarrow 10, 3 \leftrightsquigarrow 12, 5 \leftrightsquigarrow 10, 5 \leftrightsquigarrow 12$. ❑

The simplicity of our proof is due to our "smart numbering" of the squares. If the numbers of squares are standard (lexicographical), then establishing an isomorphism with the connection graph we provided can be much less transparent. Generally, reentrant knight's tours exist for any $m \times n$ boards for $n > m > 2$ unless mn is odd, $m = 2, 4$, and $m = 3$ for $n = 4, 6, 8$. See Problem 7.4.3 for $m = 4$ and any n.

Networking approach.* For us, a network is a system of recurrence relations that eventually becomes *stable*, i.e. stops changing any further. The system below is of "non-algebraic type," which is quite frequent in modeling.

For any graph G, we place *neurons* $N(a)$ at its *edges* a, which can be *active* or *inactive*: accordingly we put $V(a) = 1$ or $V(a) = 0$. A "path" will simply be a set of active neurons, those with $V(a) = 1$. Initially, $V_0(a)$ are randomly chosen (to be either 0 or 1). Similar to the predator-prey model in Section 4.3.3 or our modeling of Covid-19, we need one more function $U(a)$, which will be zero at all edges in the beginning: i.e. $U_0(a) = 0$ for all a. The graph G is as above: the vertices are squares of the $n \times n$ board, the edges are links between those connected by 1 knight move. For any a, let v_a, w_b be its endpoints (pairs (i, j) for $1 \le i, j \le n$) such that $v_a \prec w_a$ for the lexicographical ordering "$\prec$" of (i, j); we write $v \in a$ if $v = v_a$ or $v = w_a$.

Now we assume that these functions depend on m according to the following recurrence relations:

$$U_{m+1}(a) = U_m(a) + 4 - \sum_{v_a \in b} V_m(b) - \sum_{w_a \subset b} V_m(b), \text{ so } b{=}a \text{ occurs here twice,}$$

$$V_{m+1}(a) = \big\{\, 0 \text{ if } U_{n+1}(a) < 0, \quad \text{or } 1 \text{ if } U_{m+1}(a) > 3, \quad \text{or } V_m(a) \text{ otherwise} \,\big\}.$$

The purpose is to obtain ***stable solutions*** $\{U_m, V_m\}$, those which remain unchanged when $m \ge m_0$ for some m_0, aiming at Hamiltonian cycles. If such m_0 exists (not always), then the 1st relation gives the following.

(i) Any a with $V_m(a) = 1$ for $m \geq m_0$, i.e. *active*, has exactly two active neighbors: b_1 containing v_a and b_2 containing w_a.

(ii) Any a that is *inactive* for $m \geq m_0$ (i.e. not in the corresponding path) has v_a and w_a (both) belonging to this path.

Indeed, $V_m(a) = 1$ gives that a has exactly 2 active neighbors (in the path), because a is counted twice as b in the right-hand side of the first relation. Claim (ii) follows from (i): if $V_m(a) = 0$ then 4 neighbors b of a belonging to the path are needed to make $U_{m+1}(a) = U_m(a)$. Also, we obtain that all vertices of G (squares (i, j)) belong to any stable path: the graph is connected. We took this algorithm from paper [37]. The second relation, called the *hysteresis*, is necessary (as such) to avoid oscillations of $V_m(a)$.

The paths satisfying (i,ii) are generally *not* Hamiltonian cycles. It is very much possible that they are *Hamiltonian multi-cycles*, i.e. disjoint unions of such cycles for subsets of vertices covering the whole set of vertices.

It is quite a challenge to obtain a *single* cycle using the flexibility with the initial conditions of the recurrence above. See [38].

The algorithm above works for any rectangles. For instance, the union of $\{1-2-3-4-5-6\}$ and $\{12-11-10-9-8-7\}$ from Problem 6.1.4, which is a Hamiltonian 2-cycle, can be readily obtained for the 3×4 board.

Dimer configurations. There is a connection of Hamiltonian multi-cycles to dimer configurations. Recall from the chapter on dominos and magic squares, that they are equivalent to perfect coverings by dominos. The graph $G_{M\times N}$ will have vertices that are the centers of the squares of the $M \times N$ chessboard, and the edges will be *dimers*: segments connected to the centers of neighboring squares.

So a *dimer configuration* is a set of edges (the dimers) in $G_{M\times N}$ such that their endpoints do not intersect for different edges and constitute the set of all vertices (the centers of all squares). See Fig. 6.7: 2 examples of perfect coverings of 6×8 shapes and their dimer presentations.

Given two dimer configurations, their *superposition* is defined as the union of dimers in both of them minus the intersection, the dimers that belong to both configurations. This is used in physics; the number of connected components of such a superposition is of importance. We will apply this construction only when there is no intersection of dimer configurations, as in Fig. 6.7. The corresponding path is then a Hamiltonian multi-cycle.

---

[37] Y. Takefuji, K.C. Lee, *Neural network computing for knight's tour problems*, Neurocomputing 4 (1992), 249–254.

[38] I. Parberry, *Scalability of a neural network for the knight's tour problem*, Neurocomputing, 12 (1996), 19–33.

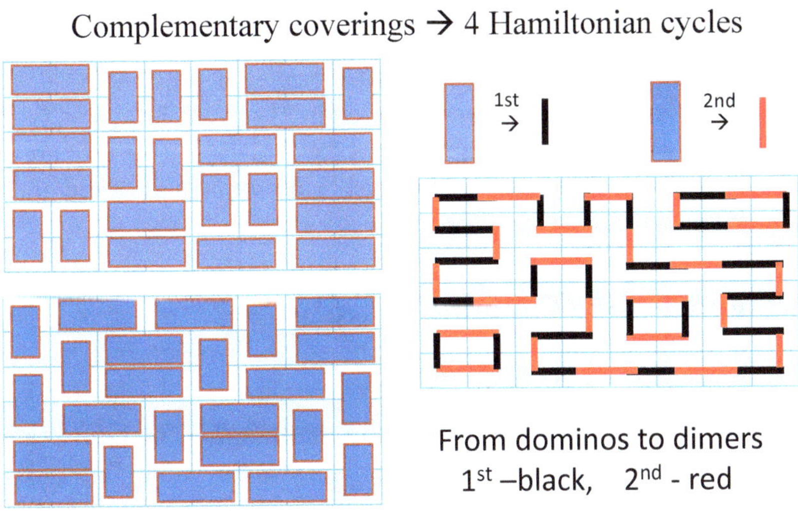

Figure 6.7: A superposition of dimer configurations.

Problem 6.1.5 *Assuming that MN is even, let C be a Hamiltonian multi-cycle C in $G_{M\times N}$. Any of its cycle has an even number of edges. We fix some initial edges in each of them; if the distance from an edge in a cycle to the corresponding initial edge is even we call it* even, *and* odd *otherwise. The distance is 0 for the initial edges, 1 for their neighbors and so on.*

(i) Then the sets of even/odd edges form two dimer configurations, *denoted by D_0 and D_1, which do not intersect (do not have coinciding dimers). Also, their superposition, a union in this case, is the whole C.*

(ii) Vice versa, if two dimer configurations D_0, D_1 for "$M\times N$" do not have coinciding dimers, then their union can be naturally (in a unique way) considered as a Hamiltonian multi-cycle.

Solution. **(i)** Any two edges in a Hamiltonian multi-cycle C share a vertex if and only if they are neighboring in one of its cycles. So D_0 and D_1 are really dimer configurations: every vertex (the center of a square on the board) will belong exactly to 1 dimer in D_0 and 1 in D_1. Obviously, D_0 and D_1 do not intersect and their union is C.

(ii) Let D_0 and D_1 be two *disjoint* dimer configurations: having no dimers in common. Each of them "covers" all vertices (centers of the squares), so every vertex "belongs" to exactly one dimer from D_0 and to exactly one from D_1. So we can make a natural *alternating sequence* of dimers: from D_0 to D_1, from D_1 to D_0, and so on.

Namely, given a dimer a in D_0, it will be placed between the dimers b, b'

from D_1 that share vertices with a; there are exactly two. Then we take the remaining vertices of b, b' (not those from a), find the (unique) dimers in D_0 containing them, place them before/after b, b', and continue.

Every vertex will occur exactly once in the resulting path: the vertices connected by the corresponding edges (the dimers in this sequence). It is a disjoint union since every vertex belongs to a unique pair of dimers, one from D_0 and one from D_1. ❑

6.2 Extended Catalan numbers

6.2.1 Non-intersecting chords

Leonard Euler was the first to study Catalan numbers systematically (1751), though some formulas for them were known before him. The following approach has solid physics connections: they occur when counting certain Feynman diagrams and in matrix models.

Consider *configurations of n non-intersecting chords in a circle.* More exactly, you place $2n$ different points on a circle and count the number h_n of the ways to connect them by segments avoiding intersections. Obviously, any such set will give the same h_n, so the latter depends only on n. This is actually one of several (equivalent) definitions of *Catalan numbers* (see below).

One has $h_1 = 1$, $h_2 = 2$:  and (for 4 points on a circle). As an exercise, check that $h_3 = 5$. The approach to finding the recurrence relation for h_n is actually no different from what we did in Section 4.1: we consider what can happen with some fixed element (a point in this case) and then reduce the "dimension" of the problem.

First of all, the count of chords does not depend at all on the positions of the points; they can be uniformly distributed or belong to a very small sector. Moreover, we can consider some "curved-chords" between points: portions of any curves instead of straight-line segments, assuming that they do not intersect. Thus, the problem is clearly *topological*, about counting some special planar graphs.

Second, we fix one point, A_1, and go through the list of its possible connections with the other $(2n-1)$ points. It can be connected with $A_2, \ldots, A_{2n}$, but not with *odd* A_i, since it must split the remaining $2n-2$ points into 2 *even* groups. See Fig. 6.8, where $n = 5$.

The remaining chords of our configurations must *not* intersect the segment $[A_1A_i]$, so we proceed independently with connecting points $A_2, \cdots, A_{i-1}$ before A_i by non-intersecting chords, and with the points $A_{i+1}, \cdots, A_{2n-1}$ after it. Here i must be even: $i = 2m$. Thus, the number of the extensions of $[A_1A_{2m}]$ to a configuration, is $h_{m-1}h_{n-m}$. Recall that it does not matter

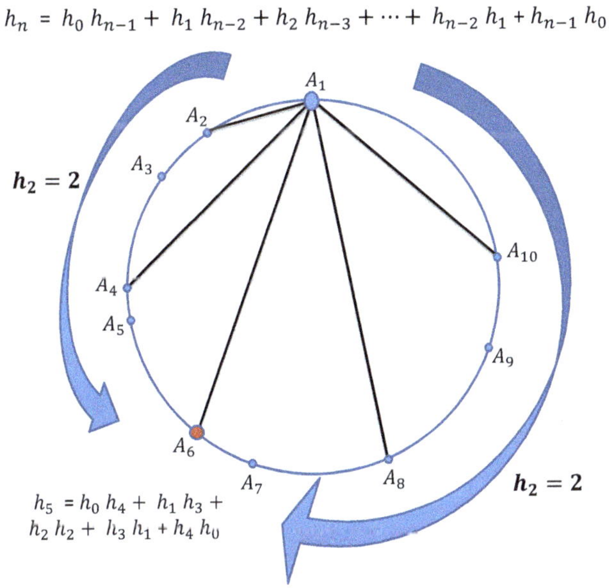

Figure 6.8: Catalan numbers via chords.

if the points are evenly spaced around the whole circle or clustered into some sectors. Finally, for $h_0 = 1$ for the sake of uniformity:

$$h_n = h_0 h_{n-1} + h_1 h_{n-2} + h_2 h_{n-3} + \cdots + h_{n-2} h_1 + h_{n-1} h_0. \tag{6.2.1}$$

This is shown in Fig. 6.8: for the chord $[A_1 A_6]$, the number of "continuations" is $h_2 h_2 = 2 \cdot 2 = 4$, the middle term in the sum for h_5; the second half of this sum repeats the first. One has: $h_3 = 1 \cdot h_2 + h_1 h_1 + h_2 \cdot 1 = 5$, $h_4 = 1 \cdot h_3 + h_1 h_2 + h_2 h_1 + h_3 \cdot 1 = 2(5+2) = 14$, and so on.

Problem 6.2.1 *Let us define the* generating function *of h_n: $g(x) = x(h_0 + h_1 x + \cdots + h_n x^n + \ldots)$. Deduce from (6.2.1) that $g(x) = x(1 + x + 2x^2 + 5x^3 + \ldots)$ and that $(g(x))^2 = g(x) - x$. Check that up to the fourth order (ignoring terms x^m with $m \geq 4$):*

$$(g(x))^2 = x^2(1 + x + 2x^2 + \ldots)^2 = x^2(1 + 2x + x^2 + 5x^2 + \ldots)$$
$$= g(x) - x = x(1 + x + 2x^2 + 5x^3 + \ldots) - x = x^2 + 2x^3 + 5x^4 + \ldots .$$

By solving this quadratic equation in terms of $g(x)$, obtain that $(1-4x)^{1/2} = 1 - 2g(x)$, where we use the expansion $(1+x)^{1/2} = 1 + x/2 - x^2/8 + \ldots$ (this is for those who know it). Why is the series in x that is the second solution of the quadratic equation $(g(x))^2 = g(x) - x$ irrelevant here? ❑

We will not use the formula from this problem and omit some details here. What we really need is the following alternative approach to counting h_n.

6.2.2 Back to random walks

Let us break the circle above between any two points and *unfold* it: the chords then become semi-circles. See Fig. 6.9 for $n = 4$. We continuously deform the (broken) circle to make it a segment, together with the chords. Keeping them non-intersecting is the key. The semi-circles can be non-intersecting "curved-semi-circles" connecting the corresponding points A_i.

We interpret each semi-circle as a pair of events: its beginning shown by $\uparrow$ will be "creation", its end shown by $\downarrow$ will be "annihilation." Accordingly we put $a_i = \pm 1$ for these events, where $i = 1, \ldots, 2n$. Counting the number of semi-circles crossing each of the vertical lines placed *between* the events, we obtain the sequence $b_0 = 0, b_1 = a_1, b_2 = a_1 + a_2, \ldots, b_i = a_1 + \cdots + a_i, \ldots, b_{2n} = a_1 + \cdots + a_{2n} = 0$. By construction, b_0, b_{2n} must be 0, automatically, and $a_1 = +1$, $a_{2n} = -1$.

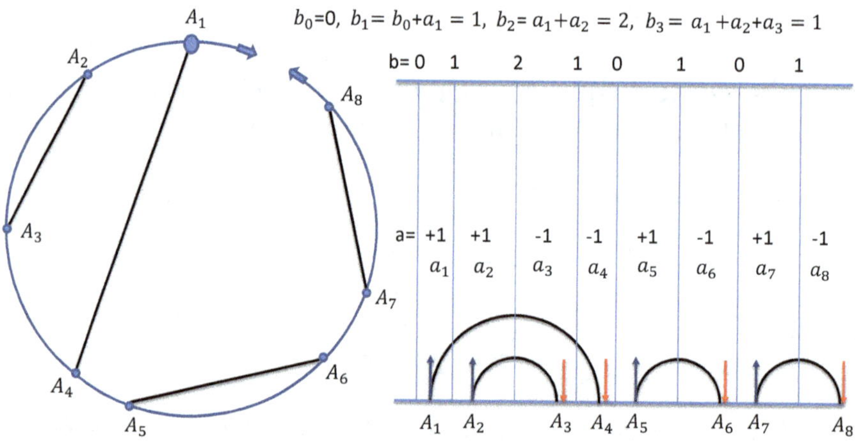

Figure 6.9: Unfolding the circle.

The sequence of semi-circles is fully determined by $\{a_1, a_2, \ldots, a_{2n}\}$. We perform the creations and annihilations of the semi-circles following this sequence. This is not ambiguous. Namely, the semi-circles created or annihilated upon $a_i = \pm 1$ must be the lowest (the closest to the x-axes) to avoid the intersections.

Not all sequences of ± 1 come from such configurations of non-intersecting semi-circles. It is no problem to create a new semi-circle upon the signal

$a_i = +1$, but we cannot terminate a semi-circle upon the signal $a_i = -1$ if we have none. Algebraically, we must impose the conditions $b_i \geq 0$ for all i; then it will always be doable. We come to a classical entirely algebraic definition of the ***Catalan numbers***, and its coincidence with the definition of h_n.

Definition 6.2.2 *The Catalan numbers C_{2n} for $n \geq 1$ count the number of sequences $\{a_1, a_2, \ldots, a_{2n}\}$ with the values ± 1 such that $a_1 + \cdots + a_i \geq 0$ for any $i = 1, \ldots, 2n$, and with the additional constraint $a_1 + \cdots + a_{2n} = 0$. As we have checked by the "unfolding procedure": $C_{2n} = h_n$.*

The condition $a_1 + \cdots + a_{2n} = 0$ seems quite artificial in this definition. If we remove it, it is not necessary to assume that the sequences of ± 1 are always of even length. If the final sum is not a zero, then geometrically, we have some "active" semi-circles left at the end. For applications below, the condition $\sum_{i=1}^{2n} a_i = 0$, will be abandoned, and we will allow any number of semi-circles to exist in the beginning. However, we will mostly continue to consider an even number of a_i for the sake of concreteness.

Definition 6.2.3 *For any $k, m \geq 0$ and $n = 1, 2, \ldots$, the **extended Catalan numbers** $C_{2n}^{2m \to 2k}$ count the number of sequences $\{a_1, a_2, \ldots, a_{2n}\}$ with the values $a_i = \pm 1$ such that*

$$2m + a_1 + \cdots + a_i \geq 0 \text{ for } i = 1, \ldots, 2n \text{ and } 2m + a_1 + \cdots + a_{2n} = 2k. \qquad \lrcorner$$

The following interpretation in terms of random walks is now immediate. We consider $a_i = \pm 1$ as "coin flipping" and present these sequences as graphs for the number of "heads" minus the number of "tails." They were constantly used in Section 2.5.1; see Fig. 2.9.

For instance, Fig. 6.10 shows two paths satisfying Definition 6.2.3, where the black one is exactly for the sequence $\{a_i\}$ from Fig. 6.9. The second (blue) one is with $m = k = 1$; generally, these values can be arbitrary non-negative.

From now on, we will call such paths *good*. Assume that the area below the x-axis is a lake (or worse), then random walks/paths avoiding this area are certainly "good". Accordingly, *bad* paths will be those crossing the x-axes (not just touching it, which is "good"). The notation for the numbers of bad paths will be $B_{2n}^{2m \to 2k}$ ("B" stands for "bad"). They play a key role in the following fundamental theorem, which has a very interesting "geometric" proof (below): *André's reflection method.*

Note that $C_{2n}^{2m \to 2k} = 0$ if $|2k - 2m| > 2n$ and $C_{2n}^{2m \to 2k} = \binom{2n}{n+k-m}$ if $n \leq k + m$. The first is obvious: it is impossible to reach $2k$ from $2m$ by $2n$ steps (in any direction) for such n. As for the second claim, the smallest

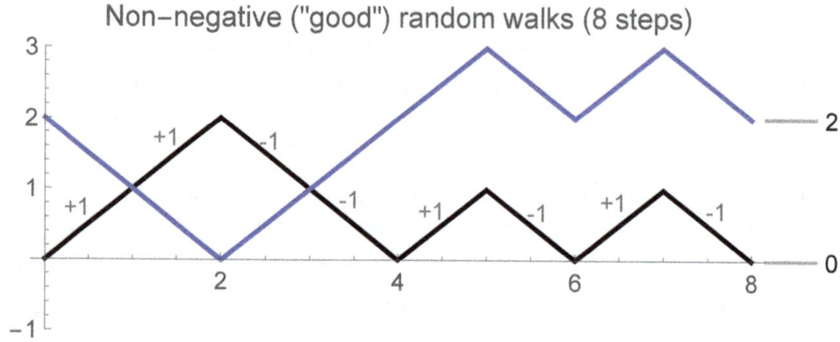

Figure 6.10: Two "good" paths.

number of steps needed for a path to go from $2m$ to $2k$ visiting the area (strictly) below zero is $2m+2k+2$. Therefore any path is good if $n \leq k+m$.

The count of the number of *all paths* of type "$2m \to 2k$" was done in Section 2.5.1; see Fig. 2.9. A path is uniquely determined if we know the group of $n+m-k$ indices i from $\{1, 2, \ldots, 2n\}$ such that $a_i = 1$. Indeed, then $\sum_{i=1}^{2n} a_i = (n+m-k) - (n-m+k) = 2m-2k$, which is the required "jump." We obtain that the number of *all* such paths is $\binom{2n}{n+m-k} = \binom{2n}{n+k-m}$.

6.2.3 The main theorem

Theorem 6.2.4 *Let $C_{2n}^{2m\to 2k}$ and $B_{2n}^{2m\to 2k}$ be the numbers of* good *and* bad *paths from $2m$ to $2k$ in the sense of Definition 6.2.3. Accordingly, $A_{2n}^{2m\to 2k}$ will be the number of* all *such paths, good and bad, where we allow m and k to be negative. Then $A_{2n}^{2m\to 2k} = \binom{2n}{n+k-m}$, and the following identities hold:*

$$C_{2n}^{2m\to 2k} + B_{2n}^{2m\to 2k} = A_{2n}^{2m\to 2k} = \binom{2n}{n+k-m},$$

$$B_{2n}^{2m\to 2k} = A_{2n}^{-2m\to 2k+2} = \binom{2n}{n+k+m+1},$$

$$C_{2n}^{2m\to 2k} = \binom{2n}{n+k-m} - \binom{2n}{n+k+m+1}. \tag{6.2.2}$$

In particular, we have the following celebrated "product formula": $C_{2n} = h_n = C_{2n}^{0\to 0} = \binom{2n}{n} - \binom{2n}{n+1} = \frac{1}{n+1}\binom{2n}{n}$.

Proof. The key is a "geometric" 1-1 correspondence between *bad* paths $\mathcal{B}$ of type "$2m \to 2k$" and *all* paths $\mathcal{A}$ of type "$-2m \to 2k+2$". The map f is described in Fig. 6.11. Given a bad path $\mathcal{B}$ (red), its (black) image $f(\mathcal{B})$ is defined as follows:

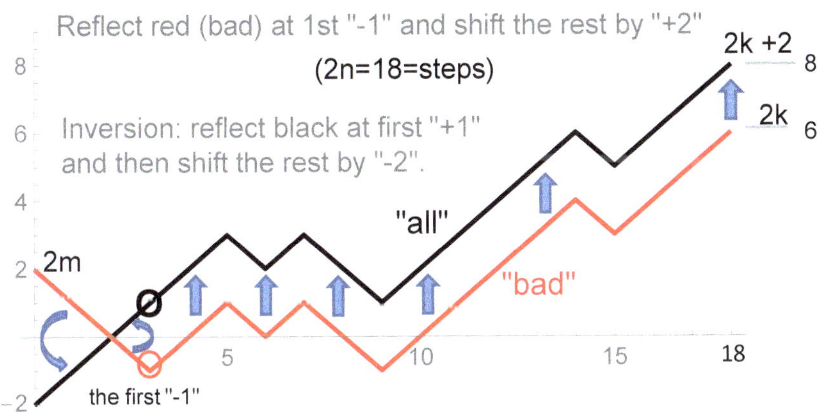

Figure 6.11: The procedure *reflect and shift.*

(1) find the first i such that $2m + a_1 + a_2 + \cdots + a_i$ is $= -1$ (the first bad point), (2) reflect in the x-axis the portion of $\mathcal{B}$ until this i (including it), and (3) shift by $+2$ the remaining portion of $\mathcal{B}$.

By construction, the image is of type "$-2m \to 2k+2$". The inverse map g is as follows: given a path $\mathcal{A}$ of this type, find the first $x = i$ when the y-coordinate is $+1$, reflect until this point, and shift by -2 the rest of $\mathcal{A}$.

The key here is that the value $y = +1$ can *always* be found for any continuous curve from $y = -2m$ to $y = 2k+2$ if $m, k \geq 0$; this is basic calculus. Thus, g is well defined for *any* $\mathcal{A}$, and obviously $gf(\mathcal{B}) = \mathcal{B}$, $fg(\mathcal{A}) = \mathcal{A}$ for any $\mathcal{A}$ and $\mathcal{B}$ considered above. Using that *good+bad= all*, we obtain (6.2.2).

Note that one can reflect here in the line $y = 1$; then no shift will be necessary. The formula will remain the same (check).

The justification of the "product formula" for $C_{2n}^{0\to 0}$ is straightforward: $C_{2n}^{0\to 0} = \binom{2n}{n} - \binom{2n}{n+1} = \frac{(2n)!}{n!n!} - \frac{(2n)!}{(n+1)!(n-1)!} = \frac{(2n)!}{n!n!}(1 - \frac{n}{n+1}) = \frac{1}{n+1}\binom{2n}{n}$. ❑

Notice that $C_{2n}^{2m\to 2k} = C_{2n}^{2k\to 2m}$. Indeed, the representations of $\{a_i\}$ by the graphs as in Fig. 6.10 or Fig. 6.11 are obviously $m \leftrightarrow k$-symmetric: we can "read" these graphs backward, from $2k$ to $2m$. The formula for $C_{2n}^{2m\to 2k}$ in terms of binomial coefficients from the theorem clearly has the same symmetry.

Let us provide a couple of calculations to connect the "product formula" for C_{2n} with h_n we calculated using the chords and the recurrence from (6.2.1): $h_3 = 2h_2 + h_1 h_1 = 5 = \frac{6\cdot 5\cdot 4}{4\cdot 3!}$, $h_4 = 2(h_3 + h_2 h_1) = 14 = \frac{8\cdot 7\cdot 6\cdot 5}{5\cdot 4!}$.

The "product formula" for C_{2n} was found by Euler in 1751. It was proven around 1758. Our proof of (6.2.1) is due to Johann Segner (1757). The name

"Catalan numbers" was assigned to the "Euler-Segner numbers" more or less accidentally (much later), but this name stuck. Catalan himself called them Segner numbers. Actually, they were used for the first time by Sharabiin Myangat (Qing Court in China; he was Mongolian, as his name indicates) in some trigonometric identities in the 1730s. We mostly use in this chapter the formula $C_{2n} = \binom{2n}{n} - \binom{2n}{n+1}$; this one is due to Catalan.

Here and below see Richard Stanley's book, which includes an appendix by Igor Pak on their history [39]. See also notes [40], and [41].

Let us extend the theorem to any number of steps, even and odd.

Proposition 6.2.5 *Let $C_n^{m\to k}$ be the number of* good *n*-paths *from m to k, which are the sequences $\{a_1, a_2, \ldots, a_n\}$ with $a_i = \pm 1$ satisfying the inequalities $m + a_1 \geq 0, m + a_1 + a_2 \geq 0, \ldots, m + a_1 + \cdots + a_n = k$. This number counts the number of non-intersecting semi-circles between points $X_1, \ldots, X_m, A_1, \ldots, A_n, Y_1, \ldots, Y_m$ such that all points are used as endpoints, no X_i and X_j are connected by semi-circles, and no Y_i and Y_j are connected by semi-circles. The number $C_n^{m\to k}$ is nonzero only if $n + k - m$ is even and $|k - m| \leq n$; then $C_n^{m\to k} = \binom{n}{(n+k-m)/2} - \binom{n}{1+(n+k+m)/2}$.*

Proof. An example of the interpretation of $C_n^{m\to k}$ via the semi-circles is provided in Fig. 6.12. The general case is completely parallel to what we did above. We leave the details to the reader. Similarly, the justification of the formula for $C_n^{m\to k}$ is absolutely the same as the one for $C_{2n}^{2m\to 2k}$. Note that, technically, $n = 0$ is allowed in the interpretation via semi-circles and in this formula; then $k = m$ and $C_0^{m\to m} = 1$, which includes $m = 0 = k$. ❑

Going back from semi-circles to chords, let us now obtain the recurrence for the extended Catalan numbers, We will generalize the relation in (6.2.1): $h_n = h_0 h_{n-1} + h_1 h_{n-2} + h_2 h_{n-3} + \cdots + h_{n-2} h_1 + h_{n-1} h_0$.

For the sake of concreteness only $C_n^k = C_n^{0\to k} = C_n^{k\to 0}$ will be considered for $k \geq 0$, where we naturally set $C_n = C_n^0$. Thus $n \geq k$, $n - k$ must be even, and $h_n = C_{2n}$ in the notation used above.

One has: $C_n^n = 1$, $C_0^k = 1$, $C_n^1 = C_{n+1}$, and $C_n^2 = C_{n+2} - C_n$. In the latter, C_n^2 equals the number of all good $(n+2)$-paths minus the number of those with Y_1 and Y_2 connected by a semi-circle, which gives $-C_n$.

It makes sense to check the relation $C_n^2 = C_{n+2} - C_n$ using the formula from the theorem. One has: $C_{2n}^2 + C_{2n} = \binom{2n}{n+1} - \binom{2n}{n+2} + \binom{2n}{n} - \binom{2n}{n+1} =$

---

[39] R. Stanley, *Catalan numbers*, Cambridge University Press, 2015.

[40] R. Stanley, *Catalan numbers*, `https://math.mit.edu/~rstan/transparencies/china.pdf`.

[41] D. Armstrong, *Catalan numbers: from EGS to BEG* (2015), `https://www.math.miami.edu/~armstrong/Talks/catalan_at_mit.pdf`.

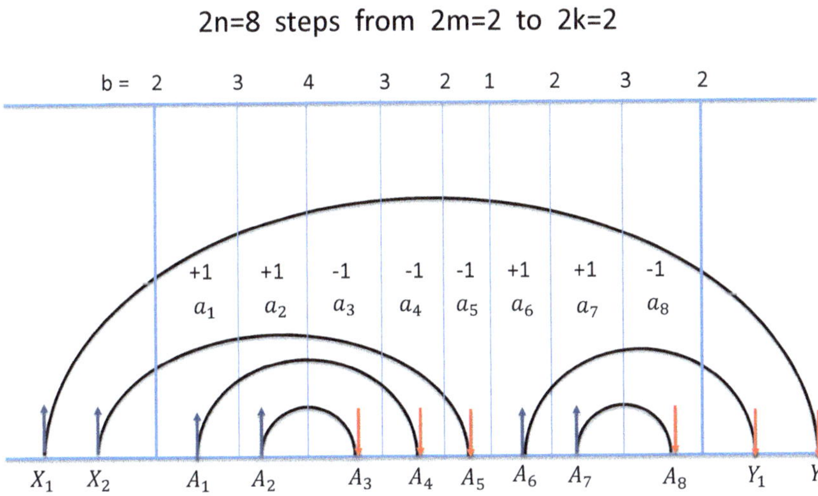

Figure 6.12: Semi-circles for extended Catalan numbers.

$\binom{2n+1}{n+1} - \binom{2n+1}{n+2}$, and, indeed (using the Pascal triangle relation):

$$C_{2n+2} = \binom{2n+2}{n+1} - \binom{2n+2}{n+2} = \left(\binom{2n+1}{n} + \binom{2n+1}{n+1}\right) - \left(\binom{2n+1}{n+1} + \binom{2n+1}{n+2}\right).$$

Problem 6.2.6 * *For $k \geq 1$ and $n \geq k$ such that $n-k$ is even, we have two types of recurrences:*

$$C_n^k = \overbrace{C_{n-1}^{k-1}C_0 + C_{n-3}^{k-1}C_2 + C_{n-5}^{k-1}C_4 + \cdots + C_{k+1}^{k-1}C_{n-k-2} + C_{k-1}^{k-1}C_{n-k}}^{(n-k)/2+1},$$

$$C_n^k = \overbrace{C_{n-k-2}C_k^k + C_{n-k-4}C_{k+2}^k + \cdots + C_2C_{n-4}^k + C_0C_{n-2}^k + C_{n-1}^{k-1}}^{(n-k)/2+1}.$$

Solution. Both justifications follow the proof of formula (6.2.1) presented in Fig. 6.8. Let us use the semi-circles instead of chords. In the second formula, A_1 can be connected with A_{n-k-2}, with A_{n-k-4} and so on until A_2. The first A here is A_{n-k-2}, because the last k points are needed to ensure the connections with $Y_1, \ldots, Y_k$.

The corresponding numbers of the configurations *inside* the semi-circle from A_1 to A_{n-k-2}, A_{n-k-4}, etc., are C_{n-k-2}, C_{n-k-4}, $\ldots$. They must be multiplied by C_k^k, C_{k+2}^k, and so on. Additionally, A_1 can be connected with

Y_k, the last Y, which gives C_{n-1}^{k-1}. Connections of A_1 with Y_i for $i < k$ are impossible, since they would result in intersections of chords.

In the first formula, Y_k can be connected with $A_{n-k}, A_{n-k-2}, \ldots, A_1$, and we arrive at the corresponding products. The number of the Y-points reduces to $k-1$ in this approach. ❑

Technically, both formulas work for $k=0$ if we set $C_n^{-1}=0$ and $C_{-1}^{-1}=1$. The second formula becomes exactly that for h_n from (6.2.1); the first formula is trivial for $k=0$: we obtain $C_n = C_n$.

The following relation is a direct variant of $C_{n+2} = C_n^2 - C_n$ for C_{n+2}^k with $k \geq 1$. The recurrence will be in terms of $C_n^{k \pm u}$ for $u = 0, 2$.

Problem 6.2.7 *For $n \geq 2, k \geq 1$ and even $n-k \geq 0$,*

$$C_n^1 = C_{n-2}^3 + 2C_{n-2}^1 \quad \text{for} \quad k=1, \text{ and } \; C_n^k = C_{n-2}^{k+2} + 2C_{n-2}^k + C_{n-2}^{k-2} \quad \text{for } k \geq 2.$$

Solution. Let us analyze the last two a_i in *good paths* for C_n^k. Such sequences $\{a_1, \ldots, a_n\}$ can be obtained as follows from $\{a_1, \ldots, a_{n-2}\}$ for the types "$0 \to k \pm u$", where $u = 0, 2$:

(a) from "$0 \to k$" by adding $a_{n-1}=+1$ and $a_n=-1$,
(b) from "$0 \to k$" by adding $-1, +1$, if $k \geq 1$, and
(c) from "$0 \to k+2$" by adding $-1, -1$ at the end,
(d) from "$0 \to k-2$" by adding $+1, +1$, for $k \geq 2$.

This gives the required recurrences. ❑

6.3 Applications

6.3.1 Playing roulette

There are two common types of roulette: American and European. The American wheel has 38 "pockets," including zero and double zero. We will only consider the simplest 50–50 bets: odd/even, red/black, or high/low. Zero and double zero are consider neither/nor: the house makes all its money on those bets when 0 or 00 comes up. So the house receives \$2 from every \$38, and the *house edge* is $2/38 = 1/19 \approx 5.26\%$. This may sound acceptable ... if you play once.

If you play once with \$100, the chances of winning \$100 are $9:19$, so it is $10:19$ to lose. If you play 5 times with \$20 each time, the average loss will be absolutely (!) the same. It is instructional to calculate the corresponding distribution of probabilities and confirm this (obvious) fact "mathematically."

Problem 6.3.1 *What is the probability of winning \$(20k) for $0 \le k \le 5$ with the American wheel if you play 5 times and your bet is \$20? Calculate the average casino's gain under this scenario.*

Solution. Assuming that you lose m times, where $m = 0, 1, 2, 3, 4, 5$, the corresponding gain is $\$20((5-m)-m)=\$20(5-2m)$, so it is negative for $m = 3, 4$, or 5 losses. The probability of gaining $\$20(5-2m)$ is $p^{5-m}q^m\binom{5}{m}$, where $p = \frac{9}{19}, q = \frac{10}{19}$ are the probabilities of winning or losing per 1 spin. Indeed, if the sequence of 5 spins is with m losses and $(5-m)$ wins, then we need to multiply the probability of each particular sequence of this type, which is $p^{5-m}q^m$, by their total number with m losses. Accordingly, the average profit for the casino will be \$20 times

$$G = \sum_{m=0}^{5} (5-2m)\, p^m q^{5-m} \binom{5}{m}.$$

As above, we count each event with its probability multiplied by the corresponding gain.

The calculation of G is an exercise on Newton's binomial and its derivative. We differentiate $(x+y)^n = \sum_{m=0}^{n} \binom{n}{m} x^m y^{n-m}$ with respect to x and multiply the result by x: $nx(x+y)^{n-1} = \sum_{m=0}^{n} \binom{n}{m} m x^m y^{n-m}$. Then we substitute: $n \mapsto 5, x \mapsto p,\ y \mapsto q$, and use that $p + q = 1$. Finally, $G = 5 - 2 \cdot 5p = 5(1-2p) = 5(q-p) = \frac{5}{19}$. This is what you can expect (statistically) to lose and the casino to gain (times \$20) under the strategy "playing 5 times with \$20." *The loss appears the same as under the strategy "playing once with \$100"*, which is obvious without any calculations! ❑

So statistically your *average* gain does not change if you split \$100 into portions and play with them separately. However, such a split provides protection against losing everything at once, if this is a consideration. The best protection of this kind is of course the strategy "$\$1 \times 100$": your loss will be very close then to \$100/19.

The following strategy is the "most aggressive." You begin with \$100 and your total bet is always everything you have at the moment; split it between many tables to make the loss close to $\frac{1}{19}$th of your bet. Let us assume that you play 19 times. Then you will have (statistically) after this game $(1-1/19)^{19} \approx 1/e \approx 0.368$ times \$100. So you will lose \$63.20 from \$100, which is very profitable for the casino.

This is the whole point of roulette: they charge you not too much for 1 spin, but 100 rounds (spins) of roulette take "only" about 3-6 hours, statistically. So they charge 5.26% for about 5 minutes of their service: quite a profit! The longer you stay at the table, the more you lose.

The European roulette is a bit "better": 37 pockets including (one) zero. The casino's edge is about $\frac{1}{37} \approx 2.7\%$, ... but Monte Carlo is more expensive than Las Vegas. This wheel is more common for online roulette games.

Many people come to casinos mostly for fun, not to play much. They are prepared to lose some money, though possibly no more than \$100 or so. For such "occasional players," the following two questions must be addressed.

(A) What are the chances of surviving $2n$ spins beginning with \$1?
(B) Assuming that they survived, what gain can be expected?

The 2nd question is mostly about measuring your personal "luck." If your actual gain is significantly (and constantly) greater than the average among those who survived, then consider becoming a professional gambler. Question (A) is of some practical importance to casinos. They do not really care about your luck ... with $q = 10/19$, but need to know your chances of surviving depending on what you are going to spend; they basically know how much. This is of course for "real roulette" (to be addressed later).

Ideal roulette. We begin with "ideal roulette", without 0 and 00; so it will be equivalent to tossing an ideal coin for us. For instance, let $2n = 200$. Any reasonable player would expect minuscule chances of surviving with so many rounds. Is this so?

Corollary 6.3.2 *Assume that someone comes with \$1 (1 chip) and the bet is always \$1 on "red/black" with zero house edge. Then the chances of surviving* $2n$ *rounds are* $P^0_{2n} = \binom{2n}{n}/2^{2n}$. *If they begin with an initial amount* $\$(1+2m)$ *(adding \$1 is a technicality) but with the same \$1-bet, then the chances are* $P^{2m}_{2n} = (\binom{2n}{n-m} + \cdots + \binom{2n}{n+m})/2^{2n}$.

Proof. We will employ formula (6.2.2): $C^{2m\to 2k}_{2n} = \binom{2n}{n+k-m} - \binom{2n}{n+k+m+1}$. We have: $P^{2m}_{2n} = C^{2m\to *}_{2n}/2^{2n}$, where $*$ means that we are supposed to consider the sum of these C-numbers over all possible outputs: $0, 2, 4, \ldots$. Then this sum, the total number of good paths, must be divided by the number of all paths, good and bad, which is 2^{2n}.

For $m = 0$, one has:

$$C^{0\to *}_{2n} = C^{0\to 0}_{2n} + C^{0\to 2}_{2n} + \cdots + C^{0\to 2n}_{2n} = \left(\binom{2n}{n} - \binom{2n}{n+1}\right)$$

$$+\left(\binom{2n}{n+1} - \binom{2n}{n+2}\right) + \cdots + \left(\binom{2n}{n+n} - \binom{2n}{n+n+1}\right) = \binom{2n}{n}.$$

Obviously, all terms but the first one will cancel each other, which is called "telescopic summation" in calculus. The last term is $\binom{2n}{n+n+1} = 0$. The case of any $2m$ is absolutely parallel: $C^{2m\to *}_{2n} = C^{2m\to 0}_{2n} + C^{2m\to 2}_{2n} + \cdots + C^{2m\to 2n}_{2n} =$

$\left(\binom{2n}{n-m} - \binom{2n}{n+m+1}\right) + \left(\binom{2n}{n-m+1} - \binom{2n}{n+m+2}\right) + \ldots = \binom{2n}{n-m} + \binom{2n}{n-m+1} + \binom{2n}{n+m}$, where the term $-\binom{2n}{n+m+1}$ from the first quantity cancels the first term in $\binom{2n}{n+m+1} - \binom{2n}{n+3m+2}$, and so on. ❑

Let us use the approximation $\binom{2n}{n}/2^{2n} \approx 1/\sqrt{\pi n}$ from (2.5.2): P^0_{200} appears to be about 5.6%, so one from 18 people will survive under 200 spins, much greater than anyone could imagine ... unless one has read this chapter. Even $P^0_{1000} \approx 2.5\%$ is far from zero.

The reason for this phenomenon is basically as follows. It is very risky, 50-50, in the beginning, since we have only one chip. However, if you are lucky enough to win 5 to 10 chips, then your chances of playing long enough significantly increase. This is of course very qualitative.

6.3.2 Conditional gains

Calculating your average gain seems significantly more involved. This is of course *conditional*, only among those who survived $2n$ rounds (starting with \$1). The average gain among all players is 0 for an ideal roulette: all such games are ***zero-sum games***.

Beyond roulette, if you run a startup, the chances of surviving is your first concern. However, if you survive, then the profitability is the key. If your gain is smaller than the average gain among those who survived, then you are good (you survived!), but probably not too good.

We will consider only $m = 0$, i.e. those playing with \$1 initially. The average gain in this group over $2n$ rounds will be denoted T_{2n}. The bet is always \$1 (one chip). Similar to the Pick-6 Lotto, we have:

$$T_{2n} = \Big(\sum_{k=0}^{n} (2k)\, C_{2n}^{0\to 2k}\Big) \,/\, C_{2n}^{0\to *}.$$

This is the sum of the gains $2k = 0, 2, 4, \ldots$ multiplied by the probabilities of the events "winning $2k$" for those who played $2n$ times. The corresponding probabilities are $C_{2n}^{0\to 2k}(2k)/C_{2n}^{0\to *}$. The denominator here is *not* 2^{2n}: only "$2n$-lucky" people (started the game with \$1) are counted.

This calculation is doable but somewhat involved algebraically. Fortunately, there is a bypass, based on the following general *principle*:

The ideal roulette is a zero-sum game: the total gain of the winners equals what the others lost. Equivalently, there can be no winning strategies.

The first claim is obvious (gains=losses). The second may not be that obvious, though they are *absolutely* equivalent. We provide an algebraic interpretation of this principle in Exercise 6.4.39 for the people with the initial amount \$(2m + 1) who play roulette $2n$ times. The total gain of those who survived, calculated via the probabilities, does coincide with the loss of those who did not, which is an interesting combinatorial identity. Though the principle is self-evident without any justifications.

Using this principle, consider *all* players who began the game with \$1 (technically, with $m = 0$) and played $2n$ times. We have: $A_{2n}^{0\to *} = 2^{2n} = C_{2n}^{0\to *} + B_{2n}^{0\to *}$, using the notation from Theorem 6.2.4; A and B stand for *all* and *bad*, and C for *good*. Since $C_{2n}^{0\to *} = \binom{2n}{n}$, one has: $B_{2n}^{0\to *} = 2^{2n} - \binom{2n}{n}$.

We of course know how much the B-people lost: $\$(B_{2n}^{0\to *})$, since everyone who did not survive $2n$ rounds lost \$1. Thus, by the very definition of T_{2n}:

$$\mathrm{Good}_{2n} \cdot T_{2n} = \mathrm{Bad}_{2n} \cdot 1, \quad \text{and}$$

$$T_{2n} = \frac{2^{2n} - \binom{2n}{n}}{\binom{2n}{n}} = \frac{2^{2n}}{\binom{2n}{n}} - 1 \approx \sqrt{\pi n} - 1. \tag{6.3.3}$$

Now, let us answer question (B) before Corollary 6.3.2: what gain can be expected assuming that you "managed" to survive $2n$ rounds?

Using a computer: $T_{200} \approx 16.7467$. The calculus approximation is 16.7245; so they coincide for practical purposes. Thus, your gain (for the \$1 you started with) can be expected to be about \$16.75. This is of course for the ideal version of roulette, without any house edge. Let us discuss real roulette.

6.3.3 Adding house edge

The formula for T_{2n} which we obtained is one of the most fundamental in the theory of random processes. Before considering playing *real* roulette, let us comment on a different aspect of (the same) T.

Low altitude targets. One of the optimization problems with anti-aircraft defense systems is to determine which *missile launch facilities* are positioned the best for the interception of a *target*.

Let us begin with modeling the target's moves (an aircraft or another missile). Its safest strategy is almost always to fly as close as possible to the ground: low-altitude targets are difficult to detect and intercept. Then, fast changes in direction and altitude by the target, especially when such maneuvers are close to the ground, are optimal. They force the guided missiles attacking the target to attempt to follow these moves. The focus is on the changes of altitude. There is some special terminology here: "rolling scissors" (up-down-up-down and so on). If this happens close to the ground, the missiles can quickly lose altitude trying to respond to such moves and, eventually, crash into the ground.

With guided missiles, delays with navigation and the time necessary for changing the course make this scenario, crashing into the ground, quite likely.

This is for one missile; if several of them attack the target simultaneously, there is not much the target can do.

Thus, to model the missile paths, we assume that it experiences almost random (for an observer) moves up and down due to the course corrections. These moves respond to the maneuvers of the target: the more random they are the better for the target. This makes these paths quite similar to our random walks; the randomness of this kind is here "by design." The *bad paths* represent the missiles that crashed. Accordingly, T_{2n} calculated above can be interpreted as the average altitude of the randomly-moving missiles (up by 1, down by 1 and so on randomly) among those that did not crash before $2n$ "steps". The "steps" are up-down corrections of the course, assuming that they are by ± 1 and uniformly distributed. Let us comment on it.

The most likely output after $2n$ steps of the random walks without any restrictions concerning the ground (high altitudes) and with equally probable moves up and down is zero. Considering the effect of the ground (a lake, a concrete wall, the number of chips you have), the average naturally increases. It must be counted only among the "good" trajectories and now depends on $2n$. In the considered example, this is the *optimal altitude for the interception* from a missile launch facility that is $2n$ steps away from the target.

It is clear in this example that some statistical setting is necessary. The probabilities, p and q, for moving up and down are generally different: the slippage is mostly downward. Practically, we can assume that the commands "move up" and "move down" are equally probable, but some former commands are not executed. This is similar to the *real roulette*: our next topic.

American wheel. Let us assume that the probabilities of moving up and down for the paths of type "$0 \to 2k$" are p and q. For ideal roulette: $p = 1/2 = q$. For the American wheel, the roulette we will consider: $p = \frac{9}{19}, q = \frac{10}{19}$. As we have already discussed, our chances of winning (upon betting on "red") are $18/38 = 9/19$, since only 36 pockets are available for the players from 38; the house possesses zero and double zero. For the European wheel, they are $p = 18/37, q = 19/37$.

Corollary 6.3.3 *With the American roulette, the chances of surviving during $2n$ rounds, starting with \$1 and playing each round (spin) with \$1 (1 chip), are*

$$\mathcal{P}_{2n} = (pq)^n \sum_{k=0}^{n} \left(\frac{p}{q}\right)^k \left(\binom{2n}{n+k} - \binom{2n}{n+k+1} \right).$$

The average gain $\mathcal{T}_{2n}$ among those who survived is

$$\mathcal{T}_{2n} = (pq)^n \sum_{k=0}^{n} 2k \left(\frac{p}{q}\right)^k \left(\binom{2n}{n+k} - \binom{2n}{n+k+1} \right) / \mathcal{P}_{2n}.$$

Also, $\mathcal{S}_{2n} = 1 - \mathcal{P}_{2n}(1 + \mathcal{T}_{2n})$ *is the casino's gain from an average person determined to play* $2n$ *times, including those in this category who lost before* $2n$ *(all have \$1 to begin with). Upon multiplication by* 100*, it is the casino's profitability in the category of such players, which depends on* $2n$*.*

Proof. The calculation is parallel to that without p and q. If $p = 1/2 = q$, we obtain the old formulas.

If the players play once, then the probability $\mathcal{P}$ of surviving is simply p, and the average gain among lucky players is $\mathcal{T} = \$1$. So the casino earns $\frac{1}{19}$ of every dollar, which is the house edge. The formula for $\mathcal{S}$ gives the same: $\mathcal{S} = 1 - \mathcal{P}(1 + \mathcal{T}) = 1 - 2\frac{9}{19}$. If $n = 1$, i.e. they play twice, then $\mathcal{P} = p$ and $\mathcal{T} = 2p^2/p = 2p$ (why?); this matches the formula above for $\mathcal{T}$ when $n = 1$. Thus, $\mathcal{S} = 1 - p(1 + 2p) \approx 7.8\%$ for this category of players.

Generally, the average gain is the sum $\sum_g g p_g$, where p_g is the probability of receiving g. Under the strategy "$2n$ *rounds unless you are out of money before* $2n$," you "invest" \$1 (what you have in the beginning) and *your* average "return" per \$1 will be $\mathcal{P}_{2n}\mathcal{T}_{2n} - (1 - \mathcal{P}_{2n}) = \mathcal{P}_{2n}(1 + \mathcal{T}_{2n}) - 1 < 0$. Accordingly, the return for the casino in this category of players will be $1 - \mathcal{P}_{2n}(1 + \mathcal{T}_{2n}) > 0$. We performed similar calculations for Pick-6 Lotto in Section 2.4.3. ❑

We provide below the table of real and ideal data for $2n = 200$ and $2n = 20$. Surprisingly, the chances of surviving the real roulette with the initial amount \$1 (and the bet=\$1) over 200 rounds are "still" about 2%, and they become significantly more acceptable, about 13.5%, for $2n = 20$, which is probably a reasonable number of rounds for those who just want to spend "several hours" in a casino.

rounds	wheel	chances	gain	house
200	real	1.96855%	\$12.8543	72.73%
200	ideal	5.63485%	\$16.7467	00.00%
20	real	13.5561%	\$4.23146	29.08%
20	ideal	17.6197%	\$4.67546	00.00%

.

The house will take about 30% of your money after 20 spins under the strategy "always play with \$1." By contrast, you will lose $1 - (1 - \frac{1}{19})^{20} \approx \frac{2}{3}$ of your money if you "invest" everything you won before each time. You can play at 1 table or at several tables, which does not matter. The greater the number of tables (with the same starting amount) the closer your loss ($\approx 2/3$) to the mathematical average. The strategy to play *always* with \$1 appeared reasonably defensive, though it of course provides no real protection against the house edge. You reduce a lot what you pay to the house under this strategy, ...but your profit will be smaller too. It will be around \$4.23 per \$1 if you survive 20 rounds (13.5% do survive).

Problem 6.3.4 *Assume that the chances of winning and losing are p and $q = 1-p \geq p$ and we begin from $2m = 0$ (practically, with \$1). Calculate the probability p_{2n+2} that we will be back to 0* for the first time *at $2n+2$ for $n \geq 0$. Using Problem 6.2.1, calculate that the probability P of touching zero again at any $2n$ (then we stop) is $\frac{1-|q-p|}{2} = p$. Here $p = \frac{9}{19}$: the wheel is American.*

Solution. The 1st move must be up in the graphic interpretation, and then we need to stay at this level or beyond it until the $(2n+1)$th step, when we must be at the level 1. Then we move down to 0. The number of such paths is $C_{2n} = \frac{1}{n+1}\binom{2n}{n}$ (not C_{2n+2}). So $p_{2n+2} = \frac{1}{n+1}\binom{2n}{n}(pq)^{n+1}$. The total probability is $P = \sum_{n=0}^{\infty} \frac{1}{n+1}\binom{2n}{n}(pq)^{n+1}$. We can use $g(z)$ for $z = pq$ for $g(z) = z(\sum_{n=0}^{\infty} C_{2n}z^n)$ from Problem 6.2.1. Using the relation $g^2 = g - z$ there, we obtain that $P^2 = P - z$ and $P = \frac{1 \pm \sqrt{1-4z}}{2}$. The plus-sign here is irrelevant, since the probability cannot be greater than 1. Thus: $2P = 1-\sqrt{1-4z} = 1-\sqrt{1-4p(1-p)} = 1-|1-2p| = 1-|q-p| = 2p$. See also (4.4.21).

The coincidence $P = p$ is actually obvious without any calculations. Let us explain this. The chances that the first step is up are p. Otherwise, all remaining paths appear below the starting level and therefore will all be excluded. If $p < q = 1-p$ and you are at 1 or at any given level, then you will eventually become below zero with a probability 1 (the casino will take all your money if you play long enough because $q > p$), i.e. you will touch 0 somewhere. This definitely proves that $P = p$. ❑

A more rigorous and exact analysis is based on Hoeffding's inequality. Let us consider a random walk with N steps with no restrictions concerning ups and downs. The probability of going below the starting level somewhere before $N = 2m+1$ is *at least*

$$Q = \sum_{k=0}^{m} \binom{N}{k} p^k q^{N-k} = 1 - \sum_{k=m+1}^{N} \binom{N}{k} p^k q^{N-k} = 1 - \sum_{k=0}^{m} \binom{N}{k} q^k p^{N-k},$$

where we use that $(p+q)^N = 1$ and replace k by $N-k$ in the last transformation. Indeed, all the paths counted here go below the starting level, but there can of course be those that go below and then go above this level. The latter are not counted, so we only have an inequality: "at least."

Let $m = (q-\varepsilon)N$ for $\varepsilon > 0$; recall that $q > 1/2$. Then $Q \geq 1 - \exp(-2\varepsilon^2 N)$, which is *Hoeffding's inequality.* It "exponentially" tends to 1 as $N \to \infty$. This immediately results in $P = p$.

The case $p = 1/2 = q$ is exceptional in this approach. To address it, we take a limit $p \to 1/2$. When p is sufficiently close to $1/2$, we obtain that P is greater than $1/2 - \delta$ for any $\delta > 0$. However, $2P \leq 1$, because $2P$ is the probability for the path to return to zero from above or from below for $p = q$. So $P = p = 1/2$ in this case too. ❑

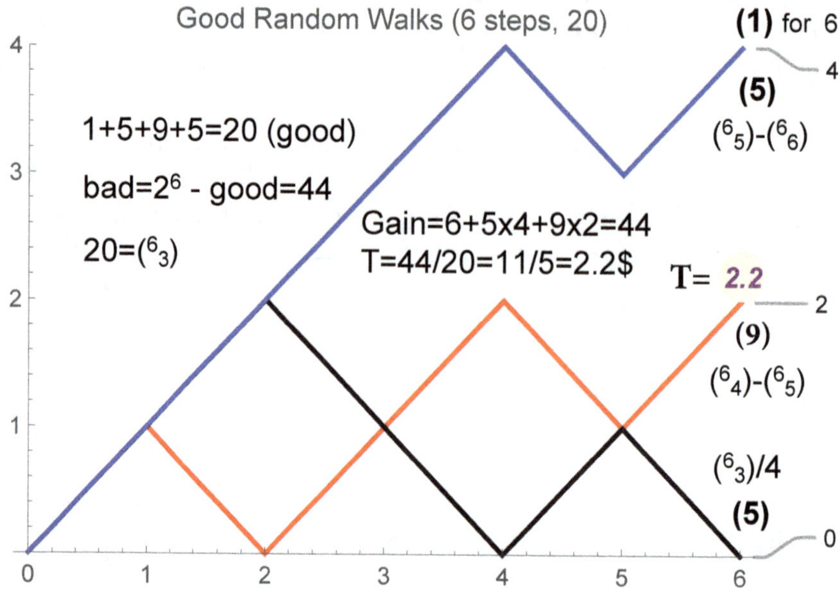

Figure 6.13: Good paths and the target.

6.3.4 Pricing options *

We will begin with an overview of the formulas above for $2n = 6$. Let us go back to $p = 1/2 = q$ and summarize in Fig. 6.13 everything we obtained for $2n = 6$ steps. Here the numbers of *good* paths can be computed directly, but the formulas help even in such a small case. The endpoints can be $2k = 0, 2, 4, 6$; the corresponding number of good paths is $5, 9, 5, 1$. The most frequent is 2. Using formula (6.3.3), the average of $2k$ over good paths is $T = 2.2$, naturally not far from $2k = 2$.

This T is the *average gain* among those who survived 6 spins of ideal roulette with \$1, or the optimal *interception altitude* if the missile facility is 6 points away from the target. Here we start with \$1 and make the bet always \$1 under the roulette interpretation. Under the 2nd interpretation, let this distance be 6 miles (with 6 navigation points) and the missile goes up or down by 0.5 miles between the navigation points with equal probabilities. Then this altitude is $1.1 = 2.2 * 0.5$ miles.

The total number of *good paths* is $1+5+9+5=20$. Here $1,5$ are for $2k=6,4$, which is obvious without any formulas. Indeed, there is only one path from 0 to 6, which is "up" six times, and there are 5 possibilities for one step-down from this path to obtain $2k=4$. The number of good paths equals $\binom{2n}{n}=\binom{6}{3}$, as we know; this is 20 indeed.

The number of *bad paths* is $2^{2n}-\binom{2n}{n}=64-20=44$. As always, we need to decide how to count them: until the first -1 or until the end. We know that only the latter is correct here! Using the roulette interpretation, good paths win $1\cdot 6+5\cdot 4+9\cdot 2+5\cdot 0=\$44$, which is exactly what bad paths lose. Thus, ideal roulette is a *zero-sum game* if the stake is always \$1 and you have only \$1 (i.e. $2m=0$). It is of course a zero-sum game under *any* strategy.

This is fully applicable to playing stock markets: there can be no winning strategy, but you follow certain "discipline" adjusted to your limitations. This is not a zero-sum game. You pay broker commissions, though the main expense in professional trading is the *spread*, the difference between "bids" and "asks". The spread can be relatively small for heavily-traded stocks. It is much smaller than 5.25% for the American wheel, but the trade volumes are huge for the most popular stocks or indices and the trades are very frequent.

Let us now discuss *pricing derivatives*, assuming that stock market is a zero-sum game (no charges for trades). This will be very basic, mostly to demonstrate the universality of Catalan numbers.

Generally, you can post any bet, not only of that financial nature: "a certain event will happen and when." This becomes then a *derivative*, which people can buy. If this event happens, the buyers of your derivative will receive from you some fixed amount per 1 derivative (they can buy many). Obviously, you must charge them when they buy: the *price of the derivative* is the percentage of the amount they purchase to compensate what you may pay if the event happens. This is of course always about probabilities. The more likely the event, the greater the price of the derivative: if the probability is 100% then the price is 100%. Abusing the terminology, we will call derivatives *options*; the real options are derivatives of a specific kind. The following example is instructional, but not too far from real option pricing models.

Pricing options. Let us discuss the following situation. Assuming that the oil price is \$40 per barrel now, your company decides to be protected against its increase to \$50 or beyond, which can really harm your business. So you need to find someone, who will take *your risks instead of you* ... for some reward.

Thus, you need a gambler to manage this, unless you want to become a gambler yourself. So these people, with their specific skills, appear necessary for this and many similar tasks, including playing stock markets.

The gentlemen who played with you the game "14 vs. 1" in the airplane (Section 2.5.3) may be the right person. Indeed, he agreed to do this for 5%, which does not seem too much for peace of mind (he says). What is the actual cost of this kind of protection?

The standard American option is that some amount, say \$1M, will be paid to you by the option dealer if some event occurs any time within the next 3 months. In this example, the event is "oil price reaches \$50."

The only thing needed to start the calculation of the price of this option (derivative, to be precise) is the average change of the oil price per day; let us "make" it $\pm\$0.5$. With commodities, this uniformity assumption is not that unreasonable. This can be different for stocks and other instruments: the distribution of the changes of the share-prices is generally normal.

Let us present this situation graphically: the x axis will be for time measured in days, and the y-axis for the oil price in dollars.

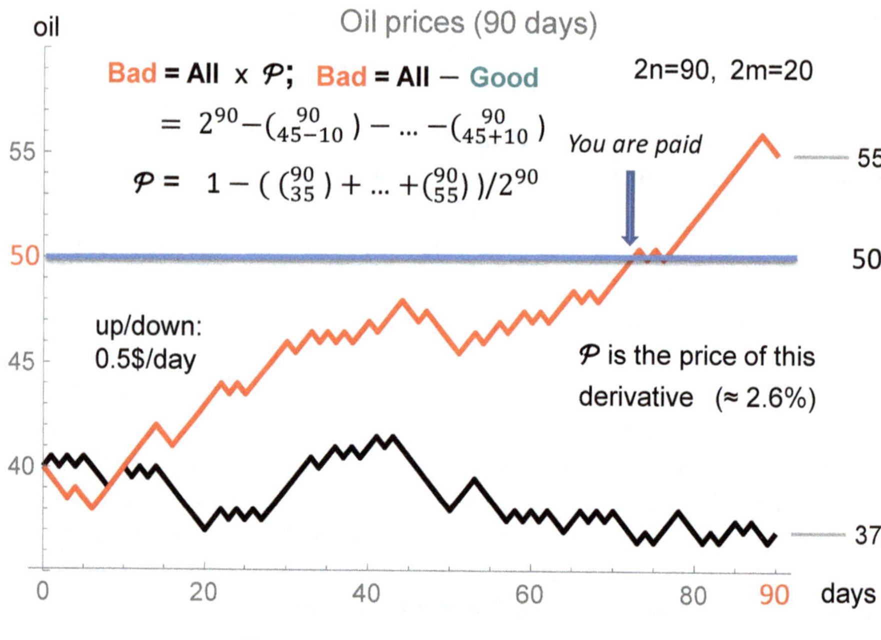

Figure 6.14: Pricing derivatives.

Fig. 6.14 is actually the same one we used for extended Catalan numbers, but now "the lake" is *the ceiling* ("limit" in the figure) at the level \$50; the starting level is \$10 lower. Since the fluctuations are ± 0.5, the number of steps to the ceiling is $2m = 20$. So $m = 10, n = 45$, and we can use the extended Catalan numbers.

For the price of this option, denoted by $\mathcal{P}_{2n}^{2m}$, one has All $\cdot\,\mathcal{P}_{2n}^{2m} = \text{Bad} \times 1$. Here 1 is the unit; for instance, $1 = \$1\text{M}$. Using Corollary 6.3.2:

$$\mathcal{P}_{2n}^{2m} = 1 - P_{2n}^{2m} = \frac{2^{2n} - C_{2n}^{2m\to *}}{2^{2n}} = 1 - \frac{\binom{2n}{n-m} + \cdots + \binom{2n}{n+m}}{2^{2n}}. \tag{6.3.4}$$

Finally: $\mathcal{P}_{2n}^{2m} = 0.0263\ldots$ for $2n = 90$ and $2m = 20$; use your computer. We obtain that your company must pay 2.63% for this kind of protection; this is the percent of the amount it wants to be paid off if the price of oil reaches \$50 from \$40 any time during 3 months. Obviously, you will need to pay for the dealer's service, but 5% requested by your "old friend" is obviously excessive, even worse (proportionally) than the bet 5 : 1 you agreed upon while playing "14 vs. 1" with him. Recall that the fair bet turned out to be 3.7 : 1. What else can you expect from the gentleman in Fig. 6.15? However, now this is about real risks and real money: you absolutely need to check and double-check this!

Figure 6.15: Your old friend ("14 vs. 1").

Discussion. We use here Theorem 6.2.4. Employing the calculus approximation from (2.5.2):

$$\mathcal{P}_{2n}^{2m} \approx 1 - \int_{-m/\sqrt{n}}^{m/\sqrt{n}} \frac{e^{-x^2}}{\sqrt{\pi}} dx, \tag{6.3.5}$$

but this is *very* approximate for small n.

The cheapest option is when $m \geq n$: $\mathcal{P}_{2n}^{2n} = 0$. In this case, the price of oil simply cannot cross the "ceiling" (when your company is paid). The most expensive one is when $m = 0$. Then $\mathcal{P}_{2n}^{0} = 1 - \binom{2n}{n}/2^{2n} \approx 1 - \frac{1}{\sqrt{\pi n}}$. It cannot be 1 because the oil price can go down and never increase above \$40 for 90 days. The price of this option will be about 92% (check).

We note that if one trades 2 times a day and the change of oil price is assumed $\pm\$0.25$ per trading point ($0.25 = 0.5/2$), then the price of the corresponding option drops about 10 times. It will be about 0.22%; here we use the formula for $n = 90$ and $m = 20$. Such a dramatic decrease is actually surprising! One can expect a similar options price because $40 \cdot 0.25 = 20 \cdot 0.5$, but this is very wrong: the binomial and normal distributions are very non-linear and such "linear approximation" is not applicable. See Exercise 6.4.47.

The black curve provided in Fig. 6.14 is really random (coin flipping). The red curve, the one intersecting the ceiling, was actually produced under the chances $2:1$ for up : down. It simply takes too long to obtain the intersection under $1:1$, so we accelerated the process. The probability of this intersection for equal chances to go up or down is the same $\mathcal{P}^{2m}_{2n}$; this is an equivalent definition of the option price. Thus, only 2.63 curves from 100 will cross the ceiling for $n = 45, m = 10$,

Exercise 6.4.40 provides the formula for the probability of the event "crossing the ceiling," where the probabilities of moving up and down are arbitrary, p and q. As an exercise, make $p = 1/3$ and $q = 2/3$. i.e. "2:1", to see that the probability of this intersection is about 88% for $n = 45$, $m = 10$. Thus only 1 from about 10 will stay below \$50 for such paths.

The standard assumption is that the distribution of daily price changes is *normal*. Equivalently, the assumption is that it is *uniform* but for some smaller interval of time, for instance for 1 hour instead of 1 day. The equivalence is due to the Stirling approximation.

Generally, uniform distributions are for any number of outputs with equal probabilities. This is rolling dice instead of tossing a coin; see Section 3.5.3. However, $\pm u$ with equal probabilities is sufficient for modeling. One simply needs to diminish the time-unit, for instance switch to 10 min instead of 1 hour. Here u must be properly adjusted: see Exercise 6.4.47. *This is basically* **Central Limit Theorem**, *where combinatorics ends and statistics begins.*

This chapter concludes the core themes of this book. The challenge was making it essentially sufficient for a motivated reader with minimal mathematical experience to study introductory combinatorics, modeling and number theory almost independently of any other sources and creating a base for advanced levels within these fields and their various applications.

6.4 Exercises

Exercise 6.4.1 $^{\heartsuit}$ *Show that in any (finite)* simple graph *(by definition, without multiple edges and ↻) with at least two vertices have some coinciding degrees.*

Exercise 6.4.2$^{\heartsuit}$ *Show that if all degrees of a graph are greater than 1, then it has a* cycle: *a reentrant walk with no repeated vertices. If all degrees are even, then any* path *(a walk with no repeated vertices) can be extended to a cycle.*

Exercise 6.4.3$^{\diamondsuit}$ *A square in graph theory is a quadrilateral. Assume that a connected simple graph G does not contain any* ***induced subgraphs*** *that are squares or paths with 4 vertices. Induced subgraphs are subsets of the set of vertices of G with the edges between their vertices that are those from G (all). Prove that there exists a* ***center*** *in G: a vertex adjacent to any other vertex.*

Hint: Remove one vertex v without destroying the connectivity, consider the corresponding induced subgraph G', and take its center c' (induction). If c' is not a center in G, let v–d–c' be a 2-path. If G' is complete, then d is a center of G. Otherwise, take d' in G' not adjacent to d and consider $v-d-c'-d'$.

Exercise 6.4.4$^{\heartsuit}$ *For a simple connected graph G, check that if any 3 of its vertices have 0, 1 or 3 edges in total between them (no 2!), then it must be a complete graph K_n, where n is the order of G.*

Hint: Use that any *induced subgraph* defined in Exercise 6.4.3 has the same property, and proceed by induction with respect to the order of G.

Exercise 6.4.5$^{\heartsuit}$ *Let G_1 and G_2 be subgraphs in a complete graph K_n, and F_1 and F_2 be obtained by removing their edges (not their vertices) from K_n. Prove that if G_1, G_2 are isomorphic graphs then so are F_1, F_2. For instance, removing any m* disconnected *edges from K_n results in the same graph up to isomorphisms, as well as removing all edges of any* path *in K_n (a walk without loops) of length m.*

Hint: Extend the isomorphism between $G_{1,2}$ to an automorphism of K_n, and then restrict it to $F_{1,2}$.

Exercise 6.4.6$^{\diamondsuit}$ *The* convex hull *of any set of points in the plane is the smallest convex polygon containing these points. By considering the convex hull of 5 points where any 3 do not belong to the same line, prove that one can always find 4 of them such that the corresponding quadrilateral is convex and the 5th point is outside of it (this problem is due to Esther Klein).*

Hint: The convex hull can be either a triangle or a quadrilateral.

Exercise 6.4.7$^{\diamondsuit}$ *In a simple connected graph G (without multiple edges and $\circlearrowleft$), let H_1 and H_2 be two paths (without cycles) and of the maximal length possible in G (both). Then they have at least one vertex in common.*

Hint: If H_1 and H_2 do not intersect then connect them by any shortest possible path and use it to construct a longer path in G (without cycles).

Exercise 6.4.8$^\heartsuit$ *A simple graph G (without multiple edges and ↻) is called* ***bipartite*** *if its vertices are a disjoint union $U \cup V$ of two sets U and V, and the edges are allowed only between the vertices of U and those of V. Prove that G is bipartite if and only if every closed walk is of even length.*

Hint: One can assume that G is connected. The "only if" part is obvious. If any closed walk in G is of even length, then fix a vertex v and define U, V to be the sets of vertices at even and odd distances from v. The distance between v and w is the smallest length of the paths between them. Then the existence of an edge between any vertices u and w in U (or in V) is impossible.

Exercise 6.4.9$^\heartsuit$ *Let G be a* bipartite *graph defined in Exercise 6.4.8. Prove that such G is a union of its Hamiltonian cycles if and only if $|U| = |V|$.*

Hint: Any path in G alternates between U and V. So any cycle (without repetitions of the vertices) has an equal number of vertices from U and from V. By removing such a cycle, the remaining graph G' remains bipartite, and $|U'| = |V'|$ if and only if $|U| = |V|$. Continue by induction.

Exercise 6.4.10$^\diamondsuit$ *In graph theory, the* **mesh** *$M_{m,n}$ is the set of points (i, j), where $1 \le i \le m, 1 \le j \le n$, taken as the vertices, which are connected by an edge when and only when the corresponding distance is 1. Check that it is* bipartite *and follow Exercise 6.4.9 to prove that it has a* Hamiltonian cycle *if and only if mn is even and $m, n > 1$.*

Hint: Fig. 1.7 (Gomory's Theorem) provides a Hamiltonian cycle if $m > 1$ and n is even: plot it in $M_{m,n}$ avoiding the 2 forks there. Let V_ϵ be a set of vertices $\{(i, j)\}$ such that the parity of $(i + j)$ is $\epsilon = 0, 1$. The vertices can be connected only if they are in different V_ϵ, which proves that $M_{m,n}$ is a bipartite graph. The existence of a Hamiltonian cycle gives that $|V_0| = |V_1|$. However, $|V_0|$ =Ceiling$[mn/2]$, so mn must be even to ensure that $|V_0| = |V_1|$.

Exercise 6.4.11$^\spadesuit$ *Consider the N-puzzle for $N = mn–1$ in the $m \times n$ frame (m is the number of rows) with N tiles labeled by $1, 2, \ldots, N$. Label the squares of this frame by $\{1, 2, \ldots, N, \emptyset\}$ naturally, left-to-right and down the frame. Initially, place the tiles correspondingly. The moves are as in the 15-puzzle from Section 2.3.1. Follow the steps below to prove that any permutation from*

$\mathbf{A}_N$ *(with $\emptyset$ added at its end) can be obtained by such moves from the initial configuration, i.e. is* solvable, *if n is even and $m > 1$. Do the following.*

(a) *Check that the 3-cycle $(N, N-n+1, N-n)$ in $\mathbf{A}_N$ can be obtained this way.* (b) *Apply Exercise 6.4.10 to construct a Hamiltonian cycle HC in $M_{n,m}$ after removing the last 2 boxes in the last row (N and $\emptyset$).* (c) *This HC results in a solvable cycle of length $N-2$ in $\mathbf{A}_N$ preserving $N-n+1, N-n$.* (d) *Then follow Problem 2.3.7. [A variant of this method works for odd $m, n > 1$. The solvability for $m, n > 1$ and even permutations with $\emptyset$ at the end is due to W. Johnston and W. Story (1879); R. Wilson generalized it to any graph [42].]*

Exercise 6.4.12 $^\diamondsuit$ *(i) Let G be a simple graph of order n: with n vertices and no multiple edges (including $\circlearrowleft$). If G has 2 connected components, then the number of its edges is no greater than $\frac{(n-1)(n-2)}{2}$. Also, among such G, the latter number occurs only for a union of the complete graph with $(n-1)$ vertices, K_{n-1}, and one isolated vertex.*

(ii) Prove by induction that if the number of edges in a simple graph G is greater than $\frac{(n-m)(n-m-1)}{2}$ for $m \geq 1$, then the number of its connected components cannot be greater than m.

Exercise 6.4.13 $^\clubsuit$ *For a simple graph of order n (with n vertices), let $d_1 \geq d_2 \geq d_3 \geq \cdots \geq d_n$ be a sequence of its degrees for the vertices $v_1, v_2, \cdots, v_n$. Prove that $\sum_{i=1}^{m} d_i \leq m(m-1) + \sum_{i=m+1}^{n} \min\{m, d_i\}$ for any $1 \leq m < n$. [Actually, these inequalities are sufficient for $\{d_i\}$ to be a sequence of degrees for some graph due to P. Erdős and T. Gallai.]*

Hint: The left-hand side of this inequality counts the number of pairs $\{v, e\}$, where e is an edge from any vertex v to the group of the first m vertices. These pairs can be for v within this m-group or not. The number of edges of the first type is no greater than $m(m-1)/2$ and they are counted twice; we obtain $m(m-1)$. The edges of the second type are to v_i for $i > m$; so no greater than m of such edges are possible (there are only m vertices in this m-group), and this number is obviously no greater than d_i.

Exercise 6.4.14 $^\heartsuit$ *A country with n citizens decided to live according to the rule "the enemy of my enemy is my friend." If A is an enemy of B then B is an enemy of A (this is a symmetric relation), and A is a friend of A.*

(i) Prove that this rule cannot be followed any longer for such a country after it reaches the stage when each citizen has more than $n/2$ enemies.

[42] A. Archer, *A modern treatment of the 15 puzzle*, `https://www.cs.cmu.edu/afs/cs/academic/class/15859-f01/www/notes/15-puzzle.pdf`.

(ii) Find an example when every citizen has exactly $n/2$ enemies for even n but the rule still holds. Check that there is only one such example for any even n up to graph isomorphisms.

(iii) It suffices to assume in (i) that the total *number of (unordered) pairs of enemies is greater than $n^2/4$ for any n, even or odd. This is* Turán's Theorem*, so try to prove it.*

Exercise 6.4.15$^\heartsuit$ *A maximal* **independent set** *in a connected simple graph (without multiple edges and ↺) is a maximal subset of its vertices where any two are not connected by an edge.*

(i) Prove that the number of its vertices is no smaller than $N/(D+1)$, where N is the total number of vertices and D is the maximal degree in this graph, which estimate can be reached for complete graphs.

(ii) Check that (i) gives that the maximal number of pan-nonattacking queens on the $n\times n$ chessboard (nonattacking vertically, horizontally and in any pandiagonals and anti-pandiagonals) is no smaller than $n^2/(4n-2)$ for odd n and no smaller than $n^2/(4n-3)$ for even n.

(iii) Thus, at least 2 pan-nonattacking queens can be placed on the 4×4 chessboard; check that it is no greater than 2.

Hint for (ii): Any queen attacks the same number of squares due to the cyclic translations. Namely, $4n-3$ for odd n and $4n-4$ for even n.

Exercise 6.4.16$^\diamondsuit$ *(i) Color the edges of the complete graph K_6 with two colors, black and white. Check that there exists its* induced subgraph *K_3 with all black edges, or such a subgraph with all white edges. Construct an example of the coloring of K_5 when such K_3 does not exist.*

(ii) For any positive integers r, s let the **Ramsey number** *$R(r,s)$ be the minimal number n such that K_n with* any *configuration of black and white edges* either *has white-only induced subgraph K_r or black-only K_s. Prove by induction on $r+s$ that $R(r,s)\le R(r-1,s)+R(r,s-1)$ for $r\ge 2$.*

(iii) Use (ii) and $R(2,s)=s$ (justify this) to check by induction that $R(r,s)\le\binom{r+s-2}{r-1}$. This results in $R(3,3)=6$, which matches (i).

Hints: (i) Any vertex has at least 3 edges from it of the same color.
(ii) Given a vertex v in K_n with $n=R(r-1,s)+R(r,s-1)$, let M be the number of black edges from v and N the number of white ones; so $n=M+N+1$ and either $M\ge R(r-1,s)$ or $N\ge R(r,s-1)$; continue by induction.

Exercise 6.4.17$^\heartsuit$ *For a simple connected graph G of order n, the* **chromatic number** *$\chi(G)$ is the minimal number of colors necessary to color its vertices* *in such a way that any two adjacent vertices are colored differently. Check that it is n only for complete graphs, and that $Ceiling[n/m] \le \chi(G) \le n$, where m is the largest size of* independent sets *from Exercise 6.4.15, i.e. the greatest number of vertices where no 2 are adjacent.*

Hint: Let c_i be the number of vertices colored by color i satisfying the required condition for $1 \le i \le k$ for some k. Then $c_i \le m$ and $n = \sum_{i=1}^{k} c_i \le k \cdot m$.

Exercise 6.4.18 ◇ *For a chromatic number $\chi(G)$ of a simple connected graph G defined in Exercise 6.4.17, prove that $\chi(G) \le d_{max} + 1$ for the maximal degree d_{max} of G (among all its vertices). For this, order the vertices, $v_1, v_2, \ldots, v_n$, color v_1 by color 1, and then always use the smallest possible color number for v_i that is not present in the vertices among those in $v_1, v_2, \ldots, v_{i-1}$ not adjacent to v_i. This procedure is called a* ***greedy algorithm****.*

Hint: There are no greater than d_{max} vertices adjacent to any v_i among $v_1, v_2, \ldots, v_{i_1}$. Thus, at least one number in the sequence $1, 2, \ldots, d_{max} + 1$ has not been used before for coloring, and the minimal such number will be taken for v_i in this algorithm. [Actually, $\chi(G) \le d_{max}$ unless for complete graphs and cycles with an odd number of vertices (R.L. Brooks)].

Exercise 6.4.19 ♡ *A tree is a connected graph without cycles. Count the number of non-isomorphic trees with 6 vertices.*

Hint: Without vertices of degree ≥ 3, it is a "road". With only one vertex of degree ≥ 3, there are 3 possibilities, and one with 2 vertices of degree 3.

Exercise 6.4.20 ◇ *Let the set of vertices V of a simple connected graph G be a disjoint union of subsets V_i, and G_i $(1 \le i \le k)$ be the corresponding* induced subgraphs *defined in Exercise 6.4.3. Assume that G_i have Hamiltonian cycles H_i. We connect H_i with H_j by one edge for every parallel pair of edges in H_i and H_j correspondingly. By parallel edges, we mean opposite edges of some "squares" (4-cycles) in G. Let $\mathcal{G}$ be the resulting graph, and $\mathcal{G}'$ its reduction to a simple graph: replacing multiple edges between H_i by one edge.*

(i) If $\mathcal{G}$ is connected, take in it any ***spanning tree*** *$\mathcal{T}$, a connected subgraph without cycles containing all of its vertices. In particular, only one edge between H_i and H_j for $i \ne j$ is taken in $\mathcal{T}$. Given $\mathcal{T}$, check that there is a unique way of constructing a Hamiltonian cycle in G by replacing the pairs of parallel edges between H_i and H_j associated with the edges of $\mathcal{T}$ with the complementary pair of parallel edges in the corresponding "squares" of G.*

(ii) Exercise 6.4.11 states that the mesh graph *$G = M_{m,n}$ for even n and $m > 1$ has Hamiltonian cycles; its vertices are the boxes of the $m \times n$ board, and the edges are (only) between neighboring boxes.*

Let $\mathbf{m}=\sum_i m_i$, $\mathbf{n}=\sum_j n_j$, *where* n_j *are supposed even and* $m_i>1$. *Partition* $M_{\mathbf{m},\mathbf{n}}$ *by the corresponding vertical and horizontal lines into* N *subgraphs* $G_{ij} = M_{m_i,n_j}$, *and take one Hamiltonian cycle* H_{ij} *in every* G_{ij}. *We define a graph* $\mathcal{G}$ *with* H_{ij} *as its vertices, which are connected by the edges as in (i).*

Let us fix a spanning tree $\mathcal{T}'$ *in* $\mathcal{G}'$. *Check that any neighboring* H_{ij} *are connected by at least* 2 *edges, and the trees* $\mathcal{T}$ *mapped onto* $\mathcal{T}'$ *in* $\mathcal{G}'$ *result in at least* 2^{N-1} *different Hamiltonian cycles in* $M_{\mathbf{m},\mathbf{n}}$.

Hint for (i): Begin with some H_i and order naturally (along H_i) the pairs of (neighboring) vertices in H_i, corresponding to the edges from H_i to H_j connected with H_i in $\mathcal{T}$ (only one pair is selected in $\mathcal{T}$ for any i, j). Then, do the same in every such H_j, ordering the pairs of vertices for the edges from H_j to H_k in $\mathcal{T}$, starting from the pair for i, j, and so on. As a result, we obtain a unique cycle visiting every such pair of vertices twice: on the way from H_i to some H_j and on its way back from H_j to H_i, and so on. Here we use the edges parallel to those already taken to go back from H_j to H_i.
Hint for (ii): Use that any Hamiltonian cycle in $M_{m,n}$ must contain all 8 edges at its corners, and that any N-tree has $(N-1)$ edges (Problem 7.4.5).

Exercise 6.4.21 ◇ *In the Hamiltonian multi-cycle from Fig. 6.7 for* $M_{6,8}$, *find all pairs of neighboring parallel edges between its 4 cycles* H_i *in the sense of Exercise 6.4.20. Plot the corresponding graph* $\mathcal{G}$ *with* H_i *as its vertices. Check that there are* 24 *choices for the spanning tree* $\mathcal{T}$ *in* $\mathcal{G}$ *and take one of them (from 24) and construct the corresponding Hamiltonian path in* $M_{6,8}$ *(different from the tree used in Fig. 6.16). Finally, plot the corresponding pair of dimer configurations (perfect non-overlapping coverings by dominos) corresponding to odd and even edges in the cycle in* $M_{6,8}$ *you constructed.*

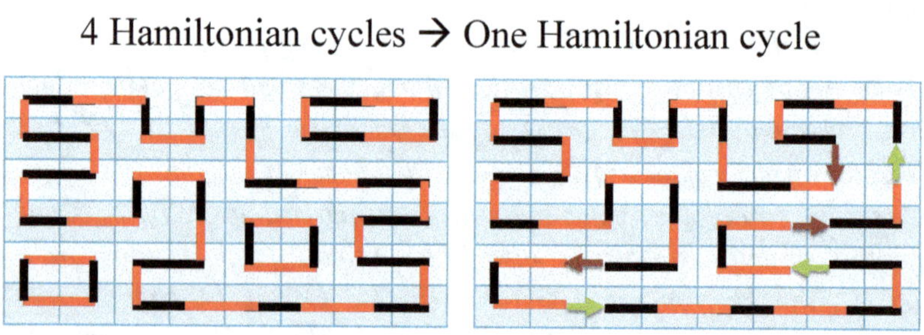

Figure 6.16: From a multi-cycle to a single one.

Exercise 6.4.22 ◇ *(i) Find explicit isomorphisms, i.e. proper numerations of the vertices, for the* 3 *isomorphic graphs in Fig. 6.17.*

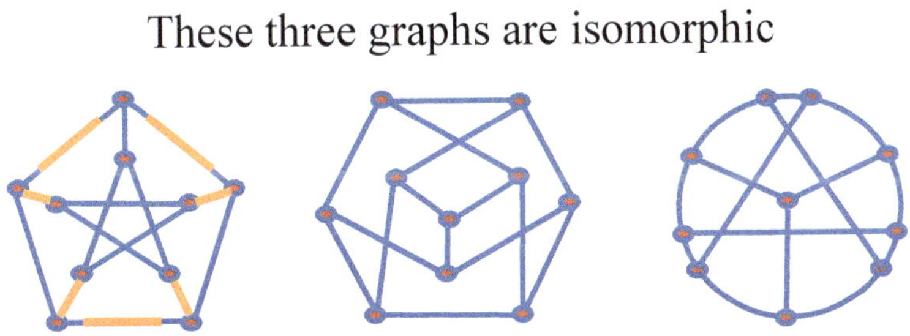

Figure 6.17: Petersen graphs.

(ii) Find a (full) Hamiltonian path *in the first one and prove that it has no Hamiltonian* cycle.

Hint for (ii): Finding a Hamiltonian path is simple. Let us assume that a Hamiltonian *cycle* exists. See the first graph in Fig. 6.17. Among the 5 edges connected to the external pentagon, called "circle" below, and the "star" inside, either 2 or 4 of them must be taken in a Hamiltonian cycle (an even number). Taking only 2 of them is impossible because either their 2 endpoints in the "circle" or their 2 endpoints in the "star" are not neighboring (not connected by an edge), which is impossible (check). If 4 of these edges are taken, then rotate the cycle and assume that the top edge between the "star" and the "circle" is ***not*** in the Hamiltonian cycle (it is not gold in the figure). Starting the cycle with the top vertex, the top two edges in the "circle" must be taken (why?). After taking them, we enter the "star": the corresponding 4 edges are gold in the figure. Then, the next two edges in the "circle" toward the bottom cannot be taken (accordingly, not gold). This gives that the bottom edge of the circle must be taken (gold). So we need only to consider the 5 edges inside the *star*. Check that there are only two ways to proceed with them and neither gives the required Hamiltonian circle.

Exercise 6.4.23$^\heartsuit$ *For any $m, n > 1$, let $K_{m,n}$ be the graph with two groups of vertices $[m]$ and $[n]$, where every one from the first group is connected with each one from the second by 1 edge, and there are no other edges. Prove that there are $n!m!$ automorphisms of $K_{m,n}$ if $m \neq n$, and $2(n!)^2$ of them if $n = m$.*

Hint: Any subset of disconnected vertices (with no edges between any of them) maps to such a subset. There are only 2 such maximal subsets, $[m]$ and $[n]$, in $K_{m,n}$. By maximal, we mean that there exists no greater subset of this kind. Thus, permutations within these 2 groups are the only automorphisms except for $n = m$, where we can additionally transpose these groups.

Exercise 6.4.24 ♣ *(i) Find all* 8 *non-isomorphic graphs with* 5 *vertices and with at least* 7 *edges, disregarding those with multiple edges. Justify your answer. You can use Exercise 6.4.12 to assume that all such graphs are connected.*

(ii) Which of these 8 *are planar? Use* Kuratowski's Theorem: *A graph G is non-planar if and only if it contains a* ***subdivision*** *of $K_{3,3}$ or K_5. A subdivision of a graph is a replacement of some edges $A_1 - A_2$ by $A_1 - B_1 - \cdots - B_m - A_1$ for additional vertices $B_1, \ldots, B_m$ (connected only with their neighbors from this path). Recall that K_5 is the complete graph with* 5 *vertices; $K_{3,3}$ is discussed in Exercise 6.4.23. They are presented in Fig. 6.18.*

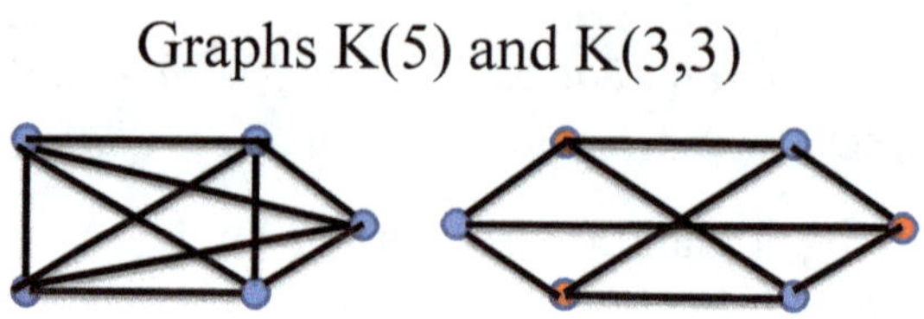

Figure 6.18: $K(5)$ (left) and $K(3,3)$ (right).

Exercise 6.4.25 ◇ *(i)* Euler's Theorem. *For any planar connected graph G: $v - e + f = 2$, where v is the number of vertices, e is the number of edges, and f is the number of faces including one infinite (unbounded) outer region.*

(ii) Check that $3f \le 2e$ for any planar connected graphs that are simple (no multiple edges and ↺). Construct a counterexample if the simplicity is not assumed here. Use this and (i) to prove that $e \le 3v - 6$ for planar graphs.

(iii) Apply (i) and (ii) to verify that the graphs $K_{3,3}$ and K_5 from Exercise 6.4.24(ii) are not planar. They are in Fig. 6.18.

Hint for (i): Check it for a tree G (then $f = 1$). If G is not a tree, remove an edge from one of its cycles, which does not change $v - e + f$. Continue doing this until you obtain a tree.

Hint for (ii): Each face has at least 3 edges due to the simplicity of G, and every edge contributes exactly to two faces. Then use Euler's formula.

Hint for (iii): If $K_{3,3}$ has a planar presentation, then any face there is bounded at least by 4 edges (check). By (i): $f = 2 + e - v = 2 + 9 - 6 = 5$. Thus, there must be at least $4 \cdot 5/2 = 10 > 9$ edges, since each of them can belong to no greater than 2 faces. For K_5: $e = 10 > 3v - 6 = 3 \cdot 5 - 6$.

Exercise 6.4.26 ♠ *(i) A simple connected planar graph G is called* ***maximal planar*** *if adding any new edge between its vertices makes it non-planar.*

Check that all faces must be triangles, *i.e. cycles $A-B-C-A$, for such G, including the face that is the outer region, which gives that $3f = 2e$ (obviously), and $e = 3v - 6$, resulting from Exercise 6.4.25(ii).*

(ii) For a maximal planar G from (i), use that $\sum_a deg_a = 2e$, where the summation is over all vertices a, to check that $\sum_a (6 - deg_a) = 12$. For instance, $3(6-2) = 12$ for a triangle. Then check that there are at least 4 vertices in G with $deg < 6$ if G is not a triangle, i.e. if $v > 3$.

(iii) For a simple connected maximal *planar G, present its outer region as the complement to a convex curved triangle and deform this triangle together with G inside to make it an* SLT, *a straight line triangle. Then apply (ii) to prove by induction with respect to v that all other edges can be made line segments too without changing the isomorphism type of G. This proves* Fáry's Theorem: *any simple planar graph can be drawn using line segments as its edges (with crossings only at the vertices).*

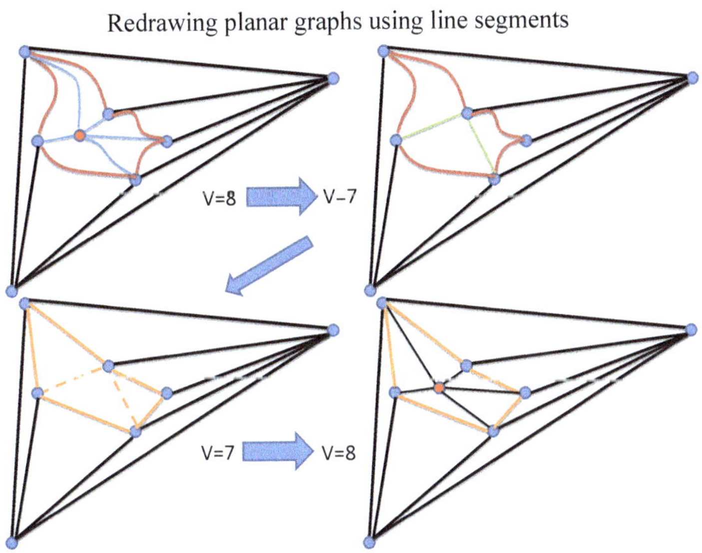

Figure 6.19: Step (iii) of the proof of Fáry-Wagner Theorem.

Hint for (iii): To make the outer curved triangle an SLT, apply a homothety for the center inside it with the variable factor (depending on the angle).
(a) Assume that $v \geq 4$ and find a vertex x different from the 3 vertices of the outer triangle and with no greater than 5 neighbors. Then remove x, consider any triangulation of this polygon formed by its neighbors, and apply the induction claim to redraw the resulting G' with segments as the edges.
(b) This curved polygon then becomes a geometric polygon, which contains

a point x inside it connected with all its ≤ 5 vertices by line segments. This is the so-called "Art Gallery Theorem," easy to check for $3, 4, 5$ vertices.
(c) Finally, delete the triangulation of the polygon obtained via the induction step, add point x above to G' as the additional vertex and connect it by the segments (edges) to the vertices of this polygon. Fig. 6.19 shows (a, b, c).

Exercise 6.4.27 ♣ *(i) Prove that the area of a triangle in the plane is $\frac{1}{2}$ if it has 3 vertices (x_i, y_i) for $i = 1, 2, 3$ with integer coordinates and no such* integer points *inside it or on its boundary (minus the vertices).*

(ii) Pick's Theorem. *For a simple polygon P (that does not intersect itself) with vertices at integer points, let A be its area, I the number of interior integer points, and B the number of integer points on the boundary. Prove that $A = I + \frac{B}{2} - 1$: use (i) and Euler's formula (there are other approaches).*

Hint for (i): Let us rotate this triangle 3 times by π at the middle points of its 3 sides, then do the same with the resulting 3 triangles, then with resulting 6 and so on until they tile the whole plane. Since there are no integer points inside these triangles, all of them are their vertices, and every integer point belongs exactly to 6 triangles. Thus, the number of these triangles approaches (1/2) of the number of the integer points they contain (in total) as $N \to \infty$, where N is the number of these operations.
Hint for (ii): Apply Euler's formula from Exercise 6.4.25(i). Any P can be subdivided into triangles from (i) with $I + B$ vertices in total and $2A+1$ faces due to (i); "+1" is because we add the face outside of P. The number of edges is $\frac{6A+B}{2}$ (check). Finally, $2 = (I + B) - \frac{6A+B}{2} + (2A + 1) = I - A + \frac{B}{2} + 1$.

Exercise 6.4.28 ♢ *(i) By analyzing their complete subgraphs, determine that the graphs 1,2, and 4 in Fig. 6.20 are non-isomorphic.*

(ii) Provide the permutations of the vertices establishing isomorphisms between graphs 2 and 3, and between graphs 1 and 5.

Hint: The 1st and the 5th contain complete subgraphs K_m for $m = 2, 3, 4$ and both are isomorphic to the one presented by 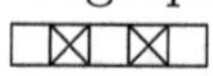 (vertices=intersections), where the left and the right vertical edges are identified. The 2nd and the 3rd contain only K_2. They are isomorphic: do transpositions in 4 neighboring pairs (top, bottom, left, right). The 4th contains K_2, K_3 and no K_4.

Exercise 6.4.29 ♡ *(i) Justify that all graphs in Fig. 6.20 have Eulerian circuits and construct an explicit Eulerian circuit for the first one.*

(ii) Use Theorem 6.1.3 due to Ore to check that all these graphs have Hamiltonian cycles. Directly: find one such cycle in the first graph.

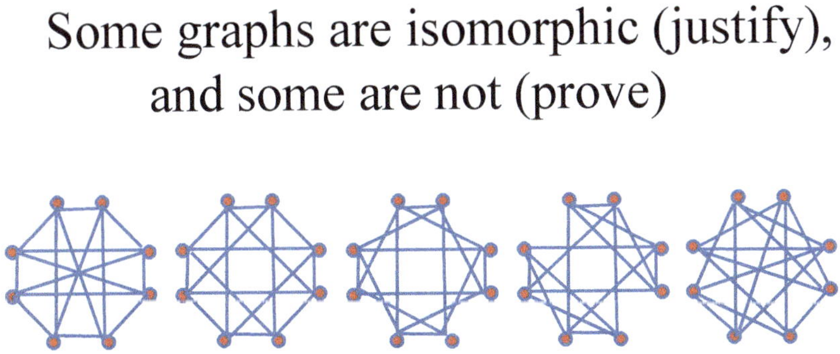

Figure 6.20: Isomorphic or non-isomorphic?

(iii) Add all *intersection points in this figure to the sets of vertices of these graphs as extra vertices (some intersections are multiple). New edges will then be (minimal) portions of old edges between new vertices, which are old vertices and the intersections. Will this graphs have Eulerian circuits?*

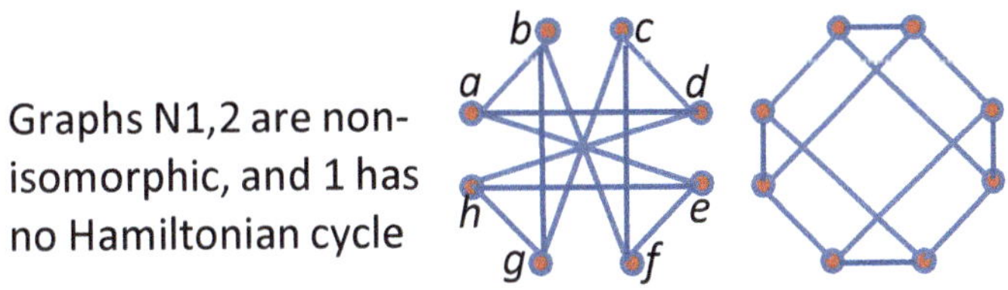

Figure 6.21: Two non-isomorphic graphs.

Exercise 6.4.30 $^\diamond$ *(i) Justify that the graphs in Fig. 6.21 are non-isomorphic.*

(ii) Prove that the 1st one has no Hamiltonian cycle.

(iii) Use Kuratowski's Theorem stated in Exercise 6.4.24 to check that graph 1 is non-planar.

Hints: (i) Count the number of edges in 1 and 2 that belong only to one "square" (4-cycle).
(ii) Check that if any of the diagonals in 1 is not included in such a cycle, then it cannot exist; the diagonals ae and ad are sufficient to examine. Thus, *all* diagonals must be present in this cycle, which gives that *only* these diagonals can occur. This readily leads to a contradiction.

(iii) The corresponding partition of $K(3,3)$ inside this graph is $a - b - f - \{c\} - d - \{h\} - g - e - a$. Here "$-$" are the edges; also, a is connected with d, b is connected with g and f is connected with e. So this is $K(3,3)$ with 2 added vertices $\{c\}, \{h\}$.

Exercise 6.4.31 $^\heartsuit$ *For any* odd *number of points in space, connect each of them with the remaining points by line segments. Let the* vertices *be the initial points and all intersection points of these segments (not only the initial points). These intersections split the initial segments into subsegments (with no vertices inside), which will be the* edges. *Prove that the resulting graph has an Eulerian circuit if no triple of* initial *points is collinear (no 3 initial points lie on one line). Prove that it has an Eulerian non-closed trail (from "A" to "B") if exactly one such triple is collinear.*

Exercise 6.4.32 $^\diamondsuit$ *As in Corollary 6.3.2, let P_{2n}^{2m} be the probability of surviving when playing ideal roulette $2n$ times with \$$(1+2m)$ (\$1 per 1 round).*

(i) For $n = 900$, check numerically that $P_{2n}^{2m} \approx 0.5, 0.75$ for $m = 14, 24$. Generally, confirm experimentally that $P_{2n}^{2m} \approx 0.5$ for $m \approx C\sqrt{n}$, where $C \approx 0.48$ for $n \gg 0$; for instance, $m = 381$ for $n = 640000$ (check).

(ii) Use the Stirling formula from (2.5.2) and that $\int_{-0.6745}^{+0.6745} \frac{e^{-x^2/2}}{\sqrt{2\pi}} dx \approx 0.500007$ to obtain that $C \approx 0.6745/\sqrt{2} \approx 0.476944$ for $n \to \infty$.

Exercise 6.4.33 $^\heartsuit$ *Deduce from the formula $C_{2n} = \frac{(2n)!}{(n+1)!\,n!}$ the recurrence relation $C_{2n} = \frac{4n-2}{n+1} C_{2n-2}$ for $n \geq 1$.*

Exercise 6.4.34 $^\diamondsuit$ *Consider the following arrangements of $2n$ different numbers:*

$$\begin{bmatrix} g_{11} & < & g_{12} & < & \cdots & < & g_{1n} \\ g_{21} & < & g_{22} & < & \cdots & < & g_{2n} \end{bmatrix},$$

where $g_{11} < g_{21}, g_{12} < g_{22}, \ldots, g_{1n} < g_{2n}$.

(i) Check that these inequalities provide that all these numbers are different. Order them naturally and replace them by ± 1 according to the row where they belong: $+1$ for the first, and -1 for the second. Then justify that the total number of such tables equals the Catalan number $C_{2n} = h_n$.

(ii) Prove that the same procedure identifies the extended Catalan numbers $C_{2n}^{0\to 2k}$ with the number of the following **standard Young tableaux:** *arrangements of $2n$ different numbers in* 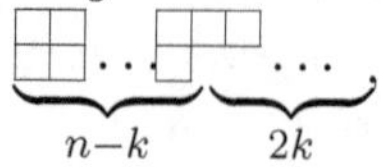 *, satisfying the same inequalities as above in 2 rows and $n-k \geq 0$ columns.*

Exercise 6.4.35 ♠ *Assuming that $k+m \le n$, consider the following* ***standard skew Young diagrams:*** 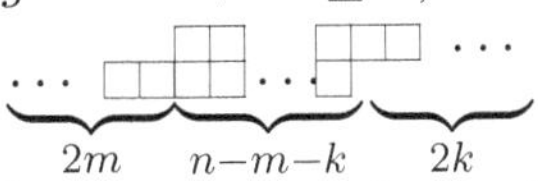 *. The corresponding* tableaux *are the arrangements of the numbers $1, 2, \ldots, 2n$ in its boxes, increasing in the rows to the right ($\rightarrow$) and down in the columns ($\downarrow$).*

(i) Given a tableau, form the sequence $a_i = \pm 1$ for $1 \le i \le 2n$: $a_i = +1$ if i is in the first row, and -1 otherwise. Check that this assignment establishes a one-to-one correspondence between such tableaux and the good paths *in the definition of the extended Catalan numbers $C_{2n}^{2m\to 2k}$.*

(ii) Generalizing, check that $C_n^{m\to k} = \binom{n}{(n+k-m)/2} - \binom{n}{(n+k+m)/2+1}$ counts the number of standard tableaux for the diagram 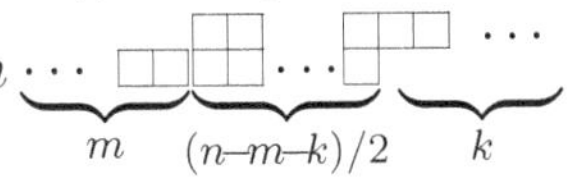 *, where we assume that $k+m \le n$ and that $n-m-k$ is even.*

Exercise 6.4.36 ♢ *Find all standard Young tableaux for the diagram* 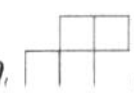 *, which are the arrangements of $1, 2, 3, 4$ in its boxes defined in Exercise 6.4.35; the entries must increase for the directions $\rightarrow$ and $\downarrow$. For such tableaux, let $a_i = +1$ for $1 \le i \le 4$ if i is in the 1st row and -1 otherwise. Check that the corresponding sequences are exactly those in the definition of $C_4^{1\to 1}$, i.e. such that $1+a_1+a_2+a_3+a_4 = 1$ and $1+a_1+a_2$, $1+a_1+a_2+a_3 \ge 0$.*

Answer: $\begin{matrix}12\\34\end{matrix}, \begin{matrix}13\\24\end{matrix}, \begin{matrix}14\\23\end{matrix}, \begin{matrix}23\\14\end{matrix}, \begin{matrix}24\\13\end{matrix},$
$\{++--\}, \{+-+-\}, \{+--+\}, \{-++-\}, \{-+-+\}$.

Exercise 6.4.37 ♡ *Calculate the probability P_{2n} that a random walk that starts at 0 and at each step moves $+1$ or -1 with equal probability will return to 0 at least once during $2n$ steps. Check that $P = \lim_{n\to\infty} P_{2n} = 1$. [This probability P is still 1 for two-dimensional random walks with 4 steps $(\pm 1, \pm 1)$, but becomes smaller than 1 in dimension 3 due to Pólya: about 0.34.]*

Hint: For $2n$ steps, the number of paths strictly above or strictly below 0 is $2\sum_{k=1}^{n} C_{2n}^{0\to 2k} = 2\binom{2n}{n+1} \approx 2\frac{2^{2n}}{\sqrt{\pi n}}$. So $1 - P_{2n} = 2\binom{2n}{n+1}/2^{2n}$, which is approximately $\frac{2}{\sqrt{\pi n}}$ due to the Stirling formula (2.5.2) and tends to 0 as $n \to \infty$.

Exercise 6.4.38 ♢ *Assume that a movie ticket is \$10. Among N people who want to buy the tickets and prefer to pay cash, $N\alpha$ can pay \$10 exactly and $N\beta$ have only twenty-dollar bills for some $0 < \alpha < 1$ and $\beta = 1 - \alpha < \alpha$. The ticket counter can only have change for a twenty if it has a ten in the box, so can serve only certain sequences of people.*

(i) Justify that there are $\binom{N\alpha+N\beta}{N\alpha}\frac{N\alpha-N\beta+1}{N\alpha+1}$ *manageable lines of people if the ticket office has 0 cash initially. Equivalently, the probability* P_N *that the ticket counter can do this is* $\frac{N\alpha-N\beta+1}{N\alpha+1}$, *which tends to* $1-\beta/\alpha$ *as* $N\to\infty$.

(ii) Check that the probability P_N *becomes*

$$1-\frac{N\beta(N\beta-1)\cdots(N\beta-k)}{(N\alpha+1)\cdots(N\alpha-k+1)},$$

if the ticket counter has initially k *ten-dollar bills for* $N\beta>k\geq 0$, *and check that this probability tends to* $1-\left(\frac{\beta}{\alpha}\right)^{k+1}$ *as* $N\to\infty$.

Hint: Follow the formula for $C_{2n}^{2m\to 2k}$ and prove that the total number of possible manageable lines of N people is $\left(1-\frac{N\beta(N\beta-1)\cdots(N\beta-k)}{(N\alpha+1)\cdots(N\alpha-k+1)}\right)\binom{N}{N\alpha}$.

Exercise 6.4.39 $^\diamondsuit$ *Recall the formulas from Theorem 6.2.4 and Corollary 6.3.2:*

$$C_{2n}^{2m\to 2k}=\binom{2n}{n+m-k}-\binom{2n}{n+k+m+1}$$
$$\text{and }\ C_{2n}^{2m\to *}=\binom{2n}{n-m}+\cdots+\binom{2n}{n+m}.$$

Use them to check directly *the following. The total gain* $\sum_{k=0}^{\infty}(2k-2m)C_{2n}^{2m\to 2k}$ *of the people who played roulette with the initial* $\$(1+2m)$ *and "survived"* $2n$ *rounds equals exactly* $(2m+1)(2^{2n}-C_{2n}^{2m\to *})$, *which is the amount lost by those who did not "survive" (with* $\$(1+2m)$ *and playing* $\leq 2n$ *rounds).*

Hint: Using the identity $\sum_{r=0}^{\infty} r\binom{2n}{r}=2n2^{2n-1}$ and changing the summation variable k to $u=n+m-k$ and $v=n+m+k+1$: $\sum_{k=0}^{\infty}(2k-2m)C_{2n}^{2m\to 2k}=2\sum_{u=0}^{n+m}(n-u)\binom{2n}{u}-2\sum_{v=n+m+1}^{\infty}(v-n-2m-1)\binom{2n}{v}=2\Big(n2^{2n}-(2n)2^{2n-1}+(2m+1)\sum_{v=n+m+1}^{\infty}\binom{2n}{v}\Big)=(2m+1)\Big(2^{2n}-\big(\binom{2n}{n-m}+\cdots+\binom{2n}{n+m}\big)\Big)$.

Exercise 6.4.40 $^\heartsuit$ *Extend Corollary 6.3.3 to check the following. Let* $p, q=1-p$ *be the probabilities for* ± 1, *where the random path with* $2n$ *steps begins from* $2m$ $(m\geq 0)$. *Check that the probability* Π_{2n}^{2m} *of such paths being* bad, *i.e. intersecting the* x*-axis, is*

$$\Pi_{2n}^{2m}=\sum_{k=0}^{m+n}p^{n+k-m}q^{n-k+m}\left(\binom{2n}{n+k-m}-\binom{2n}{n+k+m+1}\right).$$

(i) Find that $\Pi_{90}^{20}=0.0263017$ *for* $p=\frac{1}{2}=q$, *which coincides with* $\mathcal{P}_{90}^{20}$ *obtained via (6.3.4), and that it becomes* $\Pi_{90}^{20}=0.884183$ *for* $p=\frac{1}{3}, q=\frac{2}{3}$.

(ii) Calculate the probability $\mathcal{P}_{20}^{4}$ *of losing when playing roulette* $2n=20$ *times under "\$1 per round" with the initial \$5* $(2m=4)$ *for the ideal roulette, and the probability* Π_{20}^{4} *for the American wheel:* $p=9/19, q=10/19$.

Answer for (ii): $\mathcal{P}_{20}^4 = 0.263$, $\Pi_{20}^4 = 0.337$.

Exercise 6.4.41 ♣ *(i) For $n \ge m \ge 0$, consider the paths from 0 to m with n steps that are ± 1 and 0 (not only ± 1). Check that the number A_n^m of such paths equals $\sum_{k=0}^{n-m} \binom{n}{k}\binom{n-k}{(n-k+m)/2} = \sum_{k=0}^{n-m} \frac{n!}{k!(\frac{n-k+m}{2})!(\frac{n-k-m}{2})!}$ provided that $n-k+m$ is even in this sum. Also, $A_n^m = \sum_{k=0}^{[(n-m)/2]} \frac{n!}{k!(k+m)!(n-2k-m)!}$; $[x] =$Floor$[x]$.*

(ii) Following the procedure "reflect and shift" for Catalan numbers, check that $M_n^m = A_n^m - A_n^{m+2}$ for the Motzkin number *M_n^m, which is that of* good paths*: non-negative (i.e. with the partial sums no smaller than 0). They are called* ***Motzkin paths****; those from 0 to 0 are called standard. For instance, $M_1^0 = 1$, $M_2^0 = A_2^0 - A_2^2 = 3 - 1 = 2$, $M_3^0 = A_3^0 - A_3^2 = 7 - 3 = 4$, $M_4^0 = 9$.*

Exercise 6.4.42 ◇ *(i) Deduce that $M_n^m = \sum_{k=0}^{[(n-m)/2]} \binom{n}{m+2k} C_{m+2k}^{0\to m}$ from the definitions of the Motzkin and Catalan numbers; see Exercises 6.4.41 and 6.4.35.*

(ii) Following Fig. 6.8 check that $M_n^0 = M_{n-1}^0 + \sum_{k=0}^{n-2} M_k^0 M_{n-k-2}^0$, for $n \ge 1$, where $M_0^0 = 1$. For instance, $M_5^0 = 21$.

Exercise 6.4.43 ◇ *The* ***Schröder numbers*** *S_n are similar to Motzkin numbers M_{2n}^0: for the steps $(1, \pm 1)$ (as for Motzkin paths) but with $(2,0)$ instead of $(1,0)$. Equivalently, they are lattice paths from the SW corner $(0,0)$ of the $n \times n$ grid to the NE corner (n,n) using only steps $(0,1), (1,0)$ and $(1,1)$ that do not rise above the SW–NE diagonal. These steps are as in Problem 2.5.2 on the "king's moves", but the diagonal condition is added. Check that $S_1 = 2, S_2 = 6$. Similar to Exercise 6.4.42(ii), prove that $S_n = 3S_{n-1} + \sum_{k=1}^{n-2} S_k S_{n-k-1}$ for $n \ge 2$ and $S_0 = 1$. For instance, $S_3 = 22, S_4 = 90$.*

Hint: The coefficient 3 in the recurrence is because the starting point in Fig. 6.8 can be *not* connected with other points (isolated); then one of its neighbors must be isolated too.

Exercise 6.4.44 ♣ *(i) In the notation of Exercise 6.4.42, obtain from (ii) there that $M(t) = 1 + tM(t) + t^2 M(t)^2$ for $M(t) = \sum_{n=0}^{\infty} M_n^0 t^n$. Then check that $M(t)$ is the expansion of $\frac{1-t-\sqrt{1-2t-3t^2}}{2t^2}$ at $t = 0$. By a straightforward differentiation of the latter formula, check the relation $t(3t^2 + 2t - 1)\, dM(t)/dt + (3t^2 + 3t - 2)\, M(t) + 2 = 0$. Finally, deduce from it that*

$$(n+2)M_n^0 = (2n+1)M_{n-1}^0 + (3n-3)M_{n-2}^0 \quad \textit{for} \quad n \ge 1.$$

(ii) Similarly, use Exercise 6.4.43 to establish that $S(t) = \sum_{n=0}^{\infty} S_n t^n = \frac{1-t-\sqrt{1-6t+t^2}}{2t}$. *Then obtain the relation* $(n+1)S_n = (6n-3)S_{n-1} - (n-2)S_{n-2}$ *for* $n \geq 2$ *by differentiating the formula for* $S(t)$. *Compare with the recurrence from Exercise 6.4.33 for the Catalan numbers.*

Hint: Differentiate $t^2 M(t)$. Then $t^2(3t^2+2t-1)\,dM(t)/dt = f(t)M(t) + g(t)$ for some t-polynomials f, g. So the problem is to calculate them. Similarly, differentiate $tS(t)$ and find the corresponding f, g.

Exercise 6.4.45 ♣ *(i) Establish a 1-1 correspondence between all Motzkin paths from Exercise 6.4.41 from 0 to* $m \geq 0$ *and all* Catalan paths *(non-negative) from 0 to* m *that do not contain consecutive* $\{+1,+1,+1\}$ *anywhere. For this, find all triples* $\{+1,+1,-1\}$ *in the latter and all pairs* $\{+1,-1\}$ *among* ± 1 *not in such triples. Then replace the triples by* $+1$ *and the pairs by* 0. *Check that if* u_σ *is the number of* $\sigma = +1, -1, 0$ *in the Motzkin path, then the number of steps is* $3u_{+1} + 2u_0 + u_{-1}$ *in the corresponding Catalan one, and the number of* $+1$ *there is* $2u_{+1} + u_0$.

(ii) Now assume that a Catalan path from 0 to m *does not contain any consecutive* $\{+1,-1,+1\}$. *Mark any* $+1$ *followed by* -1 *and any* -1 not *followed by* $+1$. *Then, for any unmarked* $+1$, *find the first* -1 *after it* at the same level *(in the standard graph). If this* -1 *is marked, replace such* $+1$ *by* 0. *Finally, delete all marked* ± 1. *Check that you will obtain a Motzkin path for the same* m, *and that this algorithm gives a 1-1 correspondence.*

Exercise 6.4.46 ♠ *Use Exercise 6.4.41 to calculate the price of the "option" in Fig. 6.14 based on the Motzkin paths, i.e. under the assumption that the daily price changes are* \$0.5, *or* −\$0.5, *or* 0 *with equal probabilities (1/3).*

Exercise 6.4.47 ♣ *Let* $\mathcal{P}_k$ *be the price of the "option" in Fig. 6.14 where the oil price changes* k^2 *times a day by* $\pm\$1/(2k)$ *with equal probabilities (1/2 for* $\pm$*); the option is terminated immediately when the price of oil goes beyond* \$50. *Using formula (6.3.4), find that* $\mathcal{P}_2 \approx 3.0\%$ *and* $\mathcal{P}_{10} \approx 3.4\%$. *When* $k \to \infty$, *apply formula (6.3.5) to check that* $\mathcal{P}_k \approx 3.5015\%$.

Chapter 7

REVIEW ASSIGNMENTS

This purpose of this chapter is threefold: to practice the main topics, connect them better, and complement the corresponding chapters. The material we added is significant; these additional themes were not discussed earlier mostly not to lose focus of the main lines and keep the exposition compact.

Many of the review problems below were solved in the author's courses (mostly by students) as part of special "rehearsal classes" before the exams. Some were given to students to do at home for extra credit. Motivated readers can use some of the topics here as starting points for additional reading and research. We are more relaxed in this chapter with providing some advanced material. The corresponding sections mostly (but not always) begin with the problems that were discussed in the rehearsal classes.

7.1 Combinations, permutations etc.

7.1.1 Using Monte Carlo method

We continue the discussion of the game "14 vs. 1" in Section 2.5.3, where \$1 was paid for green boxes and \$5 for the red corner. Let us provide a direct *Mathematica* code: direct simulations of this game.

```
games = 200000; n = 7; bet = 5; GAIN=0;
Do[gain = 0; stop = 0; x = 0; y = 0;
  Do[
    If[stop == 0 && (x == n+1 || y == n+1 || (x == n && y == n)),
      stop = 1;
      If[x == n && y == n, gain = gain + bet, gain = gain - 1]];
    If[stop == 0, Do[rnd = Random[], {j, 1, x+1}];
      If[rnd < 0.5, x = x+1, y = y+1]], {i, 1, 2 n+2}];
      GAIN = GAIN + gain, {g, 1, games}];
Print["Average gain per simulation = ", 100.*GAIN/games];
```

Let $n+1=8$ be the size of the board and the number of games in one simulation is *games*= 200000. The average gain (per 1 game) is calculated in cents. Theoretically, it is approximately 25.684c. The program above seems sufficiently readable even you are not familiar with *Mathematica* but had experience with similar languages. This is ***Monte Carlo method***: random sampling to obtain numerical results.

Let us provide a sequence of typical outcomes for 10 simulations (each with 200000 games) for $n=7$ and under the bet "5:1" (in cents):

$\{26.096, 26.564, 26.765, 26.735, 25.469, 26.627, 25.472, 26.114, 26.087, 25.223\}$.
The average of these 10 average gains is 25.697, which is reasonably close to the exact (theoretical) average 25.684. However the standard deviation is high in this sequence; the accuracy is not always that good.

Importantly, we run the *Mathematica* operation `Random[]` several times, namely $x{+}1$ times, where x is the current x-coordinate. It appeared that using `Random[]` once, as it is supposed to be, gives unsatisfactory results. Almost all simulations, give then the averages > 26. For instance, the following sequence of average gains is typical with using `Random[]` once for $n=1000000$:

$\{26.493, 26.192, 26.336, 26.350, 26.263, 26.314, 26.473, 25.891, 26.027, 26.647\}$.
The average here is 26.2986, which is very far from 25.684 (for so many games). It seems that `Random[]` in the version of Mathematica we used is not exactly random: the sequences of random real numbers from 0 to 1 it provides are more "alternating" than they are supposed to be. This increases the chances to reach the red corner and the gain of your opponent.

Generally, creating "really random" sequences is a difficult problem. The trick above, with random number of `Random[]`, improves the "randomness" of what Mathematica provides for the purpose of this particular calculation. It appeared that the game "14 vs. 1" is a good test of the quality of randomness. We conclude that even very minor variations of the source of random data, intentional or accidental, can significantly influence the outcome.

7.1.2 Inclusion-Exclusion Principle

Problem 7.1.1 *Consider the reduced deck with only* 13 *cards of denominations* $1, 2, \ldots, 13$ *(no suits). Let a* hand *consist of* 4 *cards. We call two cards with consecutive denominations a* pair. *Count the number of hands that are exactly "2-pairs", i.e. 2 pairs where no 3 cards have consecutive denominations.*

Solution 1. We can always make the cards in a hand ordered, so we assume that the denominations increase. All "pairs" are $\{1,2\}$,$\{2,3\}, \ldots, \{12,13\}$, so there are 12 of them. The second "pair" will be assumed greater than the first. So we can extend $\{1,2\}$ by $\{4,5\}, \ldots, \{12,13\}$, totally by $12-4+1=9$ possible *pairs*. We avoid $\{3,4\}$ after $\{1,2\}$, since 4 consecutive denominations are not allowed. The number of extensions is 8 for $\{2,3\}$, and so on until $\{9,10\}$, where there is only one such extension. We obtain that the number of hands that are (strictly) 2-*pairs* is $9+8+\cdots+1=45$.

Solution 2. Let $\{b,c\}$ be the middle 2 cards in a hand. For $b \geq 2, c \leq 12$, all hands of type "2-pairs" are $\{b-1,b,c,c+1\}$ provided that $b+1<c$. Since b,c can take $12-2+1=11$ values, the answer is $\binom{11}{2}-10=55-10=45$, where we subtract those with $b+1=c$, corresponding to the hands with consecutive 4: from $\{1,2,3,4\}$ until $\{10,11,12,13\}$ (10 of them).

Solution 3. Represent the hands as $\{a<b<c<d\}$ we set $b=a+x, c=b+y, d=c+z$. Generally, $a,x,y,z \geq 1$ and $a+x+y+z+u=13$ for some $u \geq 0$. In this particular problem, $x=1, z=1, y \geq 2$, so $a+1+y+1+u=13$ and $a'+y'+u=13-5=8$ for $a=a'+1, y=y'+2$. Since now $a', y', u \geq 0$, we can use the R-numbers with $n=3$ (the "types") and $r=8$ (the "objects"). Thus, the answer is $R(n,r)=\binom{n+r-1}{n-1}=\binom{10}{2}=45$. This solution is universal, but not the simplest in this case. ❑

Problem 7.1.2 *Count the number $\mathcal{N}$ of different 5-hands for the deck of 13 cards (denominations only) with two* pairs *separated by a card in the middle, which means that this card does not form a triple of consecutive denominations with the 1st pair and with the 2nd one. We assume that the cards (their denominations) increase in 5-hands.*

Solution. Let us consider 3 cards in the middle: $\{b<c<d\}$. Then $b \geq 2$ and $d \leq 12$ can be uniquely extended to the pairs before b or after d. This gives $\binom{11}{3}$. However, we must exclude from $\{b,c,d\}$ the triples that contain a "pair", which can be $\{b=c-1,c\}$ or $\{c,d=c+1\}$. The number of triples of the first kind equals the number of $\{c,d\}$ such that $3 \leq c<d \leq 12$. It equals $\binom{10}{2}$. By the symmetry, it is the same for $\{c,c+1\}$, the second possible kind. However, these 2 kinds overlap: we counted $\{c-1,c,c+1\}$ twice. They correspond to fully consecutive hands ("straight flush hands"); there are 9 of them due to the inequalities $3 \leq c \leq 11$. Finally, $\mathcal{N}=\binom{11}{3}-2\binom{10}{2}+9=165-90+9=84$.

To double-check, let us follow Solution 3 above. For $\{a, b=a+x, c=b+y, d=c+z, e=d+t\}$, we have: $a \geq 1, x=1=t, y \geq 2, z \geq 2$ and $a+x+y+z+t+u=13$ for $u \geq 0$. This gives $a'+y'+z'+u=13-1-1-1-2-2=6$, $n=4$, $r=6$, and $\mathcal{N}=\binom{n+r-1}{n-1}=\binom{9}{3}=\frac{9\cdot 8\cdot 7}{3!}=12\cdot 7=84$. ❑

Problem 7.1.3 *Count the number N of the baskets with 7 items that can be purchased from the supply $5A+4B+3C+2D$.*

Solution. We have $N = Free - b_1 - b_2 - b_3 - b_4 + b_{12} + b_{13} + b_{14} + b_{23} + b_{24} + b_{34} - b_{123} - b_{124} - b_{134} - b_{234} + \ldots$. Here we count the number of presentations

$7 = x_1 + x_2 + x_3 + x_4$ with $x_i \geq 0$. Thus $n = 4$ (types) where the number of objects is $r = 7$, and

$$Free = R(n,r) = R(4,7) = \binom{7+4-1}{4-1} = \binom{10}{3} = \frac{10 \cdot 9 \cdot 8}{3!} = 120.$$

As we did in Section 3.4.3, $Bad_1 =$"$x_1 \geq 6$", $Bad_2 =$"$x_2 \geq 5$", $Bad_3 =$"$x_3 \geq 4$", $Bad_4 =$"$x_4 \geq 3$". Accordingly,

$$b_1 = \binom{10-6}{3} = \binom{4}{3} = 4, \quad b_2 = \binom{10-5}{3} = \binom{5}{3} = \binom{5}{2} = 10,$$
$$b_3 = \binom{10-4}{3} = \binom{6}{3} = 20, \quad b_4 = \binom{10-3}{3} = \binom{7}{3} = \frac{7 \cdot 6 \cdot 5}{6} = 35.$$

Only one from b_{ij} is nonzero: namely, $b_{34} = \binom{10-3-4}{4-1} = 1$. Finally, $N = 120 - 4 - 10 - 20 - 35 + 1 = 52$. To double-check, our answer must coincide with the coefficient of x^7 in the "expansion" below. It is 52 indeed:

$$\begin{aligned} &\text{Expand}[\,(1+x+x^2)(1+x+x^2+x^3)(1+x+x^2+x^3+x^4) \\ &\times (1+x+x^2+x^3+x^4+x^5)\,] = x^{14} + 4x^{13} + 10x^{12} + 19x^{11} + 30x^{10} \\ &+ 41x^9 + 49x^8 + \mathbf{52x^7} + 49x^6 + 41x^5 + 30x^4 + 19x^3 + 10x^2 + 4x + 1. \end{aligned}$$

❑

Problem 7.1.4 *Count the number $\mathcal{N}_1$ of 4-digit numbers $x_1x_2x_3x_4$ such that the sum of their digits $\Sigma = x_1 + x_2 + x_3 + x_4$ equals 19, and then determine the number $\mathcal{N}_2$ of such numbers with $\Sigma \leq 19$.*

Solution. One has: $y_1 + x_2 + x_3 + x_4 = 18$, where $0 \leq x_i \leq 9$ for $2 \leq i \leq 4$, and $0 \leq y_1 = x_1 - 1 \leq 8$; this is similar to the reduction $x_i \mapsto y_i = x_1 - 1$ for dice. The *free problem*, that without the upper bounds, is about R-numbers: $Free = \binom{18+4-1}{4-1}$. Then we have $B_1 =$"$y_1 \geq 9$", $B_i =$"$x_1 \geq 10$" for $i = 2, 3, 4$. Setting $b_i = |B_i|$, one obtains:

$$b_1 = |B_1| = \binom{(18-9)+4-1}{4-1} = \binom{9+4-1}{4-1}, \; b_{i\geq 2} = \binom{8+4-1}{4-1}.$$

All b_{ij} are zero: $9 + 10 > 18$ for the "greatest" b_{12}. So we have:

$$\begin{aligned} \mathcal{N}_1 &= Free - b_1 - b_2 - b_3 - b_4 + b_{12} \ldots \\ &= \binom{18+4-1}{4-1} - \binom{9+4-1}{4-1} - 3\binom{8+4-1}{4-1} \\ &= \binom{21}{3} - \binom{12}{3} - 3\binom{11}{3} = 615. \end{aligned}$$

Use Expand$[\,(x + x^2 + x^3 + x^4 + x^5 + x^6 + x^7 + x^8 + x^9)(1 + x + x^2 + x^3 + x^4 + x^5 + x^6 + x^7 + x^8 + x^9)^3\,]$ to double-check $\mathcal{N}_1$.

With $\mathcal{N}_2$, we have $y_1 + x_2 + x_3 + x_4 + u = 19$ for the same inequalities as above and with extra $u \geq 0$. This readily gives:

$$\mathcal{N}_2 = \binom{22}{4} - \binom{13}{4} - 3\binom{12}{4} = 5115, \text{ which is } > 9000/2 = 4500.$$

Actually, the average sum of digits is $4 + 3 \cdot 4.5 = 18.5$, where we divide $1 + \cdots + 9 = \frac{9 \cdot 10}{2}$ by 9 for the 1st digit and by 10 for the remaining 3. Thus, $\mathcal{N}_2$ can be obtained as $4500 + \mathcal{N}_1 = 4500 + 615 = 5115$. ❑

Problem 7.1.5 *(i) Among 3-digit numbers, how many have no ± 1-neighboring digits, pairs of neighboring digits in a number with ± 1 as their difference?*

(ii) Among $000, 001, \ldots, 998, 999$, how many have no two digits anywhere with a difference of ± 1? How many 3-digit numbers satisfy this condition?

Solution. **(i)** Let us count the numbers abc with $b = a \pm 1$. First of all, the number of pairs of digits ab with $b = a + 1$ is 8: $a = 1, \ldots, 8$ (the 1st digit cannot be 0). Similarly, $a = 1, \ldots, 9$ if $b = a - 1$: one more pair, 10, is added because $b = 0$ is allowed. So we have $8 + 9 = 17$ possibilities for ab (the first 2 digits in abc), and there are $17 \cdot 10 = 170$ "bad" numbers abc of this type totally: c can be arbitrary.

The second possibility is when $c = b \pm 1$. There are $2 \cdot 9 = 18$ such pairs bc. Here b and c are on equal footing. Then we multiply 18 by 9: the first digit a cannot be zero. This gives 162 for the "bad" numbers of this type.

However, some numbers were counted twice; these two groups of "bad" numbers overlap. To count them, we look at the middle digit b. For $b = 0$, it can be only 101; for $b = 1$, the choices are $210, 212$. For $b = 2, 3, 4, 5, 5, 7, 8$, the possibilities are $b \pm 1, b, b \pm 1$, so there are 4 choices for every such b. Finally, there is only 898 for $b = 9$. The total number in the intersection is therefore $1 + 2 + 7 \cdot 4 + 1 = 32$. Thus, the number of abc with at least one "pair" is $170 + 162 - 32 = 300$, and the final answer to (i) is $900 - 300 = 600$.

(ii) Now 0 is allowed anywhere. Accordingly, the number of abc with $b = a + 1$ is now 9 (for $01, 12, \ldots, 89$) times 2 (their transpositions), and times 10 for $c = 0, 1, \ldots, 9$. So we have 180, which coincides with the number of abc with $c = b \pm 1$ and the number of abc for $c = a \pm 1$. We will denote them N_{12}, N_{23}, N_{13}. The rotation $abc \mapsto bca$ identifies the corresponding sets. We use that only the numbers aaa are fixed under this rotation. Let N_1, N_2, N_3 be the numbers of abc for the corresponding pairwise intersections: $b = a \pm 1, c = a \pm 1$ for N_1, and so on. The intersection of all 3 sets is empty (check). One has: $N_2 = 1 + 4 \cdot 8 + 1 = 34$, where 1 is the number of continuations of $b = 0, 9$ and 4 is that for $b = 1, 2, \ldots, 8$. Using "Inclusion-Exclusion," we obtain that the final answer for the numbers from $\{000, \ldots, 999\}$ is $1000 - N_{12} - N_{23} - N_{13} + N_1 + N_2 + N_3 = 1000 - 3 \cdot 180 + 3 \cdot 34 = 562$.

The number of "good" (strictly) 3-digit numbers is 562 minus the number of "good" numbers in the form $0bc$. So we need to count the pairs bc such that $0bc$ is "good". The numbers of possible c for $b = 0, 2, 3, \ldots, 8, 9$ are $9, 8, 7, \ldots, 7, 8$; note that $b = 1$ is omitted: $a - b = -1$. So their total number is $9+8 \cdot 2+7 \cdot 6=67$, and there are $562-67=495$ "good" 3-digit numbers. ❑

7.1.3 Stirling numbers

Problem 7.1.6 *Check that for $p \geq 0$ and $a_i \geq 0$ (in the sum), $(x_1+\cdots+x_n)^p = \sum_{a_1+\cdots+a_n=p} \mathcal{S}_{a_1,\ldots,a_n} x_1^{a_1} \cdots x_n^{a_n}$, where $\mathcal{S}_{a_1,\ldots,a_n}$ is the number of partitions of the set of p elements into n* **ordered** *(distinguishable) boxes with $a_1, \ldots, a_n$ elements in each; a_i can be zero. The elements in boxes are not ordered (and there are no empty spaces there).*

Solution. The partitions are as follows. The first box will contain the quantities $(x_1 + \cdots + x_n)$ in $(x_1 + \cdots + x_n)^p$ where we pick x_1, the second box the quantities $(x_1 + \cdots + x_n)$ where we pick x_2, and so on. Thus, every quantity appears in some box, and we obtain the required identity. ❑

A sharper version of this formula is when we count the subsets of the $x_1, \ldots, x_n$ such that the corresponding exponents $b_1, \ldots, b_n$ in the monomials $x_1^{b_1} \cdots, x_n^{b_n}$ are nonzero. The justification is as above.

Problem 7.1.7 *Let $\mathcal{S}^{\dagger}_{b_1,\cdots,b_m}$ be the number of partitions of the same set of p elements into m ordered* **nonempty** *boxes: with $b_i > 0$ elements. Let $C_m(n) = \{1 \leq c_1 < c_2 < \ldots < c_m \leq n\}$ be a collection of m-element subsets of the set $[n] = \{1, 2, \ldots, n\}$. Then $(x_1 + x_2 + \cdots + x_n)^p =$*

$$\sum_{m=0}^{p} \sum_{\{c_i\} \in C_m(n)} \sum_{b_1+\cdots+b_m=p} \mathcal{S}^{\dagger}_{b_1,\ldots,b_m} x_{c_1}^{b_1} \cdots x_{c_m}^{b_m}, \quad \text{where } b_i > 0.$$ ❑

The triple summation here can be simplified in some cases. Let us make $x_1 = x_2 = \cdots = x_n = 1$, which is the simplest possible reduction. Using that the number of subsets $C_m(n)$ is $\binom{n}{m}$, we obtain the following formula:

$$n^p = \sum_{m=0}^{p} \binom{n}{m} S^{\dagger}(p, m), \tag{7.1.1}$$

where $S^{\dagger}(p, m)$ is the number of all partitions of p into m ordered (distinguishable) boxes, which are now all nonempty. Since n is arbitrary, this can be re-stated as an algebraic identity:

$$z^p = \sum_{m=0}^{p} \overbrace{z(z-1)\cdots(z-m+1)}^{m} \frac{S^\dagger(p,m)}{m!}. \tag{7.1.2}$$

Problem 7.1.8 ***Stirling numbers*** *of the 2nd kind, denoted by $S(p,m)$, count the number of partitions of a set of p elements into m* ***indistinguishable*** *boxes in which no box is empty. Then $S(p,m) = S^\dagger(p,m)/m!$, and*

$$S(p,m) = mS(p-1,m) + S(p-1,m-1) \ \textit{for}\ 1 \le m \le p-1,$$

where $S(p,0) = 0$ for $p > 0$, and $S(p,p) = 1$ for $p \ge 0$.

Solution. This recurrence relation is a variant of that for the binomial coefficients; its proof is similar. By *indistinguishable* boxes, we naturally mean that they can be permuted in any way. Recall that the elements are not ordered inside the boxes. For any partition B of p into m nonempty boxes: (1) the element p is alone in its box, or (2) this box contains other elements too.

In the first case, we simply remove this box and arrive at B' counted in $S(p-1,m-1)$. In the second case, removing p will result in B' with m nonempty boxes, but this $(p-1)$-configuration B' can be obtained from m different p-configurations. Such B are as follows: we add p to every box of B' (there are m of them). ❑

Check that $S(p,p) = 1 = S(p,1)$, $S(p,p-1) = \binom{p}{2}$ for $p \ge 1$, and $S(p,2) = 2^{p-1} - 1$ for $p \ge 2$. Concerning the latter, the direct combinatorial argument is as follows. Partitions of the set $[p] = \{0,1,,\ldots,p\}$ into 2 nonempty indistinguishable groups are simply unordered pair $\{S \subset [p], [p]\backslash S\}$ provided that both sets here are nonempty. This gives $2^{p-1} - 1$; we subtract 1 due to the partition $\{\emptyset, [p]\}$ of $[p]$. Exercise 3.7.15 provides a more algebraic and direct approach to this problem.

To give an example, let $p = 3$. Then: $n^3 = nS(3,1) + n(n-1)S(3,2) + n(n-1)(n-2)S(3,3)$, equivalently, one has: $z^3 = zS(3,1) + z(z-1)S(3,2) + z(z-1)(z-2)S(3,3)$. Here $S(3,2) = 2^{3-1} - 1 = S(3,3-1) = \binom{3}{2} = 3$ and, $z^3 = z + 3z(z-1) + z(z-1)(z-2)$.

7.1.4 Semi-magic squares

We call an $n \times n$ square a *semi-magic square* (SMS) if its entries are $1, 2, \ldots, n^2$, inclusive, and the vertical and horizontal sums are $n(n^2+1)/2$; i.e. there are no assumptions concerning the diagonal sums.

Problem 7.1.9 *(i) Check that the compositions* $v_{col}u_{row}$ *of any permutation* u_{row} *of the rows and any permutation* v_{col} *of the columns send SMS to SMS. Moreover, the composition order does not matter here:* $v_{col}u_{row} = u_{row}v_{col}$.

(ii) Prove that under the transformations from (i), any SMS can be uniquely transformed to the **normal form** *defined as follows. Let* m_i *be the minimal entry in row* i. *Then we impose the conditions (a)* $m_1 < m_2 < \ldots < m_n$ *and (b) the entries in the first row increase.*

Solution. **(i)** Transformations v_{col} and u_{row} transform the entries from any column or row to the entries from another column or row, which does not change the corresponding sums. The relations $v_{col}u_{row} = u_{row}v_{col}$ (the commutativity) are sufficient to check for any transpositions of the rows composed with any transpositions of the columns. Indeed, any permutation is a composition of transpositions (ij); actually, the *simple* ones $(i\,i{+}1)$ are sufficient, which we do not need here. For the composition of $(ij)_{row}$ for rows $i < j$ and $(kl)_{col}$ for columns $k < l$, let the intersection of the union of rows i, j and the union of columns k, l be $\begin{pmatrix} a & b \\ c & d \end{pmatrix}$ for some a, b, c, d. Then it will be transformed by $(ij)_{row}\,(kl)_{col}$ or by $(kl)_{col}\,(ij)_{row}$ to $\begin{pmatrix} d & c \\ b & a \end{pmatrix}$. So their order does not matter. Apart from this square (for the entries not in these rows or columns), the verification is simple: either $(ij)_{row}$ acts or $(kl)_{col}$ does.

(ii) To obtain the normal form, we permute the rows to ensure that $m_i < m_{i+1}$ for any $i = 1, \ldots, n-1$, and then permute the columns to make the entries in the 1st row increasing. Note that 1 is always the 1st entry.

For the uniqueness, it suffices to check that if $v_{col}u_{row}$ sends some SMS in the normal form to another one in the normal form then u, v are trivial (identity permutations). Indeed, this composition changes the order of $\{m_i\}$ according to u (for any v). So u must be the identity if the result of $v_{col}u_{row}$ is in the normal form. Thus $u = id$; but then v must be the identity too. ❑

Problem 7.1.10 *(i) Check up to permutations in rows and columns that any SMS of order* $n = 4$ *can be assumed either of type (A): with the first row* $1 < a < b < c$ *for* $a > 2$ *and* 2 *somewhere in the second row, or of type (B): with the first row* $1, 2, 15, 16$ *(in this order) and* 3 *anywhere in the second row.*

(ii) For $n = 4$, *let* N_A *be the total number of SMS of type (A), and* N_B *of type (B). Then the number of SMS in the normal form as in Problem 7.1.9 is* $(N_A + N_B)/2$, *and the number of* **all SMS** *of this size is* $24^2(N_A + N_B)/2$.

*(iii)** *Use a computer to check that* $N_A = 1804$, $N_B = 104$; *so the total number of semi-magic* 4×4 *squares is* $1908 \cdot 24^2/2 = 549504$. *Upon the*

division by the number of reflections and rotations, it gives 68688, *matching the total number of such magic squares up to symmetries from* [43].

Solution. Permuting the rows and the columns we can make 1 to be the first entry. If the first row contains 2 then the sum $1+2+15+16=34$ is the only possibility for this row, and we can make it $1,2,15,16$ (in this order) by permuting the last 3 columns. Otherwise, we can permute the remaining 3 rows and assume that 2 belongs to the second one.

Any non-trivial composition $v_{col}u_{row}$ of permutations u_{row}, which is u permuting the rows, and v_{col}, which is v permuting the columns, such that this composition preserves type (A) or type (B) can be only the transposition of the last two rows. Indeed, any v_{col} in this composition must fix the 1st row due to 1 there, and fix the 2nd due to 2 or 3 there. Since u_{row} for such v changes the order of the entries in the 1st row, u_{row} can only be the identity.

Finally, different pairs u, v transform any given SMS to different ones, the size of the group of all corresponding $v_{col}u_{row}$ is $(4!)^2$. There are only 2 symmetries in this group preserving the type, (A) or (B). ❑

We note that one can obtain 300 different MS satisfying the inequalities $m_1 < m_2 < m_3 < m_4$ from all SMS of size 4 in the normal form by permuting (only) the columns. The numbers of such permutations of the columns can be $0,2,4,6,8$ for any given SMS. There are, respectively, $863, 46, 37, 2, 6$ different SMS in the normal form with the corresponding numbers of such permutations in columns. To give an example, $\begin{pmatrix} \mathbf{1} & 4 & 14 & 15 \\ 13 & 16 & \mathbf{2} & 3 \\ 8 & \mathbf{5} & 11 & 10 \\ 12 & 9 & 7 & \mathbf{6} \end{pmatrix}$ has 8 such permutations:

$$(1,2,3,4), (1,2,4,3), (2,1,3,4), (2,1,4,3), (3,4,1,2), (3,4,2,1), (4,3,1,2), (4,3,2,1).$$

For instance, the following magic squares correspond to the last 4 permutations in this list:

$$\begin{pmatrix} 14 & 15 & 1 & 4 \\ 2 & 3 & 13 & 16 \\ 11 & 10 & 8 & 5 \\ 7 & 6 & 12 & 9 \end{pmatrix}, \begin{pmatrix} 15 & 14 & 1 & 4 \\ 3 & 2 & 13 & 16 \\ 10 & 11 & 8 & 5 \\ 6 & 7 & 12 & 9 \end{pmatrix}, \begin{pmatrix} 14 & 15 & 4 & 1 \\ 2 & 3 & 16 & 13 \\ 11 & 10 & 5 & 8 \\ 7 & 6 & 9 & 12 \end{pmatrix}, \begin{pmatrix} 15 & 14 & 4 & 1 \\ 3 & 2 & 16 & 13 \\ 10 & 11 & 5 & 8 \\ 6 & 7 & 9 & 12 \end{pmatrix}.$$

As we always assume, here column i (generally, the contents of box i) is moved to position a_i (generally, to box a_i) for a given permutation $(a_1, a_2, \ldots)$. The corresponding sequence $\{m_1 < m_2 < m_3 < m_4\}$ is $\{1,2,5,6\}$ for this square. These numbers are in bold there.

[43]W. Trump, `http://www.trump.de/magic-squares/howmany.html`.

7.2 Recurrences and random processes

7.2.1 Solving recurrences

Problem 7.2.1 *Assuming that a pair of rabbits gives birth to 2 pairs every month starting with their second month, find the recurrence relation for the number of pairs of rabbits h_n at the nth month and solve it: find an algebraic formula for $h_1 = 1,\ h_2 = 1$. In this case, we begin with 1 young pair of rabbits, $h_3 = 3,\ h_4 = 5, \ldots$ and so on.*

Solution. We need only to change the (time to) maturity in the Fibonacci recurrence: $f_{n-2} \mapsto 2f_{n-2}$. The recurrence becomes $h_n = h_{n-1} + 2h_{n-2}$ and $h_0 = (h_2 - h_1)/2 = 0$. Thus: $\lambda^2 = \lambda + 2,\ \lambda_{1,2} = 2, -1,\ C_2 = -C_1,\ 1 = 2C_1 - C_2 = 3C_1$, and $C_1 = 1/3$. Finally, $h_n = (1/3)2^n - (1/3)(-1)^n$. ❑

Let us now consider the population of a species, "intermediate" between rabbits and bacteria. The latter multiply and never die, and the maturity does not apply to them. If we deal with something like *drosophila,* the common fruit fly, the maturity can be ignored ("too short-lived"), but they do die. Also, we count them in thousands, so f_n is generally fractional.

Problem 7.2.2 *Solve algebraically the "drosophila recurrence" $f_n = \frac{5}{2}f_{n-1} - f_{n-2}$ for the initial $f_1 = 2, f_2 = 5$.*

Solution. We have $f_3 = 25/2-2=21/2,\ f_0 = 0$ (check). The master equation is $\lambda^2 = \frac{5}{2}\lambda - 1$ and $\lambda_{1,2} = \frac{5}{4} \pm \sqrt{\frac{25}{16} - 1} = 5/4 \pm 3/4$: $\lambda_1 = 2, \lambda_2 = 1/2$. Setting $f_n = C_1\lambda_1^n + C_2\lambda_2^n$, we obtain that $C_2 = -C_1$ and $2C_1 + C_2/2 = 2$, which gives that $\frac{3}{2}C_1 = 2$, and $C_1 = 4/3$. Finally, $f_n = \frac{4}{3}(2^n - (\frac{1}{2})^n)$. To double check: $f_2 = 5 = \frac{4}{3}(4 - 1/4)$. Obviously, only powers of 2 can be expected in the denominators of f_n due to the recurrence. The coefficient $\frac{4}{3}$ does not contradict this: $2^n - (1/2)^n = 0 \bmod 3$. ❑

Problem 7.2.3 *Let us use 6 colors, namely b, w and r, g, c, y, to color the line of n consecutive boxes. We assume that only colors r, g can be used after □ (white) and only r, g, c after ■ (black). Find the recurrence for the number h_n of such colorings, and solve it algebraically (in terms of $\lambda_{1,2}$).*

Solution. Obviously, $h_1 = 6$. One has: $h_2 = 6^2 - 3 - 4 = 29$, because w, b, c, y are not allowed after □ and w, b, y are not allowed after ■. The recurrence is as follows. If the first box is □, then only r, g are allowed after it and then we have no restrictions for box 3. If the first box is ■, then only r, g, c are allowed after it, and then there will be no further restrictions. Combining, we obtain $5h_{n-2}$ colorings in these two cases.

If the first box is not b or w, then we have 4 possibilities for the first color, r, g, c, y, and no further restrictions. We arrive at $4h_{n-1}$. Finally, $h_n = 4h_{n-1} + 5h_{n-2}$.

The master equation $\lambda^2 = 4\lambda + 5$ has roots $\lambda_1 = 5,\ \lambda_2 = -1$. Thus $h_n = C_1 5^n + C_2(-1)^n$. Using that $h_1 = 6, h_2 = 29$, we obtain that $h_0 = 1$; indeed, $h_2 = 29 = 4h_1 + 5h_0 = 4 \cdot 6 + 5$. Then we determine $C_{1,2}$ as follows:

$$C_2 = 1 - C_1,\ 5C_1 - (1 - C_1) = 6 = 6C_1 - 1,\ C_1 = \frac{7}{6}, C_2 = -\frac{1}{6}.$$

Finally, $h_n = \frac{7}{6}5^n - \frac{1}{6}(-1)^n$. To double-check this formula: $h_1 = \frac{7}{6}5 + \frac{1}{6} = 6$, and, indeed, $h_2 = \frac{7}{6}5^2 - \frac{1}{6} = \frac{7 \cdot 25 - 1}{6} = \frac{174}{6} = 29$.

Note that $7 \cdot 5^n - (-1)^n = 0 \bmod 6$, so the integrality of h_n can be easily seen directly from the formula. ❑

Problem 7.2.4 *With 5 colors A, B, C, D, W, find 2 initial values and the recurrence relation for the number k_n of colorings of n consecutive boxes by colors A, B, C, D, W avoiding AA, BB, CC, DD anywhere (WW is allowed). Solve it algebraically.*

Solution. For this kind of composing, it is convenient to do the passage from $(n-1)$ to n, what we called "growing stalagmites" in Section 4.3.2.

There are always at least 4 ways to continue coloring from k_{n-1} to k_n. If the last box is "W", there is an extra way: "W" after "W". The number of such exceptional cases is k_{n-2} and we obtain:

$$\begin{aligned}
k_n &= 4k_{n-1} + k_{n-2},\quad \lambda^2 = 4\lambda + 1,\quad \lambda_{1,2} = 2 \pm \sqrt{5},\\
k_n &= C_1\lambda_1^n + C_2\lambda_2^n,\ k_1 = 5,\ k_2 = 25 - 4 = 21,\ k_0 = 21 - 4 \cdot 5 = 1,\\
C_2 &= 1 - C_1,\ C_1\lambda_1 + C_2\lambda_2 = 5 = C_1(\lambda_1 - \lambda_2) + \lambda_2,\\
C_1 &= (5 - \lambda_2)/(\lambda_1 - \lambda_2) = (3 + \sqrt{5})/(2\sqrt{5}) = (5 + 3\sqrt{5})/10,
\end{aligned}$$

and $C_2 = (5 - 3\sqrt{5})/10\ (= C_1(\sqrt{5} \mapsto -\sqrt{5}))$. ❑

Problem 7.2.5 * *(i) Using $k_n, \lambda_{1,2}, C_{1,2}$ from the previous problem, check that $k_{n+1}/k_n - \lambda_1 = (\lambda_2 - \lambda_1)\sum_{m=1}^{\infty}(-z_n)^m$ for $z_n = \left(\frac{C_2\lambda_2^n}{C_1\lambda_1^n}\right)$, which is a convergent series. For instance, one has: $k_{n+1}/k_n - \lambda_1 = (\lambda_2 - \lambda_1)z_n + \epsilon_n$, where $|\epsilon_n| < |(\lambda_2 - \lambda_1)(z_n)^2|$ for odd n. Approximately, $|k_{n+1}/k_n - \lambda_1| \approx 0.652 \cdot 0.056^n$.*

(ii) Similarly, check that $|f_{n+1}/f_n - \frac{1+\sqrt{5}}{2}| \approx 2.236 \cdot 0.382^n$ for the Fibonacci numbers f_n. Conclude that about $3n$ terms for the Fibonacci numbers are needed to reach the same order of approximation for $\sqrt{5}$ versus the usage of the sequence k_{n+1}/k_n.

Solution. **(i)** One has:

$$\frac{k_{n+1}}{k_n} = \frac{C_1\lambda_1^{n+1} + C_2\lambda_2^{n+1}}{C_1\lambda_1^n + C_2\lambda_2^n} = \frac{\lambda_1 + \lambda_2 z_n}{1+z_n}$$
$$= \frac{\lambda_1(1+z_n) + (\lambda_2 - \lambda_1)z_n}{1+z_n} = \lambda_1 + \frac{(\lambda_2-\lambda_1)z_n}{1+z_n}.$$

Then expand: $\frac{1}{1+z} = \sum_{m=0}^{\infty}(-z)^m$. We obtain that

$$\frac{k_{n+1}}{k_n} = \lambda_1 + (\lambda_2 - \lambda_1)\sum_{m=1}^{\infty}(-z_n)^m.$$

The inequality for ϵ_n is because $z_n > 0$ for odd n. Check numerically that $\frac{C_1}{C_2} \approx -0.146$, $\frac{\lambda_2}{\lambda_1} \approx -0.056$.

(ii) For the Fibonacci numbers, the ratio is $\frac{\lambda_2}{\lambda_1} \approx -0.382$, "worse" than that in (ii). Since 0.382^3 is almost exactly 0.056, we need f_{3n+1}/f_{3n} to reach the same accuracy as with k_{n+1}/k_n for calculating $\sqrt{5}$. ❑

Problem 7.2.6 *Use Binet's formula for the Fibonacci numbers f_n and Newton's binomial to check that f_n is divisible by 5 if and only if 5 divides n.*

Solution. The following identity is complementary to (4.3.13):

$$\frac{(1+x)^n - (1-x)^n}{2} = \frac{1 + nx + \binom{n}{2}x^2 + \binom{n}{3}x^3 + \dots}{2} - \frac{1 - nx + \binom{n}{2}x^2 - \binom{n}{3}x^3 + \dots}{2}$$
$$= \frac{2nx + 2\binom{n}{3}x^3 + \dots}{2} = \sum_{m=0}^{2m+1\le n}\binom{n}{2m+1}x^{2m+1}. \tag{7.2.3}$$

Let us combine it with (4.2.8) for $n = 0, 1, \dots$:

$$f_n = \frac{\left(\frac{1+\sqrt{5}}{2}\right)^n - \left(\frac{1-\sqrt{5}}{2}\right)^n}{\sqrt{5}} = \frac{\left(1+\sqrt{5}\right)^n - \left(1-\sqrt{5}\right)^n}{2^n\sqrt{5}}. \tag{7.2.4}$$

Upon $x \mapsto \sqrt{5}$ in (7.2.3):

$$f_n = \frac{n\sqrt{5} + \binom{n}{3}5\sqrt{5} + \binom{n}{5}5^2\sqrt{5} + \dots}{2^{n-1}\sqrt{5}} \tag{7.2.5}$$
$$= \frac{n + \binom{n}{3}5 + \binom{n}{5}5^2 + \dots}{2^{n-1}} = \frac{1}{2^{n-1}}\sum_{m=0}^{2m+1\le n}\binom{n}{2m+1}5^m.$$

All terms in the last sum are divisible by 5 except for the first one, n. So its divisibility by 5 is if and only if 5 divides n. The divisibility of this sum by 2^n is far from obvious in this approach, but we already know this, ... since Fibonacci rabbits are integral. Here the field with 4 elements can be used (we omit this). Since $\gcd(2,5) = 1$, the powers of 2 do not influence the divisibility of this sum by 5, and we obtain the required. ❑

Problem 7.2.7 *(i) Any solution* $\widetilde{f}_n$ *of the classical Fibonacci recurrence is a lincar combination* $C_1 f_n + C_2 d_n$ *of the Fibonacci numbers* $f_n = 0, 1, 1, \ldots$ *for* $n = 0, 1, \ldots$ *and* $d_n = f_{n+1} = 1, 1, 2, \ldots$. *If* $\widetilde{f}_n$ *are integers, then so are* $C_{1,2}$.

(ii) Extend (i) to any recurrence $f_n = f_{n-1} + \sum_{i=1}^{k} a_i f_{n-i}$ *for* $k \geq 1$ *with integral* a_i. *Let* $f_0 = f_1 = \cdots = f_{k-2} = 0, f_{k-1} = 1$, *and we set* $f_n^j = f_{n+j}$ *for* $j = 0, \ldots, k-1$. *This set of* k *solutions generalizes the pair* $\{f_n, d_n\}$ *above.*

Solution. **(i)** Since the coefficients of this recurrence are constant (do not depend on n), the sequence $\{f_{n+m}\}$ for any m is a solution too. Importantly, f and d are not proportional to each other. Generally, if $d_n = f_{n+1} = cf_n$ for a solution f_n of any recurrence and some c (and all $n \geq 0$), then $f_{n+1} = cf_n$ and $f_n = c^n f_0$. This is obviously not the case with the Fibonacci sequence: the matrix $\begin{pmatrix} f_0 & f_1 \\ d_0 & d_1 \end{pmatrix} = \begin{pmatrix} 0 & 1 \\ 1 & 1 \end{pmatrix}$ is invertible. Moreover, its determinant is -1, and the integrality of $\widetilde{f}_1, \widetilde{f}_2$ implies that for $C_{1,2}$.

The usage of "shifted" f_n as the second solution is convenient for obtaining $\widetilde{f}_n$, especially when f_n are integral by construction. This allows us to eliminate the calculation of $\lambda_{1,2}$, $\widetilde{C}_{1,2}$ for $\widetilde{f}_n$ from scratch.

(ii) The corresponding matrix of initial values becomes

$$\begin{pmatrix} f_0 & f_1 & \cdots & f_{k-1} \\ f_1 & f_2 & \cdots & f_k \\ f_2 & f_3 & \cdots & f_{k+1} \\ \cdots & \cdots & \cdots & \cdots \end{pmatrix} = \begin{pmatrix} 0 & \cdots & 0 & 0 & 1 \\ 0 & \cdots & 0 & 1 & 1 \\ 0 & \cdots & 1 & 1 & 1+a_1 \\ \cdots & \cdots & \cdots & \cdots & \cdots \end{pmatrix}.$$

Its determinant is $(-1)^{k-1}$ due to zeros above the anti-diagonal. ❑

7.2.2 Transfer matrix

Perfect coverings of $4 \times n$.

Problem 7.2.8 *Count* $\mathcal{N} = \mathcal{N}_{4\times 4}^{\square\square}$, *the number of perfect coverings of the* 4×4 *shape by dominos. For this, split it by the (horizontal) middle line into 2 shapes of dimensions* 2×4, *and analyze the dominos intersecting this line.*

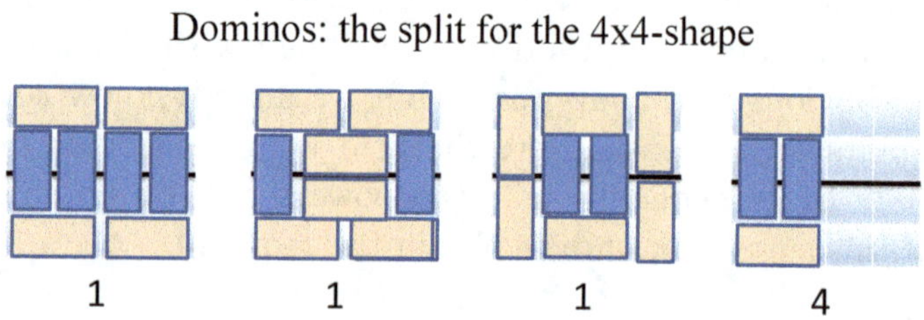

Figure 7.1: Covering 4×4.

Solution. If the horizontal middle line does not intersect the dominos, then the number of such perfect coverings is d_4^2 for $d_4 = f_5 = 5$. If it intersects 4 dominos, then the number of such coverings is 1. The latter case and the remaining 3 cases of possible intersections are in Fig. 7.1; blue dominos are those intersected. The last case must be taken with multiplicity 2; its rotation by π must be added. Finally, $\mathcal{N}=5^2+1+1+1+2\cdot 4=36$. ❑

We provided the recurrence relation for $h_n = \mathcal{N}^{\square\square}_{4\times n}$ in Problem 4.3.3: $h_{n+1}-h_n-5h_{n-1}-h_{n-2}+h_{n-3}=0$. One has: $h_4-h_3-5h_2-h_1+h_0$, where $h_0=1, h_1=1, h_2=d_4=5,\ h_3=g_4=11$. Indeed, $36-11-25-1+1=0$. The proof is in the next Problem 7.2.9.

Problem 7.2.9 * *Prove the relation* $h_n = h_{n-1}+5h_{n-2}+h_{n-3}-h_{n-4}$ *for* $h_n = \mathcal{N}^{\square\square}_{4\times n}$ *for* $n=4,5,\ldots$. *Recall that* $h_0=1, h_1=1, h_2=5, h_3=11$, *and we found that* $h_4=36$.

Solution. Setting $x_n = \mathcal{N}$ 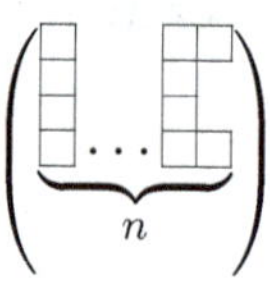, $y_n = \mathcal{N}$ 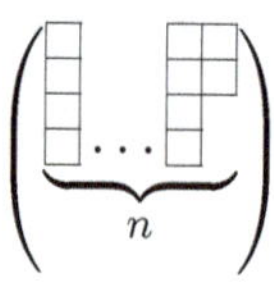 :

$x_n = \mathcal{N}$ 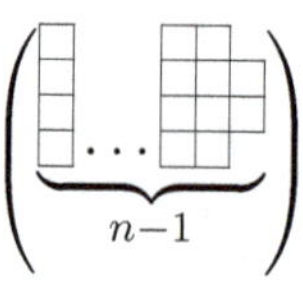 $= x_{n-2}+h_{n-2},\quad y_n = h_{n-1}+y_{n-1},$

$h_n = \mathcal{N}$ 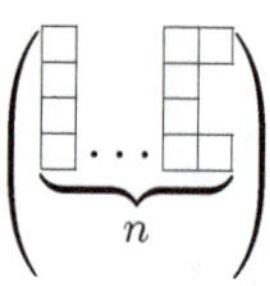 $+\,\mathcal{N}$ 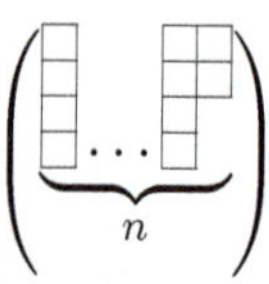 $+\,\mathcal{N}$ 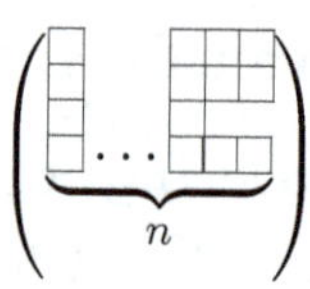

We obtain that $h_n = x_n + y_n + 2h_{n-2} + y_{n-2}$, where we will substitute:

$$x_n = h_{n-2}+h_{n-4}+h_{n-6}+\ldots,\ y_n = h_{n-1}+h_{n-2}+h_{n-3}+\ldots,$$

using the recurrence relations (above) for x_n, y_n. Then, we perform the following algebraic manipulation:

$$
\begin{aligned}
&h_n = (h_{n-2}+h_{n-4}+h_{n-6}+\cdots) + (h_{n-1}+h_{n-2}+h_{n-3}+\cdots) + 2h_{n-2}\\
&+(h_{n-3}+h_{n-4}+\ldots) = h_{n-1}+4h_{n-2}+2h_{n-3}+3h_{n-4}+2h_{n-5}\\
&+3h_{n-6}+\ldots = h_{n-1}+5h_{n-2}+2\underbrace{h_{n-3}}+3\underline{h_{n-4}}+\boxed{2h_{n-5}+3h_{n-6}+\cdots}\\
&-(\underbrace{h_{n-3}}+4\underline{h_{n-4}}+\boxed{2h_{n-5}+3h_{n-6}+\ldots}) = h_{n-1}+5h_{n-2}+h_{n-3}-h_{n-4}.
\end{aligned}
$$

In the last infinite sum, we changed $4h_{n-2} \mapsto 5h_{n-2}-h_{n-2}$ and replaced there $-h_{n-2}$ by the quantity $-(\cdots)$. Then we reduced similar terms; the boxed expressions cancel each other. The reduction of the infinite sums to some finite expressions can be surprising, but this is some general fact. ❑

The approach from Problem 4.3.4 results in the following system for the *types* $s_0 = \emptyset$, $s_1 = \boxed{XXXX}$, $s_2 = \boxed{X\ \ \ \ X}$, $s_3 = \boxed{\ \ \ \ XX}$, $s_4 = \boxed{\ \ XX\ \ }$, where we plot the first columns of "$4\times n$" horizontally and indicate by X the boxes removed there. Let $h_n^{(i)}$ be the number of perfect coverings by dominos of the remaining shape; n is the initial number of columns. For instance, $h_n^{(0)} = h_n$ (there is nothing to remove), $h_n^{(1)} = h_{n-1}$. For the vector $H_n = \{h_n^{(0)}, h_n^{(1)}, h_n^{(2)}, h_n^{(3)}, h_n^{(4)}\}$ considered as a column, the recurrence relation is

$$
H_n = MH_{n-1} \text{ for } M = \begin{pmatrix} 1 & 1 & 1 & 2 & 0\\ 1 & 0 & 0 & 0 & 0\\ 1 & 0 & 0 & 0 & 1\\ 1 & 0 & 0 & 1 & 0\\ 0 & 0 & 1 & 0 & 0 \end{pmatrix}.
$$
The characteristic polynomial of M is $Det(M - \mathbf{1}_5 x) = 1-2x-4x^2+4x^3+2x^4-x^5 = (1-x)(x^4-x^3-5x^2-x+1)$. This polynomial provides the recurrence relation for h_n due to Theorem 4.2.3. The one serving the recurrence above is the 2nd factor.

Problem 7.2.10 *Let* $d_n = \mathcal{N}^{\square\square}_{2\times n}$. *Following Fig. 7.1, split "$4\times n$" into two shapes of dimensions $2\times n$ using the middle line M there, and check that*

$$
h_n = d_n^2 + \sum_{1\le m\le n/2}\ \sum_{\{a_j\}} \binom{\frac{n-(a_1+a_2+\cdots+a_{m+1})}{2}-1}{m-1} d_{a_1}^2 d_{a_2}^2 \cdots d_{a_{m+1}}^2,
$$

where the difference $n-(a_1+a_2+\cdots+a_{m+1})$ *is even from* $2m$ *to* n*, and the summation is over all such possible* $a_j \ge 0$*. For instance,* $h_5 = 95$ *from this formula, which coincides with* $h_5 = h_4 + 5h_3 + h_2 - h_1$.

Solution. So we divide the shape $4\times n$ by its middle horizontal line M into two shapes "$2\times n$"; this line may intersect dominos. Let us consider two dominos

d, d', intersecting M assuming that there are no such dominos (intersecting M) between them. By a *block* from d to d' (d before d'), we mean a portion of "$4\times n$" between the 4-columns containing d and d', *including these columns.* The total number of dominos intersecting M is even, and we obtain a natural system of consecutive non-overlapping blocks in "$4\times n$". For instance, there are $2, 1, 1, 1$ such blocks in the covering from Fig. 7.1, respectively.

Let us check that the covering is completely determined by a sequence of independent coverings of the rectangles before, after, and between the natural blocks above. Let d be the 1st domino intersecting M (leftmost). Then the domino through the box on top of d must belong to the block from d to d' that begins with d. Indeed, it cannot intersect the rectangle before d because the number of remaining boxes to cover beyond M and before d will then be odd (non-coverable). We use that the dominos before d do not intersect M. Similarly, the dominos through the boxes below d and those above/below d' must belong to this block. So we can continue by induction.

Moreover, the blocks themselves can be covered uniquely: use that there is a unique way to cover the boxes above/below d, d'. We obtain that the total number of perfect coverings of "$4\times n$" is the product of those for the rectangles before/after/between the blocks.

Let the blocks be of the size $4\times p_i$ for some p_i, where $1 \le i \le m$: m is the number of such blocks. So we need to count the number of possibilities for $\{p_i\}$, and the number of perfect coverings of the rectangles $4\times a_j$ (they can be empty) before/after/between such blocks. Here $1 \le j \le m+1$, but some of a_j can be zero. The relation between $\{p_i\}$ and $\{a_j\}$ is $p_1+\cdots+p_m = n-(a_1+\cdots+a_{m+1})$. This gives the required formula. ❑

The previous problem and the next two are some introduction to the ***method of transfer matrices***, where we represent any shape $M\times N$ as the union of its rows, $1\times N$ shapes, and analyze possible intersections of the dominos with the "fault lines" between the rows.

Problem 7.2.11 *Let* $g_n = \mathcal{N}^{\square\square}_{3\times n}$ *from Section 1.1.1 and* $d_n = \mathcal{N}^{\square\square}_{2\times n}$, *considered above. Similar to Problem 7.2.10, split the* $3\times n$ *shape into "*$1\times n$*", the bottom row of boxes, and "*$2\times n$*", the 2 top rows of boxes; let* L *be the horizontal line between them. Use this to prove the following. Let* $\mathcal{M}$ *be the set of integers* $\{0 \le m_0 < m_1 < m_2 < \ldots < m_{k-1} < m_k = n\}$ *for* $k = 0, 1, \ldots, n$. *Then:*

$$g_{2n} = \sum_{\mathcal{M}} (d_{2m_0} 2^{m_1-m_2-1} d_{2m_2-2m_2} 2^{m_3-m_4-1} \ldots),$$

where the last term is either $d_{2n-2m_{k-1}}$ *or* $2^{n-m_{k-1}-1}$. *For instance, let us* *provide $\mathcal{M}$ and the corresponding terms for $n=3$:*

$$\begin{array}{cccccccc} \{3\} & \{2,3\} & \{1,2,3\} & \{1,3\} & \{0,3\} & \{0,2,3\} & \{0,1,3\} & \{0,1,2,3\} \\ d_6, & d_4, & d_2d_2, & d_22, & d_02^2, & d_02d_2, & d_0d_4, & d_0d_2\ . \end{array}$$

Using $d_0 = 1, d_2 = 2, d_4 = 5, d_6 = 13$, we obtain that $g_6 = 13+5+4+4+4+4+5+2 = 41$, which matches g_6 from Section 1.1.1.

Solution. Let us make the boxes in the vertical dominos black and white otherwise. We will focus on the bottom row $1\times n$ of the rectangle $2\times n$ above L, a sequence of black and white boxes.

An *alt-block* is a rectangle formed by columns of "$2\times n$" above L with the following bottom row. It must begin with a maximal black segment and end with a maximal black segment. Then, any *maximal* black segment or white segment, between them must be of even length. A *segment* is a connected sequence of boxes; it is called maximal of some color if it cannot be increased as such. A simple alt-block is for white boxes between 2 black ones:

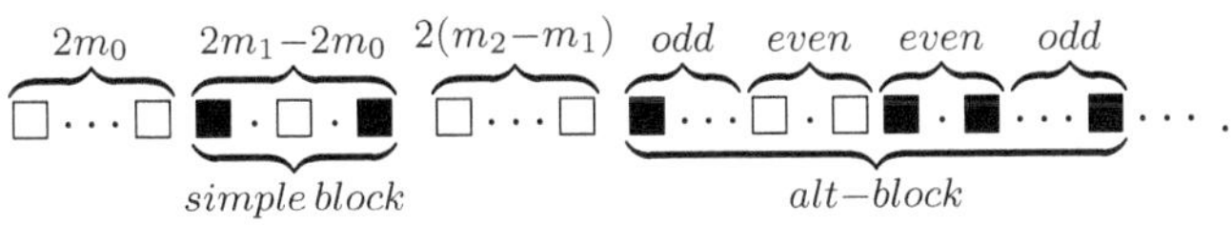

Practically, we begin with the first (leftmost) box in the bottom row and find the first maximal black segment. If it is of even length, then it is an alt-block and we continue. If its length is odd, we go to the next maximal odd black segment. The segment from the first maximal segment to the second one (including them) will be an alt-block, and we continue. Maximal white segments must be of even size to make the corresponding portions of "$1\times n$" below L coverable by dominos. This row is identical to the bottom row of "$2\times n$" : a vertical domino crossing L covers both, its box beyond L and the one below it. If the row below L is coverable with a particular combination of black boxes (excluded), then this can be done uniquely.

We obtain that "$2\times n$" above L must be a union of alt-blocks and maximal white rectangles with an even number of columns between them to ensure that the corresponding shape $1\times n$ below L is coverable. Moreover, the dominos covering the white boxes on top of the black ones at the 2 ends of a given alt-block must belong to this alt-block, and there is only 1 way to cover it. Recall that black boxes are covered by vertical dominos crossing L. See:

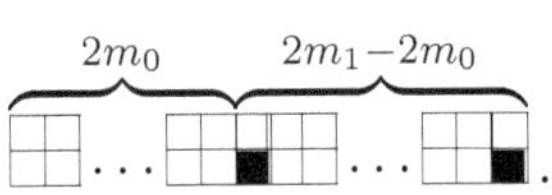

So we have 1 simple alt-block here of length $2m_1 - 2m_0$. Obviously, the only way to cover this shape is when we cover the $2\times 2m_0$ white shape by dominos, which gives d_{2m_0} choices, and then put the domino on top of ■ like this: ▆. Then we have no choice but to put dominos alternatively: at the bottom, at the top, at the bottom, and so on; the last one will be inside this block.

We define *reduced alt-blocks* by removing the 1st and the last columns of alt-blocks. Accordingly, their bottom rows are unions of maximal black and white segments of *even* length.

Consider alt-blocks of length $2m$ with a fixed complementary set in "$2\times n$". The corresponding reduced alt-blocks are of length $2m-2$, and they are 1-1 with arbitrary sets of vertical lines between the columns in $\overbrace{\square\square\cdots\square}^{m-1}$. We use that their maximal white/black segments are of *even* length, so here we divide $2m-2$ by 2. A vertical line before the first column is needed to address the initial color of the bottom row of reduced alt-blocks.

We obtain that the number of alt-blocks of length $2m$ with a fixed complementary set is 2^{m-1}, and arrive at the required formula. ❑

Problem 7.2.12 * *Let $N_n^m = N_{m\times n}^{\square\square}$ for fixed even $m > 0$ and any $n \geq 0$; we set $N_0^m = 1$. Consider arbitrary sequences s of boxes in the first column $c =$ "$m \times 1$" of "$m\times n$" with even number of boxes, including the empty set $\emptyset$ and the whole c. Let N_n^s be the number of perfect coverings of the shape X_n^s obtained by removing s from "$m\times n$". For instance, $N_n^{\emptyset} = N_n^m$ and $N_n^c = N_{n-1}^m$. We set $M_{s,s'} = 1$ for even sets s, s' if $s\cap s' = \emptyset$ and if the maximal segments in $c\setminus(s\cup s')$ are all of an even size; otherwise $M_{s,s'} = 0$.*

(i) Prove that $N_n^s = \sum_{s'} M_{s,s'} N_{n-1}^{s'}$, where $n \geq 1$ and the summation is over all possible sequences s' of even size.

(ii) Let $P(x)$ be the characteristic polynomial $Det(M-\mathbf{1}x)$ for M considered as a matrix of size $2^{m-1}\times 2^{m-1}$. Setting $P(x) = x^m - \sum_{i=1}^{2^{m-1}} a_i x^i$, one has: $N_n^m = \sum_{i=1}^{2^{m-1}} a_i N_{n-i}^m$.

(iii) Use a computer to calculate the transfer matrix *M (of size $2^{m-1}\times 2^{m-1}$) for $m = 6$ and the corresponding $P(x)$. Let $Q(x)$ be from the master equation $Q(\lambda) = 0$ of the* minimal recurrence *for N_n^6. Check that Q divides P.*

Solution. We go through all possible coverings of the boxes in the 1st column of X_n^s. The first boxes of the horizontal ones constitute exactly s' with $M_{s,s'} = 1$. Upon the removal of these dominos the shape becomes $X_{n-1}^{s'}$. If the horizontal dominos are known, then the vertical ones (strictly inside the 1st column) are uniquely determined. This proves (i). Then we use Theorem 4.2.3.

We will omit the matrix M for $m = 6$, which is of order 64. However, let us provide the corresponding characteristic polynomial $P(x)$, which equals : $x^{32}(1-x^2)(1-5x+6x^2-x^3)(1+6x+5x^2+x^3)(1-x-2x^2+x^3)^2(1-2x-x^2+x^3)^2(-1-2x+x^2+x^3)^2(-1-x+2x^2+x^3)^2$. The product $(1-x^2)(1-5x+6x^2-x^3)(1+6x+5x^2+x^3) =$

$x^8 - x^7 - 20x^6 - 10x^5 + 38x^4 + 10x^3 - 20x^2 + x + 1$ results in the *minimal* recurrence relation for N_n^6, obtained numerically after Problem 4.3.4. ❑

Problem 7.2.13 * *Transfer matrix method. (i) For even $m > 0$, prove that N_n^m coincides with the first entry of the matrix power M^n, which is $(M^n)_{\emptyset,\emptyset}$.*

(ii) Use a computer to find that $N_8^8 = 12988816 = 2^4(901)^2$ and calculate N_n^8 for $n \le 50$. Then check for $50 \ge n \ge 16$ the recurrence relation:

$$\begin{aligned} N_n^8 &= N_{n-1}^8 + 76N_{n-2}^8 + 69N_{n-3}^8 - 921N_{n-4}^8 - 584N_{n-5}^8 \\ &+ 4019N_{n-6}^8 + 829N_{n-7}^8 - 7012N_{n-8}^8 + 829N_{n-9}^8 + 4019N_{n-10}^8 \\ &- 584N_{n-11}^8 - 921N_{n-12}^8 + 69N_{n-13}^8 + 76N_{n-14}^8 + N_{n-15}^8 - N_{n-16}^8. \end{aligned}$$

Solution. Claim (i) follows from Problem 7.2.12(i). The recurrence from (ii) can be justified for *any* $n \ge 16$ as follows. One calculates the characteristic polynomial $P(x)$ for $m = 8$ from Problem 7.2.12(ii). Then the polynomial $x^{16} - x^{15} - 76x^{14} - 69x^{13} + 921x^{12} + 584x^{11} - 4019x^{10} - 829x^9 + 7012x^8 - 829x^7 - 4019x^6 + 584x^5 + 921x^4 - 69x^3 - 76x^2 - x + 1$ is its (irreducible) factor. This and the numerical checks of the recurrence for sufficiently many n are sufficient for the proof. The passage to the recurrence is as above: $x^{16} - x^{15} - 76x^{14} - 69x^{13} + \ldots$ corresponds to $N_n^8 = N_{n-1}^8 + 76N_{n-2}^8 + 69N_{n-2}^8 + \ldots$. ❑

Let us provide the polynomial corresponding to the recurrence for $m = 10$: $x^{32} - x^{31} - 285x^{30} - 411x^{29} + 18027x^{28} + 20689x^{27} - 472275x^{26} - 271027x^{25} + 6149853x^{24} + 471319x^{23} - 42303393x^{22} + 10402780x^{21} + 157353820x^{20} - 58545372x^{19} - 335484428x^{18} + 123321948x^{17} + 429447820x^{16} - 123321948x^{15} - 335484428x^{14} + 58545372x^{13} + 157353820x^{12} - 10402780x^{11} - 42303393x^{10} - 471319x^9 + 6149853x^8 + 271027x^7 - 472275x^6 - 20689x^5 + 18027x^4 + 411x^3 - 285x^2 + x + 1$.
Thus, $N_n^{10} = N_{n-1}^{10} + 285N_{n-2}^{10} + 411N_{n-3}^{10} + \ldots$.

Comments. (a) Note that $\lim_{n\to\infty} N_{n+1}^m / N_n^m = \lambda_1$, where $\lambda_1 \approx 9.007$ for $m = 8$ and ≈ 16.094, for $m = 10$. Accordingly, $\lambda_1^{2/m} \approx 1.732, 1.743$; the limit of $\lambda_1^{2/m}$ as $m \to \infty$ is $e^{2G/\pi} \approx 1.792$ for Catalan's constant $G = 1 - \frac{1}{3^2} + \frac{1}{5^2} - \frac{1}{7^2} + \ldots$.

(b) The fact that such polynomials are always of degree $\frac{m}{2}$ or smaller follows from Kasteleyn's formula. See formula (2) in Richard Stanley's paper [44]. At the end, the conjecture is discussed there that they always have distinct roots, equivalently, $\frac{m}{2}$ is the actual depth of the recurrence for N_n^m. Jeffrey Lagarias proved that such polynomials are *irreducible* over rational numbers if $m + 1$ is a *prime number* $\neq 7$, which readily gives that their roots are simple.

[44] R. Stanley, *On dimer coverings of rectangles of fixed width*, Disc. Appl. Math. 12 (1985), 81–87.

(c) The "reciprocity" $N^m_{-n-2} = (-1)^{\frac{mn}{2}} N^m_n$ is due to Stanley. See also James Propp's remarks in [45]. The symmetries of M from Problem 7.2.13 provide this relation, though Stanley obtained them from Kasteleyn's formula.

7.2.3 Modeling reinfections *

A very convenient way to introduce this distribution is via *reinfections*. Continuing Exercises 2.6.2 and 2.6.3, let us assume that the probability of being exposed to the infection during one day is β, the duration of 1 epidemic cycle is N days, and the immunity period is L. The latter means that after being infected on day x, an individual cannot be reinfected on days $x+1,\ldots,x+L$. Otherwise, all exposed individuals are assumed to become infected.

This is a lattice variant of *Matérn process II* in dimension one; see [46]. The immunity interval can of course continue *after* day N, which significantly impacts the formulas below. Thus, we consider sequences of r non-overlapping segments (no points in common) of the length $L+1$ with integer endpoints in the segment $[1,N]$, where the last segment is allowed to go beyond N. If these segments are within $[1,N]$, then the probability of one such configuration is $\beta^r(1-\beta)^{N-Lr-r}$. Let $P(r,N,\beta) = \binom{N-Lr}{r}\beta^r(1-\beta)^{N-Lr-r}$.

The following formula is from Exercise 2.6.3. Let $\pi_r = \pi_r(N,L)$ be the probability to be infected r times during N days. Then

$$\pi_r = P(r,N,\beta) + \beta\sum_{i=1}^{L} P(r-1,N-i,\beta),\ r=0,1,\ldots\ . \tag{7.2.6}$$

The summation here is when the last interval goes beyond the right endpoint. Indeed, it can be $[N-L+1,N+1]$, $[N-L+2,N+2]$, ..., $[N,N+L]$. To give an example, let $L=1$. Then $\pi_r = \beta^r(1-\beta)^{N-2r}\left(\binom{N-r+1}{r}+\beta\binom{N-r}{r}\right)$.

Problem 7.2.14 *For the probability $\pi_r(N,L)$ to be infected r times during N days from (7.2.6), let $G(t,u) \stackrel{\text{def}}{=} \sum_{N=0}^{\infty} t^N \pi_r(N,L)t^N u^r$, where we fix L. Then:*

$$G(t,u) = \frac{1}{1-t} + \frac{\beta\,(u-1)\,t}{(1-t)(1-(1-\beta)t-\beta u t^{L+1})}. \tag{7.2.7}$$

Solution. *First,* let $G^\circ_N = \sum_{r=0}^{\infty} P(r,N,\beta)u^r = \sum_{r=0}^{\infty} u^r\beta^r(1-\beta)^{N-Lr-r}\binom{N-Lr}{r}$. This is the generating function for tiling the shape "$1\times N$" with $(L+1)$-minos,

[45] J. Propp, *A reciprocity theorem for domino tilings*, The Electron. J. Combin. 8:1 (2001), R 18.

[46] I. Cherednik, *"Discrete Poisson hardcore 1D model and reinfections"*, arXiv:2202.09176 (2022).

sequences of $(L+1)$ consecutive boxes, and with 1-minos. For $u=1, \beta = 1/2, L=1$: $G^\circ_N = f_{N+1}/2^N$ for the Fibonacci numbers f_N. Generally:

$$G^\circ_N = (1-\beta)G^\circ_{N-1} + \beta u G^\circ_{N-L-1},\ G^\circ_0 = 1,\ G^\circ_i = (1-\beta)^i \text{ for } 1 \le i \le L.$$

For instance, $G^\circ_{L+1} = (1-\beta)^{L+1} + \beta u$. Using Proposition 4.2.4 or directly:

$$G^\circ(t,u) = \sum_{N=0}^{\infty} G^\circ_N t^N = \frac{1}{1-(1-\beta)t - \beta u t^{L+1}}.$$

Second, setting $G_N = \sum_{r=0}^\infty \pi_r(N,L)u^r$, *it satisfies the same recurrence* as for G°_N, but with different initial conditions. Consider how the first box can be covered. Check directly or deduce from Exercise 2.6.3 that $G(t,u) = (1+\beta ut + \beta ut^2 + \cdots + \beta ut^L)\, G^\circ(t,u)$. Finally,

$$\begin{aligned} G(t,u) &= \frac{1-t+\beta ut(1-t^L)}{(1-t)(1-(1-\beta)t-\beta ut^{L+1})} \\ &= \frac{1-(1-\beta)t-\beta ut^{L+1}+\beta ut-\beta t}{(1-t)(1-(1-\beta)t-\beta ut^{L+1})} = \frac{1}{1-t}\left(1+\frac{\beta\,(u-1)\,t}{(1-(1-\beta)t-\beta ut^{L+1})}\right). \end{aligned}$$ ❑

For $u=0$: $\sum_{N=0}^\infty \pi_0(N,L) = \frac{1}{1-(1-\beta)t}$, which we know without any calculations. For $u=1$: $\sum_{r=0}^\infty \pi_r(N,L)t^N = \frac{1}{1-t}$, which gives a combinatorial proof of the identity $\sum_{r=0}^\infty \pi_r(N,L) = 1$ for any N, L. This equality of course holds because $\{\pi_r(N,L)\}$ are probabilities for *all* possible numbers of infections.

Formula (7.2.7) readily gives that

$$\pi_r(N,L) = d^r\left(\frac{1}{1-t}\left(1+\frac{\beta\,(u-1)\,t}{1-(1-\beta)t-\beta ut^{L+1}}\right)\right)/du^r\,(u\mapsto 0). \tag{7.2.8}$$

Performing the differentiations, obtain the following "telescopic-type" presentation of π_r and its explicit expression in terms of binomial coefficients.

Problem 7.2.15 *Let* $\Pi_r(L) = \frac{\beta^r t^{(L+1)r-L}}{(1-t)(1-(1-\beta)t)^r}$ *for* $r \ge 1$ *and* $\Pi_0(L) = \frac{1}{1-t}$. *Then* $\sum_{N=0}^\infty \pi_r(N,L)t^N = \Pi_r(L) - \Pi_{r+1}(L)$ *for* $r \ge 1$. *This gives a more direct (algebraic) justification of the relation* $\sum_{r=0}^\infty \pi_r(N,L) = 1$ *for any* N, L. ❑

Problem 7.2.16 *Provided that* $N - L(r-1) - r \ge 0$,

$$\pi_r(N,L) = \beta^r \sum_{s=0}^{m} \binom{s+r-1}{r-1}(1-\beta)^{\max\{m-L,s\}},\quad m = N-L(r-1)-r.$$

There are $L+1$ *different powers of* $(1-\beta)$ *in the resulting expression with integral positive coefficients provided that* $m \ge L$. *Namely, these powers are* $(1-\beta)^{m-L+k}$ *for* $k = 0,1,\ldots,L$ *and the coefficients depend only on* k, r *except for* $k=0$. *For* $m < L$, *these powers are* $(1-\beta)^k$, *where* $0 \le k \le m$ *and the corresponding coefficients depend only on* k, r. ❑

Continuous limit. Combinatorially, $\pi_r > 0$ for $N \geq L(r-1) + r$, i.e. for $r \leq \frac{N+L}{L+1}$. For instance, $\pi_1 = 1 - (1-\beta)^N$ for *any* L such that $L > N$. Notice that π_0 does not depend on L; this is obvious because the duration of the immunity interval does not affect those non-infected.

We will set $\beta \stackrel{\text{def}}{=} \alpha/N$ for some $\alpha \geq 0$ and assume that $\lim_{N\to\infty} N/L = \nu \geq 0$, the problem is to calculate the limits $\pi'_r = \lim_{N\to\infty} \pi_r$. Let us provide the final formulas; their justification will be the next problem.

Let us, first, assume that $r\nu \leq 1$. Then

$$\pi'_r = e^{(r\nu-1)\alpha}\left(1 + \alpha(1-r\nu) + \frac{\alpha^2(1-r\nu)^2}{2} + \ldots + \frac{\alpha^r(1-r\nu)^r}{r!}\right) \tag{7.2.9}$$
$$- e^{((r-1)\nu-1)\alpha}\left(1 + \alpha(1-(r-1)\nu) + \ldots + \frac{\alpha^{r-1}(1-(r-1)\nu)^{r-1}}{(r-1)!}\right).$$

The last value of r here is $r_\flat$ =Floor$[\frac{1}{\nu}]$; Floor$[x]$ is the integer part of x. For such r, the corresponding π'_r are nonzero. There can be one more nonzero probability: π_r are nonzero if and only if $r \leq \frac{N+L}{L+1}$. Thus, π'_r can be nonzero for $r_\sharp$ =Ceiling$[\frac{N}{L}]$, which is $r_\flat + 1$ if $\frac{1}{\nu}$ is *not* an integer and $\pi_\flat$ if it is an integer. Indeed, $\frac{N}{L} + 1 > \frac{N+L}{L+1} > \frac{N}{L}$. When $r_\sharp = r_\flat + 1$:

$$\pi'_{r_\flat+1} = 1 - e^{(r_\flat\nu-1)\alpha}\left(1 + \alpha(1-r_\flat\nu) + \frac{\alpha^2(1-r_\flat\nu)^2}{2} + \ldots + \frac{\alpha^{r_\flat}(1-r_\flat\nu)^{r_\flat}}{(r_\flat)!}\right).$$

Problem 7.2.17 * *Assume that $N \to \infty$, $\beta = \alpha/N$, $\lim_{N\to\infty} L/N = \nu$ for some $\alpha > 0$ and $0 \leq \nu \leq 1$. Then justify formula (7.2.9) for any $0 \leq r \leq r_\flat$, and the additional formula above for $r_\sharp = r_\flat + 1$ when $\frac{1}{\nu}$ is not an integer.*

Solution. We use the formula $\pi_r = P(r, N, \beta) + \beta \sum_{i=1}^{L} P(r-1, N-i, \beta)$ for $r = 0, 1, \ldots$ from (7.2.6). Recall that the summation here is when the last interval goes beyond the right endpoint.

It can be presented as follows. Let $F_r(X) = d^{r-1}\left(X^{N-Lr}\,\frac{1-X^L}{1-X}\right)/dX^{r-1}$ (the $(r-1)$th derivative), where $r > 0, L > 0$ and $N - Lr > 0$. Then

$$\pi_r = \binom{N-Lr}{r}\beta^r(1-\beta)^{N-Lr-r} + \frac{\beta^r}{(r-1)!}F_r(X \mapsto 1-\beta). \tag{7.2.10}$$

We will omit the identification of (7.2.10) with (7.2.6). Use *Leibnitz Rule*: $(FG)^{(p)} = \sum_{i=0}^{p}\binom{p}{i}F^{(p-i)}G^{(i)}$ for the derivatives of the products.

The following step is non-trivial. We can simplify F_r from (7.2.10) when it is considered in the limit. Let $\Phi_r \stackrel{\text{def}}{=} \mathcal{B}^{r-1}\left(\frac{X^{N-Lr}-X^{N-L(r-1)}}{1-X}\right)$ for the following *formal* differentiation $\mathcal{B}$.

It acts in the ring generated by X^{N-sL} and $1/(1-X)$ *treated as independent symbols* and all their polynomial combinations. We need only to know what are the $\mathcal{B}$-derivatives of these symbols; then *Leibnitz Rule* will be used: $\mathcal{B}^p(FG) = \sum_{i=0}^{p} \binom{p}{i} \mathcal{B}^{p-i}(F)\mathcal{B}^i(G)$ for $\mathcal{B}$. We set:

$$\mathcal{B}^p(X^{N-sL})=(1-s\nu)^p X^{N-sL},\ \mathcal{B}^p\Big(\frac{1}{1-X}\Big)=\frac{\beta^p\, p!}{(1-X)^{p+1}},\ p=0,1,2,\ldots .$$

This differentiation is actually $\beta\, d/dX(\cdot)$ with the following simplifications due to taking the limit. First, we replace $\binom{N-sL}{r}$ by $\frac{N^r}{r!}$ for $s \le r$ due to $r \ll N-sL$. Second, $X^{N-sL\pm p}$ is replaced by X^{N-sL} for $0 \le p \le r$ because these powers will be finally evaluated at $X = 1-\beta = 1-\frac{\alpha}{N} \to 1$.

We obtain: $\pi'_r = \frac{\alpha^r(1-r\nu)^r}{r!}\, e^{(r\nu-1)\alpha} + \frac{1}{(r-1)!}\,\Phi'_r$, where Φ'_r will denote here and below the limit of $\beta\Phi_r$ after the evaluation $1-X \mapsto \beta$, $X^{N-Ls} \mapsto e^{\alpha(s\nu-1)}$.

One has: $\Phi_r = \sum_{s=0}^{r-1} \binom{r-1}{s} \mathcal{B}^s\Big(X^{N-Lr} - X^{N-L(r-1)}\Big)\,\mathcal{B}^{r-1-s}\Big(\frac{1}{1-X}\Big) =$

$$\sum_{s=0}^{r-1} \beta^{r-s-1}\,\frac{(r-1)!}{s!}\,\bigg(\frac{\alpha^s(1-r\nu)^s X^{N-Lr} - \alpha^s(1-(r-1)\nu)^s X^{N-L(r-1)}}{(1-X)^{r-s}}\bigg).$$

Finally, $\frac{\beta^{r-s-1}}{(1-X)^{r-s}} \mapsto \beta^{-1}$, $\lim_{N\to\infty}(1-\beta)^{N-Ls} = e^{\alpha(s\nu)-1}$, and

$$\Phi'_r = \sum_{s=0}^{r-1} \frac{(r-1)!}{s!}\,\Big(\alpha^s(1-r\nu)^s e^{\alpha(r\nu-1)} - \alpha^s(1-(r-1)\nu)^s e^{\alpha((r-1)\nu-1)}\Big).$$

This gives formula (7.2.9). The calculation is similar for $\pi_{r_\sharp}$. ❑

Some numerical examples. Let $N = 750, L = 150$. The combinatorial formulas and continuous ones for $\nu = L/N = 0.2$ give essentially the same:

$$\begin{aligned}
&\pi_0 \approx 0.37,\ \pi_1 \approx 0.44,\ \pi_2 \approx 0.17,\ \pi_3 \approx 0.02 &&\text{for } \alpha = 1,\\
&\pi_0 \approx 0.50,\ \pi_1 \approx 0.39,\ \pi_2 \approx 0.10,\ \pi_3 \approx 0.01 &&\text{for } \alpha = \log(2),\\
&\pi_0 \approx 0.61,\ \pi_1 \approx 0.33,\ \pi_2 \approx 0.06,\ \pi_3 \approx 0.004 &&\text{for } \alpha = 0.5.
\end{aligned}$$

7.2.4 Hardcore random walks *

They are for random walks inside a given interval where the steps, positive or negative, can be of different lengths. Modeling reinfections is when all steps are in one direction and zero-steps are allowed. *Motzkin paths* from Exercise 6.4.41 are another example. Their continuous limits are interesting. We will consider two such limits, complementing $\{\pi'_r\}$ from (7.2.9).

Poisson-Catalan distribution. Let us assume that some physics system is subject to N random events of 3 sorts: (i) adding $1/N$ to its *energy* E

with probability $(1-\frac{\alpha}{N})$, (ii) adding L/N to E with probability $\frac{\alpha p}{N}$ and (iii) subtracting L/N from E with probability $\frac{\alpha q}{N}$. Here $p, q = 1-p \geq 0$.

This models some "slow" linear growth of E with relatively rare "significant" accidental increases and decreases of energy, for instance due to transitions to excited states and back. Another possible interpretation is when E is assumed a share-price subject to some constant trend $+1/N$ per time-unit with accidental significant $\pm L/N$ fluctuations.

Let us begin with $E = 0$ and require that the intermediate balances of additions and subtractions of L/N are always non-negative (regardless of the "slow" gains), i.e. the allowed sequences of $\pm L/N$ are Catalan paths for $r+m$ events of $+1$ and m events of -1. The total number of such steps is $r+2m$. This is a pure combinatorial condition; it ensures the positivity of E at any points and is really necessary when $L \gg 1$.

We will assume that $\lim_{N\to\infty} L/N = \nu$ for $0 < \nu < 1$ in the formulas below, making it $\nu \to 0$ in some formulas. In the next problem, $\pi'_{r,m}$ denotes π'_{r+2m} from formula (7.2.9) where $r \mapsto r+2m$ and the parameters α, ν are replaced by $\alpha_m = (1+2m\nu)\,\alpha$, $\nu_m = \frac{\nu}{1+2m\nu}$.

Problem 7.2.18 *(i) For $m \geq 0$ and $0 \leq r \leq$ Ceiling $[1/\nu]$, one has:*

$$\mathcal{P}'_{r,m} = \pi'_{r,m}\, p^{r+m} q^m \binom{r+2m}{m} \frac{r+1}{r+m+1}. \tag{7.2.11}$$

(ii) Let $\mathcal{P}^0_{r,m} \stackrel{\mathbf{def}}{=\!=} \lim_{\nu\to 0} \mathcal{P}'_{r,m}$. Then $\pi'_{r,m}$ becomes the α-Poisson distribution $\frac{\alpha^{r+2m}e^{-\alpha}}{(r+2m)!}$ in this limit, and we obtain that $\mathcal{P}^0_{r,m} = \frac{\alpha^{r+2m}p^{r+m}q^m e^{-\alpha}}{m!(r+m)!}\frac{r+1}{r+m+1}$. Furthermore, let $p = \frac{1}{2} = q$ and $\sigma^0_r \stackrel{\mathbf{def}}{=\!=} \sum_{m=0}^{\infty} \mathcal{P}^0_{r,m}$. Then $\sigma^0_r = \frac{r+1}{\alpha/2} e^{-\alpha} I_{r+1}(\alpha)$ for the hyperbolic Bessel function *defined as follows: $I_u(z) = \sum_{m=0}^{\infty} \frac{(z/2)^{2m+u}}{m!\Gamma(m+u+1)}$. Note that $I_0(z) = \sum_{m=0}^{\infty} \frac{(z/2)^{2m}}{(m!)^2}$ and $I_1(z) = dI_0(z)/dz$.*

(iii) Under $p = 1/2 = q$ and for $\nu \to 0$ as in (ii), let σ^0_* be the probability $\sum_{r=0}^{\infty} \sigma^0_r$ of obtaining a Catalan path with $m \geq 0$ (for any $r \geq 0$). Then it equals $e^{-\alpha}\big(I_0(\alpha) + I_1(\alpha)\big)$. Also, the average value $T = \sum_{r=0}^{\infty}(r+1)\,\sigma^0_r$ of $(r+1)$ with respect to the distribution $\{\sigma^0_r\}$ is $T = 1$.

(iv) Setting $\nu \to 0$, let us consider all *paths, disregarding now the Catalan condition and allowing r to be negative. Let $p_r(m)$ be the corresponding probability with m steps -1 and the output r. Then one has:*

$$p_r(m) = p^{r+m} q^m \binom{r+2m}{m} \frac{e^{-\alpha}}{(r+2m)!} \quad \text{for } r \geq -m\,, \text{ and}$$

$$P_r = \sum_{m=0}^{\infty} p_r(m) \;\text{ equals }\; e^{-\alpha}\Big(\frac{p}{q}\Big)^{r/2} I_r\big(2\alpha(pq)^{1/2}\big). \tag{7.2.12}$$

In particular, $\sum_{r=-\infty}^{\infty} P_r = 1$.

Solution. **(i,ii)** The number of sequences $\{a_i = \pm 1, 1 \le i \le n\}$ such that $\sum_{i=1}^{k} a_i \ge 0$ for any $1 \le k \le n$ and $\sum_{i=1}^{n} a_i = r \ge 0$ is as follows: $C_n^{0\to r} = \binom{n}{(n-r)/2}\frac{r+1}{(n+r)/2+1}$, where $n-r$ must be even. See Exercises 6.4.35(ii) and 6.4.38. For $n = r+2m$, we obtain $C_{r+2m}^{0\to r} = \binom{r+2m}{m}\frac{r+1}{r+m+1}$. Then we multiply the latter expression by $p^{r+m}q^m$ and by $\pi'_{r,m}$.

For $n = r+2m$, we obtain $C_{r+2m}^{0\to r} = \binom{r+2m}{m}\frac{r+1}{r+m+1}$. Then we multiply the latter expression by $p^{r+m}q^m$ and by $\pi'_{r,m}$. This proves (i); the justification of (ii) is straightforward.

(iii) Differentiating (7.2.13) with respect to t, we obtain the identity $e^{\frac{\alpha}{2}(t+\frac{1}{t})-\alpha} = \sum_{r=0}^{\infty} \mathcal{P}_r^0 \frac{t^{r+1}-t^{-r-1}}{t-t^{-1}}$. The formula $T = 1$ is when $t \to 1$. Indeed, $\frac{t^{r+1}-t^{-r-1}}{t-t^{-1}} = t^r + t^{r-2} + \cdots + t^{-r}$. To obtain σ_*^0, observe that the latter sum contains 1 for even r and t for odd r; then use (7.2.13) for $r = 0, 1$.

Actually, we do not need (7.2.13) for (iii). The total number of *all* Catalan paths with n steps is $\binom{n}{n/2}$ for even n and $\binom{n}{(n-1)/2}$ for odd n. It is $C_n^{0\to *} = \sum_{r=0}^{n} C_n^{0\to r}$, where $n-r$ is even, in the notation above. Thus, the total probability of obtaining a Catalan path is

$$\sum_{r=0}^{\infty} \sigma_r^0 = \sum_{n=0}^{\infty} e^{-\alpha}\frac{\alpha^n}{n!}\binom{n}{\text{Floor}[n/2]} = e^{-\alpha}\big(I_0(\alpha) + I_1(\alpha)\big).$$

Then, $2^n - \binom{n}{\text{Floor}[n/2]} = \sum_{r=0}^{\infty} r\, C_n^{0\to r}$ for even $n-r$, since the ideal roulette is a 0-sum game, and $2^n = \sum_{r=0}^{\infty}(r+1)\, C_n^{0\to r}$ using the formula above for $C_n^{0\to *}$. This proves directly that $T = 1$.

(iv) To allow any sequences of ± 1, we must omit the terms $\frac{r+1}{m+r+1}$. Then $\sum_{r=-\infty}^{\infty} P_r = 1$ for $P_r = \sum_{m=0}^{\infty} p_r(m)$ becomes the classical identity

$$e^{\frac{\alpha}{2}(t+\frac{1}{t})} = \sum_{r=-\infty}^{\infty} I_r(\alpha)t^r, \text{ where } I_r(\alpha) = I_{-r}(\alpha). \tag{7.2.13}$$

To see this, we represent $p_r(m) + p_{-r}(m)$ for $r > 0$ as

$$(p^{r+m}q^m + q^{r+m}p^m)\binom{r+2m}{m}\frac{e^{-\alpha}}{(r+2m)!} = (pq)^{\frac{r+2m}{2}}\left(\left(\frac{p}{q}\right)^{\frac{r}{2}} + \left(\frac{q}{p}\right)^{\frac{r}{2}}\right)\frac{e^{-\alpha}}{m!(r+m)!}.$$

Then $\sum_{r=-\infty}^{\infty} P_r - \sum_{r=-\infty}^{\infty}\big(p_r(0) + \sum_{m=1}^{\infty}(p_r(m) + p_{-r}(m))\big)$ becomes

$$e^{-\alpha}I_0(2\alpha(pq)^{1/2}) + e^{-\alpha}\sum_{r=1}^{\infty}\left(\left(\frac{p}{q}\right)^{r/2} + \left(\frac{q}{p}\right)^{r/2}\right)I_r(2\alpha(pq)^{1/2})$$

$$= \exp\left(-\alpha + \alpha(pq)^{1/2}\left(\left(\frac{p}{q}\right)^{1/2} + \left(\frac{q}{p}\right)^{1/2}\right)\right) = 1 \text{ due to } p+q = 1.$$

Since this is a full sum of probabilities, it is 1. This results in (7.2.13) for $t = \left(\frac{p}{q}\right)^{1/2}$; we can make t arbitrary (real) for proper q, p. ❑

Problem 7.2.19 *Let $p=1/2=q$ and $\nu = 0$ as in part (iii) of Problem 7.2.18. Now the starting level will be $k \geq 0$ and the corresponding probability of Catalan paths will be $\sigma_*^0\langle k\rangle = e^{-\alpha}\sum_{n=0}^{\infty}\frac{\alpha^n}{n!}C_n^{k\to *}$. Following Corollary 6.3.2:*

$$C_n^{k\to *} = \sum_{n=0}^{\infty}\left(\binom{n}{\mathrm{Floor}[\frac{n-k}{2}]} + \binom{n}{\mathrm{Floor}[\frac{n-k}{2}]+1} + \cdots + \binom{n}{\mathrm{Floor}[\frac{n+k}{2}]}\right).$$

Check that $\sigma_*^0\langle k\rangle = e^{-\alpha}\big(I_{-k}(\alpha) + I_{-k+1}(\alpha) + \cdots + I_{k+1}(\alpha)\big)$. Practically, for $\alpha = 2.2$: $\sigma_*^0\langle k\rangle \approx 0.5, 0.81, 0.95$ for $k = 0, 1, 2$. ❑

Normal approximation. We are back to the case of zero or positive steps with one $L > 0$ inside $[1, N]$, i.e. to the case of reinfections. For any combinatorial probabilities, the connection to the *normal distribution* must be always inspected. Let us do this for the probability π_r of r infections from (7.2.6). This is a good example of theoretical and practical interest.

Recall that we consider non-overlapping subintervals of length L inside $[1, N]$, where the last one is allowed to go beyond N. The formula is $\pi_r = P(r, N, \beta) + \beta\sum_{i=1}^{L} P(r-1, N-i, \beta)$, where $P(r, N, \beta) = \binom{N-Lr}{r}\beta^r(1-\beta)^{N-Lr-r}$ and $r = 0, 1, \ldots$. In contrast to the Poisson-type limit, the length of the immunity interval L is now assumed fixed. Also, the probability β of creating an L-subinterval will be now a constant: $0 < \beta < 1$.

Both conditions make sense for modeling the spread of infections. The immunity intervals are of biological nature, so L can be assumed a constant. The probably β of being infected (during 1 day) may depend on N, but it is sufficiently stable during the main part of the epidemic.

To evaluate the terms $P(r, N, \beta)$ and $P(r-1, N-i, \beta)$, we assume that $r - \beta(N - rL)$ is sufficiently small. Using Exercise 3.7.47, $p^r(1-p)^{n-r}\binom{n}{r}$ for $0 < p < 1$ tends to $\dfrac{e^{-(r-np)^2/(2np(1-p))}}{\sqrt{2\pi np(1-p)}}$ for $n \gg 0$ and $r - np = O(\sqrt{n})$. In our case, $p = \beta$ and $n = N - \beta L$. Then all P-terms in (7.2.6) coincide asymptotically for $N \gg 0$ due to $i \ll N$ and we arrive at the estimate

$$\pi_r \approx \widetilde{\pi}_r \stackrel{\text{def}}{=\!=} (1+\beta L)\frac{e^{-(r-\beta u)^2/(2u\beta(1-\beta))}}{\sqrt{2\pi u\beta(1-\beta)}}, \text{ where } u \stackrel{\text{def}}{=\!=} N - Lr, \tag{7.2.14}$$

$$u = \frac{N}{1+\beta L} + O(\sqrt{N}),\; r = \beta u + O(\sqrt{u}) = \frac{\beta N}{1+\beta L} + O(\sqrt{N}). \tag{7.2.15}$$

Here u is the number of points in $[1, N]$ not covered by the L-subintervals plus $0 \le i \le L-1$ if the last subsegment goes beyond the endpoint by i points (the edge effect). We assume that $u + Lr$ is fixed, so u and r are not independent variables. Accordingly, the term $2np(1-p)$ in the classical normal approximation is not a constant: it depends now on r. The range of r needed for the normal approximation is $r = \beta u + O(\sqrt{u})$. It can be readily recalculated in terms of N, which relation is provided in (7.2.14).

Problem 7.2.20 *Justify that the formula (7.2.14) for $\widetilde{\pi}_r$ gives a* distribution of probabilities *when $0 \le r < \infty$ is considered as a variable and $u = N - Lr$ for sufficiently large fixed N.*

Solution. This means that we need to check that $I_N \to 1$ as $N \to \infty$ for

$$I_N = (1+\beta L) \int_0^{N/L} \frac{e^{-(r-\beta u)^2/(2u\beta(1-\beta))}}{\sqrt{2\pi u\beta(1-\beta)}} dr$$
$$= \frac{1+\beta L}{L} \int_0^{N} \frac{e^{-(r-\beta u)^2/(2u\beta(1-\beta))}}{\sqrt{2\pi u\beta(1-\beta)}} du,$$

where we switch to the integration in terms of u using $dr = -\frac{1}{L}du$.

Let $u = v^2$, $du = 2vdv$, $a = \frac{1+\beta L}{L}, b = \frac{N}{L}$, $c = \beta(1-\beta)$. We will consider a, b, c as arbitrary independent positive parameters. Then

$$I_N = 2a \int_0^{\sqrt{N}} \frac{e^{-(av-bv^{-1})^2/(2c)}}{\sqrt{2\pi c}} dv.$$

Let I_∞ be its value when the upper limit is ∞. We will see that I_∞ does not depend on b, so $\lim_{N\to\infty} I_N$ makes sense and coincides with I_∞. This remark is needed to address that $b \to \infty$ when $N \to \infty$.

This is a "standard integral", but in a somewhat unusual form. Let us provide its calculation. Let $z = \frac{b}{a}v^{-1}$; then $v = \frac{b}{a}z^{-1}$ and $dv = -\frac{b}{a}z^{-2}$. Switching from v to z in I_∞, we obtain that $I_\infty = 2b\int_0^\infty \frac{e^{-(bz^{-1}-az)^2/(2c)}}{\sqrt{2\pi c}}\frac{dz}{z^2}$. Combining it with the formula above and using $w = az - b/z$:

$$2I_\infty = 2\int_0^\infty \frac{e^{-(bz^{-1}-az)^2/(2c)}}{\sqrt{2\pi c}}(a+\frac{b}{z^2})dz, \text{ and } I_\infty = \int_{-\infty}^{\infty} \frac{e^{-w^2/(2c)}}{\sqrt{2\pi c}}dw = 1.$$

Here the derivative of w, which is $a + b/z^2$, is positive for $0 < z < \infty$, and $-\infty < w < \infty$ in the considered range. This gives the required. ❑

7.3 Modular arithmetic and finite fields

We will begin with a more systematic account of the theory of finite fields, providing complete justifications. We do not need all tools available in their theory. For instance, the Frobenius automorphism will be used, but the Galois theory will not be really employed. For finite fields, it can be readily "replaced" by the usage of the cyclicity of $\mathbb{F}_q^*$, which we will justify and constantly use in this section.

7.3.1 Finite fields

Recall from Sections 5.3.3 and 5.3.4 that for every finite field $\mathbb{F}$, there exists prime p, its *characteristic*, such that $p \cdot 1 = \overbrace{1 + \cdots + 1}^{p} = 0$. The simplest field is $\mathbb{F}_p$, which is $\mathbb{Z}_p = \mathbb{Z} \bmod p$ with the standard modular addition and multiplication. Both notations, $\mathbb{F}_p$ and $\mathbb{Z}_p$, will be used. Obviously, $p \cdot 1 = 0$ in $\mathbb{Z}_p$, so p is its characteristic.

If $\mathbb{F}$ is obtained by adding a root x of some *irreducible polynomial* $P(X)$ of degree m with the coefficients in $\mathbb{F}_p$, then $|\mathbb{F}| = p^m$. The irreducible polynomials are those that are not products $P(X) = Q(X)R(X)$, where the coefficients of Q and R are from $\mathbb{F}_p$ (or from any base field), and their degrees, $deg(Q)$ and $deg(R)$, are nonzero. *It is not sufficient for irreducibility to check that* $P(X)$ *has no roots in* $\mathbb{F}_p$ unless $deg(P) = 2, 3$. In these two cases, if $P = QR$ for Q, R of degree 1 or greater, then one of Q, R has degree one and therefore has a root in $\mathbb{F}_p$. This means that the absence of roots in $\mathbb{F}_p$ is necessary and sufficient for the irreducibility of such $P(X)$. For $deg(P) \geq 4$, this is generally insufficient and the justification of the irreducibility can be much more involved.

Let us check that $|\mathbb{F}| = p^m$, where m is the *dimension* of $\mathbb{F}$ over $\mathbb{F}_p$. The field $\mathbb{F}_p$ obviously belongs to $\mathbb{F}$: it is $\{1, 2 = 1+1, 1+1+1, \ldots\}$. The dimension is the greatest number of elements $x_1, \ldots x_m$ such that no linear combination of them with coefficients in $\mathbb{F}_p$ is zero. That is $c_1x_1 + \cdots + c_mx_m = 0$ with $c_i \in \mathbb{F}_p$ if and only if $c_1 = \cdots = c_m = 0$. This property and the maximality of m gives that any x can be *uniquely* represented as $x = c_1x_1 + \cdots + c_mx_m$ for $c_i \in \mathbb{F}_p$. Thus, the size of $\mathbb{F}$ is p^m indeed.

The same argument gives that any *field extension* $\mathbb{F} \subset \mathbb{F}'$ of the field $\mathbb{F}$ with p^m elements will have p^{km} elements, where k is the dimension of $\mathbb{F}'$ over $\mathbb{F}$. Generally, the dimension of L over K divides the dimension of M over K in any "tower" of fields $K \subset L \subset M$. This is for any base field; the justification is straightforward.

Let us consider $\mathbb{F}^* = \{a \in \mathbb{F} \,|\, a \neq 0\}$, the multiplicative group of nonzero elements in $\mathbb{F}$. We have already used Lagrange Theorem 5.3.1. Let us apply it to the multiplicative group $\mathbb{F}^* = \{a \in \mathbb{F} \,|\, a \neq 0\}$, which is of order $p^m - 1$. We obtain that the order of any a in $\mathbb{F}^*$, the smallest $k > 0$ such that $a^k = 1$, divides $p^m - 1$. Thus $a^{p^m-1} = 1$. By adding 0: $x^{p^m} = x$ for any $x \in \mathbb{F}$.

The key point in the theory of finite fields is that $x \in \mathbb{F}$ constitute *all* solutions of the equation $x^{p^m} = x$ in $\mathbb{F}$ *and in any larger fields containing* $\mathbb{F}$. Indeed, no equation of degree n can have more than n roots in any field, but we already have n: all elements of $\mathbb{F}$.

An *isomorphism* between fields $\mathbb{F}$ and $\mathbb{F}'$ is a bijection $\phi : \mathbb{F} \to \mathbb{F}'$ as sets such that $\phi(x+y) = \phi(x) + \phi(y)$ and $\phi(xy) = \phi(x)\phi(y)$ for any x, y in $\mathbb{F}$. It automatically sends $0 \mapsto 0$ and $1 \mapsto 1$ (check).

Problem 7.3.1 *For any prime p and $m \geq 1$ construct a field with p^m elements by adding all roots of $X^{p^m} - X$ to $\mathbb{F}_p$.*

(i) Assuming that $X^{p^m} - X$ has all p^m roots in some field, check that the set of all roots of this polynomial is a field with p^m elements.

(ii) Prove that the field with p^m elements is unique up to isomorphisms; so the notation $\mathbb{F}_{p^m}$, which we will constantly use, makes sense.

Solution. We can add to $\mathbb{F}_p$ any root x of an irreducible polynomial $P(X)$ of degree n with the coefficients in $\mathbb{F}_p$. Let us take as $P(X)$ any *irreducible factor* $P(X)$ of $X^{p^m} - X$ over $\mathbb{F}_p$ of degree greater than 1, and switch from $\mathbb{F}_p$ to $\mathbb{F}_p^{(1)}$ obtained by adding its root to $\mathbb{F}_p$. Then take an *irreducible* factor $P^{(1)}(X)$ of degree greater than 1 of the ratio $(X^{p^m} - X)/P(X)$ with the coefficients in $\mathbb{F}_p^{(1)}$ and construct $\mathbb{F}_p^{(2)}$ by adding a root of $P^{(1)}(X)$ to $\mathbb{F}_p^{(1)}$. Continue this way until $X^{p^m} - X$ becomes a product of linear factors in some $\mathbb{F}_p^{(k)}$ (after k steps). This field can be larger than we need.

(i) Let us check that this set $\mathbb{F}$ in $\mathbb{F}_p^{(k)}$, or in any other field containing all roots of $X^{p^m} - X$ (from (ii)) is closed with respect to addition and multiplication, i.e. it is a field. This will prove (i) and (ii).

The products or ratios of such nonzero roots are roots again, which is obvious. Let us check that their $\mathbb{F}_p$-linear combinations are some roots too. If x_1, x_2 are two such roots, then so is $(ax_1+bx_2)^{p^m} = ax_1^{p^m} + bx_2^{p^m} = ax_1+bx_2$, for $a, b \in \mathbb{F}_p$. We use Newton's binomial, formula (5.4.6) to be exact. Thus, these roots form a field $\mathbb{F}$ exactly with p^m elements.

(ii) A natural proof that the field $\mathbb{F}$ we constructed does not depend on the process above up to isomorphisms is via the induction with respect to the sequences of the extensions we used. It also gives that any fields with p^m elements can be obtained in such a way. We proceed as follows.

Let $\widetilde{\mathbb{F}}$ be a field with p^m elements. Take any sequence $P, P^{(1)}, P^{(2)}, \cdots$, as above: P is an irreducible factor of $X^{p^m} - X$, $P^{(1)}$ is an irreducible factor of $(X^{p^m} - X)/P(X)$ with the coefficients in $\mathbb{F}_p[x]$ for a root x of $P(X)$, and so on. Solve the equations $P(x)=0, P^{(1)}(x)=0, P^{(2)}(x)=0, \cdots$ in $\widetilde{\mathbb{F}}$, instead of creating some abstract extensions of $\mathbb{F}_p$ as we did before. This embeds $\mathbb{F}_p^{(1)}, \mathbb{F}_p^{(2)}$ and so on into $\widetilde{\mathbb{F}}$. Thus, $\mathbb{F}$ constructed in (i),(ii) will be mapped into $\widetilde{\mathbb{F}}$, which identifies these fields: they have the same number of elements. ❑

A careful reader noticed that we have used the following general fact.

Lemma 7.3.2 *Any nonzero ring homomorphism f from a field K is injective, i.e. $f(x) \neq 0$ for any $x \neq 0$ in K. Ring homomorphisms are maps f such that $f(x+y) = f(x) + f(y)$ and $f(xy) = f(x)f(y)$ for any x, y.*

Proof. Let us assume that $f(x) = 0$ for some $x \neq 0$. Then $f(K) = f(Kx) = f(K)f(x) = 0$, i.e. f is zero identically (for the whole K). ❑

A version of this problem is the following description of $\mathbb{F}_{p^m}$:

The field $\mathbb{F}_{p^m}$ is the unique extension of $\mathbb{F}_p$ satisfying the following conditions: (a) all its elements are roots of the polynomial $X^{p^m} - X$, and (b) this polynomial has all its roots in this field.

7.3.2 $\mathbb{F}^*_{p^m}$ is cyclic

The cyclicity of $\mathbb{F}^*$ for finite $\mathbb{F}$ provides another approach to the fact that finite fields with the same number of elements, p^m, are isomorphic to each other. See part (ii) of Problem 7.3.1.

Let us demonstrate that the uniqueness of a field with p^m elements is granted if any such field $\mathbb{F}'$ contains a root x' of some *irreducible* polynomial $P(X)$ over $\mathbb{F}_p$ of degree m. Then $\mathbb{F}' = \mathbb{F}_p[x']$. Indeed, $\mathbb{F}[x']$ is of dimension m over $\mathbb{F}_p$, so it has p^m elements and must coincide with $\mathbb{F}'$.

Since $x^{p^m} = x$ for any $x \in \mathbb{F}'$, the polynomial $P(X)$ must divide $X^{p^m} - X$; otherwise use long division of $X^{p^m} - X$ by $P(X)$ to come to a contradiction with its irreducibility. However, $X^{p^m} - X$ is a product of linear factors in the field $\mathbb{F}$ we constructed above. Therefore $\mathbb{F}$ contains a root x of $P(X)$, and the map $\sum_{i=0}^m c_i(x')^i \mapsto \sum_{i=0}^m c_i x^i$ for $c_i \in \mathbb{F}_p$ embeds $\mathbb{F}'$ into $\mathbb{F}$; any nonzero ring homomorphisms of fields are embeddings (see above). Thus, its image coincides with $\mathbb{F}$: they have the same number of elements by the assumption. This is actually a "one-step variant" of the argument we used in the proof of the uniqueness in Problem 7.3.1(ii).

We will find such irreducible $P(X)$ and their roots in a very constructive way. Some basics from the theory of commutative groups are needed.

A ***commutative group*** G is a set with the multiplication $\{x, y\} \mapsto xy$, the *neutral element* 1, satisfying $1x = x$ for any x, and the division. The relations $xy = yx$, $x(yz) = (xy)z$ for any elements $x, y, z \in G$ must hold.

Such a neutral element is unique if it exists. Indeed, if there is another $\widetilde{1}$ satisfying $\widetilde{1}x = x$ for any x, then, substituting $x \mapsto 1$, we obtain $\widetilde{1}1 = \widetilde{1} = 1$. Similarly, $1/x$ is unique if it exists. By definition, $1/x = x'$ such that $x'x = 1$, and $z = x/y$ is such that $zy = x$. If there is another reciprocal $\widetilde{x}'$, then $\widetilde{x}'x'x = x' = \widetilde{x}'$. We obtain that $x/y = x(1/y)$, so it is unique too.

Proposition 7.3.3 *Let G be a commutative group G of order $n = |G|$ satisfying the following property. Given any divisor d of n, the elements satisfying the relation $y^d = 1$ in G form a subgroup of order d (with d elements); actually the inequality $\leq d$ is sufficient in (i).*

(i) Then G is cyclic. Namely, all its elements are powers of some $x \in G$, which is then called a cyclic generator *of G. Equivalently, the order of x equals n. There is only one cyclic group of order n up to isomorphisms, and the solutions of the equation $y^d = 1$ there are $x^{kn/d}$ for $0 \leq k < d$.*

(ii) This holds for the group $G = \mathbb{F}^*$ for any finite $\mathbb{F}$ or for an arbitrary finite subgroup in $\mathbb{F}^*$ for any field $\mathbb{F}$. If $\mathbb{F}$ is a field with p^m elements then $n = p^m - 1$. So all elements in $G = \mathbb{F}^*$ are powers of some element x of order $p^m - 1$, called a primitive root mod p *when $m = 1$.*

Proof. First, we check (ii): that $\mathbb{F}^*$ satisfies (i). For a field $\mathbb{F}$ with p^m elements, the number of roots of the equation $x^d = 1$ in $\mathbb{F}$ cannot be greater than d for any d. It cannot be smaller than d if d divides $p^m - 1$, because $X^d - 1$ divides $X^{p^m-1} - 1$. Indeed, all roots of the latter polynomial are in $\mathbb{F}$ and all are simple. Thus the roots of $X^d - 1$ are in $\mathbb{F}$ and are simple too. So their number is d. This argument is no different for *finite* subgroups in $\mathbb{F}^*$ in any field $\mathbb{F}$. For instance, for the field of complex numbers the roots of unity of order n form a cyclic subgroup with n elements.

We use here that if all roots of a polynomial $P(X)$ belong to some field and all are simple, then their number is $deg(P)$. The justification is by consecutive divisions of $P(X)$ by the linear factors $(X - x)$ for its roots x.

(i)* The proof that G is cyclic is constructive; we will find its cyclic generator x. Let $n = p_1^{m_1} p_2^{m_2} \cdots p_r^{m_r}$ be the prime decomposition. For any p_i, there exists at least one element g_i in G such that $g^{n/p_i} \neq 1$, because the number of elements in G satisfying $g^{n/p_i} = 1$ is $n/p_i < n$ by the assumption. We set: $h_i = g_i^{n/p_i^{m_i}}$ for $1 \leq i \leq r$.

Then $h_i^{p_i^{m_i}} = 1$ due to the Lagrange Theorem, but $h_i^{p_i^{m_i-1}} \neq 1$. Indeed, $h_i^{p_i^{m_i-1}} = \left(g_i^{n/p_i^{m_i}}\right)^{p_i^{m_i-1}} = g_i^{n/p_i} \neq 1$. So the order of h_i is exactly $p_i^{m_i}$, the maximal possible power of p_i. In particular, $h_i^{n/p_i} \neq 1$.

Let us prove that $x = h_1 h_2 \cdots h_r$ is the required element of order n, i.e. a cyclic generator of G.

If its order is smaller than n, then it divides at least one of the numbers $n/p_1, n/p_2, \ldots, n/p_r$. Let it be n/p_1. Then $x^{n/p_1} = 1 = h_1^{n/p_1} h_2^{n/p_1} \cdots h_r^{n/p_1}$. However, $h_2^{n/p_1} = 1$ because $h_2^{p_2^{m_2}} = 1$ and $p_2^{m_2}$ divides n/p_1, $h_3^{n/p_1} = 1$, and so on. Thus, $h_1^{n/p_1} = 1$, a contradiction. ❑

As an application, we obtain a "constructive proof" of the uniqueness of $\mathbb{F}$ with p^m elements up to isomorphisms. Indeed, any cyclic generator x of the group $\mathbb{F}^*$ generates *algebraically* the whole $\mathbb{F}$ over $\mathbb{F}_p$. The algebraic generators are any elements x of $\mathbb{F}$ such that linear combinations of x and their powers with the coefficients in $\mathbb{F}_p$ constitute the whole $\mathbb{F}$. Only its powers (and 0) are needed to have the whole $\mathbb{F}$ for a cyclic generator.

There are many algebraic generators that are not cyclic generators. For instance, $\mathbb{F}_{16}^*$ has 15 elements and contains x of order 5. It can belong to a smaller field only if $x^{4-1} = x^3 = 1$, which is impossible. Indeed, subfields of $\mathbb{F}_{p^m}$ are $\mathbb{F}_{p^k}$ for m divisible by k. So x is an algebraic generator. By the way, this gives that the irreducibility of $\frac{X^5-1}{X-1} = X^4+X^3+X^2+X+1$ over $\mathbb{F}_2$.

Recall that the key step in the proof of the existence and uniqueness of $\mathbb{F}_{p^m}$ was as follows: given an *irreducible* polynomial $P(X)$ of degree m over $\mathbb{F}_p = \mathbb{Z}_p$, the field $\mathbb{F}_p[x]$ has p^m elements for x such that $P(x) = 0$. We did not use that they exist for any m. Instead, we began with $\mathbb{F}_p$ and added all roots of $X^{p^m} - X$ one after another. This appeared sufficient to construct $\mathbb{F}$ with p^m elements and then prove its uniqueness up to isomorphisms.

To summarize, the following facts are fundamental in the theory of finite fields. For an irreducible $P(X)$ of degree m over $\mathbb{F}_p$ and $P(x) = 0$:

the elements of $\mathbb{F}_p[x]$ *(all of them) are roots of* $X^{p^m} - X \Longrightarrow$ (7.3.16)
$P(X)$ *divides* $X^{p^m} - X$ *and all roots of* $P(X)$ *belong to* $\mathbb{F}_p[x]$;
$\mathbb{F}_p[x]$ *is a subfield of* $\mathbb{F}_{p^{mk}}$ *for any* $k \geq 1$, *and only for such* $\mathbb{F}_q$;
moreover, $\mathbb{F}_p[x]$ *is isomorphic to* $\{y \in \mathbb{F}_{p^{mk}}$ *such that* $y^{p^m} = y\}$. (7.3.17)

Here $\mathbb{F}_p$ and $\mathbb{F}_p[x]$ can be subfields of $\mathbb{F}_q$ only for some $q = p^\ell$; the characteristics must coincide. The embedding $\mathbb{F}_{p^m} \subset \mathbb{F}_{p^{mk}}$ follows from (7.3.16) and the cyclicity theorem because $p^{mk}-1$ is obviously divisible by p^m-1. The cyclicity gives that $\mathbb{F}_{p^m}$ is a subfield of $\mathbb{F}_{p^\ell}$ *if and only if* p^m-1 divides $p^\ell-1$. Then and only then $\mathbb{F}_{p^\ell}^*$ contains a cyclic subgroup with p^m-1 elements; all roots of $X^{p^m}-X$ can be found in $\mathbb{F}_{p^\ell}$ in this case.

The divisibility of $p^\ell - 1$ by $p^m - 1$ holds if and *only* if ℓ is divisible by m; see Exercise 5.6.8. Let us reprove the only part via finite fields. Assuming that $p^\ell - 1$ is divisible by $p^m - 1$, let us embed $\mathbb{F}_{p^m}$ into $\mathbb{F}_{p^\ell}$ as above. The latter is a vector space over $\mathbb{F}_{p^m}$, which gives that $\ell = mk$, where k is its dimension. See Exercise 5.6.27.

We can employ (7.3.17) to calculate the number $N(p, m)$ of irreducible polynomials $P(X)$ of degree m up to proportionality. Let $m = p_1^{m_1} p_2^{m_2} \cdots$. Then $N(p,m) = \big(p^m - (p^{m/p_1} + p^{m/p_2} + \cdots) + (p^{m/(p_1 p_2)} + \cdots) - \ldots\big)/m$. It is the number of elements of $\mathbb{F}_{p^m}$ that do not belong to smaller subfields, i.e. the number of algebraic generators, divided by m. We use that there is only one subfield $\mathbb{F}_{p^d}$ for any d dividing m (only for such d), and apply the Inclusion-Exclusion Principle. The connection with $N(p, m)$ is as follows: (i) every such P has m roots in $\mathbb{F}_{p^m}$ and all of them are algebraic generators, (ii) non-proportional irreducible polynomials P_1, P_2 cannot have roots in common.

For instance, we need to exclude from $\mathbb{F}_{16}$ the elements from $\mathbb{F}_4$, which gives $(16-4)/4$. Such P are: $X^4+X^3+X^2+X+1$, X^4+X^3+1, X^4+X+1. They have no roots in $\mathbb{F}_2$ as well as $(X^2+X+1)^2 = X^4+X^2+1$, which is reducible.

We have already verified (ii); this is a general fact. If x is such a root, the long division of polynomials provides a smaller polynomial that has x as a root and divides both. This contradicts their irreducibility.

Galois theory generalizes (7.3.16)-(7.3.17). It is as follows for finite fields.

Problem 7.3.4 *(i) Check that the map $F : x \mapsto x^p$ is an automorphism of $\mathbb{F}_{p^m}$, the famous* Frobenius automorphism, *i.e. $F(x+y) = F(x) + F(y)$, $F(xy) = F(x)F(y)$ and the inverse of F exists. Justify that the powers F^k form a (cyclic) group of order m.*

(ii) Let x be an algebraic or cyclic generator of $\mathbb{F}_{p^m}$ such that $P(x) = 0$ for an irreducible $P(X)$ of degree m. Then x^{p^k} for $k = 0, \ldots, m-1$ constitute all roots of $P(X)$, and they are correspondingly algebraic or cyclic generators.

(iii) For an arbitrary divisor d of m, the subfield $\mathbb{F}_{q^d}$ in $\mathbb{F}_{q^m}$ is formed by (all) x such that $F^{m/d}(x) = x^{p^{m/d}} = x$, which is a reformulation of (7.3.17). In particular, $\mathbb{F}_p = \{x \in \mathbb{F}_{q^m} \mid F(x) = x\}$.

Solution. **(i)** The property $F(xy) = F(x)F(y)$ is obvious; $F(x+y) = F(x) + F(y)$ follows from Newton's binomial. See the proof of Problem 5.4.1. If $F(x) = F(y)$ then $F(x-y) = (x-y)^p = 0$ and $x = y$ ($\mathbb{F}_{p^m}$ is a field). This gives that the image of F contains p^m elements; therefore, $F(\mathbb{F}_{p^m})$ is the whole $\mathbb{F}_{p^m}$. Thus, F is an automorphism: its inverse exists. Here one can use that $F^m =$id (the identity map), so the inverse obviously exists; it is F^{m-1}.

Now, if $x^{p^\ell} = 1$ for $1 \le \ell < m$, then x belongs to $\mathbb{F}_{p^\ell}$. This cannot hold for algebraic generators, so F^ℓ cannot be the identity in $\mathbb{F}_{p^m}$.

(ii) Since F is the *identity* in $\mathbb{F}_p$, we have $F(P(x)) = P(F(x)) = 0$ (the coefficients of $P(X)$ are in $\mathbb{F}_p$) and F sends the roots of $P(X)$ to its other roots. The sets of algebraic and cyclic generators are obviously preserved under F (under any automorphism). Then use that $F^k(x) = x$ if and only if m divides k for algebraic generators x; cyclic generators are algebraic generators. ❑

As an application, we obtain that m divides $\varphi(p^m - 1)$ for the Euler function, the number of cyclic generators; see Exercise 5.6.14.

When $m = 1$, Proposition 7.3.3 gives the cyclicity of $\mathbb{F}_p^*$, which readily results in the cyclicity of $\mathbb{Z}_{p^m}^*$ for odd p and any $m \geq 1$. The corresponding generators are called *primitive roots* modulo p^m. See Exercises 5.6.16, 5.6.17. This has applications in cryptography.

The RSA algorithm (U.S. Patent 4405829). It is as follows. For 2 (very) large prime numbers $p \neq q$ such that $\gcd(p-1, q-1)$ is relatively small (at least 2 of course), let $m = (p-1)(q-1)/\gcd(p-1, q-1)$. The *key* will be $1 < e < m$ such that $\gcd(e, m) = 1$. We use that $x^m = 1 \bmod n$ for $n = pq$ and any $x \in \mathbb{Z}_n^*$ (the group of invertibles $\bmod n$). This is due to Exercise 5.6.13: $\varphi(n) = (p-1)(q-1)$. Accordingly, $(x^e)^d = x$ if $d = 1/e \bmod m$. The encryption sends $x \mapsto x^e = y$; the decryption is then $y \mapsto y^d = x^{ed} = x \bmod n$. Here, x is "our message" to be encrypted. The numbers n and e are known to *all* users, but m and the corresponding d are known only to the recipients. Finding m requires the prime factorization of n, which is very involved for large n. Generally, n can be anything, but the greater the number of its prime factors the faster its prime factorization and the simpler it is to find the "secret" d.

Problem 7.3.5 *Assuming that* $\gcd(n, 10) = 1$, *prove that the period of the decimal representation of* m/n *for* $\gcd(n, m) = 1$ *is the smallest* $u > 0$ *such that* $10^u = 1 \bmod n$, *i.e. it is the order of the number* 10 *modulo* n *(for any such* m*). The beginning of the period can be somewhere in the decimal representation. For instance,* u *is some divisor of* $\varphi(n)$*, the Euler function counting the number of invertibles in* $\mathbb{Z}_n$*. It is exactly* $p-1$ *if* $n = p$ *(a prime number) provided that* 10 *generates the cyclic group* $\mathbb{F}_p^*$.

Solution. Following Problem 5.1.2, we can assume that $1 \leq m < n$. The first "input" r in the solution of this problem is then m and we need to find the period of the sequence $m, m10, m10^2, \ldots$ modulo n.

Here $m, 10$ are invertible in $\mathbb{Z}_n$. Thus, $m10^i = m10^j$ in $\mathbb{Z}_n$ for some $0 \leq i < j$ if and only if $10^{j-i} = 1 \bmod n$, which gives the first claim. The order of 10 in $\mathbb{Z}_n^*$, the group of invertibles in $\mathbb{Z}_n$, divides the order (the number of elements) of this group; see Section 5.1.3. The cyclicity of $\mathbb{Z}_p^*$ for prime p is from Proposition 7.3.3. ❑

Artin's conjecture states that the density of primes p such that $a = 10$ generates $\mathbb{F}_p^*$ is the ***Artin constant*** $C = \prod_p(1 - \frac{1}{p(p-1)})$, where the product is over all primes p. This is for any integer $a > 0$ (with the same C) unless $a = b^m$ for $m > 1$ or $a = (4k+1)b^2$ for some integers $b, k \geq 1$ and square-free $4k+1$. Approximately, $C \simeq 0.374$. This and similar problems require the (generalized) ***Riemann hypothesis***, so they are open. See e.g. [47].

7.3.3 Various problems

Problem 7.3.6 *Are the polynomials $X^2 - 2$, and $X^2 + 1$ irreducible over the field $\mathbb{R}$ of real numbers, the field $\mathbb{Q}$ of rational numbers, and the field $\mathbb{F}_5 = \mathbb{Z}_5$?*

Solution. Modulo 5: $X^2+1 = X^2-4 = (X-2)(X+2)$, and 2 is not a square. So $X^2 - 2$ is reducible only over $\mathbb{R}$, $X^2 + 1$ is reducible only over $\mathbb{F}_5$. ❑

Problem 7.3.7 *(i) Calculate directly the number of reducible polynomials $P(X)$ over $\mathbb{F}_p$ of degree 3 that are monic: begin with X^3. (ii) Check "without calculations" that the number of irreducible ones is $(|F_{p^3}| - |F_p|)/3 = (p^3 - p)/3$. Thus, the number from part (i) is $p^3 - (p^3 - p)/3$.*

Solution. We will begin with (ii). The total number of monic polynomials of degree 3 is p^3 because they have 3 arbitrary coefficients. Every polynomial is completely determined by its roots, and the irreducibility gives that all roots are different. Let us check the latter.

We differentiate P and obtain that its multiple roots satisfy a quadratic or linear polynomial over $\mathbb{F}_p$ unless $p = 3$ and $P(X) = X^3 + a$. Using that $a = b^3$ for some b in $\mathbb{F}_3$ in the latter case, $P(X) = (X + b)^3$, which is impossible.

If $x \in \mathbb{F}_{p^3}$ is not in $\mathbb{F}_p$, then the corresponding $P(X)$ cannot be of degree 2, because there are no quadratic extensions of $\mathbb{F}_p$ inside $\mathbb{F}_{p^3}$ due to the divisibility of dimensions; see Section 7.3.1. Thus all elements of $\mathbb{F}_p^3 \subset \mathbb{F}_p$ are combined in triples, the roots of irreducible polynomials over $\mathbb{F}_p$. Therefore the number of such triples, $(p^3 - p)/3$, gives exactly the number of irreducible P of degree 3.

(i) The number of monic *reducible* polynomials is that for irreducible quadratic ones, which is $(p^2 - p)/2$ (check), times the number of linear ones $(X - x)$ with $x \in \mathbb{F}_p$, which is p, plus the number of $P(X) = (X - x_1)(X - x_2)(X - x_3)$ for any $x_1, x_2, x_3 \in \mathbb{F}_p$. The latter number is $R(p, 3) = \binom{p+2}{3} = \frac{(p+2)(p+1)p}{6}$ (3 objects from p types). We obtain $(p+2)(p+1)p/6 + (p^2 - p)p/2 = (p^2 + 3p + 2 + 3p^2 - 3p)p/6 = (2p^2 + 1)p/3$. So it is $p^3 - (p^3 - p)/3$ indeed. ❑

[47] I. Cherednik, *A note on Artin's constant*, Mosc. Math. J., 15:2 (2015), 283–291.

Problem 7.3.8 *Prove that the equation $x^3 - a = 0$ for any $a \in \mathbb{F}_p^*$ has a unique solution in $\mathbb{F}_p$ for $p \neq 1 \bmod 3$. Otherwise it has no solution or exactly 3 in $\mathbb{F}_p$. Check that the field obtained by adding an element x satisfying $x^3 = 2$ to the field $\mathbb{F}_7$ is $\mathbb{F}_{7^3}$ (the field with 7^3 elements). Calculate $(x \pm 1)^{-1}$: express them as linear combinations of $1, x, x^2$.*

Solution. The group $\mathbb{F}_p$ is cyclic with $(p-1)$ elements, so if $(p-1)$ is not divisible by 3, then the map $x \mapsto x^3$ is 1-1. So there is exactly one solution of $x^3 = a$ for such p and *any* $a \neq 0$. Note that the remaining solutions will then be roots of an irreducible quadratic polynomial.

Similarly, if a solution x exists when 3 divides $p-1$, then all solutions are xu for all u satisfying $u^3 = 1$. At least one such $u \neq 1$ is granted in $\mathbb{F}_p$. You can use the cyclicity of $\mathbb{F}_p^*$ (above) or a general fact that any G of order divisible by a prime number q contains an element of order q. So the group of such u is of order 3: this order is no greater than 3 since $X^3 - 1$ has no greater than 3 solutions in any field.

In $\mathbb{F}_7$, the cubes are only ± 1 and the polynomial $X^3 - 2$ is irreducible. Indeed, if it is a product of two, then one of them must be linear, which means that $X^3 - 2$ has a root in $\mathbb{F}_7$. Thus $\mathbb{F}_7[x] = \mathbb{F}_{7^3}$. Now let $u = x + 1$. Then $x = u - 1$ and $x^3 = 2$ becomes $(u-1)^3 = 2 = u^3 - 3u^2 + 3u - 1$. So $u^3 - 3u^2 + 3u = 3$, and $3/u = u^2 - 3u + 3 = (x+1)^2 - 3(x+1) + 3 = x^2 - x + 1$. Finally, $1/u = (x^2 - x + 1)/3 = 5(x^2 - x + 1)$. A direct calculation gives: $(1+x)(x^2 - x + 1) = x^3 + 1 = 3$. One can also do the following: $x^3 - 2 = (x^3 + 1) - 3 = (x+1)(x^2 - x + 1) - 3 = 0$ and $\frac{1}{u} = \frac{x^2 - x + 1}{3}$. Similarly, $(x-1)(x^2 + x + 1) = x^3 - 1 = 1$ and $\frac{1}{x-1} = x^2 + x + 1$. ❑

Problem 7.3.9 * *Let us assume that the polynomial $P(X) = X^2 - aX - b$ with the coefficients in some field $\mathbb{F}$ is irreducible and has a multiple root that is* not *in $\mathbb{F}$. This means that $P(X) = (X - \lambda)^2$ in the field $\mathbb{F}[\lambda]$ obtained by adding λ satisfying $P(\lambda) = 0$ to $\mathbb{F}$.*

(i) Obtain that this is possible only if $2 \cdot 1 = 0$ in $\mathbb{F}$ (i.e. when its characteristic is 2) and if not all elements in $\mathbb{F}$ are squares.

(ii) Check that if $\mathbb{F}$ is a finite *field of characteristic 2, then* all *its elements are squares, so there is no such $P(X)$ for finite $\mathbb{F}$.*

Solution. **(i)** The second root of $P(X)$ is $a - \lambda$. If $\lambda = a - \lambda$, then $2\lambda = a$. Since λ is not in $\mathbb{F}$, we obtain that $2 = 0$ in $\mathbb{F}$ in any of its extensions. Otherwise, $\lambda = a/2 \in \mathbb{F}$. We say that the *characteristic* of the field $\mathbb{F}$ is 2 in this case. Thus, $P(X) = X^2 - b$ and it has no root in $\mathbb{F}$. This simply means that b is not a square in $\mathbb{F}$. The conditions from (i) are actually necessary and sufficient for the existence of irreducible quadratic polynomials with multiple roots, but the problem is that they are necessary.

(ii) This never happens in *finite* fields of characteristic 2; all their elements are squares. To check this, we use that they have 2^m elements; we need only to know that this number is even. So the group $\mathbb{F}^*$ of nonzero elements has $2^m - 1$ elements and the preimage of 1 in $\mathbb{F}^*$ of the map $x \mapsto x^2$ can only be 1. Indeed, if $x^2 = 1$ for $x \neq 1$, then the order of $\mathbb{F}^*$ must be even due to the Lagrange Theorem. We conclude that different elements $x, y \in \mathbb{F}^*$ have different x^2, y^2: if $x^2 = y^2$ then $(x/y)^2 = 1$ and $x = y$. Therefore *all* elements in $\mathbb{F}^*$ are squares due to the Pigeonhole Principle.

Another justification of (ii) is based on the fact that the *Frobenius map* $F : x \mapsto x^p$ is a ring homomorphism in characteristic p, i.e. preserves multiplication and addition. It has a trivial kernel: if $F(x) = 0$ then $F(1/x)F(x) = F(1) = 0$ and F is zero identically. Thus, F must be an isomorphism for any finite (!) field of characteristic p, and all its elements are x^p for proper x. ❑

An example when (i) holds is the *field* of all rational functions $f(t)/g(t)$ for polynomials $f(t), g(t)$ of the variable t of any degree with the coefficients in $\mathbb{Z}_2$ (and nonzero $g(t)$). Check that this is a field. Then t is not a square, and $X^2 - t$ is an irreducible polynomial with a double root.

Problem 7.3.10* *Algebraic integers are complex numbers x satisfying a relation $R(x) = 0$ for a monic polynomial $R(X)$ (in the form X^n+ lower powers of X) with integral coefficients. Units are algebraic integers u such that u^{-1} are algebraic integers too. Lattices are spaces $\{\sum_i m_i z_i\}$ for finite sets of complex numbers z_i and with arbitrary integral coefficients m_i.*

(i) Prove that x is an algebraic integer if and only if the multiplication by x preserves some lattice L. Deduce from this that sums and products of algebraic integers are algebraic integers.

(ii) Prove that u is a unit if and only if $R(u) = 0$ for some monic $R(X) = X^n + \sum_{i=1}^{n-1} a_i X^i$ with integral a_i and $a_0 = \pm 1$, and that the product of two units u_1 and u_2 is a unit.

(iii) Given an algebraic integer x, prove that the irreducible monic polynomial $P(X)$ such that $P(x) = 0$ has integer coefficients. For this, prove that if $R(X) = P(X)Q(X)$ and the coefficients of the monic polynomials $P(X)$ and $Q(X)$ are rational, then they must be integral (Gauss). Accordingly, any unit satisfies $a_0 = \pm 1$ for the constant term a_0 of its minimal polynomial $P(X)$, which was the definition of units in Section 5.5.1.

Solution. **(i)** There are several natural steps. (a) Given an algebraic number x for monic $R(X)$ with integral coefficients, the multiplication by x preserves the lattice $\{\sum_{i=0}^{n-1} m_i x^i\}$.

(*b*) Vice versa, assume that the multiplication by x preserves some L. Dividing L by any of its element $\neq 0$, it can be assumed with 1; then L must contain all powers x^i. Consider the lattices $L_p = \{\sum_{i=0}^{p-1} m_i x^i\}$. They belong to L, and the maximal number of linearly independent elements in L_p over rationals (or integers) must reach its top value for some p°. Then for any $y \in L_p$, one has: $ay \in L_n$ for some integer $a > 0$ if $p \geq p^\circ$. Let L' be a subspace of all y in L such that $ay \in L_{p^\circ}$ for some integral $a > 0$. It is a lattice too (justify this).

(*b′*) One has $L_p \in L'$ for any p, and the quotient L'/L_{p° (elements in L' up to L_{p°) is finite. It is a finite commutative group, to be more exact. We obtain that all L_p coincide for $p \geq n$ for a certain sufficiently large n. This results in $xL_n = L_n$ and in the relation $x^n = \sum_{i=0}^{n-1} a_i x^i$ with integral a_i.

(*c*) The sum $x_1 + x_2$ and the product $x_1 x_2$ for algebraic integers x_1, x_2 preserve the lattice $\sum_{i,j} m_{ij} x_1^i x_2^j$ for $0 \leq i \leq n_1 - 1$ and $0 \leq j \leq n_2 - 1$, where n_i is the degree of the corresponding $R_i(X)$ for $i = 1, 2$. This proves (i).

(ii) As in (i), we can find a lattice L preserved by (the multiplications by) u and $v = \frac{1}{u}$. Then $uL \subset L$ and $L = uvL = u(vL) \subset uL$. We obtain that $uL = L$. As above, L can be assumed $L = \{\sum_{i=0}^{n-1} m_i u^i\}$ for integral m_i, where n is the smallest p such that u^p is a linear combination of u^i with integer coefficients for $0 \leq i < p$. Then $uL = L$ holds if and only if $u^n = \sum_{i=0}^{n-1} a_i x^i$ with $a_0 = \pm 1$. Otherwise, 1 would not belong to L. The same consideration gives that $a_0 = \pm 1$ implies that $uL = L$, which gives that $u^{-1}L = L$.

(iii) If $R(X) = P(X)Q(X)$ for monic P, Q with rational coefficients, let $f, g > 0$ be minimal integers such that $\widetilde{P} = fP$, $\widetilde{Q} = gQ$ are with integral coefficients. If $fg > 1$, then it is divisible by some prime p, and $\widetilde{P}\widetilde{Q} = (fg)R = 0 \bmod p$ (coefficient-wise). The latter is impossible because $\widetilde{P}$ and $\widetilde{Q}$ are nonzero $\bmod p$ by construction. Thus, P, Q are with integral coefficients. The last touch: if $a_0 = b_0 c_0$ is ± 1, where a_0, b_0, c_0 are constant terms of R, P, Q, then $b_0 = \pm 1$ (and so is c_0) because they are integers. ❑

Continued fractions via $SL(2, \mathbb{Z})$. Here $SL(2, \mathbb{Z})$ is the group of all 2×2 matrices of determinant $Det = 1$ with integer coefficients. This is a group with respect to multiplication of matrices; the unit element is the unit matrix. The inverse matrices always exist if the determinant is nonzero. If it is 1, then they are from $SL(2, \mathbb{Z})$. We claim that there is fundamental relation between the matrices from $SL(2, \mathbb{Z})$ with non-negative entries and with continued fractions for rational numbers.

Problem 7.3.11* *We set* $\tau_+ = \begin{pmatrix} 1 & 1 \\ 0 & 1 \end{pmatrix}$, $\tau_- = \begin{pmatrix} 0 & 1 \\ 1 & 0 \end{pmatrix}$. *For integers* $a, b > 1$ *with* $\gcd(a, b) = 1$, *let* $[b_0, b_1, \ldots, b_m] = [b_0, b_1, \ldots, b_m - 1, 1]$ *be the short* *(canonical) and* long *continued fractions for* a/b. *See Problem 5.5.5. We set*

$$T_{a,b} = \tau_+^{b_0}\tau_-^{b_1}\cdots\tau_+^{b_{m-1}}\tau_-^{b_m} \text{ for odd } m, \text{ and } T_{a,b} = \tau_+^{b_0}\tau_-^{b_1}\cdots\tau_+^{b_m-1}\tau_- \text{ for even } m.$$

(i) Check that $T_{a,b} = \begin{pmatrix} a & y \\ b & x \end{pmatrix}$ *for integers* x, y *such that* $ax - by = 1$, $0 < x < b$ *and* $0 < y < a$. *These conditions uniquely determine* (x, y).

(ii) Deduce from (i) that any matrix $\begin{pmatrix} a & y \\ b & x \end{pmatrix}$ *of* $Det = 1$ *with positive integer* x, y *is* $T_{a,b}\tau_+^k$ *for* $k \geq 0$. *The remaining integer matrices of* $Det = 1$ *with any non-negative* a, b, x, y *are* $\tau_-^m\tau_+^k$ *or* $\tau_+^m\tau_-\tau_+^k$ *for* $m, k \geq 0$.

Solution. Check directly that if $M = \begin{pmatrix} a & y \\ b & x \end{pmatrix}$ is such that $a, b, x, y > 0$, $a \geq y$ and $b \geq x$, then $\tau_\pm M$ satisfies the same inequalities. Since $a, b > 1$ and $Det(M) = 1$ for $M = T_{a,b}$, we obtain $0 < x < b$ and $0 < y < a$ from (i).

From the definition: $M = T_{a,b}$ ends with $\tau_-^{b_1}$ for $a < b$ and with $\tau_+^{b_0}$ for $a > b$. The matrix $\tau_+ M = \begin{pmatrix} a+b & y+x \\ b & x \end{pmatrix}$ results in $a/b \mapsto a/b + 1$ and $b_0 \mapsto b_0 + 1$. If $a > b$, then $b_0 > 0$ and $T_{a+b,b}$ is, indeed, $\tau_+ T_{a,b}$. If $a < b$, then $[0, b_1, b_2, \ldots]$ changes to $[1, b_1, b_2, \ldots]$, which gives $T_{a+b,b} = \tau_+ T_{a,b}$ too.

Similarly, $\tau_- M = \begin{pmatrix} a & y \\ a+b & y+x \end{pmatrix}$, which changes $\frac{a}{b} \mapsto \frac{a}{a+b}$. The corresponding continued fraction becomes $[0, b_1+1, b_2, \ldots]$ for $a/b = [0, b_1, b_2, \ldots]$ if $a < b$. Since $M = T_{a,b}$ ends with $\tau_-^{b_1}$ in this case, we obtain the required. If $a > b$, then the continued fraction for $\frac{a}{a+b}$ becomes $[0, 1, b_0, b_1, \ldots]$, where $a/b = [b_0 > 0, b_1, \ldots]$, which corresponds to $\tau_- T_{a,b}$, as it was claimed. ❑

Nonattacking queens and Latin squares. We will practice here modular methods in their theory.

Problem 7.3.12 *Use the formula* $j = ai + b \bmod 13$ *from Proposition 5.2.5 to extend the 2 queens in Fig. 7.2 to a configuration of pan-nonattacking queens in the square* 13×13, *where we use the indices* $0 \leq i, j \leq 12$ *for the rows and columns. Plot them. Which conditions for the positions of the initial queens are necessary and sufficient for such an extension to exist?*

Solution. We have $3a + b = 6, 11a + b - 3$. Thus, $8a = -3 = 10$ and $a = 5/4 = 1 + 1/4 \bmod 13$. The multiples of 4 are as follows: $4, 8, -1, \ldots$. We obtain $4 \cdot 3 = -1$, $a = 1 + 1/4 = 1 - 3 = -2 = 11 \bmod 13$. Therefore, $b = 6 - 3a = 6 + 6 = -1 = 12 \bmod 13$. Finally, we construct the queens using the formula $j = -2i - 1 \bmod 13$. Actually, the pattern of this configuration is clear "combinatorially" after the first 2 or 3 queens.

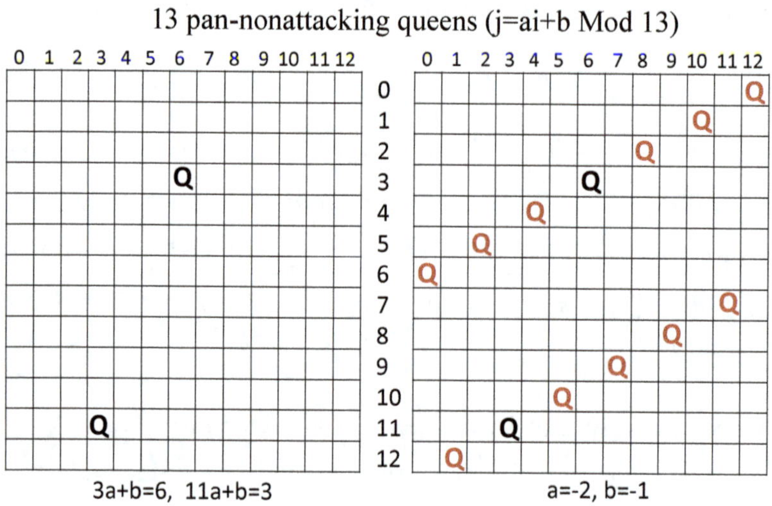

Figure 7.2: Pan-nonattacking queens through 2.

Generally, we have $ai_1+b=j_1, ai_2+b=j_2$ for the queens at $(i_1,j_1),(i_2,j_2)$, which gives that $a=\frac{j_i-j_2}{i_1-i_2}$. The condition $a=\pm1$, means *exactly* that these two queens either belong to the same pandiagonal (for $+1$) or to the same anti-pandiagonal (for $a=-1$). Otherwise $a\neq\pm1$ and, as we proved above, *all* a queens obtained via the formula $j=ai+b \bmod n$ will be pan-nonattacking.

So the necessary and sufficient condition for our formula to work is simply that 2 given queens must not belong to the same pandiagonal or anti-pandiagonal. This (combinatorial) condition is obviously necessary for such an extension; we proved that it is sufficient. ❑

Problem 7.3.13 *Construct* 3 *pairwise orthogonal Latin squares using the formula* $a'_{ij}=ri+j$ *in* $\mathbb{F}_4$ *for* $r\neq0$, *where the row and column indices* i,j, *and the corresponding entries are from* $\mathbb{F}_4$.

Solution. Let $a'_{ij}=ri+j$ in $\mathbb{F}_4$ for $r=\widehat{2},\widehat{3},1$ from Section 5.3.2. The ones for $r_1=\widehat{2}, r_2=\widehat{3}$ were calculated there, so we need only to add $r_3=1$, the addition table for $\mathbb{F}_4$:

1	0	1	$\widehat{2}$	$\widehat{3}$
0	0	1	$\widehat{2}$	$\widehat{3}$
1	$\widehat{2}$	$\widehat{3}$	0	1
$\widehat{2}$	$\widehat{3}$	$\widehat{2}$	1	0
$\widehat{3}$	1	0	$\widehat{3}$	$\widehat{2}$

,

2	0	1	$\widehat{2}$	$\widehat{3}$
0	0	1	$\widehat{2}$	$\widehat{3}$
1	$\widehat{3}$	$\widehat{2}$	1	0
$\widehat{2}$	1	0	$\widehat{3}$	$\widehat{2}$
$\widehat{3}$	$\widehat{2}$	$\widehat{3}$	0	1

,

3	0	1	$\widehat{2}$	$\widehat{3}$
0	0	1	$\widehat{2}$	$\widehat{3}$
1	1	0	$\widehat{3}$	$\widehat{2}$
$\widehat{2}$	$\widehat{2}$	$\widehat{3}$	0	1
$\widehat{3}$	$\widehat{3}$	$\widehat{2}$	1	0

.

❑

These 3 MOLS can be used for the error correction code for the alphabet with 12 symbols. The standard English keyboard contains 26 letters, 42 digits and symbols, *Shift*, and *Space*. By adding two emoji symbols to represent these, we make the total $26+42+2+2=72=9\cdot 8$. Thus, the field $\mathbb{F}_9$, with the corresponding 8 modular MOLS, is sufficient for optimal transmission of any (English) text (plus 2 emoji $\ddot{\smile}$, $\ddot{\frown}$) through any noise under some natural assumptions.

Problem 7.3.14 *Represent the elements from the field* $\mathbb{F}_9$ *as* $a\lambda_1+b\lambda_2$, *where* λ_1 *and* λ_2 *are assumed linearly independent over* $\mathbb{F}_3=\{0,1,2\}$, *and* a,b *are from* $\mathbb{F}_3$. *For* $r\neq 0$ *not in* $\mathbb{F}_3$ *construct the corresponding Latin square* LS'_r *by the formula* $(i,j)\to ri+j$ *in* $\mathbb{F}_9$. *Order the elements of* $\mathbb{F}_9$ *as follows:* $\{\{0,\lambda_1,2\lambda_1\},\{\lambda_2,\lambda_2+\lambda_1,\lambda_2+2\lambda_1\},\{2\lambda_2,2\lambda_2+\lambda_1,2\lambda_2+2\lambda_1\}\}$. *Check that* LS'_r *then satisfies the Sudoku condition: any* 3×3 *block from* 9 *for the above partition of* $\mathbb{F}_9$ *in the rows and columns contains all numbers from* $\mathbb{F}_9$.

Solution. For i,i',j,j' from any block, $ri+j=ri'+j'$ implies that $r(i-i')=j'-j$. Upon the division by λ, we obtain that r must belong to $\mathbb{F}_3$ unless $i=i'$ and $j=j'$. Thus all numbers from $\mathbb{F}_9$ will appear in this 3×3 block. We leave the Sudoku application to the reader. ❑

We will now generalize (the purely combinatorial) Exercise 5.6.45: the addition table of $\mathbb{Z}_4$ considered combinatorially as a Latin square has no orthogonal mate. An application is Exercise 2.6.39: there are no pandiagonal Latin squares for *even* n. They do not exist even if only pandiagonals (or only anti-pandiagonals) are required to have no repetitions. Following Corollary 5.2.2, LS where pandiagonals have no repetitions exist for any *odd* n.

Problem 7.3.15 *For an even* n, *the addition table of* $\mathbb{Z}_n$, *which is a Latin square* L *with the entries* $0,1,\ldots,n-1$, *has no orthogonal mate* L'.

Solution. Assume that a Latin square L' is orthogonal to L. Take *any* k $(0\le k\le n-1)$, and consider the boxes in L' with the entries k. Let their row and column numbers be $(i_1=0,j_1),(i_2=1,j_2)\ldots,(i_n=n-1,j_n)$; then $(j_1,j_2,\ldots,j_n)$ is a permutation of $(0,1,\ldots,n-1)$ since L' is Latin. The numbers $a_1,a_2,\ldots,a_n$ in these boxes considered in L must be all different too; otherwise, there will be repetitions in the corresponding bi-square $L\vee L'$, namely, in the sequence of bi-numbers $\langle a_1,k\rangle,\langle a_2,k\rangle,\ldots,\langle a_n,k\rangle$.

Now, $S\stackrel{\mathbf{def}}{=\!=}\sum_{m=1}^{n}(i_m+j_m)=2\frac{(n-1)n}{2}=(n-1)n$, which is 0 modulo n. Using that $a_m=i_m+j_m \bmod n$, we obtain that $S \bmod n=\sum_{m=1}^{n}a_m=n(n-1)/2=n(n/2)-n/2=n/2 \bmod n$. However, it is 0 due to the first calculation, a contradiction. ❑

7.3.4 Quadratic Reciprocity *

There are many contributions to this celebrated fact: Euler, Legendre, Gauss (the first 8 proofs of it), and so on. In our book, Pisano periods are an application. Also, it is related to the theory of Pell's equations $x^2 - dy^2 = \pm p$ for odd prime p. Namely, d must be assumed a square in $\mathbb{Z}_p$ when this equation is considered modulo p. We mention that the combinatorial part of the proof we provide below is related to the Frobenius coin problem.

Problem 7.3.16 *The Gauss Lemma. For an odd prime number p, let a be any positive integer such that $\gcd(a,p) = 1$. Consider the sequence $\mathcal{A}$ of the residues* mod p *of the numbers $a, 2a, \ldots, \frac{p-1}{2}a$. Let $\mathcal{V} = \{v_1, \ldots, v_n\}$ be the subsequence of $\mathcal{A}$ formed by the residues that are greater than $\frac{p}{2}$; so $n = |\mathcal{V}|$ is its cardinality. We treat here the residues as numbers from 0 to $p-1$.*

(i) Prove that a mod p *is a square in the field $\mathbb{Z}_p$, a* quadratic residue, *if and only if $n = |\mathcal{V}|$ is even.*

(ii) Use (i) to check that 2 mod p *is a square for an odd prime number p if and only if $p = \pm 1$* mod 8.

Solution. **(i)** First of all, $ia \neq ja \bmod p$ for $1 \leq i \neq j \leq p-1$ due to $\gcd(a,p) = 1$. Let $\mathcal{U} = \{u_1, \ldots, u_m\}$ be the subsequence of $\mathcal{A}$ formed by the residues smaller than $\frac{p}{2}$ (considered as numbers). One has: $\mathcal{A} = \mathcal{U} \cup \mathcal{V}$ because $p/2$ is not an integer (p is odd), and $m + n = |\mathcal{A}| = (p-1)/2$. Then, $ia \neq p - ja \bmod p$ for any $1 \leq i, j \leq \frac{p-1}{2}$ because $i + j \leq p-1$. This means that $\mathcal{A}$ and $p - \mathcal{A}$ do not intersect, where $p - X$ is the set $\{p - x\}$ for x from a given set X.

Using that $\mathcal{A} \cap (p - \mathcal{A}) = \emptyset$ and that $p - \mathcal{V} \subset p - \mathcal{A}$, we obtain that the sequence $\mathcal{C} = \{u_1, v_2, \ldots, u_m, p - v_1, p - v_2, \ldots p - v_n\}$ has $(p-1)/2$ different elements. Since all of them are from 1 to $\frac{p-1}{2}$ by construction, the sequence $\mathcal{C}$ is a permutation of $\{1, 2, \ldots, \frac{p-1}{2}\}$.

Let us calculate now the product of the elements of $\mathcal{A}$ modulo p in two different ways: directly due to its definition and then using that $\mathcal{C}$ is a permutation of $\{1, 2, \ldots, \frac{p-1}{2}\}$. One has:

$$a \cdot (2a) \cdots \left(\frac{p-1}{2}a\right) = a^{\frac{p-1}{2}} \left(\frac{p-1}{2}\right)! = u_1 \cdots u_m \cdot v_1 \cdots v_n \bmod p$$

$$= (-1)^n u_1 \cdots u_m \cdot (p - v_1) \cdots (p - v_n) \;=\; (-1)^n \left(\frac{p-1}{2}\right)! \bmod p.$$

Then we divide both sides by $(\frac{p-1}{2})!$ and use that a is a square mod p (a quadratic residue) if and only if $a^{\frac{p-1}{2}} = 1$. The latter is due to Euler. It readily follows from the cyclicity of the group $\mathbb{Z}_p^*$; see Exercise 5.6.18(i).

(ii) For $a=2$: $\mathcal{A}=\{2,4,\ldots,p-1\}$. First, let $p=1 \bmod 4$. Then $\mathcal{U}=\{2,\ldots,\frac{p-1}{2}\}, \mathcal{V}=\{\frac{p+3}{2},\ldots,p-1\}$. So $m=\frac{p-1}{4}=n$, and 2 is a square modulo p in this case if and only if $p=1 \bmod 8$. Second, let $p=3 \bmod 4$. Then $\mathcal{U}=\{2,\ldots,\frac{p-3}{2}\}, \mathcal{V}=\{\frac{p+1}{2},\ldots,p-1\}, m=\frac{p-3}{4}, n=\frac{p+1}{4}$. We conclude that $2 \bmod p$ is a square if and only if $p=-1 \bmod 8$ in this case. ❑

Problem 7.3.17 *(i) We set $N_{a,b}=[\frac{a}{b}]+[2\frac{a}{b}]+\cdots+[\frac{(b-1)}{2}\frac{a}{b}]$ for odd b. Recall that $[\alpha]$ is the integer part of α. For $a,p,n=|\mathcal{V}|$ from Problem 7.3.16, check that $n=N_{a,p}+(a-1)\frac{p^2-1}{8} \bmod 2$. If a is odd or $p=\pm 1 \bmod 8$, then $n=N_{a,p} \bmod 2$; otherwise $n=N_{a,p}+1 \bmod 2$.*

(ii) ***Quadratic Reciprocity.*** *Let $a=q$, $b=p$ for any two distinct odd prime numbers p,q. For $p=3 \bmod 4=q$, the claim is that $p \bmod q$ is a square in $\mathbb{Z}_q$, a quadratic residue in $\mathbb{Z}_q$, if and only if $q \bmod p$ is not a square in $\mathbb{Z}_p$. For any other combination of $p,q \bmod 4$, one has : $p \bmod q$ and $q \bmod p$ are squares or non-squares simultaneously.*

Solution. **(i)** Recall that $ia \bmod p = ia-\lfloor\frac{ia}{p}\rfloor p$ for any i, where $[z]=$Floor$[z]$. One has: $a+2a+\cdots+\frac{p-1}{2}a=\frac{p^2-1}{8}a$, and $\frac{p^2-1}{8}a+N_{a,p}\,p=u_1+\cdots+u_m+v_1+\cdots+v_n$. The latter sum is $np+u_1+\cdots+u_m+(p-v_1)+\cdots+(p-v_n)$ modulo 2, which is $np+1+\cdots+\frac{p-1}{2}=np+\frac{p^2-1}{8}$. Thus, $np=(a-1)\frac{p^2-1}{8}+N_{a,p}p$ and $n=(a-1)\frac{p^2-1}{8}+N_{a,p}\,p \bmod 2$ because p is odd. Finally, use that $p^2=1 \bmod 16$ for $p=\pm 1 \bmod 8$ and only for such p.

We note here that $N_{a,p}=0$ for $a=2$ and any odd p; so $n=0 \bmod 2$ precisely for $p=\pm 1 \bmod 8$, which matches Problem 7.3.16(ii).

(ii) Generally, for any a and odd b such that $\gcd(a,b)=1$, $N_{a,b}$ is the number of *positive* differences $xa-yb$ for $x=1,2,\ldots,\frac{b-1}{2}$, $y=1,2,\ldots$. This gives that $N_{a,b}+N_{b,a}=\frac{a-1}{2}\frac{b-1}{2}$ if both, a and b, are odd. Indeed, $xa-yb$ cannot be 0 in this range due to $\gcd(a,b)=1$, so this difference can be either positive or negative (positive by swapping a and b).

Using (i) for odd prime p,q, we conclude that both, $p \bmod q$ and $q \bmod p$, are squares or non-squares simultaneously if and only if $N_{a,b}+N_{b,a}=\frac{a-1}{2}\frac{b-1}{2}$ is even. Otherwise $p=3 \bmod 4=q$ and only one of them is a square. ❑

Some examples are in Exercise 5.6.19. Let us mention a connection of (ii) with Problem 7.5.5 below (the Frobenius coin problem). The *set of gaps* there is $G_{b,a}=\{xa-yb>0$ such that $1\le x\le b-1,\ 1\le y\le a-1\}$. It is similar to what we considered above, but we imposed the relations: $1\le x\le\frac{b-1}{2}$, $1\le y\le\frac{a-1}{2}$.

Problem 7.3.18 *The* ***Legendre symbol*** *$\left(\frac{m}{p}\right)$ is ± 1 if m is a square modulo odd prime p and -1 otherwise, provided that $\gcd(m,p)=1$. Accordingly, Problem 7.3.17(i) can be stated as $\left(\frac{p}{q}\right)\left(\frac{q}{p}\right)=(-1)^{\frac{p-1}{2}\cdot\frac{q-1}{2}}$ for distinct odd* *prime numbers p, q. The* **Jacobi symbol** *for odd $n, m > 0$ is defined as $\left(\frac{m}{n}\right) = \left(\frac{m}{p_1}\right)^{a_1}\left(\frac{m}{p_2}\right)^{a_2}\cdots\left(\frac{m}{p_\ell}\right)^{a_\ell}$, where $\gcd(m,n) = 1$ and $n = p_1^{a_1}p_2^{a_2}\cdots p_\ell^{a_\ell}$ is the prime decomposition; this symbol is 0 otherwise.*

Note that the Legendre symbol is always defined for m relative to an odd prime p while the Jacobi symbol extends this concept to any n. Check that $(-1)^{\frac{ab-1}{2}} = (-1)^{\frac{a-1}{2}+\frac{b-1}{2}}$ for odd a, b, and prove that $\left(\frac{m}{n}\right) = (-1)^{\frac{m-1}{2}\cdot\frac{n-1}{2}}\left(\frac{n}{m}\right)$. Also, $\left(\frac{m}{n}\right) = 1$ if m is a square in $\mathbb{Z}_n^*$, but not "if and only if" unless $n = p^a$ for odd prime p and odd a. ❑

7.4 Graphs, trees, line graphs

7.4.1 Trails and paths

Problem 7.4.1 *Does the ornament in Fig. 7.3 have an Eulerian circuit, E-circuit for short, considering the intersections as vertices and the corresponding smallest portions of arcs between them as edges?*

Solution. It has an E-circuit since it is made from intersecting circles, though some consideration is necessary near the border. The degrees are 12 or 8: all even. Intersections of any closed curves have even degrees. ❑

Figure 7.3: Ornament: Flower of Life.

Problem 7.4.2 *Does the bearded face in Fig. 7.4 have an Eulerian circuit or trail, considering the intersections as vertices and the corresponding smallest portions of the curves between them as edges?*

Solution. It has only one non-closed curve: from the nose to the ♡. All others are closed, so all intersections,, except two, have even degrees. Also, it is connected. Thus, an Eulerian trail, from nose to ♡, exists (but not an Eulerian circuit). ❑

Figure 7.4: Does it have an E-circuit of type (I) or (II)?

Problem 7.4.3 *Prove that closed knight's tours visiting any square exactly once do not exist on the* $4\times n$ *board for any* n.

Solution. The squares can be black (b) or white (w) with respect to the standard colorings of chessboards. Also, the squares belonging to 2 middle n-rows of "$4\times n$" will be called *interior* (i), and *exterior* (e) otherwise; to be exact, interior/exterior in their 4-columns. Any square on the board is one of these: (be), (bi), (we), (wi). Knight's moves change (b) $\to$ (w), (w) $\to$ (b), and (e) $\to$ (i). However, the moves (i) $\to$ (i) are possible. Since all fields must be visited and the numbers of the interior and exterior squares coincide, the required tour must be alternating: (e) $\to$ (i) $\to$ (e) $\to$ (i) $\to\cdots$. If (i) $\to$ (i) occurs at least once, then the number of (e) will be smaller, which is impossible in the *closed* tour. Open tours can exist. Thus, (i) $\to$ (e) always, and only 2 types from 4 can be present in any such tour: (be), (wi) or (bi), (we). This is very far from being Hamiltonian. ❑

Problem 7.4.4 *For any positive integer r, s, t let the corresponding* ***Ramsey number*** *$R(r, s, t)$ be the minimal number n such that K_n with* each *configuration of black, white and red edges contains either a white induced subgraph K_r, or a black induced subgraph K_s, or red such K_t. Use the formula $R(3,3) = 6$ from Exercise 6.4.16 to prove that $R(3,3,3) \leq 17$ (in fact, it equals 17).*

Solution. Assume that there exists a 3-coloring K_{17} without monochromatic induced subgraphs K_3. Given a vertex v (any) in K_{17}, consider the induced subgraphs B, W, R formed by the vertices connected with v only by black, white or red edges. Neither contains edges of the same color (which would result in a monochromatic K_3). Thus, these subgraphs are 2-colored, and $|B|, |W|, |R| \leq 5$ due to Exercise 6.4.16. We obtain that $17 = |B| + |W| + |R| + 1 \leq 15 + 1 = 16$, a contradiction. ❑

Using that $R(3,3,3) = 17$, we obtain in a similar way that $R(3,3,3,3) \leq 2 + 4 \cdot 16 = 66$ in the natural notation. It can be somewhat improved: $R(3,3,3,3) \leq 64$ (A. Sanchez-Flores). Finding exact values of Ramsey numbers is generally an extremely complicated problem.

7.4.2 Labeled trees

Connected graphs without cycles are called *trees*; a tree with n vertices (of order n) is called an n-tree. So trees are simple graphs.

Problem 7.4.5 *Prove that the number of edges of any connected graph G of order n (with n vertices) is no smaller than $(n-1)$, and that it is exactly $(n-1)$ if and only if G is a tree.*

Solution. Proceed by induction with respect to n. Any connected G with a loop remains connected if an edge from this loop is removed (without touching the vertices). This proves the inequality $> (n-1)$ for the number of edges if G is not a tree. In the case of an n-tree G, take any vertex of degree 1 (they exist) and remove it together with the edge from it. The corresponding, G' is then an $(n-1)$-tree by the induction hypothesis, and so is G. ❑

By a *labeled n-tree*, we mean a tree with n vertices (of order n) labeled by the numbers $1, 2 \ldots, n$; different vertices must receive different labels. The following is *Cayley's formula.*

Problem 7.4.6 * *Prove that the number of labeled n-trees for $n > 1$ considered up to isomorphisms is n^{n-2}. Equivalently, n^{n-2} is the number of* spanning subtrees *in the (labeled) complete graph K_n. Spanning subtrees of a graph are those containing all of the vertices of the original graph.*

Solution. We will use the expansion of $(x_1 + x_2 + \cdots + x_n)^{n-2}$, so there is a direct link to *Stirling numbers*. Let us use graphs to describe the monomials in this expansion. Calling them *words* is more appropriate actually, since the order of the x-variables matters. They are $x_{i_1}x_{i_2}\cdots x_{i_{n-2}}$, where x_{i_1} is taken from the first quantity in the product ($1 \le i_1 \le n$), x_{i_2} from the second, and so on. They can be presented combinatorially as follows: $\begin{pmatrix} 1\ 2\ \cdots n-2 \\ i_1 i_2 \cdots i_{n-2}\end{pmatrix}$. Accordingly, we obtain a graph G: $\{1 \to i_1,\ 2 \to i_2,\ \ldots,\ (n-2) \to i_{n-2}\}$. So it has n vertices, which are the indices of x_i, and $(n-2)$ edges. The latter are *from* the vertices $1, 2, \ldots, n-2$ to any vertices (can be $n-1$ or n). The edges have natural directions here; such graphs are called *directed*.

There are no edges *from* vertices $n-1$ and n by construction. If any of them are missing in a given monomial, then there will be no edges *to* it, and it will be an isolated vertex of G describing this monomial.

We will construct the corresponding labeled n-tree Γ, which means it will be directionless and without cycles. The graphs G above generally have cycles, so they must be "eliminated."

Represent cycles in G (if any) as $C_i = \{a_i \to b_i \to c_i \to \cdots \to a_i\}$, starting with their smallest vertices and with respect to the directions of the edges. Any vertex can belong to no greater than one cycle; let m be their number. So a_i is the minimal possible vertex in C_i for $1 \le i \le m$. The cycles in G will be ordered with respect to a_i: C_{i+1} is after C_i if $a_{i+1} > a_i$.

Begin with vertex $(n-1)$ and connect it with b_1, the second vertex in the 1st cycle. Then go from b_1 to a_1 along the 1st cycle, i.e. via c_1, connect a_1 with b_2 by a *new edge* (not from G), and then delete the *old edge* between a_1 and b_1. Next, go from b_2 to a_2 via c_2 in the 2nd cycle, connect a_2 with b_3 by a new edge, and then delete $a_2 \to b_2$. Continue in the same manner until a_m is reached (in the last cycle). Then connect it with vertex n by a new edge, and, finally, delete the edge $a_m \to b_m$.

There are two exceptional situations. If the cycle is $a_i \to a_i$ (i.e. x^i is taken from the ith quantity in the product), then connect $a_{i-1} \to a_i \to b_{i+1}$ and delete $a_i \to a_i$. Only cycles in G are involved in this procedure. However, if there are no cycles in G, then add 1 edge: connect $(n-1)$ with n.

Finally, the resulting graph Γ contains all vertices and it is connected (by construction). It has exactly $n-1$ edges. Indeed, we deleted m old edges, but added $m+1$ new ones. Since the initial number of edges was $n-2$, this number becomes $n-1$ in Γ. Thus Γ is a tree: Problem 7.4.5. The directions of edges, initial or added, are disregarded in Γ.

In the opposite direction, i.e. from Γ to G, we connect vertices $(n-1)$ and n by a *unique* path P in Γ (it is a tree). Let the vertex after $(n-1)$ in this

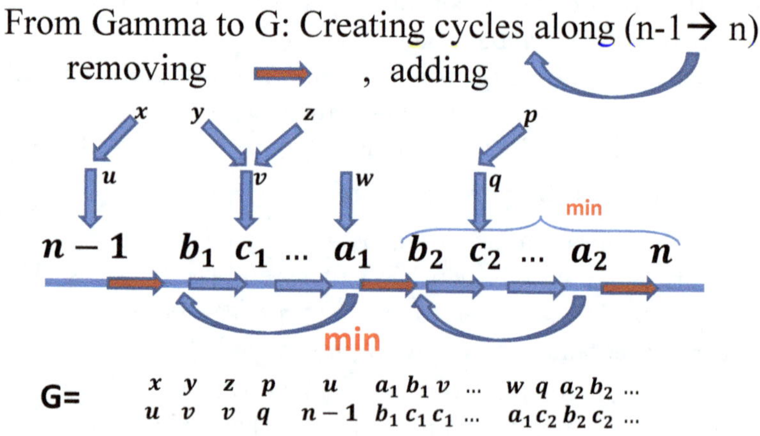

Figure 7.5: From tree Γ to directed G.

path be b_1, and the minimal vertex in this path be a_1. The first cycle in G will be: $a_1 \to b_1$, then all the way from b_1 to a_1 along P. To construct the 2nd cycle, we disregard the 1st one and continue in the same manner with the remaining vertices of Γ. Namely, let b_2 be the remaining neighbor of a_1 in P (one of its neighbors was used in the 1st cycle), and a_2 the minimal vertex apart from the first cycle. As above, we create the 2nd cycle in G, disregard it, create the 3rd one, and continue until vertex n is reached.

The edges that are not from P remain untouched. We obtain that vertices n and $(n-1)$ can have only arrows *to* them, and that any other vertex has exactly one edge *from* it. This is sufficient to *uniquely* supply the edges of G constructed above with the directions. See Fig. 7.5.

Thus, the construction described above establishes a 1-1 correspondence between the monomials (words, to be exact) and labeled n-trees. Namely, the map is $x_{i_1}x_{i_2}\cdots x_{i_{n-2}} \to G \to \Gamma$. Finally, let us take $x_1 = \cdots = x_n = 1$; this results in the formula n^{n-2}. The usage of the complete graph K_n is a simple reformulation. ❑

Examples. Consider all of the labeled trees for $n = 3$; they are: $\{1-2-3\}$, $\{1-3-2\}$, and $\{2-1-3\}$, (3 of them.) The number of such trees is 16 for $n = 4$, and such numbers grow rapidly for $n > 4$.

Let $n = 8$. Then G corresponding to the monomial $x_4x_1x_4x_8x_3x_8$ will be $\{1 \to 4,\ 2 \to 1,\ 3 \to 4,\ 4 \to 8,\ 5 \to 3,\ 6 \to 8\}$.

There are no cycles here; so no edges will be removed and only one edge must be added: $7 \to 8$. This will be Γ corresponding to G upon ignoring the directions of edges.

For $n = 8$ and the monomial $x_4x_2x_5x_1x_3x_7$, the graph G is as follows:

$$1 \to 4,\ 2 \to 2,\ 3 \to 5,\ 4 \to 1,\ 5 \to 3,\ 6 \to 7.$$

The corresponding ordered sequence of cycles is $\{a_1 = 1 \to b_1 = 4 \to 1\}$, $\{a_2{=}2 \to b_2{=}2\}$, $\{a_3{=}3 \to b_3{=}5 \to 3\}$. Accordingly, we

add: $n{-}1{=}7 \to b_1{=}4,\ a_1{=}1 \to b_2{=}2,\ a_2{=}2 \to b_3{=}5,\ a_3{=}3 \to n{=}8,$

and delete : $a_1 = 1 \to b_1 = 4,\ \ a_2 = 2 \to b_2 = 2,$ and $a_3 = 3 \to b_3 = 5.$

Finally $\Gamma = \{7 - 4 - 1 - 2 - 5 - 3 - 8 - 6\}$, i.e. it is a path from 7 to 6.

Problem 7.4.7 *Let $F(n,k)$ be the numbers of labeled forests with n vertices that are disconnected unions of k subtrees such that the vertices $1, 2, \ldots, k$ belong to different subtrees. Prove that $F(n,k) = kn^{n-k-1}$.*

Solution. Surprisingly, this generalization of Problem 7.4.6 is simpler to check, but its proof below is more algebraic. We follow [48]. This formula is due to Cayley too (1889), but without indications on its justification.

The *secondary vertices* will be those connected with the *primary vertices* $1, 2, \ldots, s$ in the corresponding subtrees. There are $\binom{n-k}{j}$ options for j secondary vertices (totally): the only restriction is that $1, 2, \ldots, s$ are not allowed. If vertex 1 and all edges from it are removed, then we obtain a *forest* with $n - 1$ vertices and $k + j - 1$ subtrees. Doing this for all primary vertices:

$$F(n,k) = \sum_{j=0}^{n-k} \binom{n-k}{j} F(n{-}1, k{+}j{-}1), \text{ where } F(1,1){=}1,\ F(n,0){=}0,\ n \geq 1.$$

Then we proceed by induction on n, k. Let us assume that $F(n-1, s) = s(n{-}1)^{n-s-2}$ for $n > 1$ and $1 \leq s \leq n{-}1$. Then $n{-}1{-}(k{+}j{-}1){-}1 = n{-}k{-}j{-}1$ and by switching to $i = n{-}k{-}j$ from j we obtain that $F(n,k) =$

$$\sum_{j=0}^{n-k} \binom{n-k}{j} (k + j - 1)(n - 1)^{n-k-j-1} = \sum_{i=0}^{n-k} \binom{n-k}{i} ((n-1) - i)(n-1)^{i-1}$$

$$= \sum_{i=0}^{n-k} \binom{n-k}{i} (n-1)^i - \sum_{i=0}^{n-k} \binom{n-k}{i} i(n-1)^{i-1} = n^{n-k} - (n{-}k)n^{n-k-1}$$

$$= n^{n-k} - (n-k)n^{n-k-1} = (n - n + k)n^{n-k-1} = kn^{n-k-1}.$$

Here the formula for $(1 + x)^{n-k}$ and its x-derivative are used. ❑

---

[48] L. Takács, *Note on Cayley's formula for counting forests*, Journal of Combinatorial Theory, Series A 53 (1990), 321–323.

7.4.3 Line graphs

For these graphs, Hamiltonian cycles can be obtained in a relatively simple way. For every *simple* connected graph G (without multiple edges and ↺), let $L(G)$ be the corresponding ***line*** (or ***derived***) ***graph***, with the vertices that are edges of G. Two vertices in $L(G)$ are *adjacent*, connected by an edge there, if and only if the corresponding edges in G share a vertex. Since G is simple, such a vertex is unique. Hassler Whitney proved that only two *non-isomorphic* G have *isomorphic* $L(G)$, namely, the complete graph K_3 with 3 vertices (geometrically, △), and the one of order 4, where the edges are from one of them to the remaining 3 (geometrically, Y); the notation is $K_{3,1}$. The proof is somewhat involved.

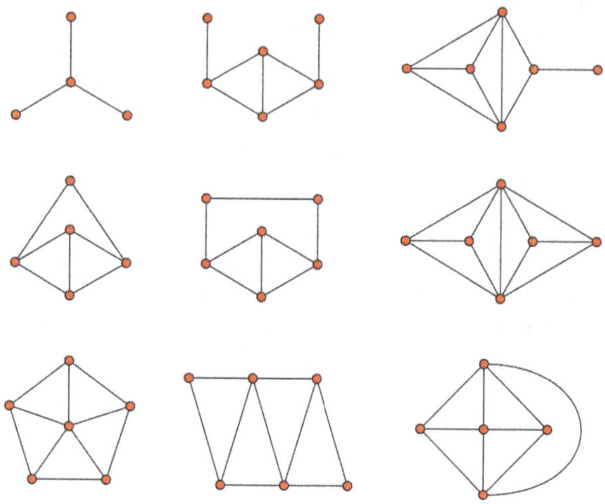

Figure 7.6: 9 obstacles to being a line graph.

L. Beineke and, independently, N. Robertson proved that line (or derived) graphs are those that do not contain these 9 as ***induced subgraphs***. A subgraph H of some G is called *induced* if all edges from G between the vertices from H are its edges.

Fig. 7.6 is from [49], originally from [50]. Here and in the examples below all intersections in geometric presentations are considered as vertices, and the edges are the shortest portions of the corresponding curves between them.

[49] `https://commons.wikimedia.org/wiki/File:Forbidden_line_subgraphs.svg`.

[50] L. Beineke, *Characterizations of Derived Graphs*, Journal of Combinatorial Theory, 9 (1970), 129–135.

Problem 7.4.8 *For any vertex v in G, all edges from it form a complete subgraph $L_v(G)$ in $L(G)$. We will assume that G has at least 3 vertices.*

(i) Check that every $L(G)$ is a union of its complete subgraphs $L_v(G)$ for $deg_v \geq 2$, that every edge of $L(G)$ lies in exactly one of these complete subgraphs, and every vertex of $L(G)$ belongs to no greater than 2 of them.

(ii) Moreover, $L_v(G)$ for $deg_v \geq 3$ are maximal *complete subgraphs in $L(G)$, i.e. do not belong to greater complete subgraphs, and maximal complete subgraphs with > 3 vertices in $L(G)$ are in the form $L_v(G)$ for some v in G.*

Solution. **(i)** By construction, every edge of $L(G)$ corresponds to a pair of edges in G from 1 vertex. So it lies only in 1 of such complete subgraphs. Every edge of G belongs only to L_v for its two endpoints v. At least one of them is of degree ≥ 2 because G is connected and has ≥ 2 vertices.

(ii) If $deg_v \geq 3$ for v in G, then any edge not through v that "intersects" all the edges through v must have at least 3 endpoints, which is impossible. By intersecting, we mean having vertices in common. So the corresponding $L_v(G)$ are maximal. Similarly, all edges in G corresponding to the vertices in any maximal complete subgraph M in $L(G)$ must be pairwise intersecting. If these edges are not through one vertex and there are at least 3 of them, then there will be a triangle formed by some 3 of them. If the number of such edges is greater than 3, then there exists an edge intersecting (having a common vertex with) all these 3, which is impossible. ❑

Problem 7.4.9 *Use the previous problem to check that any of the graphs in Fig. 7.6 cannot be represented as $L(G)$ for any G.*

Solution. The solution of the problem is by inspection. Any representation of each of these 9 graphs G_i as a union of complete subgraphs K_2 (a segment), K_3 (a triangle), or K_4 either

(*a*) has at least 1 vertex belonging to 3 such subgraphs, or

(*b*) has at least one of the edges belonging to two of them.

Both situations contradict (i) in Problem 7.4.8. The following reduction to (a) or (b) is by a case-by-case verification; let us describe the main steps.

First of all, subgraphs K_4 can occur only in the middle-right graph. There are 2 of them and both must be in the form L_v due to (ii); they do share an edge, so this is situation (b). Otherwise, only K_2 and K_3 can occur.

Second, the presentation of the 1st graph, G_1, in this figure as a union of complete subgraphs can only be the union of its 3 segments. Then the middle vertex belongs to all 3, which results in (a), a contradiction.

Third, to avoid this in G_i for $i > 1$, there must be some triangles in their corresponding unions such that any subgraph in G_i isomorphic to G_1 has at least one edge belonging to one of these triangles. However then 1 vertex will belong to 3 complete subgraphs (in any such union) unless there are 2 triangles there with a common edge. So we arrive at either (a) or (b). ❑

We provide several examples of $L(G)$ in Fig. 7.7. The one in the middle of the first line contains  as an *induced* subgraph, so it is *not* in the form $L(G)$ for any G (check this directly).

However, its extension $\widetilde{G}$ right below this one is $L(G)$ for the following G. Take the bottom-right graph and inscribe it in a circle. The resulting one has 12 edges, matching the number of vertices of $\widetilde{G}$. Check that its $L(G)$ is $\widetilde{G}$.

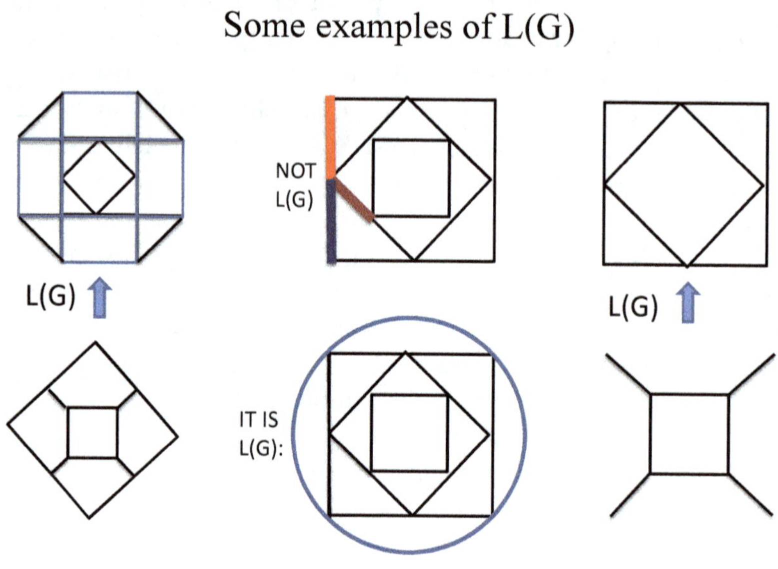

Figure 7.7: Some line and non-line graphs.

Problem 7.4.10 *(i) Every $L(G)$ has a Hamiltonian cycle, a reentrant Hamiltonian path, if G has an Eulerian circuit. However, not "if and only if": find a counterexample.*

(ii) Applying (i), prove that $L(G)$ with all even degrees (in $L(G)$) and with at least one even deg_v in G always have Hamiltonian cycles.

Solution. **(i)** Follow the Eulerian circuit in G and connect the vertices in $L(G)$ for the corresponding neighboring edges. This will be a required cycle.

An example of "not only if", is $G = K_4$, the complete graph with 4 vertices. It is the tetrahedron:  without the middle intersection. Then $L(G)$ is a square with 4 vertices (the sides of the initial square) connected with 2 additional vertices for the diagonals. This is the octahedron. It has a Hamiltonian cycle (check). The tetrahedron has no E-circuit or E-trail.

(ii) If $e = [v, v']$ is an edge in G considered as a vertex in $L(G)$, then $deg_e = deg_v + deg_{v'} - 2$. So if at least one v in G has an even degree in G and if all deg_e in $L(G)$ are assumed even, then all deg_v must be even in G. Recall that G is connected. Then use Euler's Theorem in G and apply (i). ❑

7.5 Catalan numbers revisited

Let us begin with practicing the construction from Theorem 6.2.4. We mostly consider the even number of steps; it will be odd in the following problem.

Problem 7.5.1 *Consider sequences $\{a_1, a_2, ..., a_m, a_{m+1}, ..., a_{2n+1}\}$ for $n \geq 0$ with $a_i = \pm 1$ such that $a_1 + a_2 + \cdots + a_{2n+1} = 3$. A sequence is called* bad *if there exists m such that $a_1 + \cdots + a_m = -1$. Then we take the smallest possible m and transform it to the sequence $\{c_1 = -a_1, c_2 = -a_2, ..., c_m = -a_m, c_{m+1} = a_{m+1}, ..., c_{2n+1} = a_{2n+1}\}$. This is the procedure "reflect and shift" used in the calculation of $C_{2n}^{0\to 2k}$, naturally adjusted to $2n+1$ steps.*

(i) Check that $c_1 + \cdots + c_{2n+1} = 5$, and this procedure is a bijection between bad *$\{a_i\}$ and* all *representations $c_1 + c_2 + \cdots + c_{2n+1} = 5$ with $c_i = \pm 1$.*

(ii) Calculate the number N of representations $a_1 + a_2 + \cdots + a_{2n+1} = 3$ that are good, *i.e. such that $a_1, a_1 + a_2, a_1 + a_2 + a_3, \ldots \geq 0$.*

Solution. The one-to-one correspondence is exactly the same as for $C_{2n}^{0\to 2k}$. In this notation, $N = C_{2n+1}^{0\to 3}$. The number of $a_i = +1$ must be $n+2$ now; indeed, $(n+2) - (2n+1-n-2) = 2n+3-2n = 3$. So $N = \binom{2n+1}{n+2} - \binom{2n+1}{n+3} = \frac{(2n+1)!}{(n+2)!(n-1)!} - \frac{(2n+1)!}{(n+3)!(n-2)!} = \binom{2n+1}{n+2}\left(1 - \frac{n-1}{n+3}\right) = \binom{2n+1}{n+2}\left(\frac{4}{n+3}\right) = \frac{4\cdot(2n+1)!}{(n+3)!(n-1)!}$. More generally, $C_{2n+1}^{0\to 2k-1} = \frac{2k\cdot(2n+1)!}{(n+1+k)!(n+1-k)!}$ for $k = 1, 2, \ldots$. ❑

7.5.1 Counting triangulations

Recall that our initial approach to Catalan numbers was based on non-intersecting chords between $2n$ points. The following classical interpretation of Catalan numbers is different, though its justification is similar to that presented in Fig. 6.8.

Problem 7.5.2 *Consider the number of triangulations H_n of a convex $(n+2)$-polygon. A triangulation is its representation as a union of non-intersecting triangles with the vertices in those of the polygon; they are allowed to have common sides and vertices. See Fig. 7.8.*

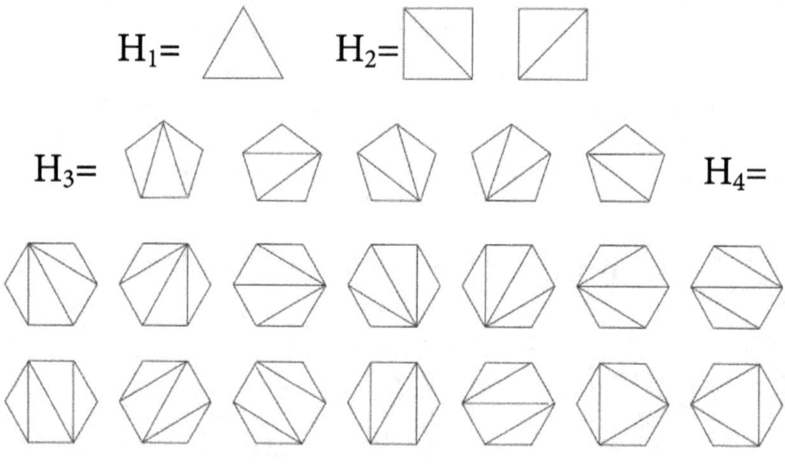

Figure 7.8: Catalan numbers as triangulations.

(i) Prove that H_n equals $h_n = C_{2n}$, the nth Catalan number; see Fig. 6.8 for the recurrence for h_n.

(ii) Check that H_n is the number of ways to plot $(n-1)$ non-intersecting diagonals (except at the vertices) inside this polygon.

Solution. **(i)** The initial 3 values of H_n coincide with h_1, h_2, h_3. The recurrence relation is exactly the same as for h_n in (6.2.1):

$$H_n = H_0H_{n-1} + H_1H_{n-2} + H_2H_{n-3} + \cdots + H_{n-2}H_1 + H_{n-1}H_0, \; H_0 = 1,$$

and the proof is parallel to that for h_n presented in Fig. 6.8. We fix any *side* (not a vertex) of the polygon. It belongs to a unique triangle, so it is connected by a triangle with exactly one of the vertices of the polygon. By removing this side, we obtain two subpolygons: the one with $m+2$ vertices (and $m+2$ sides) before this triangle and the remaining one, after it, with $(n+3)-(m+2) = n-m+1 = (n-m-1)+2$ vertices. Here $m = 0, 1, \ldots, n-1$. The cases $m = 0, m = n-1$ are those when this side forms a triangle in this particular triangulation with the side before it and the side after it.

Then we triangulate both polygons, before and after, independently and obtain, indeed, that $H_n = \sum_{m=0}^{n-1} H_m H_{n-m-1}$. Since the initial values of H_n coincide with those for h_n, both sequences coincide *"identically"* (for any n).

(ii) * This part is elementary but a little bit involved combinatorially.

Let us justify that the triangulations are exactly any collections of $(n-1)$ non-intersecting diagonals. The triangulations are obviously *maximal* collections of non-intersecting diagonals, i.e. those that cannot be increased by adding one more diagonal. So we need to check that such maximal collections are those with $(n-1)$ diagonals. We will do this by induction.

Let us split the polygon into two by any diagonal from such a maximal collection. In each of these subpolygons, the corresponding diagonals form a maximal non-intersecting family; otherwise we could increase a non-maximal one there and thus increase the initial one too (which is impossible due to the assumption that it is maximal).

Let the number of vertices in these 2 subpolygons be $m+2$ and $n-m+2$ with $m = 1, \ldots$; the sum of these numbers must be $(n+2)+2 = n+4$. By induction, the number of diagonals will correspondingly be $m-1$ and $n-m-1$ in the resulting families. So the number of the diagonals in the initial collection equals $(m-1)+(n-m-1)+1 = n-1$ $(+1$ because of the diagonal used to split the polygon). This proves that the maximal collections are always of the size $n-1$.

Vice versa, we assume now that we have $(n-1)$ non-intersecting diagonals and will prove that they form a maximal collection in the above sense. We argue by induction, employing the same split as above. Then either the corresponding 2 families have exactly $m-1$ and correspondingly $n-m-1$ diagonals or one of them has a greater number of diagonals. In the first case, both families are *maximal* by induction, and so is the initial collection. The second case is impossible because the number of non-intersecting diagonals simply cannot be greater than $N-1$ in any $(N+2)$-polygon: we proved above that any *maximal* collection contains $(N-1)$ diagonals. ❑

7.5.2 Rational Catalan numbers

We will present an alternative proof of the formula $h_n = C_{2n} = C_{2n}^{0\to 0} = \frac{1}{n+1}\binom{2n}{n}$, which better clarifies the denominator here. It goes via *rational (slope) Catalan numbers.* The definition is as follows. We consider all integer up-right paths in the rectangle $a \times b$ for $b \geq a \geq 1$ from $(i=0, j=0)$, the bottom-left corner, to $(i=a, j=b)$, the top-right one. We deviate here for the usage of (i,j) in grids and frames in this book. The diagonal will then be from $(i=0, j=0)$ to $(i=a, j=b)$. The steps are $\Uparrow$ and $\Rightarrow$: by one unit. If the paths are strictly above or touching the diagonal, they are called ***Dyck paths***. The corresponding ***rational Catalan number*** is the number of such paths, denoted by $C(a,b)$. Obviously, $C(n,n) = C_{2n}$.

Problem 7.5.3 *Establish an isomorphism between the Dyck paths in the rectangles $n \times (n+1)$ and $n \times n$; the latter are counted by C_{2n}.*

Solution. Obviously, the last two steps in Dyck paths in "$n \times (n+1)$" must

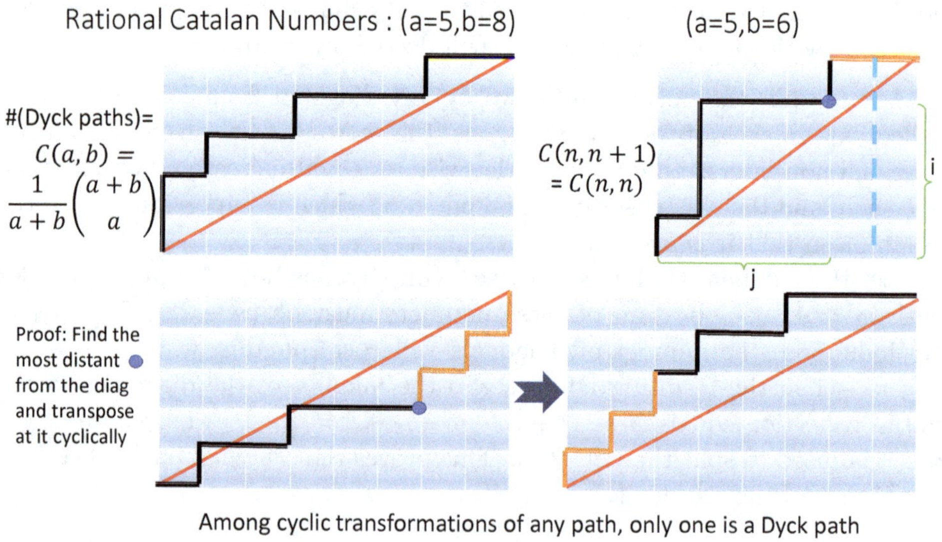

Figure 7.9: Formula for rational Catalan numbers.

be $\Rightarrow$ and $\Rightarrow$ again. Thus, we can cut off the last column and need only to check that any (i, j) with $1 \le i \le n, 1 \le j < n$ such that $\frac{i}{j} > \frac{n}{n+1}$ satisfy the inequality $\frac{i}{j} \ge 1$. We use that $j < n$. ❑

Examples for $a = 5, b = 8$ and $a = 5, b = 6$ are in Fig. 7.9; two upper diagrams and the last one are Dyck paths. The bottom-left one is not. The upper-right path is a Dyck path in "$n\times(n+1)$" for $n = 5$ restricted to "$n\times n$".

Problem 7.5.4 *Assuming that* $\gcd(a, b) = 1$ *for positive integers* a, b *such that* $a < b$, *prove that* $C(a, b) = \frac{1}{a+b}\binom{a+b}{a}$.

Solution. We follow M.T.L. Bizley (1954); the formula was announced by H.D. Grossman (1950). Firstly, the diagonal of the rectangle $a \times b$ has only two integer points, $(i = 0, j = 0)$ and $(i = a, j = b)$ due to the condition $\gcd(a, b) = 1$. Moreover, there can be no two integer points (i, j) inside this rectangle such that the segment connecting them is parallel to the diagonal.

Let us consider all "up-right" paths from $(0, 0)$ to (a, b), not only Dyck paths. For any $1 \le n \le a + b$, let (i_n, j_n) be the corresponding points: the vertices of this path, from $(i_1, j_1) = (1, 0)$ or $(i_1, j_1) = (0, 1)$ to $(i_n, j_n) = (a, b)$. Given a path and $1 \le m \le a + b$, we move its portion for $m, m + 1, \ldots, a + b$ to the origin: algebraically, the translation is $(i, j) \mapsto (i - i_m, j - j_m)$.

The image of the diagonal under this (or any) translation is parallel to it. Thus, if we add the portion for $1, \ldots, m - 1$ of this path after the one translated above, it will be some path in the same rectangle. Algebraically, this move is the translation $(i, j) \mapsto (i + a - i_m, j + b - j_m)$.

By doing this for all m, we obtain the *cyclic orbit* of the initial path: $a+b$ *different* paths. The key is that *this orbit contains exactly one path, which is a Dyck one* (above the diagonal). See the bottom of Fig. 7.9.

It is necessary and sufficient to take m such that (i_m, j_m) is below the diagonal and the most distant from the diagonal among such vertices of the path. It is $(i = 2, j = 6)$ in the lower-left path in the figure. Such m is unique due to the gcd $= 1$ condition for a, b. Indeed, the integer points of the diagonal are only its ends. When we translate the diagonal to (i_m, j_m), the integer points there can come only from integer points of the diagonal.

We obtain the required: the number of Dyck paths is the total number of all paths, which is $\binom{a+b}{a}$, divided by $a + b$. ❑

Using the previous problem: $C_{2n} = C(n, n) = C(n, n+1) = \frac{1}{2n+1}\binom{2n+1}{n} = \frac{1}{2n+1}\frac{(2n+1)!}{n!(n+1)!} = \frac{(2n)!}{n!(n+1)!} = \frac{1}{n+1}\binom{2n}{n}$. We see that the usage of rational Catalan numbers clarifies the denominator $(n + 1)$ in this formula.

7.5.3 Frobenius coin problem

Rational Catalan numbers and the corresponding Dyck paths are closely related to the following classical problem, a great instance of the connections between combinatorics and number theory (this book is full of them).

Assume that we have a lot of coins with denominations a and b. What amounts can be obtained using them? More formally, find the set $\Gamma = \Gamma_{a,b}$ of integers $n \geq 0$ represented as $n = xa + yb$ for integers $x, y \geq 0$. Equivalently, describe *the set $G_{a,b}$ of gaps*: the numbers $n > 0$ that do not belong to $\Gamma_{a,b}$. This is about the solvability of a simple Diophantine problems with 2 variables. As such, it attracted a lot of attention from many mathematicians. The following problem was managed by James Sylvester in 1862.

Problem 7.5.5 *(i) Assume that* $\gcd(a, b) = 1$ *for* $a, b \geq 1$. *Prove that all numbers* $n \geq (a - 1)(b - 1)$ *belong to* $\Gamma_{a,b}$.

(ii) Check that the map $g \mapsto \gamma = ab - a - b - g$ *establishes a one-to-one correspondence between* g *from the set of gaps* $G_{a,b}$ *and* $\gamma \in \Gamma_{a,b}$ *such that* $\gamma < (a - 1)(b - 1)$.

(iii) Deduce from (ii) that the number of gaps g *is exactly* $\frac{(a-1)(b-1)}{2}$.

Solution. **(i)** The numbers ℓb for $\ell = 0, 1, \ldots, a - 1$ represent all remainders modulo a. We claim that the gaps are exactly the numbers $\ell b + ka > 0$ for $k < 0$ for any such ℓ. Indeed, any number can be represented as $\ell b + ka$ for some k and $0 \leq \ell \leq a-1$. If $k \geq 0$, then it belongs to $\Gamma_{a,b}$. Let $k < 0$. Assume that $\ell b+ka = xa+yb$ for some $x, y \geq 0$. Then $(x-k)a = (\ell-y)b$ for $x-k > 0$, $\ell - y = (x-k)a$ due to $\gcd(a, b) = 1$, and $\ell = y + (x-k)a \geq a$, which is impossible. Thus $G_{a,b} = \{\ell b + ka > 0 \text{ such that } 1 \leq \ell \leq a-1, k < 0\}$. The greatest number here, the last gap, is obviously $(a-1)b - a = ab - a - b$. All numbers n after this one are in $\Gamma_{a,b}$.

(ii) One has: $g = \ell b+ka \mapsto b(a-1)-a-\ell b-ka = (a-1-\ell)b+(-1-k)a$, where $a-1-\ell \geq 0$ and $(-1-k) \geq 0$ due to (i). Thus, the image of this map is exactly the set of all γ until $b(a-1)-a+1$; the latter one is for the smallest gap, which is $g = 1$. The claim from (iii) readily follows from (ii). ❑

Geometric interpretation.* There is another, more geometric, approach to $G_{a,b}$ and counting the gaps from (iii). This is how *Dyck paths* (and rational Catalan numbers) enter the coin problem. We mostly follow [51].

Consider the set $\{(i, j)\}$ for $0 \leq i \leq a$ and $0 \leq j \leq b$; the bottom-left vertex of the corresponding $a \times b$-rectangle is $(0, 0)$, the upper-right one is (a, b). Let the unit boxes in this rectangle be $B_{i,j}$ for $0 \leq i \leq a-1$, $0 \leq j \leq b-1$, *where* (i, j) *is the bottom-left corner of* $B_{i,j}$.

Problem 7.5.6 *Let us insert* $ib-(j+1)a$ *into box* $B_{i,j}$*; so the first entry is* $-a$ *for* $B_{0,0}$*, and the last entry is* $(a-1)b - ab = -b$ *for* $B_{a-1,b-1}$.

(i) Prove that $G_{a,b}$ *is exactly the set of the entries in the boxes in the rectangle* $a \times b$ *that are strictly above the diagonal, i.e. with bottom-right corners* $(i, j+1)$ *satisfying the inequality* $i/a > (j+1)/b$*. Actually,* $i/a \geq (j+1)/b$ *is sufficient here because the condition* $\gcd(a, b) = 1$ *is imposed.*

(ii) Moreover, the entries $ib-(j+1)a$ *of any boxes not above the diagonal are non-positive for any* $i \geq 0, j \geq 0$ *(not only within this rectangle). Also, if* $i \geq a$ *and* $0 \leq j \leq b-1$*, i.e. for the boxes in the strip beyond "*$a \times b$*", then the corresponding entries are all different and constitute the set* $\Gamma_{a,b}$.

(iii) Check that $\pm\frac{ib-(j+1)a}{\sqrt{a^2+b^2}}$ *is the distance from* (i, j) *to the line* $i = at, j = bt$ *(for any* t*), where* $+$ *is for* (i, j) *above the diagonal and* $-$ *otherwise. So the smaller the value of* $ib-(j+1)a$ *the smaller the distance to the diagonal.*

Solution. **(i)** This follows directly from Problem 7.5.5. We set $i = \ell$ and $j = -1-k$ in the presentation $n = \ell b + ka$ for the gaps there. Then the inequality $-\ell\frac{b}{a} \leq k \leq -1$, describing the gaps, becomes $-i\frac{b}{a} \leq -j-1 \leq -1$ and $ib \geq (j+1)a \geq 1$.

(ii) Obviously, $bi-a(j+1) \leq 0$ if $i/(j+1) \leq b/a$, which holds if the box is below the diagonal or intersects it. The second claim is direct $bi-a(j+1) = b(a+x) - a(j+1) = bx + a(b-j-1) = bx + ay$ for $x = i-a \geq 0$ and

[51] E. Gorsky, M. Mazin, *Compactified Jacobians and q,t-Catalan numbers, I*, Journal of Combinatorial Theory, Series A 120 (2013), 49–63.

$y = b - j - 1 \geq 0$. Here x, y can be arbitrary non-negative integers. If $xa + yb = x' + y'b$ here, then $x = x'$ and $y = y'$ due to the conditions $0 \leq y, y' \leq a-1$ and our permanent assumption $\gcd(a, b) = 1$.

We leave the calculation of the distance from (iii) to the reader. See two examples for Part (iii) in Fig. 7.10, where $a{=}5, b{=}8$ and $a{=}5, b{=}6$. ❑

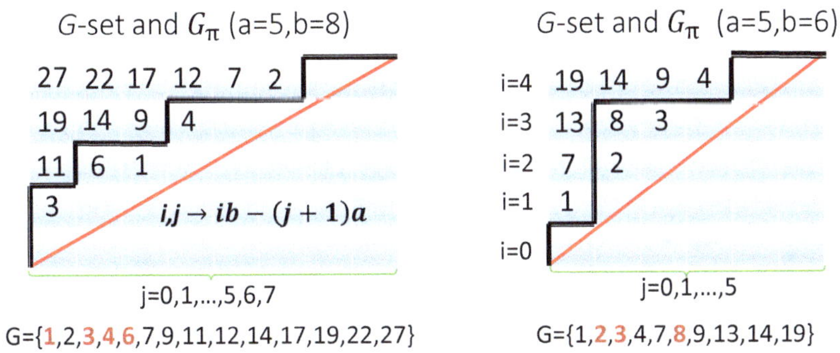

Figure 7.10: Red numbers cannot be obtained for a, b.

The total number of entries is $(a-1)(b-1)/2$, and $ab-a-b$ is the greatest entry; see Problem 7.5.5(iii). Let us re-prove these facts geometrically.

Pick's Theorem from Exercise 6.4.27 gives that the number of interior integer points (i, j) in the triangle $\Delta((0,0),(a,0),(a,b))$ (above the diagonal) is 1 plus its area minus the number of integers boundary points divided by 2. This is exactly the number of boxes above the diagonal. We obtain that it equals $1 + \frac{ab}{2} - \frac{a+b+1}{2} = \frac{(a-1)(b-1)}{2}$. This is not too difficult to count directly.

Problem 7.5.7 *(i) For any set of coins $a, b, c, \ldots$ (integers ≥ 2) assuming that* $\gcd$ *of all of them is 1, let Γ be the set $\{xa + yb + zc + \ldots\}$ for integers $x, y, z \geq 0$, and G the set of gaps, the numbers that cannot be represented this way. We set $\delta = |G|$ and denote by κ the greatest gap. Prove that $\kappa \leq 2\delta - 1$ and that $\kappa = 2\delta - 1$ implies that the map $\sigma : g \mapsto \kappa - g$ identifies G with $[\Gamma] = \Gamma \cap [\kappa]_+$, where we set $[\kappa]_+ = \{0, 1, \ldots, \kappa\}$.*

(ii) Check that $\kappa = 2\delta - 1$ for the following triples $\{a, b, c\}$: $\{4, 6, 2m+1\}$ for $m \geq 1$, and $\{2m, 2m+1, 2m+2\}$ for $m \geq 2$. However, this does not hold for $\{3,4,5\}$, $\{5,6,7\}$, $\{7,8,9\}$, and so on.

Solution. **(i)** The images $\sigma(\gamma)$ for $\gamma \in [\Gamma]$ cannot belong to Γ. Indeed, this would give that κ is a sum of 2 elements from Γ and therefore is from Γ too. Thus, $\sigma([\Gamma]) \subset G$ and $\sigma(\sigma([\Gamma])) = [\Gamma] \subset \sigma(G)$. Therefore $|[\Gamma]| \leq |G| = \delta$, and $2\delta \geq \kappa + 1$, since $[\kappa]_+ = [\Gamma] \cup G$. The equality here means precisely the coincidence of the sets $\sigma([\Gamma])$ and G.

(ii) The notation will be G_m, Γ_m, δ_m, and κ_m for both families. For the first one, we proceed by induction. Problem 7.5.5 covers the case of $m = 1$. For $m = 2$, we have $G_2 = 1, 2, 3, 7$, $\delta_2 = 4$, $\kappa_2 = 7 = 2\delta_2 - 1$. When going from m to $(m+1)$ for $m \geq 2$ we change $c = 2m+1$ to $c = 2m+3$ and the set of gaps becomes a union of 1 and G_m shifted by 2; we use that $4, 6$ linearly generate all even numbers greater than 2. This increases δ_m by 1 and κ_m by 2. So $\delta_m = 4+(m-2) = m+2$, and the equality $\kappa = 2\delta - 1$ holds for $m+1$.

The 2nd family. Γ_m is the union of $\{2mx, 2mx+1, \ldots, 2mx+2x\}$ for $x = 1, 2, \ldots$. Accordingly, the set of gaps will be the union of $\{1, 2, \ldots, 2m-1\}$, $\{2m+3, 2m+4, \ldots 4m-1\}$ and so on. To be exact, the union of the sets $\{2mx + 2x + 1, \ldots, 2m(x+1) - 1\}$ for $x = 1, 2, \ldots$. Their total number will be $\delta_m = (2m-1) + (2m-3) + \cdots + 1 = m^2$. The last gap is for $x = m-1$: $2m(x+1) - 1 = 2m^2 - 1$, which is $2\delta_m - 1$. The example from (ii) of a family with $\kappa < 2\delta - 1$ is similar; we leave it as an exercise. ❑

Exercise 5.6.7 provides an example with $a = 6, b = 9, c = 20$, when $2\delta = 44$; this is the problem of *McNugget numbers.*

Problem 7.5.8 *For $G = G_{a,b}, \Gamma = \Gamma_{a,b}$ as above and a Dyck path π in "$a \times b$", let G_π be the set of the numbers $ib - (j+1)a$ for the boxes $B_{i,j}$ above the diagonal but below π, including those touching π. In particular, the set G is associated in this sense with the path π that is the left side of this rectangle continued by its top side: $\uparrow\!\rightarrow$.*

(i) Let Δ_π be $\mathbb{Z}_+$, the set of all non-negative integers, where we remove the set G_π. Check that it has the following properties: $\Gamma \subset \Delta_\pi$ and $\Gamma + \Delta_\pi = \Delta_\pi$, where $\Gamma = \Gamma_{a,b}$ and $\Gamma + \Delta_\pi$ is understood as the set of all pairwise sums.

(ii) Vice versa, prove that any set Δ of non-negative integers such that $\Gamma \subset \Delta$ and $\Gamma + \Delta = \Delta$ is Δ_π for some Dyck path π. Thus the number of Δ such that $\Gamma \subset \Delta$ and $\Gamma + \Delta = \Delta$ equals $C(a, b) = \frac{1}{a+b}\binom{a+b}{a}$.

Solution. **(i)** By construction, $G_\pi \subset G$, and Δ_π is obtained from Γ by adding the entries $a_{i,j} = ib - (j+1)a$ of the boxes $B_{i,j}$ below π and above the diagonal. If $B_{i,j}$ is of this kind, then so are $B_{i+1,j}$ and $B_{i,j-1}$ if they remain in this rectangle and above the diagonal. Indeed, the corresponding entries are $a_{i,j} + b$, and $a_{i,j} + a$, so they belong to Δ_π.

(ii) The conditions $n + a, n + b \in \Delta$ if $n \in \Delta$ are necessary and sufficient for the inclusion $\Gamma + \Delta \subset \Delta$. Paths are the sequences of unit steps $\rightarrow$ and $\uparrow$. Equivalently, $\{B_{i,j}$ is above $\pi\}$ implies that $\{B_{i+1,j}, B_{i,j-1}$ are above $\pi\}$. Here any such paths in the plane are considered, not only Dyck paths. The corresponding changes of the entries are precisely $+b$ and $+a$.

Given Δ we consider *all* points (i, j) with non-negative integers i, j such that $0 \leq j \leq a-1$ and $ib - (j+1)a$ belong to Δ. Then the above conditions are satisfied and Δ is formed by all boxes above some path π. The condition $\Delta \subset \mathbb{Z}_+$ implies that the path π corresponding to Δ must be above the diagonal. The condition $\Gamma \subset \Delta$ gives that all the boxes above the rectangle are beyond π. See Problem 7.5.6(ii). Thus π is inside this rectangle. ❑

The entries below the paths in Fig. 7.10 form the corresponding G_π. They are shown in red in $G_{a,b}$. From the viewpoint of the coin problem, the amounts above π are assumed "obtainable," and any number of coins can be added to them. We do not consider them as coins. You can use only one of them, and then pay the rest using the coins (any number of them).

This problem has far-reaching applications. For those interested in relations to algebraic and arithmetic geometry: it describes the cell decomposition of the Compactified Jacobian for the singularity $X^a = Y^b$. This is a particular case of the so-called Fundamental Lemma, which is for any plane curve singularities. It was quite a challenge in number theory and related representation theory. Finally, it was proved by Ngô Bao Châu in full generality. Not much is known beyond plane curve singularities. The condition $\kappa = 2\delta - 1$ from Problem 7.5.7 is equivalent to the Gorenstein condition for the corresponding singularity. This is a natural challenge: plane curve singularities are Gorenstein, but not the other way around. We omit the definitions; this paragraph was just to put Problem 7.5.8 into perspective.

Let us interpret combinatorially the numbers $N_{a,b}$ used in the proof of *Quadratic Reciprocity* from Section 7.3.4.

Problem 7.5.9 *(i) Let the rectangle $R =$ "$a \times b$" for* $\gcd(a,b) = 1$ *be as above. Assume that a is odd and take the subrectangle $R' =$ "$\frac{a+1}{2} \times b$" formed by the first $\frac{a+1}{2}$ rows counting from the bottom up. Then the number of the boxes in R' above the diagonal in R is $N_{b,a} = [b/a] + [2b/a] + \cdots + [b\frac{(a-1)}{2}]$, which was an important ingredient of Problem 7.3.17.*

(ii) The combinatorial quadratic reciprocity *is the identity $N_{a,b} + N_{b,a} = \frac{a-1}{2}\frac{b-1}{2}$ for odd a, b, which can be interpreted as follows. Remove the top-left corner "$\frac{a-1}{2} \times \frac{b-1}{2}$" from R. The number of the remaining boxes in R beyond the diagonal is then $N_{a,b} + N_{b,a}$, which is $\frac{(a-1)(b-1)}{2}$, the total number of such boxes minus $\frac{(a-1)(b-1)}{4}$ due to the corner. So it is $\frac{a-1}{2}\frac{b-1}{2}$ indeed.*

(iii) Let p be an odd prime number. Then $b \bmod p$ is a quadratic residue, *i.e. a square in $\mathbb{Z}_p$, if $N_{p,b}$ and $(b-1)\frac{p^2-1}{8}$ are even, or if both are odd. Otherwise it is not a square. For instance in Fig. 7.10: $a = p = 5$, $N_{5,8} = 4$ for $b = 8$ (the numbers in the boxes there are $1, 3, 6, 11$), and $N_{5,6} = 3$ for $b = 6$ (the numbers are $1, 2, 7$). Here $(b-1)\frac{p^2-1}{8}$ is odd and we obtain that $8 \bmod 5 = 3$ is not a square in $\mathbb{Z}_5$, but $6 \bmod 5 = 1$ is a square.* ❑

We omit the proof: it is straightforward from the considerations above. Part (iii) is the Gauss Lemma.

Combinatorially, this kind of reciprocity can be stated for any k, ℓ such that $0 < k < a$, $0 < \ell < b$, and $\frac{a-k-1}{\ell+1} < \frac{a}{b} < \frac{a-k}{\ell}$. These inequalities mean that the top-left corner " $k \times \ell$ " does not intersect the diagonal, but the corner " $(k+1) \times (\ell+1)$ " intersects it. Geometrically: take any box crossed by the diagonal and consider the top-left corner "before" it, excluding this box.

Then the combinatorial reciprocity is the transposition of a and b. Let the number of the boxes beyond the diagonal in R without the top k rows be $N_{a,b}(a-k, b)$. Similarly, let the number of boxes in R without the left ℓ columns be $N_{a,b}(a, b-\ell)$, which equals $N_{b,a}(b-\ell, a)$ upon the reflection swapping a and b. Then $N_{a,b}(a-k, b) + N_{a,b}(a, b-\ell) = \frac{(a-1)(b-1)}{2} - k\ell$. This construction is used in the theory of plane curve singularities.

In (iii) above: $k = \frac{a-1}{2}$, $\ell = \frac{b-1}{2}$, and $\frac{a-k-1}{\ell+1} = \frac{a-1}{b+1} < \frac{a}{b} < \frac{a+1}{b-1}$.

To finalize, let us connect this construction with the Dyck paths in " $a \times b$ ". Recall that touching the diagonal is allowed in their definition if $\gcd(a, b) \neq 1$. Let (p, q) be an integer point such that $0 \le p \le a$, $0 \le q \le b$ and the triangle with the vertices $(0,0), (p,q), (a,b)$ contains no *inner* integer points. Provided that $\gcd(a, b) = 1$ and (p, q) is above the diagonal, we claim that the combinations of any Dyck paths in the bottom-left "$p \times q$" with those in the upper-right " $(a-p) \times (b-q)$ " are Dyck paths in "$a \times b$". This is an easy exercise, though with some potential. For instance, all Dyck paths are such for $(a, b) = (n, n+1)$, $(p, q) = (n, n)$, which gives that $C(n, n+1) = C(n, n)$. Note that $k = b - p, \ell = q$ satisfy the conditions for k, ℓ above.

7.5.4 Skew Young diagrams *

We will discuss here a multi-dimensional generalization of Catalan numbers. Arbitrary Young diagrams occur in this way, and even arbitrary skew Young diagrams, which are among the main objects of representation theory of the *symmetric group*. The extended Catalan numbers in this book are associated with 2-row diagrams. Let us extend them to any number of rows.

Consider now paths of any length with r possible steps that will be e_i for $e_1 > e_2 > e_3 > \cdots > e_r$. The steps e_i can be positive or negative. We assume that e_i are ***generic***: only these inequalities will be imposed and used below.

Higher Catalan numbers are defined for the vectors: $\vec{m} = \sum_{i=1}^{r} m_i e_i$ and $\vec{k} = \sum_{i=1}^{r} k_i e_i$, where we assume that $m_i \ge m_{i+1}, k_i \ge k_{i+1}$ for any i, i.e. these sequences of integers are non-increasing, and that $k_i \ge m_i$ for all i. For any concrete choice of $\{e_i\}$ some of their linear combinations with positive and negative coefficients can "accidentally" become positive, but we do not consider such paths "good". That is, e_i are assumed undetermined; only the inequalities above are imposed. We arrive at the following definition.

Definition 7.5.10 *Good paths are sequences $\{e_{i_1}, e_{i_2}, \ldots, e_{i_n}\}$ for the indices $i_1, i_2, \ldots, i_n$ from 1 to r such that the coefficients $b_i(s)$ of $e_1, e_2, \ldots, e_r$ in any partial sums $\vec{m}+e_{i_1}+\cdots+e_{i_s}$ for $s = 1, \ldots, n$ are non-increasing with respect to i for any s: $b_i(s) \geq b_{i+1}(s)$, i.e. in the same sense as for $\vec{m}$ and $\vec{k}$. If $\vec{m}+e_{i_1}+\cdots+e_{i_n} = \vec{k}$, then $C_n^{\vec{k}\to\vec{m}}$ is the number of good paths.*

In the case of $r = 2$, $b_1(s)$ is the number of e_1 used until point s, and $b_2(s)$ is the number of e_2. So *good paths* are exactly the sequences of ± 1 such that the number of $+1$ until any s plus m_1 is no smaller than the number of -1 until s plus m_2 and $m_i + b_i(n) = k_i$ for $i = 1, 2$. Thus $b(s) = b_1(s) - b_2(s)$ are the sums we used in Definitions 6.2.2 and 6.2.3. We obtain that the number of good paths for $\vec{m} = (m, 0)$ and $\vec{k} = (k, 0)$ for $m, k \geq 0$ with n points s is $C_n^{m\to k}$ in our standard notation.

If e_i are (daily) changes of some share-price, and its initial value is $p + \sum_{i=1}^{r} m_i e_i$, then these conditions mean that the price will stay at the levels $\geq p$ for n days. This will be for *any* e_i satisfying the inequalities above. For fixed values of e_i, there can be of course other paths with the prices always no smaller than p. So the equivalence of the inequalities $\geq p$ with the conditions from the definition is for *arbitrary* decreasing $\{e_i\}$. For $r = 2$, one can take $e_1 = +1$ and $e_2 = -1$ to have the exact equivalence.

We associate with $\vec{m}, \vec{k}, n$ the following *skew Young diagram* μ:

m_1 □	$\cdots$ □□ $\cdots$	□□ k_1+1
m_2 □□	$\cdots$ □□ $\cdots$	□ k_2+1
m_3 □□□□	$\cdots$ □□ $\cdots$	□ k_3+1
$\cdots$	$\cdots$	$\cdots$.

Let us use j to number the boxes in these rows. So the top row of boxes is for j satisfying $m_1 < j \leq k_1$, the next one is for $m_2 < j \leq k_2$, and so on. We will "normalize" m_i here by the condition $m_r = 0$, using a proper horizontal translation. The number of boxes in μ will be n.

For instance, 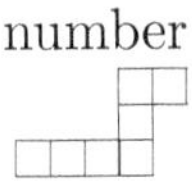 corresponds to $m_1 = 3, m_2 = 3, m_3 = 0$ and $k_1 = 5, k_2 = 4, k_3 = 4$; $n = 7$ in this case.

Then we put the numbers $1, 2, \ldots, n$ in the boxes of μ making them increasing in the rows ($\to$), and in the columns ($\downarrow$). Such configurations of numbers inside μ are called ***standard skew Young tableaux*** of shape μ.

We will associate with any tableau the following sequence of the numbers $i_1, i_2, \ldots, i_n$, where $1 \le i_s \le r$: i_1 is the row number of 1 in this tableau, i_2 is the row number of 2, and so on. The column numbers of $1, 2, \ldots, n$ will be disregarded.

Problem 7.5.11 *For any* tableau *and the corresponding sequence* $i_1, i_2, \ldots, i_n$, *consider the sequence* $\{e_{i_1}, e_{i_2}, \ldots, e_{i_n}\}$.

(i) Check that $\vec{m} + (e_{i_1} + e_{i_2} + \cdots + e_{i_n}) = \vec{k}$, *and all partial sums here,* $\vec{m} + e_{i_1}$, $\vec{m} + e_{i_1} + e_{i_2}$ *etc., satisfy the positivity conditions as for* $\vec{m}, \vec{k}$, *i.e. the sequences* $\{e_{i_1}, e_{i_2}, \ldots, e_{i_n}\}$ *are* good paths.

(ii) Vice versa, given a skew Young diagram μ, *check that an arbitrary good path such that* $\vec{m} + (e_{i_1} + e_{i_2} + \cdots + e_{i_n}) = \vec{k}$ *can be obtained from a unique standard skew Young tableau of shape* μ. ❑

So *good paths* are in 1-1 correspondence with standard Young tableaux. The number of standard Young tableaux for a given skew Young diagram can be counted, for instance, by the relatively recent *Naruse hook-length formula*. See e.g. [52]. Let us explain this a bit. The usual Young diagrams are with $m_1 = 0 = m_2 = m_3 = \ldots$. The formula for the number of standard Young tableaux in this case is classical: $\mathcal{C}_n^{\vec{k}\to\vec{m}} = n!/\prod_{\ell=1}^{n} h_\ell$, where h_ℓ is the number of boxes in the hook $\begin{array}{l}\boxed{\ell}\to\square\\ \downarrow\\ \square\end{array}$, the number of boxes from box ℓ in its column and its row *after it* (i.e. following $\downarrow$ and $\to$). Box ℓ is counted. The enumeration of boxes in μ is arbitrary; let us do it lexicographically.

For instance, h is 1 for any right corners in any diagram, and it is the number of rows plus the number of columns minus 1 for the first box. To provide an example, the h numbers for the Young diagram that is the rectangle $2 \times p$, are as follows: $\begin{array}{|ll|}\hline p{+}1 & \cdots 432\\ p & \cdots 321\\ \hline\end{array}$. We place them respectively. This readily gives the standard formula for the Catalan numbers: $C_{2p} = (2p)!/((p{+}1)!p!)$. Here $\vec{m} = (0,0)$ and $\vec{k} = (p{+}1, p{+}1)$. Recall that $2m = m_1 - m_2 = 0$ and $2k = k_1 - k_2 = 0$ when switching to $C_{2n}^{2m\to 2k}$ ($2n$ is the number of boxes). Interestingly, the procedure "reflect and shift" works for general $\mathcal{C}_n^{\vec{k}\to\vec{m}}$, but leads to a different, Frobenius-type, formula.

[52] A. Morales, I. Pak, G. Panova, *Hook formulas for skew shapes I. q-analogues and bijections*, Journal of Combinatorial Theory, Series A 154 (2018), 350–405.

Appendix A

Card Games as Cognitive Models

A.1 Combining bridge and poker

A sizable part of this book is devoted to combinatorics of card games and focused mostly on the corresponding probabilities. We will try to put games of chance into perspective. The card game *pont*, which we will describe, arguably provides a toy model for *momentum* risk-taking and decision-making toward general-purpose *artificial intelligence* (*AI*). This is a stand-alone section, an *epilogue* of some sort. A natural question is, how much can games of chance help us to study real processes of decision-making?

A.1.1 Games as concepts

Games reflect life in various ways, potentially preparing us for real challenges; we discuss only "intellectual games." They do not always follow real processes, but certainly borrow their concepts from *nature*. To give an example, airplanes are very different from birds, but the *concept* of flying is obviously from nature. Similarly, artificial intelligence systems borrow a lot from human intelligence, though the final implementations can be quite different.

Many faces of decision-making are reflected in the games we play. Some include "timing," some do not. For instance, solving puzzles and playing chess are of analytic type; they are not really "momentum" unless in tournaments. Poker's bidding is a great model of real-time momentum decision-making. However, the *poker* risks are too straightforward: directly based on the ranks of the hands. Also, psychological factors, important in poker, are too much game-specific, though "poker face" is a concept well beyond poker.

Bridge-type games provide more subtle risks: the strength of the hand is the key but the card play is equally important. They are of analytic kind; in this respect, bridge is similar to *chess*. Chess and bridge involve risks, but changes of the hands and other features of poker are closer to *momentum risk-taking*, which always involves significant uncertainties.

Some combination of poker-style risks with the auction and play typical for contract games can be expected as a suitable cognitive-behavioral model for decision-making. The card play must be complex enough and require analytical skills, as in bridge. Otherwise, our brain will use only a fraction of its decision-making capacities. The nature of the risks is actually not that important, but they must not be too straightforward and simplistic; this component is weak in poker. However, real risks always include "timing" and "momentum forecasting", which is weak in bridge.

Pont: basic principles. *Pont* is a combination of *bridge* and the so-called *Russian preference* (played in various European countries) with some poker-style features. "Pont" is French for "bridge", but *bridge* was actually derived from earlier "biritch", a transliteration of an old Russian word for a herald. *Pont* utilizes a standard deck of 52 cards or a smaller one of 36 cards. The auction is quite different from that of bridge: the hands may be updated while bidding, which resembles *draw poker.* Also, the bidding does not use the order of suits and the player who starts the auction has no advantage. The winner of the auction, the *declarer*, may increase the size of the hands, from 6 to 9 cards, and then determines the *contract*: the trump and the obligation to take a certain number of tricks. The contract must be no smaller than the bid. Increasing the size of the hands after the end of the auction adds certain poker-style "time-evolution" to bridge.

Following suit and using trump cards is similar to bridge-type games. The scoring system is simpler than that of bridge. The bidding process is as close as possible to the poker bidding in *poker pont*, a poker variant of *pont.* As in bridge, the declarer's award is based on the value of the contract depending upon whether or not it was made. The game can be for 2, 3, 4 players, 2 partnerships, or for 1 versus computer. All variants, including the one for 2 players (with 36 cards), are reasonably dynamic and playable.

We paid a lot of attention to making the game playable, which was a challenge since it uses unusual *fractional bids*. We think that it reflects as much as it possibly can for a card game, playing the real stock markets. Generally, playable games have stronger roots in our life and psychology.

The auction and bidding seem fundamental principles of any intelligence (not only with humans) for almost any risk-taking. There are always different choices, and there must be auction-type procedures to pick the most optimal ones. From this perspective, poker and contract card games serve well humankind as a risk-taking playground because they obviously capture something important about human cognition. See [53].

Stock market connections. Modeling stock markets is a classical challenge. Their "agents" look only to their own interests and to market prices by design, though investing is more involved than that. It is a very interactive process, a clear subject for behavioral sciences. Stock market is a game due to its solid rules and protocols, but with very real risks.

Momentum trading provides a universal example of risk-taking, and can be modeled. By *momentum*, we mean relatively short-term and mostly real-time decisions, which are frequently based on incomplete information; long-term investing is different. We of course consider the stock market as an end-in-itself, disregarding its economic and societal purpose.

Pont is a certain model of momentum trading. The *pont bidding table* is actually a simplified linear version of that used in a real trading system described in [54], which was nonlinear; *pont* was introduced there.

Our *bids* are some counterparts of the forecasts of share-prices. They can be verified only upon the termination of the taken positions (declared contracts), as in card games. In the same way, the number of cards per hand and the number of tricks taken reflect respectively the duration of the position and the return. The downplay and *misère* resemble selling short, though there is some stretch here. The suits are substitutes for the time-horizons of investments or the considered equities. They are on equal grounds in *pont* in contrast to other bridge-type games. The stronger the suit in a hand, the more reasons to make it a trump, which is very universal everywhere.

A.1.2 Pont: full description

Pont uses a standard card deck of 52 cards for 4 players, or a smaller four-suit deck of 36 cards (from the ace down to the 6) when there are two or three players. In the case of 4 players, they may divide themselves into two partnerships; here the whole deck is used too. The dealers are changed clockwise after each game. The cards are dealt singly in the clockwise order and placed face down, giving each player six cards. After the players pick up their hands, the dealer starts the auction by making the bid or passing.

Auction. A *bid* is a fraction N/D with the denominator D from 6 to 8 and the numerator N no larger than D. Generally speaking, the bid is the expected number of tricks to be taken (N) divided by the final number of cards per hand (D). The latter may be from 6 to 9, but D is < 9 in the bids.

[53] D. Parlett, *A history of card games*, Oxford University Press, 1991.

[54] I. Cherednik, *Artificial intelligence approach to momentum risk-taking*, International Journal of Financial Studies 9:4 (2021), 1–42

The fraction N/D must be no smaller than 3/6 for 3 or 4 individual players, and no smaller than 4/6 for 2 players or partnerships. The fractions 4/8, 7/7, 8/8 are excluded. The bids 3/6, 4/7, and 5/8 are not allowed for 2 players, but are accepted for 3 or 4 players.

The auction proceeds clockwise with each player either making a bid that is not lower than the previous ones of other players; for instance, 4/6, 5/7, 6/8, 5/6, 6/7, 7/8, 6/6 may be claimed after 4/6 or saying "pass". If a player has already passed, his/her further bidding is forbidden after the first bid is made.

Passing is allowed after bidding only if there are other players who did not pass; also, the last remaining (surviving) player may not pass. The round of bidding continues until the *last bid*, when a player (who then becomes the *closer*) repeats his/her previous bid for the first time, or simply says "close". If the others (two opponents for the team variant) pass after this, the closer becomes the *declarer*. Otherwise there is no declarer.

More rounds are necessary if all players pass or at least two of them claim the same bid. To start the next round, the dealer *upgrades* the cards, giving out an additional card per hand face down. Then each player picks up the card and after this removes one card from his/her hand by laying it face down. The hands must be 6 again. Then the closer (or the dealer if all pass) claims first, repeating or enlarging his/her last bid, and the auction continues following the same rules until the first repetition. Those who passed during the previous rounds do not bid, unless all players passed. The cards may be upgraded no greater than 2 times.

Taking no tricks. If all players pass after the second upgrade (which is the final), the dealer leads to start the *downplay* notrump, where the players are trying to win the smallest number of tricks. Also, the closer starts the downplay if two or more players (or both teams, if applicable) do not pass after the second upgrade, which is the last, but claim *coinciding* bids, i.e, neither of them is the winner.

At the end of the downplay, the numbers of tricks taken will be diminished by the minimum number of tricks taken; so this will be zero for some. Then this number will be subtracted from the corresponding scores. In the case of 2 players, this number must be divided by two before subtracting. For example, if 2 players took 4 and 2 tricks, the first will lose $\frac{4-2}{2} = 1$ points.

A player may claim *misère*, which means that no tricks will be taken by him/her. This may be done only before the first upgrade, and can be then beaten by 6/7 or higher for 2 players (teams), and by 5/6 or higher for 3 or 4 players. *Misère* is played notrump. The declarer makes the opening lead by placing the card on the table face up. If there are 3 or 4 individual players, then all cards are placed face up on the table after this. It is the same for partnerships, but the partner does not participate laying his/her cards face down; so only the declarer and the opponent team (with all hands face up) will continue. The *misère* contract is defeated if one of the opponents finds the way where the declarer takes at least one trick.

Play. After the auction, the declarer may *increase*, asking the dealer to deal out 1 card per player face down. The procedure can be repeated several times, but the maximal number of cards per hand must be no greater than 9. The declarer picks up the cards every time. The others keep the additional cards face down and pick them up after the declaration of the contract.

Then the declarer announces the *contract*, choosing the trump suit or notrump (which is allowed), and stating the minimal number of tricks to be taken (including the partner's tricks for the partnerships). The denominator "D" equals the number of cards per hand after the last increase: from 6 to 9.

The number of tricks to win cannot be smaller than the final number of cards per player (after the last increase) times the fraction from the last declarer's bid. The bid "*misère*" can be changed by the declarer by the contracts 6/7, 7/8, 8/9, 6/6, 7/7, 8/8, 9/9. It is the same for 2,3,4 players, and the partnerships. Also, the last bid 6/8 can be changed by *misère* if there were no upgrades and increases. The partner's hand is discarded *face down* when playing *misère* in the team variant.

The declarer starts the *play* trying to take enough tricks to fulfill the contract or take no tricks for *misère*. For partnerships, the declarer may ask the partner anytime during the play to place all his/her cards face up on the table and then plays both of the partnership hands (unless in *misère*).

All players have to follow suit if they can. Otherwise they must trump if they can. Only the declarer may lead a trump. Other players may do this only if they have no other cards in the corresponding suit left. The play lasts until the declarer (with the partner if applicable) takes the declared number of tricks or until the contract is defeated.

Scoring. At the end of the play, the declarer's *score* goes up by the *value* of the contract, the number of tricks from the contract minus 3 for 2 players (or partnerships) and minus 2 for 3-4 individual players, if the contract was made. Otherwise this value is subtracted from the score.

If the last bid *before the first upgrading* is more or equal than 5/6, then this value goes up by one, called *premium* (when adding or subtracting). The same *premium* is added to *misère*, treated as 5/6 when calculating the score (3 points for 2 players/teams and 4 points for 3,4 individual players).

A fulfilled contract with the fraction $N/D = 1$, gives 1 *bonus point* for 2 players (partnerships) and 2 bonus points for 3-4 individual players. For 3 or 4 individual players, successful contracts 5/6, 6/7, 7/8, 8/9, 9/10, and *misère* add 1 bonus point to the declarer's score. In contrast to the premium, the bonus is not subtracted from the score if the contract fails.

There is another, somewhat more involved, variant of the *pont score system* with more "punishment" for defeated contracts. The play goes then until the end. If the number of tricks taken is less than it was declared; then the score of the declarer is diminished by the value of the contract multiplied by the number of tricks missed. Say, if the declarer took the necessary tricks but one, the score becomes smaller by the value of the contract. This score system is for experienced players.

Finally, the *rewards* will be proportional to the scores of the players diminished by their arithmetic mean, that is the total of all scores divided by the number of players. The partners may redistribute the total partnership reward (the sum of their rewards). The standard recommended way is as follows. If both rewards are positive or both are negative, then it is the same as for individuals. If the first reward is positive, the second is negative and the total is negative, then the first partner does not pay. If the total is positive here, then the second partner receives nothing (and pays nothing).

Bidding table. The following table is the list of bids in increasing order and the corresponding minimum contracts for different hands.

THE MAIN BIDDING TABLE

3–4 individual players			contracts	2 players(partnerships)		
names	**b p**	**bids:**	tricks / cards	**:bids**	**p b**	*names*
1		**3/6:**	3/6, 4/7, 4/8, 5/9	: —		—
1+1		**4/7:**	4/6, 4/7, 5/8, 6/9	: —		—
1+2		**5/8:**	4/6, 5/7, 5/8, 6/9	: —		—
2		**4/6:**	4/6, 5/7, 6/8, 6/9	**:4/6**		*1*
2+1		**5/7:**	5/6, 5/7, 6/8, 7/9	**:5/7**		*1+1*
2+2		**6/8:**	5/6, 6/7, 6/8, 7/9	**:6/8**		*1+2*
m	* *	**m/6:**	...., 6/7, 7/8, 8/9	**:5/6**	*	*2*
3	* *	**5/6:**	...., 6/7, 7/8, 8/9	**:m/6**	*	*m*
3+1	* *	**6/7:**	6/6, 6/7, 7/8, 8/9	**:6/7**	*	*2+1*
3+2	* *	**7/8:**	6/6, 7/7, 7/8, 8/9	**:7/8**	*	*2+2*
4	** *	**6/6:**	6/6 ,7/7, 8/8, 9/9	**:6/6**	* *	*3.*

The number of stars in this table gives the number of points to be added to the score as the premium **p**, which is for declaring the contract after the *first* round of the auction. The bonus points are the number of stars in column **b** there, which is for making the contract. Recall that in contrast to the premium, the bonus is not subtracted from the score if the contract fails.

The usage of the *names* of the bids is convenient while bidding. The name consists of the value of the bid and the number of additional cards (after +); the premium and bonus are disregarded. For instance, bid *2+2*, corresponding to **6/8** for 3,4 players, results in 2+2=4 points. For 2 players, bid *1+2*, corresponding to **6/8**, gives 1+2=3 points.

By *m*, we mean *misère*; (*m* = **m/6**) has the same list of admissible contracts as **5/6** but it is ranked higher for 2 players (or 2 partnerships) and is ranked lower for 3 or 4 individuals. Recall that the *misère* contract may be declared and played if the highest bid is **6/8** or smaller; **m/6** must be placed after **6/8** and can be beaten by **5/6** or higher for 3-4 players, and by **6/7** by 2 players/partnerships. We omit **m/6** in the middle column of contracts.

A.1.3 Variants, AI-aspects

Basic Pont (BP). The simplest version of the game is *basic pont*, which is played without *misère*, and without "premium". The table is also simplified by dropping the bids of denominator 8, the bids *x+2*.

BASIC BIDDING TABLE

3–4 individuals			contracts	2 players(teams)		
names	b	**bids:**	tricks / cards	**:bids**	b	*names*
1		**3/6:**	3/6, 4/7, 4/8, 5/9	: —		—
1+1		**4/7:**	4/6, 4/7, 5/8, 6/9	: —		—
2		**4/6:**	4/6, 5/7, 6/8, 6/9	**:4/6**		*1*
2+1		**5/7:**	5/6, 5/7, 6/8, 7/9	**:5/7**		*1+1*
3	*	**5/6:**	5/6, 6/7, 7/8, 8/9	**:5/6**		*2*
3+1	*	**6/7:**	6/6, 6/7, 7/8, 8/9	**:6/7**		*2+1*
4	**	**6/6:**	6/6, 7/7, 8/8, 9/9	**:6/6**	*	*3.*

Poker Pont (PP). Another variant is *poker pont* for 2, 3, 4 individual players. It follows the table of *basic pont*, without bonuses.

bids:	contracts	**5/7:**	5/6, 5/7, 6/8, 7/9
3/6:	3/6, 4/7, 4/8, 5/9	**5/6:**	5/6, 6/7, 7/8, 8/9
4/7:	4/6, 4/7, 5/8, 6/9	**6/7:**	6/6, 6/7, 7/8, 8/9
4/6:	4/6, 5/7, 6/8, 6/9	**6/6:**	6/6, 7/7, 8/8, 9/9.

The *PP-betting* procedure is as follows. As in poker, each player puts up *ante* (one chip or more) to form a *pool*, which consists of the *pot* and the sectors, one for each player. A player always puts chips in the corresponding sector. The dealer starts *betting*, adding chips to the pool or putting nothing, which is *passing*.

The player on the dealer's left may *pass*, or *call* by putting the same, or *raise* by adding extra chips of his/her own. Other players continue clockwise until all have finally called any raises. A player may raise after passing if the latter was before the first raise. Passing is allowed after raising or calling if there are other players who did not pass.

The first player who calls without adding is the *closer*. If all other players pass, the closer becomes the *declarer*. If there is no declarer, the dealer upgrades cards and the closer starts another round of betting by raising or doing nothing. Those who passed before may not bid if any player raised after him/her. There are *optional upgrades* in *poker pont*. A player who upgrades puts 1 chip to his/her sector per 1 new card dealt to him/her and then removed 1 card face down. The number of upgrades is no more than 3 (for 4 rounds of betting). If still all pass after the last upgrade then the *ante* goes to the pot, the dealer is changed clockwise, and a new game starts.

The process of *PP-play* is as follows If there are two or more players who put the same number of chips (regardless of the chips for upgrades, which may be different), then the closer begins one round of bidding among those players only. It is as in *basic pont*; the declarer is a player claiming the highest bid. If there is no declarer after this round, there will be no more upgrades, the dealer moves all chips except the *ante* from the sectors to the pot, and the next dealer starts a new game.

The declarer may increase several times (but no more than 3), adding a chip per increase to the pool. After the declaration of the contract all opponents pick the cards and *respond* or *pass* clockwise starting with the first on the declarer's left. One must add 1 chip per each increase (totally, the current number of cards per hand minus 6) to the pool to *respond* and therefore become an *active opponent*. Other opponents are *passive*. However all participate in the card play, which follows the standard *pont* rules.

The declarer leads and wins the pool (including the pot) when making the contract. If the contract is defeated then all opponents, active and passive, take chips back from their sectors and the active opponents divide the declarer's chips and the pot among themselves proportionally to the number of tricks taken. The fractions are ignored and the remaining chips (if any) go to the pot. If there are no active opponents, the declarer takes his/her chips (not the pot!) back even in the case of the failure. The contract has to be the minimal possible for the current number of cards per hand. Namely, 3/6,

4/7 for 3-4 players, 4/6, 5/7 for 2 players, and 6/8, 6/9 for either. It may not be lower than the last bid if there has been at least 1 round of bidding.

Additional rules. Extra penalties can be added for breaking the rules. The opponents may decide to diminish the declarer's or partnership's score by the value of the contract if the declarer (partnership) made a mistake against the rules when playing. Vise versa, in the case of the opponent's mistake, the declarer has the right to consider the contract to be fulfilled and the other player(s) may decide to subtract its value (or its doubled value) from the score of the opponent whose fault it is. In *poker pont*, the contract is considered to be defeated in the case of a declarer's mistake. If it is an opponent's mistake, the chips from the pool are distributed as if the contract were defeated, and the opponent who made a mistake gives this very number of chips to the declarer. These are basics, to be developed by players.

As to the partnerships, a general recipe is to at least repeat the bid of your partner if you have 2 sure tricks for any trump, but it depends, For instance, this can be with "A A", or "A K" in the same suit, or "A" in one suit and "K Q" in another. If you pass, but the opponents do not, it may stimulate your partner to pass or claim *misère*. So it makes sense to bid if you can count on 3 (or more) tricks upon declaring your trump, especially if you have many honors and the hand is good for the increase. This is common sense.

A computer version. This is basically the variant for two players where one of them is a computer. Generally, the computer is supposed to take into account all information about the cards of the player available during the play and consider all possible choices. This requires a lot of practical statistical analysis, which can be very difficult (too many choices). Since it can be complicated to achieve, we suggest that the computer mostly does what people do, but diminishes its bids and contracts vs. what human players would do. Basically, computer plays "defensive" unless sophisticated programs are used.

The simplest *one-way version* is when computer never bids and has no score. Then the player either determines the contract (without upgrades) and then plays following the standard rules or passes and subtract 2 points from his/her score. This variant follows *basic pont*, the simplest for writing code. In this case, the bidding scale and the admissible contracts start from 4/7, considered as *0+1* and resulting in 1 point. Generally, *0+k* means the fractional bid $(3+k)/(6+k)$ and is counted as k points for $k=1,...,6$. Let us provide (very) basic examples of the computer strategy.

Generally, the computer is trying to win the trick led by the player with the smallest possible card or to play the lowest card otherwise. If it has no proper suit and no trump left, the card can be the lowest (from the shortest suit, if there are several cards of the same rank). However, the suits where the player has no cards according to the information during the process of play are considered the best in this case.

When the computer leads, then the suit where the player (possibly) has no cards is the first choice. If the highest card (one of them if there are several of the same rank) has the *adjacent* one in the same suit, then this can be the second choice for leading. For instance, the pairs "A K" or "K Q" are adjacent. The third choice can be when the next card in this suit (in the computer's hand) is lower by 4 or more; for instance, the pair "A 10" is one such. Otherwise, the suit can be the longest and the card the highest among the longest suits. The longest suit where the two highest cards are adjacent is considered the most suitable in this case. If there are several choices of this kind, then the computer decides randomly. These are very basic hints; the actual computer program can be significantly more developed.

Toward general AI systems. We did a lot of modeling in this book. The most ambitious modeling projects aim at artificial intelligence systems (*AI*). They are quite a few *AI*; for specific tasks. General purpose *AI* are the most challenging; decision-making is of this kind. *Pont* reflects well some features of momentum risk-taking, *MRT* for short. Here "momentum" means real-time management that is mostly based on the trends and results in relatively short-term decisions. It is expected that *MRT* can be modeled. It was introduced in [55]. In some sense, it is a development of [56]; see also [57]. Though *momentum trading* was the main source of *MRT*.

Driving a car is a typical example of *MRT*. Permanent acquiring basic visual (and other) information is a must here. This is very resource consuming, much greater than *MRT* itself. The actual beginning of *MRT* is when our brain identifies *events*. They can be conditions of the road, road signs, pedestrians, neighboring cars, navigation matters and so on. All such *categories* of potential events are supposed to be analyzed constantly even if the current actions are supposed to be in one particular category. The priorities are determined via auction-type procedures in our brain, which can be similar to those in *pont*. A special feature here is that there are sharp time limits for the driver's decisions, though this is so common for *MRT*.

Only some *events* will reach the level of *signals*. The separation of the

[55] I. Cherednik, *Artificial intelligence approach to momentum risk-taking*, International Journal of Financial Studies 9: 58 (2021).

[56] D. Kahneman, *Thinking, fast and slow*, New York: Farrar, Straus and Giroux, 2011.

[57] J. Bargh, M. Chen, L. Burrows, *Automaticity of social behavior: Direct effects of trait construct and stereotype activation on action*, J. Pers. Soc. Psychol, 71:2 (1996), 230–244.

signals from "noise" requires prior experience. By noise, we mean the events that hardly require special consideration and actions in a particular situation. In our brain, they are those that do not reach sufficient *action potentials*. The action potentials correspond to minimum allowed bids in *pont*.

After the signals are determined (there can be several of them), our brain estimates and compare their priority. This goes via "bidding" of some sort. "Bids" are of course much more sophisticated in our brain vs. *2-bids* in *pont*. However, they are likely to be of numerical nature and *fractional*; the denominator can be *time-to-action* for driving. Finally, one signal is identified as the top priority, and its analysis and necessary actions begin. There can be two or more "winners", but one of them usually leads. If the significance of the selected event later fades, then a switch to other signals is considered and a new round of bidding begins.

Presumably, "protection functions" are constantly created in our brain, which is parallel to the usage of $p(t)$ in Section 4.5.3 on modeling Covid-19. The function $p(t)$ is not always "protection"; generally, it is the one that directly impacts our actions. For instance, it is "price function" in stock markets. The significance of a market news diminishes if this news is "priced in": the corresponding price approaches some target. This is general: if $p(t)$ increases, the importance of the event for us diminishes.

In our brain, $p(t)$ is supposed to reflect the importance of the event, which can fade, the change of its significance due to our actions, and the allocation of the resources of our brain for this particular task. When the resources reach some projected amount, the problem is assumed to be fully addressed, though the resources can be added if this projection changes.

This may not be too different from the role of *random access memory (RAM)* in computers, which is about 1% of the total memory and valuable. The core of the *MRT* "block" in our brain cannot be too large.

Generally, *MRT* is a combination of the pre-analysis described above, the corresponding actions and the post-analysis of the results of these actions. The "final cost" and "final outcome" of our actions will be stored in our memory in some form; it will influence our future *MRT* in such and similar situations. Anyway, such components occur in the networks we design.

The steps we outlined must be somehow present in any *autonomous vehicle (AV)* projects. If some of them are omitted, it will be a car with an advanced driver assistance system, but not a self-driving car.

Conclusion. We see that *pont* and similar games capture some key aspects of *MRT*. Importantly, our experience with *MRT* in some special situations, including games, can improve our *general* risk-taking skills. This seem *almost regardless of the field*, which is surprising. It indicates that the decision-making of this kind goes via a certain universal "risk-taking block" in our brain. If so, modeling this block can be of obvious importance for *general-purpose artificial intelligence systems*, for modeling any intelligence.

According to "The National Artificial Intelligence Research and Development Strategic Plan (2019 update)" by the National Science and Technology Council (USA): *"parallel sustained efforts are required to fully realize the vision of general-purpose AI-systems that exhibit the flexibility and versatility of human intelligence in a broad range of cognitive domains." MRT* is certainly in this category. Common sense reasoning and problem solving, probabilistic reasoning, combinatorial optimization, decision making and human-machine interaction are among the critical directions prioritized there. They write in the plan: *"most AI researchers believe that general AI is still decades away."*

Human intelligence will always be quite a challenge for modeling, and not only because our brain contains about 100 billion neurons. The complexity of the processes in our brain is astonishing. However, we can try to model some segments of decision-making, especially those of momentum type. The universality of *MRT* is a strong hint that it is governed by some "source code" in our brain, which hopefully can be successfully studied and modeled. *Pont* is designed to capture some features of this code. Generally, *MRT* seems a Bessel-type process due to following natural assumptions:

(a) our brain constantly evaluates the priorities of any tasks, which are some numerical "bids", and compare them via "auctions";

(b) it constantly generates "protection functions" used to allocate and adjust the resources needed for their analysis and our actions;

(c) forecasting is a necessary part of this process, which includes auto-termination of a task if its importance begins to diminish;

(d) "pruning the branches" as shown in Fig. 4.7(b) takes place: omitting some aspects of a task and possible directions of its analysis.

Here (a,b,c) are necessary to restrict the resources used for a particular task; (d) is a likely mechanism for this, resulting in a Bessel-type saturation. Some limit must be specified, though this can worsen our analysis and actions.

Games of chance shaped very much combinatorics during several thousand years of its history. In contrast to real life, the events there have exact probabilities, though not always straightforward to calculate. The actions of players add a lot of randomness beyond mathematics and make the games what they are supposed to be: practicing risk-taking skills in an artificially created environment. The best we can do when designing games, including *pont*, is to make the basic events reasonably distributed; this is combinatorics. The top level here is designing *AI*-systems utilizing the power of modern computers; this is a great new challenge, where combinatorics will be present.

Printed in the USA
CPSIA information can be obtained
at www.ICGtesting.com
LVHW021339040824
787142LV00002B/72

9 789811 265396